THE CERTIFICATE LIBRARY

* * *

PHYSICS

THE CERTIFICATE LIBRARY

THIS SERIES of books has been designed to cover the various syllabuses of the Ordinary Level Examination for the General Certificate of Education. The books are equally suitable for all students who wish to improve their standards of learning in the subjects of the various volumes, and together they form an invaluable reference library.

The Editor of each book is an experienced Examiner for one or more of the various Examination Boards. The writers are all specialist teachers who have taught in the classroom the subject about which they write; many of them have also had experience as Examiners.

Each volume is comprehensive and self-contained and therefore has far more than an ordinary class textbook: each has a full treatment of the subject; numerous illustrations; questions on the chapters with answers at the end of the book; and a Revision Summary for each section of the book.

These summaries give the information in a concise and compact way so that examination candidates may easily revise the whole contents of the book—they can soon see how much they readily know, and how much they have forgotten and need to study again. The facts in these summaries serve, therefore, as a series of major pegs upon which the whole fabric of the book hangs.

Advice is given on answering the examination paper and there are questions of examination standard together with suggested answers.

Series Executive Editor
B. E. COPPING, B.A.HONS.

COLOUR—THE VISIBLE ELECTROMAGNETIC RADIATIONS

FIG. 1. *Impure continuous spectrum of white light.*

FIG. 2. *Pure continuous spectrum of white light.*

FIG. 3. *Line spectrum of light from a sodium vapour lamp.*

FIG. 4. *Line spectrum of light from a mercury vapour lamp.*

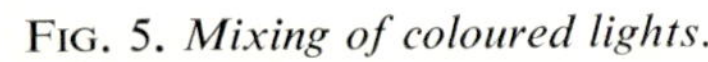

FIG. 5. *Mixing of coloured lights.*

FIG. 6. *Mixing of coloured pigments.*

THE CERTIFICATE LIBRARY

★ ★ ★

PHYSICS

Edited by

D. W. SCOTT, M.A.

Director, Centre for Science Education, University of Bristol; formerly Physics Master, Shrewsbury School

Contributors:

J. R. BAKER, M.A., B.D., A.Inst.P.

Lecturer in Education, University of Leicester

P. HAMBLETON, B.Sc.Hons.

Formerly Head of Science Department, City of Bath School

D. V. CLISH, M.A., A.Inst.P.

Head of Science Department, St. Luke's College, Exeter

J. A. BEESTON, M.A.

Head of Physics Department, Lancing College

H. A. MAYOR, M.A., D.Phil.

Head of Science Department, King Edward's School, Birmingham

W. E. MATTHEWS, M.A.

Formerly Senior Physics Master, Shrewsbury School

T. DUNCAN, B.Sc., A.Inst.P.

Senior Lecturer in Education, University of Liverpool

THE GROLIER SOCIETY LIMITED

LONDON

First published 1965
Reprinted 1968
Revised 1971

ISBN 0 7172 7715 1

Made and printed in Great Britain by
Odhams (Watford) Ltd., Watford, Herts

PREFACE

THE AIM of this book is to provide a course in physics which, while covering the syllabus requirements for the Ordinary Level examinations of the General Certificate of Education, also presents the basis of a living science directly related to our own age. In consequence, the authors had in mind not only future G.C.E. candidates, be they working at home or in school, but also the older student who in seeking to educate himself wishes to have some knowledge of the basic concepts of physics and their relation to everyday life.

To achieve this aim care has been taken to ensure that the text is authoritative, up-to-date, and readable. Recommendations made by the Association for Science Education in its reports on Science and Education have been taken into consideration so that much discovered in the last sixty years has been included in this book, while other material which in that time has lost its interest and significance has been excluded.

Physics is often described as the basic science, the science that has to be understood before any other science can be studied. That this is so can be appreciated when we consider the topics dealt with under the title of physics.

Firstly, movement and changes in movement, the study of which was the historical origin of physics. From these ideas the ideas of force and inertia were developed and these have remained immensely important to this day. Secondly, the idea that all matter is made of very small particles and that these particles are in constant motion has provided us with a fundamental picture of the physical world which can be used to describe a great many phenomena. These topics lead on to the idea of energy which is of such importance that physics is often described as a study of energy in its different forms. Thirdly, physics includes electricity which is of increasing importance not only in everyday life but also in the development of other sciences.

With these basic topics firmly in mind a fuller appreciation of the phenomena observed in such fields as heat, light, sound, and atomic energy will be gained.

Like all sciences physics defines as precisely as possible the terms it uses; but unlike most sciences, in which the terms are nearly always new words, physics often uses ordinary words as terms: unless the meanings attached to these words (such as *work*, *pressure*, *beam*, *current*) are thoroughly understood, difficulty will be encountered in understanding physics. Notes should be made of these words and, at the same time, diagrams should be copied out and all examples of calculations should be carefully worked through.

D. W. SCOTT

Note to 1971 edition: In this revision the recommendations made in the ASE report *S.I. Units, Signs, Symbols, and Abbreviations* have been adopted, though the simpler, and in the editor's view most necessary, approach to the concept of temperature has been used.

CONTENTS

CONTENTS

FIGS. 1 AND 2. *The stars* (above) *are millions of millions of kilometres away. The crystal* (below) *is magnified 2 million times. Physics explores the workings of the Universe between and beyond these two extremes.*

FUNDAMENTAL IDEAS

CHAPTER 1

WHAT IS PHYSICS?

As we progress through this book we shall meet many quantities, such as time, mass, space and temperature. Such quantities are the basis of physics and as we meet them we shall define them, but physics itself we shall not attempt to define, for it is far more important that you get the feel of what physics is all about, rather than memorize some three-line definition of what it is.

Man has existed on earth for a million years or so, and from the beginning has performed actions based on what we would now call principles of physics. Without knowing the reason behind their actions, these early men knew the direction a stone had to be thrown in order to hit a running animal; they knew that if they threw the stone directly at the animal they would miss, for when the stone arrived, the animal would no longer be in its original position. They knew also the best position of their bodies to lift something heavy; that it was easier to roll a large stone than to slide it; that by striking two flints together, sparks could be produced. Such things they learnt by trial and error. Such things could be learnt today from the principles of physics.

The earliest civilizations that we know of existed some ten thousand years ago in the Near and Middle East. By this time man had accumulated much experience learnt by preceding generations by trial and error. As a result of this accumulated experience, cities were built containing comparatively large

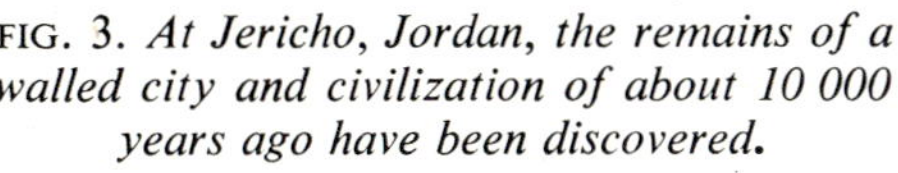

FIG. 3. *At Jericho, Jordan, the remains of a walled city and civilization of about 10 000 years ago have been discovered.*

buildings, surface drainage systems were installed, and simple machines such as primitive ploughs and carts were in use.

The Sun Revolves Round the Earth?

During the eight thousand years that followed, man started to think more about abstract things; about the principles that lay behind the everyday things around him. Arab astronomers, for example, attempted to explain the movement of the stars and planets across the sky, work that was to be carried on by the Greeks and Romans. By the second century A.D., a Greek called Ptolemy championed the theory that the earth was the centre of the universe, a theory that was to be accepted for over a thousand years.

FIG. 4. *Ptolemy and his Universe of which the earth is the centre.*

In Greece also, the study of mathematics was started by such famous men as Euclid and Pythagoras. Others, such as Archimedes, were designing machines more complex than any known before. Principles of physics were being discovered, theories were being formulated. It was a period in history when men were beginning to seek knowledge for the sake of knowledge. But too much reliance was being put on theory and not enough on experiment. It was inevitable that many wrong conclusions were reached.

With the fall of the Roman Empire, the Dark Ages began in Europe. They were to last for a thousand years. The advance of knowledge was halted.

The old ideas were kept alive primarily by the monks in their monasteries, but neither time, energy nor desire were available for the pursuit of new ones. The monks manipulated the old ideas to fit their religious beliefs and that knowledge of the physical world, which had been passed on from generation to generation, was now subordinated to religion.

Towards the end of the fourteenth century the minds of men began to reawaken. The desire for learning was being reborn, and in Italy the first stirrings of the Renaissance were felt. In the circumstances, a head-on collision between the long-established theories of the Church and the new-found theories of the Renaissance was unavoidable.

The Earth Revolves Round the Sun

In 1530 a Polish priest called Copernicus made public some new ideas he

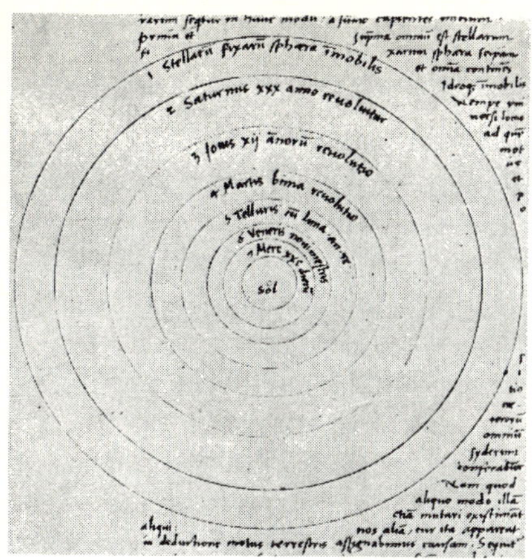

FIG. 5. *Sketch from Copernicus' notebook showing the planets revolving round the sun.*

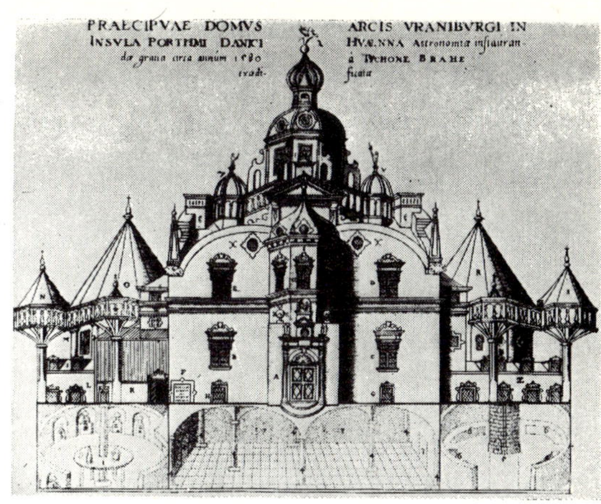

FIG. 6. *Brahe's observatory in Denmark where observations were made that confirmed the theory of Copernicus.*

had formulated on the structure of the universe. Copernicus had rejected the old theory of Ptolemy and had come to the conclusion that the earth rotated round the sun. The theory was not new, as Copernicus was the first to admit, having already been advanced by some members of the school of Pythagoras in the fifth century B.C.; but this in no way detracts from the greatness of Copernicus' achievement for by advancing his new theory of the universe, he had to go against not only what seemed common sense based on much careful observation, but against an idea that had been firmly established as truth for 1300 years and from which many other ideas had grown.

By the 1580's the Danish astronomer Tycho Brahe was making detailed observations of the planets which appeared to confirm the revolutionary ideas of Copernicus; further confirmation was not long in coming. The German scientist Kepler by studying the observations of Brahe was able to calculate the orbits of the planets, while in Italy, Galileo observed for the first time the four inner satellites of the planet Jupiter; the satellites were seen to be revolving round the planet like a miniature Copernican solar system. Notice how theory and experiment worked together in the rebirth of an old idea, setting it this time on a firm basis.

The new theory conflicted violently with the long-established belief that the earth was the centre of the universe. Its implications, that the earth was in fact an insignificant planet amongst countless others in an infinitely large universe, were too much for a world that, through a sense of supreme self-importance, believed itself to be the central point of creation. But not all minds were closed to the new ideas. The Italian philosopher Giordano Bruno attempted to incorporate the new scientific facts into a fresh view of religion. As a reward for his effort, Bruno was arrested by the Inquisition and in Rome in 1600 was burnt at the stake for heresy.

FIG. 7. *Isaac Newton (1642-1727).*

Nor were scientists immune from the persecution of the Church. Despite the growing popularity of the new theory, Galileo, now an old man of seventy with his life's work mostly behind him, was arrested by the Inquisition in 1633. Galileo maintained that it was wrong to interpret the Word of God in such a manner that it contradicted the Works of God; yet under threat of torture he publicly recanted his belief in the Copernican universe. It was too late however to stop the scientific revolution that was sweeping Europe. Men wanted to know about things and, as a result of the work carried out by such pioneers as Galileo, they would now pursue that knowledge from a firm basis of experiment.

Newton's Theory of Gravity

In the year of Galileo's death, 1642, Isaac Newton was born in Lincolnshire. He was to make many important contributions towards the advancement of both physics and mathematics, none more so than his theory of gravity, one of the great landmarks in science.

The theory of the Copernican universe, expanded and substantiated by both Kepler and Galileo, showed that the earth revolved round the sun. Yet the question of why it did so, the problem of what held the planets in their orbits, although greatly troubling the minds of the day, had not been answered.

With the idea of gravity already formulating in his mind, Newton turned to the original calculations of Kepler. Studying and reworking them he eventually postulated the idea of a force called gravity to explain the observed behaviour of the planets. The theory of gravity had been born. When Newton died in 1727, he left the world of science far richer than it had ever been before; discovery was to follow discovery at an ever-increasing rate.

By the beginning of the nineteenth century we find Michael Faraday launching out into ten years of arduous experiments from which he would discover the relationship between magnetism and electricity. The start of the modern era had been reached. And yet in 1867, the year of Faraday's death, probably no more than one-quarter of the information given in this book was known.

FIG. 8. *Albert Einstein (1879-1955).*

The theories of the Greeks, found and lost and found again; the revolutionary thinking of such men as Copernicus; the vision of such men as Kepler and Newton; the self-sacrifice of Bruno; the tough determination of Galileo; the relentless hard work of Faraday and countless other men—all this and much more had put science on a firm basis from which it would now advance so rapidly that the world would be carried with it, and the life of man changed beyond recognition.

In 1898 the French physicist Pierre Curie and his Polish wife Marie announced their discovery of the radioactive element radium. Much work on radioactivity was to follow; by the turn of the century many men such as J. J. Thomson, Lord Rutherford and Niels Bohr were devoting their energies to a new field, that of atomic physics.

With the turn of the century came also a change in approach to science. For a long time each branch of this vast subject had tended to become more and more isolated from the rest. Oblivious to the work going on outside their own small sphere, many scientists were often reaching totally different solutions to the same problem. These conflicting answers were by and large ignored but with the beginnings of the modern era such watertight compartments began to break down. Physicists began working with chemists, chemists with mathematicians and mathematicians with astronomers.

Another Theory of Gravity

In 1905 the world of science was shaken to an extent never before experienced; it was the year that the twenty-six-year-old German-born scientist, Albert Einstein, published his special theory of relativity.

Einstein insisted that scientific results obtained by observation must be consistent with the properties of the means of observation. This is not an easy idea to grasp but it is an important one. Let us consider a simple illustration of it.

An observer travelling in a train *A* wishes to measure the speed of a second train *B* which is overtaking in the same direction. By observing the number of carriages that pass him in one minute and knowing the length of each carriage, the observer can calculate the speed of *B* relative to *A*. But the

FIG. 9. *The devastating physical and moral effects of the first nuclear bombs exploded in 1945 were a terrible confirmation of the mass/energy relationship stated by Einstein forty years before. The photograph shows an underwater test explosion of a nuclear bomb in the Pacific.*

observer wishes to know the speed of *B* relative to the ground, he must therefore take into account "the properties of the means of observation," in other words, the speed of the train *A* in which he is travelling. This speed the observer can measure by counting the number of telegraph poles that pass him in one minute. Knowing that the telegraph poles are equally spaced, and the distance between each pole, the observer can calculate the speed of *A* relative to the telegraph poles. Since the telegraph poles are stationary relative to the earth, this would also be the speed of *A* relative to the earth. Hence, by simple addition, the speed of *B* relative to the earth can also be found.

Einstein applied this logical point of view on a universal scale, using it to take a fresh look at the established principles of physics. In the process he was led to the idea that the mass of a body was dependent on its speed relative to an observer; Einstein calculated that as the speed of a body increased, so its mass increased. This, and much else, led him to postulate a relationship between energy and mass, a relationship that was to receive spectacular confirmation some years later in the release of atomic energy. The new theory threw doubt on the classical ideas of time, space and mass, which had been universally accepted since Newton. Yet, at the same time, the new theory applied to mechanics and electrodynamics was able to clear away a conflict of ideas which had been growing up between these two fields of science.

The fact that the classical ideas applied separately to experimental mechanics and electrodynamics led to a conflict in results, whereas Einstein's ideas did not, gained a certain rapid acceptance for the new theory. Other scientists remained sceptical but Einstein had not finished with the subject and in 1919 he published his general theory of relativity.

Even greater doubt was now thrown on the classical ideas; Einstein was offering a new theory of gravitation to replace that of Newton. Certain observed facts, such as a peculiarity in the orbit of the planet Mercury, could not be explained using Newton's theory, whereas with Einstein's theory they could. To some people at the time, however, the theory of Newton was established without a doubt for all time, and their minds were closed to the possibility of improving upon it; they were tending to confuse theory with fact by failing to realize that theories are only attempts to explain observed facts and are not facts in themselves.

From his theory of gravitation Einstein was able to predict facts for which no evidence had at that time been observed. One such prediction involved the bending of light rays by gravitation, a phenomenon that was eventually observed by a group of British scientists studying an eclipse of the sun in 1919. Although this was by no means conclusive evidence, it caused something of a sensation at the time, winning a rapid acceptance for Einstein's theories. Since 1919 much more supporting evidence has been discovered and today Einstein's theories of relativity are firmly established on a level with Newton's theory of gravitation, and are among the most important and brilliant in the entire field of science.

Perhaps, after what we have just said, you will be surprised to find that there is no further mention of Einstein's theory of gravitation in this book, whereas there is of Newton's theory. We use theories to help us understand observed facts, and for understanding the facts mentioned in this book, Newton's theory is perfectly adequate. It is only in extremely advanced physics that certain facts occur demanding the application of Einstein's theories before those facts can be understood.

Progress takes place in science only when experiment and theory go hand in hand. As we have seen, sometimes one comes first and sometimes the other, but no theory is acceptable unless experiment confirms it. If experimental facts conflict with theory, either the theory is wrong or the facts have been observed inaccurately; history is full of either possibility and only the continuous checking of both fact and theory will eventually give the answer. As science progresses more and more facts are discovered and perhaps in consequence old theories will have to be changed. It is for the possibility of this change that the mind of a scientist must always remain open and ready.

CHAPTER 2

BASIC QUANTITIES

As an introduction to this book we need to get a picture of the extent of the ground that we must cover. We need to think about the basic quantities that must be measured; about the most important ideas and theories that past generations of scientists have built up; about the basic phenomena that we must study. This having been done we will have a framework in which to place each small piece of work as it occurs. Without such a framework much of our work in physics would remain meaningless and in consequence, dull.

As a foundation for such a framework we need first to consider the basic quantities that must be measured in physics: time, space, mass, and temperature.

TIME

We have all used clocks and watches and have lived with a sense of time every day of our lives. What then is the problem? Just this: how do we measure time accurately over the range from millions of years (wanted by geologists and astronomers) to a millionth of a millionth of a second (wanted by the nuclear physicist)?

Over the ages man has developed more and more accurate timepieces for use in our day-to-day life, but it is only in the last fifty years or so that he has had any success in measuring either the very long or the very short periods of time we have just mentioned. Man has always used the day as his basic unit, now subdivided into the hour, and the further subdivisions of the minute and the second. The year is another unit that is fixed by nature like the day, but it is not as easy to use as a standard. Hour-

Fig. 10. *Sundials once served as clocks.*

glasses, sundials and candles served man for a long, long time. Then at the beginning of the seventeenth century the idea of regulating a clock with a pendulum occurred to the Italian scientist Galileo as he sat in the cathedral at Pisa watching a chandelier swinging in a draught. He noticed that the time of swing hardly depended on how much the chandelier was swinging and, also, that all the chandeliers swung at the same speed.

You can make similar observations for yourself. Make a pendulum out of a small heavy object and a piece of cotton. Tie one end of the cotton to the object and the other end to a rigid support and then set it swinging (Fig. 11).

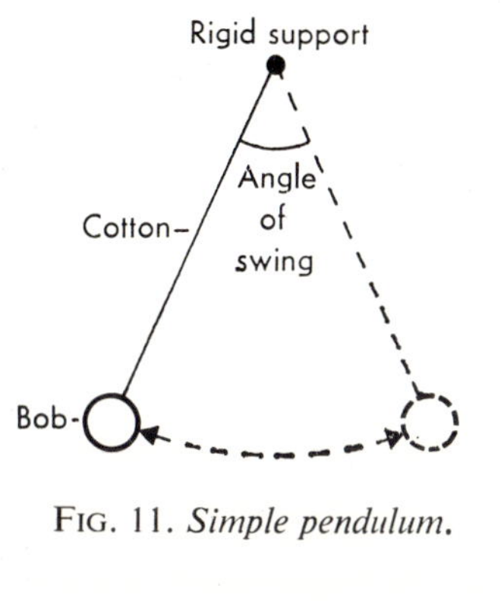

FIG. 11. *Simple pendulum.*

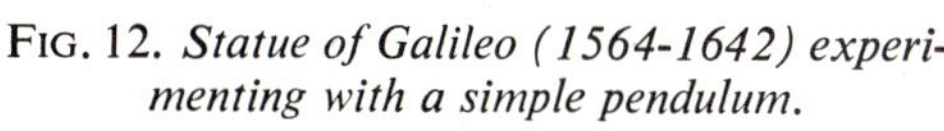

FIG. 12. *Statue of Galileo (1564-1642) experimenting with a simple pendulum.*

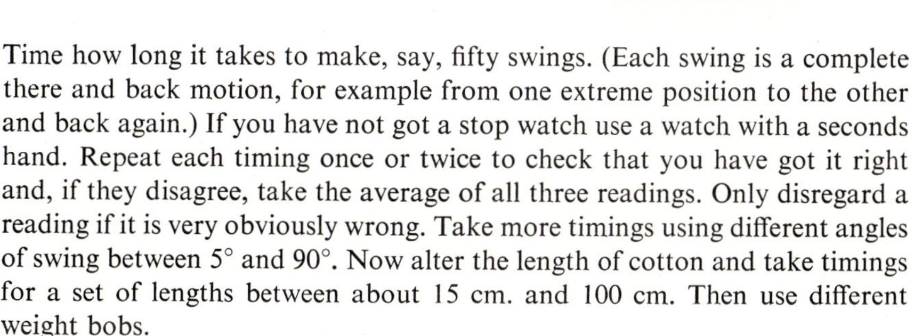

Time how long it takes to make, say, fifty swings. (Each swing is a complete there and back motion, for example from one extreme position to the other and back again.) If you have not got a stop watch use a watch with a seconds hand. Repeat each timing once or twice to check that you have got it right and, if they disagree, take the average of all three readings. Only disregard a reading if it is very obviously wrong. Take more timings using different angles of swing between 5° and 90°. Now alter the length of cotton and take timings for a set of lengths between about 15 cm. and 100 cm. Then use different weight bobs.

You should get readings similar to those shown in Table 1. Try to spot the relationship between the time period for one swing T and the length of the cotton l.

Table 1. Pendulum Readings

(With the angle of swing 10° and bob of weight 50 g)

Length in cm	10	25	50	100	200
Time for 20 swings in seconds	12·6	20·1	28·6	40·3	56·8
	12·6	20·1	28·8	39·9	56·9
	12·7	20·1	28·8	40·1	57·3
Average time for 20 swings	12·6	20·1	28·7	40·1	57·0
Average time for one swing	0·63	1·01	1·44	2·01	2·85

(With length of pendulum 50 cm and bob of weight 50 g)

Angle	10°	20°	30°	40°	60°
Time for 20 swings in seconds	28·6	28·8	29·0	28·7	29·2
	28·8	29·0	28·8	28·9	29·1
	28·8	28·7	28·8	28·8	29·3
Average time for 20 swings	28·7	28·8	28·9	28·8	29·2
Average time for one swing	1·44	1·44	1·45	1·44	1·46

(With length of pendulum 50 cm and angle of swing 10°)

Mass of bob in g	10	20	50	100	200
Time for 20 swings in seconds	29·0	29·2	28·6	28·8	28·9
	28·9	29·1	28·8	28·9	28·7
	29·0	29·1	28·8	28·9	28·9
Average time for 20 swings	29·0	29·1	28·7	28·9	28·8
Average time for one swing	1·45	1·46	1·44	1·45	1·44

Plot a graph of T against $\sqrt{l}$ (Fig. 13). You will find that this graph is a straight line through the origin, showing that T is directly proportional to $\sqrt{l}$.

$$T \propto \sqrt{l}$$

You should also find that your values of T hardly change when you use different angles of swing or different bobs.

This is the principle of physics which lies behind pendulum clocks. Such clocks were at first worked by weights; later they were driven by wound-up springs; while today in most mechanical clocks (as in watches) the pendulum has been replaced by a balance wheel, the oscillations of which are controlled by a "hair spring" and driven by a "main spring." The latest type of such watches will adequately meet all our needs in the laboratory, measuring accurately to 0·2 of a second over a period of an hour or so.

In physics the unit that we shall use most is the second. How is this defined if we want it to be as accurate as it is possible to be? The simple answer is that it is $\frac{1}{86,400}$ of a day. The snag appears when we realize that scientists

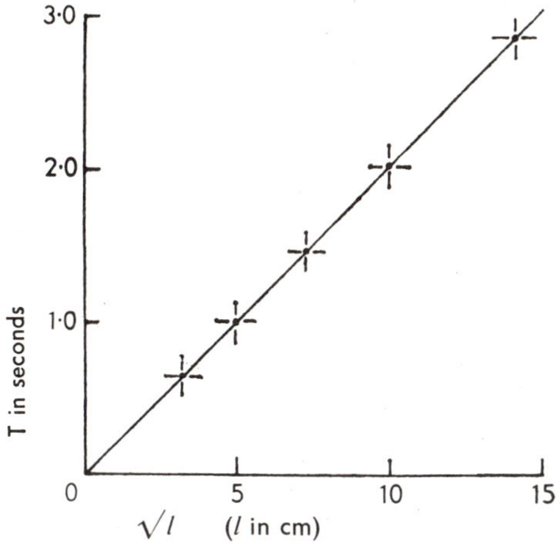

FIG. 13. *Graph for a simple pendulum.*

need very accurate astronomical instruments to be able to measure the length of the day and, worse than this, that the length of the day is always varying, only very slightly, but still varying. Another difficulty is that we can only compare two estimates of the length of the day by clocks whose accuracy is unknown.

In order to try to measure millionths of a second, when a stop clock could not be used because the actual starting and stopping would take many times as long as the interval we wish to measure, scientists have developed electrical and atomic methods. The most accurate are the atomic clocks using peculiarities of the ammonia molecule or of the caesium atom. Such is the accuracy of these clocks that they are never out by more than one second in thirty years or more. So the second is now defined in terms of the properties of the caesium atom because scientists find these easier to measure very accurately than the length of the day.

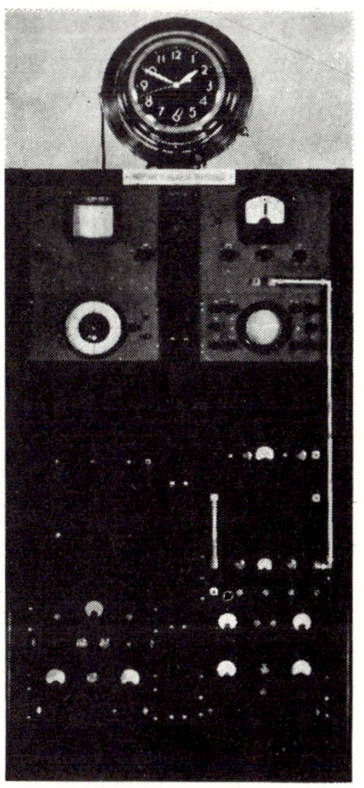

FIG. 14. *This atomic clock uses the vibrations of the ammonia molecule to measure time with an accuracy of one part in 100 million.*

Very long periods of time pose a different problem for here the physicist is trying to measure into the past, the age of a rock for example. Anything to measure this is hardly a clock in the normal sense of the word. The physicist, however, can answer problems such as this by making use of radio-activity. Certain elements, like uranium, and certain types of other elements (radio-active isotopes), like carbon-14, are likely to change into different elements. Each change takes place at a fixed rate. Therefore, if we can find out by chemical analysis how much is left and how much is changed, we can calculate for how long the change has been going on. The physicist has been able to measure back about 2000 million years by this method The same method can be used in certain cases to help the archaeologist in dating remains in tombs and elsewhere. Greater ages, for example the age of the earth or the sun, can be estimated by calculations based on similar principles.

In this course you will learn something about a number of these methods of measuring time; as well as the pendulum, the electric clock and radio-activity will be mentioned. Notice how different branches of physics (mechanics, electricity, atomic physics) have combined in order to measure the same thing, "Time."

You will see examples, like this one, of the help given by one branch of science to another, at intervals throughout the course.

Notice, too, how easy it is to measure intervals for everyday life in comparison with measurements of either the much smaller or the much larger; this is a theme that will also recur.

SPACE

As with time, we know all about ordinary measurements of length, area and volume. Again the problem is one of size: the millions of millions of millions of kilometres between one galaxy of stars and another; the millionths of millionths of millionths of a centimetre which is the order of size of the nucleus of an atom.

In laboratory work only the range from a fraction of a millimetre up to a few metres is used. Notice that we have not quoted these lengths in feet and inches. In physics we use the metre as our basic unit. The reason for this is mainly arithmetic convenience, in that all the other units are multiples of 10, 100, 1000 and so on. The standard metre used to be the length of a piece of metal kept near Paris but now, as with time, scientists find it easier to work very accurately from the properties of atoms of a particular element, in this case krypton, and the metre is defined in terms of these properties. In deciding on our methods of measuring lengths we must consider the accuracy that we require. For our purposes an accuracy of within 1 per cent will be quite sufficient; we would be doing very well

FIG. 15. *An electron microscope is here being used to examine the structure of a metal. An image of the metal, magnified many thousands of times, can be seen on the circular screen.*

indeed if all our measurements were as accurate as this. For distances between 5 cm and a few metres, a metre rule is suitable. This will give us a reading to the nearest millimetre; in other words a reading accurate to within half a millimetre, and half a millimetre is 1 per cent of 5 cm. Between 1 cm and 5 cm we can use vernier calipers and between 1 mm and 1 cm we can use a micrometer screw gauge.

These two simple instruments, which are fully described later in the book, will give readings to within 1 per cent over the ranges mentioned.

For distances smaller than these, a microscope giving high magnification is used, while for still smaller distances, indirect methods such as the electron microscope are employed. These methods, like atomic clocks, involve principles too complicated to be discussed here.

To measure distances longer than those on which measuring tapes can be used, the method of *triangulation* is employed. Fig. 16 shows the basis of this method. If we measure the distance x and the angles A and B we can find the distance to the house either by calculation or by drawing a diagram to scale. The problem in this case is how to measure the angles accurately. This method is used in surveying and has even been used to find the distance to the nearest star. In this last instance the baseline (the distance x in Fig. 16) is the diameter of the earth's orbit round the sun; this is obtained by taking

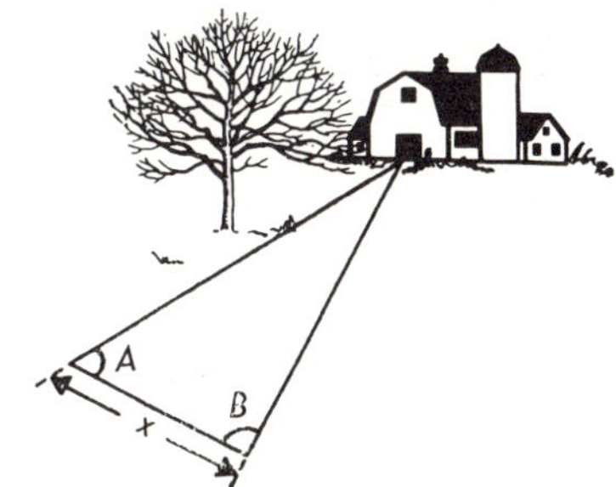

FIG. 16. *Triangulation.*

FIG. 17. *Andromeda nebula, about 1·5 million light years away.*

readings at six-month intervals. Large, accurate telescopes are used to measure the angles: the accuracy is such that the system would be capable of measuring the angle between lines drawn from you to the two sides of a tennis ball hung up some miles away! The value obtained for this distance is so great that it takes light four years to get from the star to us, and light travels at 300 000 000 metres per second. (This is about 186 000 miles per second.) Light takes about 8 minutes to reach the earth from the sun and just over 1 second to earth from the moon.

For distances still greater than these, complicated methods of comparison are used. There is, for example, no direct method of measuring the distance of the Andromeda nebula away from the earth, nor the distances of countless other groups of stars. The light from the Andromeda nebula seen on earth left the nebula 1 500 000 years ago: we have already said that light travels at a speed of 3×10^8 metres per second, so you can work out the distance for yourself.

The measure of the motion of a body involves the measurement of both time and distance. The *average speed* of a moving body is the distance moved, divided by the time it took to move that distance. If the time interval is small enough, we can call this the speed of the body at that moment. If the speed is not changing, we say that the body is moving with *uniform speed.*

$$\textit{Average speed} = \frac{\textit{Distance}}{\textit{Time}}$$

In physics we often require to know the direction in which a body is moving, as well as its speed; this combination of speed and direction is called *velocity*, which may be defined as the *distance travelled in unit time in a given direction.* Quantities such as velocity, acceleration, force, having both magnitude and direction, are called *vector quantities.* Quantities such as speed, mass, volume, having magnitude but no direction, are called *scalar quantities.*

While we may speak of a body moving with uniform speed in a circle, it would not be correct to say that the body moved with uniform velocity in a circle.

Movement around a circle represents a constant change in direction; a body moving with uniform velocity does not change its direction, but moves in a straight line.

The usual units that we shall use for measuring motion are m/s (metres per second).

For the measurement of very high or very low velocities, special techniques are required. One such technique involves high-speed photography. Film is exposed at intervals of a fraction of a second so that a fast-moving object, such as a bullet, forms a series of images on the photographic plate. The interval between exposures and the positions of the images can then be used to calculate the velocity of the bullet.

When a car starts from rest its velocity increases; it is said to be *accelerating*. Acceleration is the increase in velocity in unit time and is measured in units of metres per second per second, i.e. m/s^2.

$$\textit{Acceleration} = \frac{\textit{Increase in velocity}}{\textit{Time}}$$

MASS

The mass of an object is a measure of the amount of material in the object; it is also a measure of how difficult it is either to start the object moving or, once it is moving, to change its velocity either in magnitude or direction.

Consider three tin cans lying on a path: one is empty; one is full of paint; the third is full of lead. If you were to kick each tin can in turn, the first would fly away from your foot, the one full of paint would roll away slowly, while the one full of lead would probably cause considerable damage to your toes. The empty tin has less mass than the one full of paint and this, in turn, has less mass than the one full of lead.

The weight of an object exists because the earth attracts any object near to it. This "*gravitational force*" depends on the mass of the earth, the position of the object, and the mass of the object. If you were to take the object up in a space ship, its *weight* becomes less as its distance from the earth increases; its *mass*, however, remains constant. In practice we measure the weight of an object, we measure how heavy it is and we do this by comparing it with a standard on some form of balance.

The unit of mass used in physics is the kilogramme. The standard is a mass of platinum kept near Paris, the mass of which has been defined as 1 kilogramme. Note that 1 gramme is very nearly the mass of 1 cubic centimetre of water.

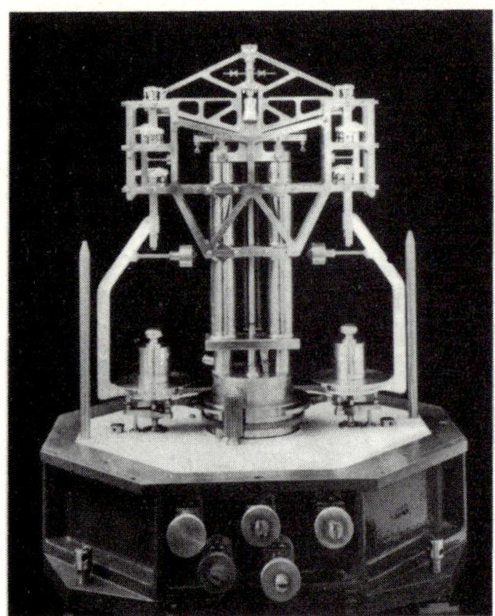

FIG. 18. *High precision balance accurate to 1 part in 1000 million.*

The measurement of mass in a laboratory is often performed on a *beam balance* (Fig. 19). This consists of a beam, or balance arm, carrying scale pans at either end, and balanced in the middle on a knife-edge. The knife-edge is normally made from some very hard material such as agate, balancing on a small block of the same material. When the balance is not in use, the beam is usually removed from the knife edge by means of a simple lever mechanism. This saves unnecessary wear on the knife-edge. To the centre of the beam is attached a pointer which swings over a scale.

Provided that this type of balance is correctly set up, the gravitational force on two equal masses will make the pointer come to rest on the zero mark of the scale, whether one mass is placed in the right-hand pan and the other in the left, or vice versa. The setting-up of a balance involves making sure that it is level and that the pointer is reading zero when the unloaded balance arm is lifted on to the knife-edge. Having accurately set up a balance, it can then be used to compare an unknown mass with standard masses. These standards will not have been compared directly with the standard kilogram kept near Paris, but there will be an unbroken chain of comparison back to that standard. However, because of the small inaccuracies that are bound to be involved in each of these comparisons, it is very unlikely that our laboratory standards will be quite accurate. They should, however, be accurate enough for us to measure 0·01 g, which is 1 per cent of 1 g.

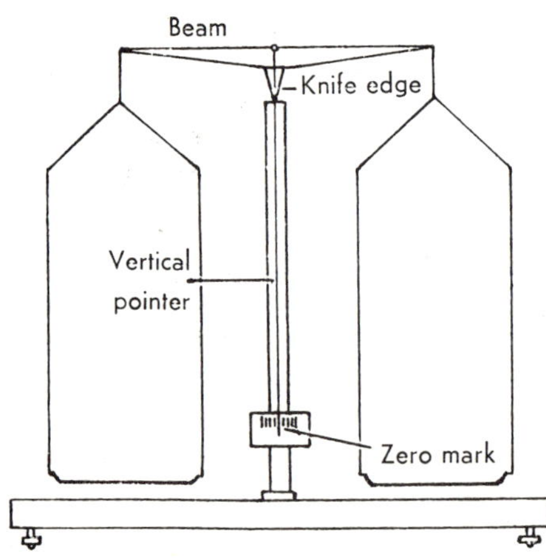

FIG. 19. *Principle of the beam balance.*

Although the beam balance is used to compare masses, the whole process is often referred to as "weighing", and one has got to remember that this is strictly wrong and is just a short way of saying "finding the mass of . . ."

For masses of 100 g or more, provided we do not wish to find a small difference between two such masses, a balance that will give us readings accurate to the nearest gramme is adequate. Such a balance is one which utilizes the effect of the weight of a mass in stretching (extending) a spring. The scale of the *spring balance* is marked on at the time of manufacture according to the effects of standard masses upon the spring. Alternatively the spring balance may be calibrated in units of force (newtons). The spring balance measures forces but, provided we stay on the surface of the earth, the weights of two bodies are proportional to their masses and so it is also fair to calibrate the balance in units of mass. A spring balance is often more convenient and quicker to use than the beam balance but it is not so accurate.

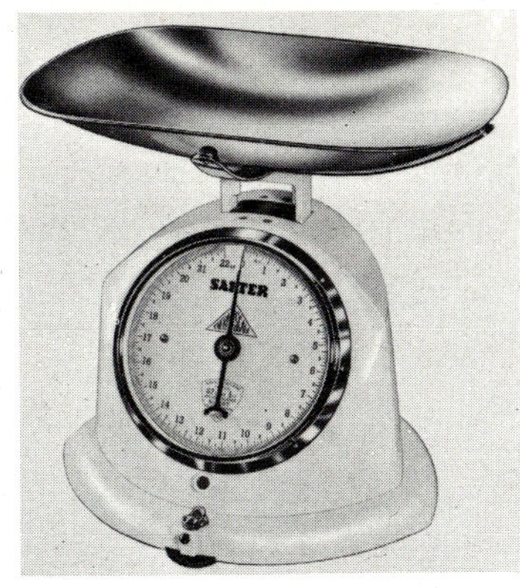

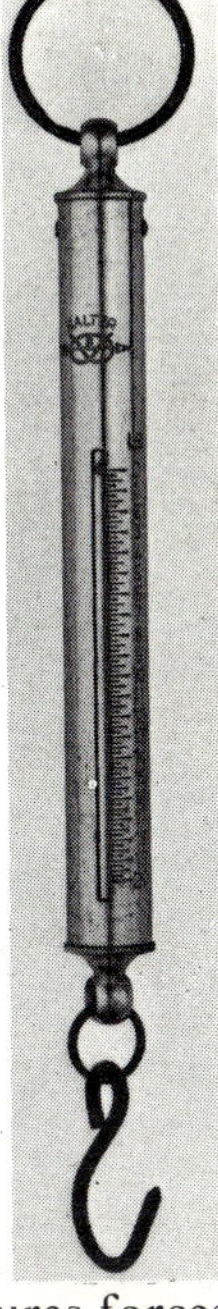

FIG. 20. *Two types of extension spring balance. The type above is sometimes called, misleadingly, a compression balance, but it still depends on the extension of a spring, by means of a system of levers.*

Once again it is the very large and very small measurements that cause difficulties; how to measure the mass of the sun, or that of an electron. There are methods, and we shall learn more of them later in the book. For the moment, however, let us only say that the theory of gravitation applied to certain observations of the solar system allows us to arrive at an answer for the first measurement, whereas the fact that electrons move in circles when placed in a strong magnetic field allows us to arrive at the second.

Density

In Fig. 21, *A* is a block of wood 36 cm long, 10 cm wide and 2 cm deep,

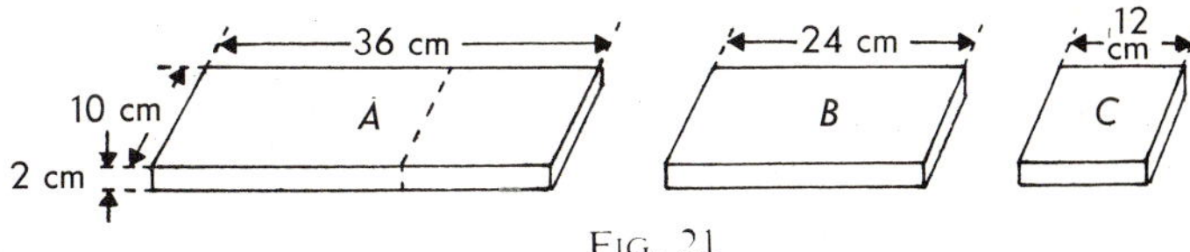

FIG. 21

its volume is therefore 36 cm × 10 cm × 2 cm or 720 cm³. When *A* was weighed on a balance, a measurement of 240 g was recorded. Block *A* was then cut in two, 12 cm from one end as shown, giving a larger block *B* and smaller block *C*. The masses of these two blocks were then found by weighing and their volumes by calculation. A table was drawn up:

	Mass (g)	Volume (cm^3)	Mass/Volume (g/cm^3)
A	240	720	$\frac{1}{3}$
B	160	480	$\frac{1}{3}$
C	80	240	$\frac{1}{3}$

You will see from the table that although the masses and volumes of each body are different, the ratio of mass/volume is constant in all three cases. This ratio is called the *density* of the bodies; it is a property of the substance from which the bodies are made and since the three bodies we are considering are all made from the same type of material, their densities are all equal.

Measurement of density usually involves the separate measurement of the mass and volume of the piece of material under investigation. It is, however, often convenient to compare substances with water instead of actually measuring their density. In these cases we measure the *relative density*.

$$\text{Relative density} = \frac{\text{Mass of a volume of the substance}}{\text{Mass of the same volume of water}}$$

Since, if we use cubic centimetres and grammes the density of water is numerically very nearly one, the relative density of a substance will be the same figure as the density. The density, though, has units of g/cm^3, while the relative density is a ratio and therefore has no units. The methods used in the laboratory may be grouped under five headings.

1. *To find the density of a regular solid:*
   (*a*) Weigh it.
   (*b*) Measure its dimensions and calculate its volume.

2. *To find the density of an irregular solid that does not dissolve in water:*
   (*a*) Weigh it.
   (*b*) Place it in a measuring cylinder that is half full of water. The difference in the two volume readings will be the volume of the solid (Fig. 22).

3. *To find the relative density of a solid available only in small volumes, that does not dissolve in water:*
   (*a*) Weigh a density bottle empty (Fig. 23).
   (*b*) Weigh the bottle with some of the solid in it.

(*c*) Fill the bottle up with water, the solid still being in the bottle. The water should completely fill the bottle and the narrow hole in the stopper. The outside of the bottle and the top of the stopper should be dried. Weigh again.

(*d*) Empty out the solid and fill completely with water. Weigh again.

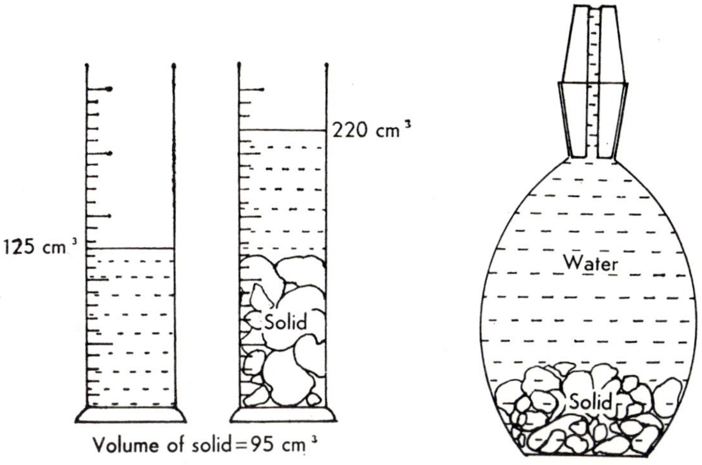

FIG. 22. FIG. 23.

Specimen results using lead shot

Mass of bottle =	24·50 g
Mass of bottle plus lead shot =	144·32 g
Mass of bottle plus lead shot plus water =	183·76 g
Mass of bottle full of water =	74·52 g
∴ Mass of lead shot =	119·82 g
Mass of water to fill bottle =	50·02 g
Mass of water over the lead shot =	39·44 g

Hence the mass of water which would occupy the same volume as the lead shot = 10·58 g

$$\therefore \frac{\text{Mass of lead shot}}{\text{Mass of same volume of water}} = \frac{119{\cdot}82}{10{\cdot}58}$$

∴ Relative density of lead shot = 11·3

4. *To find the relative density of a liquid:*

(*a*) Weigh a density bottle empty.

(*b*) Fill the bottle with water, dry the outside and weigh again.

(*c*) Fill the bottle with the liquid under investigation and weigh again.

From the results, the ratio of mass of liquid to fill bottle/mass of water to fill bottle, is the relative density of the liquid.

5. *To find the density of air:*
   (*a*) Weigh a 500 cm³ flask, which is fitted with a bung and tubing (Fig. 24).
   (*b*) Pump out the air and tighten the clip to keep the vacuum. Then reweigh.
   (*c*) Check the volume of the flask by filling with water and transferring the water to a measuring cylinder.

Specimen results

Mass of flask full of air = 118·23 g
Mass of flask without air = 117·57 g
Volume of flask = 500 cm³
Density of air = Mass of air/Volume of air
= 0·66/500
= 0·0013 g/cm³

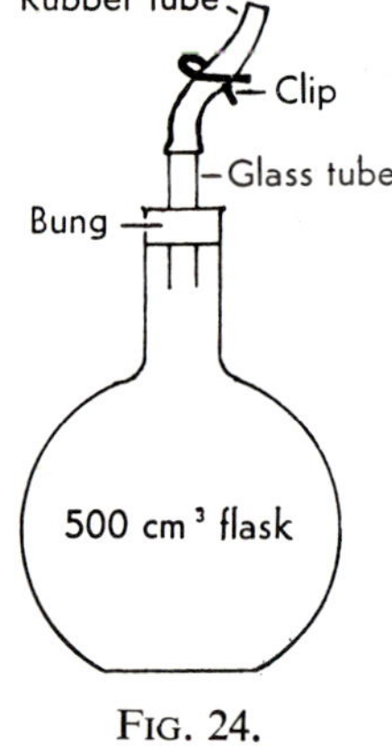

FIG. 24.

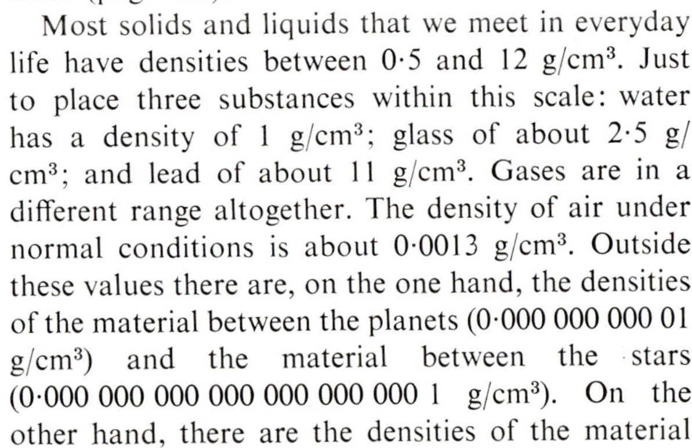

There are also methods for measuring density which depend on the application of Archimedes' Principle. These methods are discussed later in the book (page 113).

Most solids and liquids that we meet in everyday life have densities between 0·5 and 12 g/cm³. Just to place three substances within this scale: water has a density of 1 g/cm³; glass of about 2·5 g/cm³; and lead of about 11 g/cm³. Gases are in a different range altogether. The density of air under normal conditions is about 0·0013 g/cm³. Outside these values there are, on the one hand, the densities of the material between the planets (0·000 000 000 01 g/cm³) and the material between the stars (0·000 000 000 000 000 000 000 1 g/cm³). On the other hand, there are the densities of the material at the centre of a star (up to 100 000 g/cm³), and the actual nucleus of any atom (100 000 000 000 000 g/cm³). These very low and very high densities have been calculated by complicated and indirect means which are beyond the scope of this book.

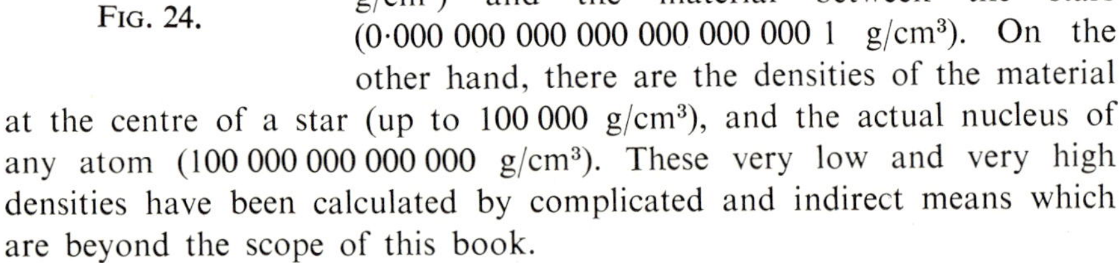

Force

In explaining the term weight, we used the word force. Let us now consider what we mean by force. A brief answer is "force is that which will tend to move or to change the shape of any body."

Press a drawing pin into a board; put your finger into a piece of Plasticine;

kick a ball; push hard against a brick wall; pull a heavy box over a rough surface; in each of these cases a force is being exerted. We have different names for the different types of force that we meet: the force due to the attraction of the earth we call *gravity*; the force trying to stop us move a heavy object over a rough surface is called *friction*; the force in a picture cord holding the picture up against the force of gravity is called a *tension*. These are some of the most common examples.

What will actually happen when a force is exerted on a body depends on all the forces acting on that body. The force that you apply may change the shape or size of the body (pressing a piece of Plasticine); it may just balance other forces such as friction (trying to push a heavy object over a rough surface but not succeeding) or gravity (holding something in your hand to stop it falling to the floor); it may overcome such forces as friction or gravity, it will then alter the speed or direction of motion of the body (kicking a ball).

Now that we have defined units of mass, length and time we can define units of force as well. The unit of force is called the newton (symbol N) and it is defined as that force which would cause a mass of 1 kilogramme to accelerate at 1 metre per second per second. The weight of a mass of 1 kilogramme, i.e. the force exerted by gravity on it, is about 10 newtons (more accurately 9·81 N). You will learn more about the measurement of forces later (page 75).

Energy and Work

Closely associated with the idea of force is the idea of energy and work. Whenever a force makes something move, work is done by the force and energy is transferred. It should be noted that the phrases "energy used" and "work done" mean exactly the same thing.

We can think out the relationship between work done, force and distance, like this: put a kilogramme weight on the floor; lift it on to a table, say 1 m from the floor; a certain amount of work will be done. The force is about 10 newtons and the distance 1 m. Now lift a second kilogramme weight from floor to table; there is no difference between this and the first weight so the same amount of work must be done. Now put them both down and then lift them up together; we must now be doing the same amount of work as when we lifted them separately, i.e. the work done is proportional to the force. Now lift them both from the table to a cupboard top at 2 m; we are lifting them another 1 m and so will be doing the same amount of work as lifting them from the floor to the table. Now put them both on the floor and then lift them both together straight on to the cupboard top; we must now have done the same work as lifting them in two stages and twice the work of either stage. Hence the work done is proportional to the distance moved. Therefore work done is proportional to force × distance and we can choose our units of work so that:

$$\textit{Work done} = \textit{Force} \times \textit{Distance moved}$$

With the force measured in newtons and the distance in metres, then the work done is in joules.

Note that anything which is capable of doing work is said to possess energy. This energy may appear in many different forms: *kinetic energy* by virtue of the motion of a mass; *potential energy* by virtue of position such as of height or of being wound up; *chemical energy* which can be released in a chemical reaction; *nuclear energy* which can be released in nuclear reactions; *electrical energy; heat energy; light energy*—all of these are different forms of energy. Energy can be changed from one form to another but as we shall see later, it cannot disappear.

Power

Power is the rate of doing work:

$$\textit{Power} = \frac{\textit{Work done}}{\textit{Time taken}}$$

With the work done measured in joules and the time in seconds, the power is in watts.

You can find your own mechanical power by carrying out two simple experiments:

(1) *Lifting weights*

Lift some heavy masses from floor to table. (Someone could be putting them back on the floor for you to pick up again while the experiment is in progress.) Time yourself lifting a certain number of masses. Here is an example:

Masses of 12 kg lifted 10 times in 18·0 s
Height lifted each time = 1 m
Force required to lift a 12 kg mass is about 120 N

$$\begin{aligned}\text{Total work done} &= \text{Force} \times \text{Distance moved}\\ &= 120 \times 10 \times 1\\ \text{Power} &= \text{Work done/Time taken}\\ &= \frac{120 \times 10 \times 1}{18}\ \text{watts}\\ \therefore \text{Power of boy} &= 67\ \text{watts}\end{aligned}$$

2. *Running up stairs*

Run up a long flight of stairs timing yourself from bottom to top. Also measure the height of the stairs and weigh yourself.

Mass of boy = 50 kg
Height climbed = 40 steps each 15 cm high
Time taken = 8·0 s
Weight of a boy of mass 50 kg is about 500 N
Height of each step is 0·15 m
Total work done = Force × Distance
= 500 × 40 × 0·15 J
Power = Work done/Time taken
∴ Power of boy = 375 W

TEMPERATURE

The last of the basic quantities that we must be able to measure is temperature.

It is probable that you think that you know what you really mean by temperature, but you may find it difficult to put it into words and to distinguish it from heat. *Heat is a form of energy while temperature is the degree of hotness;* this is the short way of saying it. If two things have the same temperature then, if they are put in contact, neither will gain heat at the expense of the other. On the other hand, if two things of different temperatures are put in contact heat will flow from one to the other so that the cold one will become warmer and the hot one will become cooler. Remember that the hotter body does not necessarily possess more heat energy than the cooler. Throw a piece of red hot iron into the Atlantic Ocean: the iron will certainly lose heat but, equally certainly, the Ocean possesses far, far more heat than the iron.

In order to compare temperatures we must first invent a scale. There are two scales that are used in everyday life in Britain: the *Celsius* (*or Centigrade*) *scale*, which is the one that we shall always use in scientific work; and the *Fahrenheit scale*. On the Celsius scale the freezing point of water is called 0 degrees (0°) and the boiling point of water is called 100 degrees (100°). The Fahrenheit scale, invented about 1720, called the lowest temperature that could be reached at that time 0°, and blood temperature 96°. The odd figure of 96 was chosen because it can be divided by a large number of smaller figures. We can, however, now reach temperatures far below 0° F and so the scale is now defined as having the freezing point of water at 32° F and the boiling point of water at 212° F. This gives a temperature scale very nearly the same as the old one but with easy "fixed points" to refer to. On this scale the normal blood heat is now taken to be 98·4° F. This Fahrenheit scale is gradually going out of use.

Thermometers

The second necessity for the comparison of temperatures is an instrument that will give a uniformly changing reading as the temperature is increased from the freezing point of water to the boiling point of water. In the laboratory a mercury or alcohol thermometer is usually used. As the liquid in the thermometer is heated, it expands and moves up the fine tube of the thermometer. For a mercury thermometer we can thus mark on the 0° C point when it is in melting ice, and the 100° C mark when it is in the steam just above the boiling water, and then divide the stem into 100 equal divisions between these two points. A mercury thermometer can read between about −30° C and 300° C and an alcohol thermometer between about −100° C and 80° C. Below these lower limits the liquids would soon freeze and above the higher limits the liquids would soon boil.

Outside these ranges other methods have to be used. Going down to lower temperatures electrical methods are employed as far as −200° C while the expansion of gases can be used as low as −260° C. In the last few degrees before the lowest possible temperature is reached we have to use complicated indirect methods. This lowest possible temperature, when any matter would have no heat energy at all and would be completely still, is about −273·2° C. Later in this book (page 175) you will meet a third temperature scale, which is the really scientific scale, which has the lowest possible temperature (absolute zero) as zero or 0 kelvin (0 K). The freezing point of water is about 273·2 K, for one kelvin difference in temperature is the same as 1 degree C difference in temperature. Going up to higher temperatures electrical methods can be used up to about 1000° C and then we use the colour of the light given out by the hot body: the hotter the object the whiter the light given out because of its hotness. Above 2000° C or 3000° C we have to resort to other methods, such as calculating the amount of heat given out and working out the temperature from equations that have been discovered linking these two quantities. We know, in this way, that the temperature of the surface of the sun is roughly 6000° C and that other stars have surface temperatures of up to 25 000° C. Right at the top end of the scale comes the temperature at the centre of stars; in the case of the sun it has been estimated that the value is 20 000 000° C.

We often have to measure heat energy as well as temperature. The unit that we use is the same as for mechanical energy, the joule. Measuring heat is not just a matter of using a meter; we must either do experiments and calculations, or calculations by themselves, to find out how much heat is involved in any process.

Various methods of measuring heat energy by experiment and calculation will be discussed in detail later on in this book (page 154).

Anders Celsius, 1701-44, a Swedish scientist, whose scale had 0° for the melting point of ice and 100° for the boiling point of water—the Centigrade scale.

Lord Kelvin, 1824-1907, a British scientist, who originated the absolute scale with —273° C or 0 K for absolute zero: ice melts at 273 K, water boils at 373 K.

Fig. 25. *Two men who invented temperature scales.*

TABLE OF SIZES

Degree of size	*Time (s)*	*Space (m)*	*Speed (m/s)*	*Mass (kg)*	*Density (g/cm^3)*	*Temperature (°C)*
10^{30}				Mass of Sun		
10^{25}		Distance to furthest observed galaxy				
10^{24}				Mass of Earth		
10^{17}		Distance to nearest star				
10^{14}					Nuclear matter	
10^{11}						
10^{8}			Speed of light			
10^{7}	One year	Diameter of Earth		*Queen Elizabeth 2*		Inside star
10^{5}	One day				Inside star	
10^{4}			Speed of Earth round Sun			Surface of star
10^{3}		One kilometre	Speed of rifle bullet			Melting point of copper
10^{2}	$1\frac{2}{3}$ minutes					Boiling point of water
10^{1}					Lead	Spring morning
10^{0}		One metre	Walking pace	$1000cm^3$ of water	Water	0°C Ice melts
10^{-2}				A 5p piece		—183°C Oxygen liquefies
10^{-3}	Wing beat of fly	Thickness of pane of glass	Snail's pace		Air	
10^{-4}		Thickness of piece of paper				—273·2°C Absolute zero
10^{-5}				Postage stamp		
10^{-9}		Thickness of thin soap bubble				
10^{-12}		Distance between atoms in a solid				
10^{-13}				Red blood corpuscle		
10^{-17}		Diameter of proton				
10^{-22}	Smallest time scale known in atom					
10^{-24}					Inter-stellar space	
10^{-28}				Proton		
10^{-31}				Electron		

EXPRESSION OF LARGE AND SMALL FIGURES

There is a shorthand method of writing very small and very large figures. We can write, for example, 2×10^9 for 2 000 000 000. Here 10^9 means 10 multiplied by itself nine times, i.e. 1 000 000 000. Similarly 10^4 means 10 multiplied by itself four times, i.e. 10 000.

You will recall that we have already used this method in expressing the units of different quantities. For example we measure length in centimetres (cm), area in square centimetres (cm^2) and volume in cubic centimetres (cm^3).

For small figures we use a minus sign to indicate that the number must be divided by the power of ten, e.g. we can write 2×10^{-9} for 0·000 000 000 2 or $\frac{2}{1\,000\,000\,000}$. Note that $10^0 = 1$.

Let us now use this notation to compile a table comparing the degree of size of some of the quantities that we have been discussing. In compiling such a table you will notice that temperature, unlike all the other quantities, becomes negative and not fractional in the lower portion of the table. Notice also that things directly concerned with our everyday life occur about half-way down the table: the galaxies of stars are roughly the same number of times bigger than us as the atoms are smaller.

CHAPTER 3

ELEMENTS, ATOMS AND MOLECULES

BEFORE we go on to study the physics of everyday things, some idea of how everything is made up is necessary. So let us now review the evidence for the existence of a limited number of fundamentally different substances from which the entire physical world is constructed: such substances are known as *elements*. Let us also consider the idea that the elements are themselves constructed from incredibly small basic particles in the form of *atoms*.

Evidence for Elements

By an element, we mean a substance which can neither be split into different substances by normal physical means (dissolving, boiling, freezing, shaking, magnetizing, etc.), nor by normal chemical means. The simple evidence for the existence of these elements has been discovered mainly by chemists who, over the years, have found substances which can neither be split up into any simpler substance, nor formed from other substances.

When two or more substances are mixed together without any chemical action taking place, the resulting product is said to be a *mixture*. For example, iron filings and sulphur can form a mixture from which the iron filings can be removed by a magnet; salt and sand can form a mixture from which the salt can be washed out. If, however, a substance cannot be split up by any normal physical means, but can be made from, or split up into, other substances by chemical means, then that substance is called a *compound*. For example, iron sulphide is a compound formed by heating a mixture of iron filings and sulphur: after heating, the iron can no longer be separated by means of a magnet, but only by chemical means; zinc and sulphuric acid mixed together give off hydrogen which is an element, so that one of them, at least, must be a compound containing hydrogen.

In 1789 the French chemist Lavoisier published a list in which he named 23 substances correctly as elements; the list also contained such substances as oxides, alkalis, and even light, similarly named, but incorrectly so. Lavoisier's list was, however, the first to be reasonably accurate; of the 28 elements which had been obtained in a pure form at that time, 23 had been identified by Lavoisier as elements. By 1850 the number of named elements had risen to 59, and by 1900, to 82. There were no longer any incorrect inclusions in the list.

Today we know that there are 92 naturally occurring elements on earth. Scientists have recently manufactured about a dozen more elements, all of which are unstable, and they are likely to manufacture others in time.

The chemist, on purely chemical evidence, cannot be sure that there is no way of splitting up a substance. If after thorough investigation he has found no way of splitting up the substance, he can only assume that the substance is an element, but he cannot be sure. The elements identified in this manner were noticed, after some time, to have distinctive chemical properties which fitted them into several separate groups. By the beginning of this century the groups themselves had been fitted together to form a table, known as the *periodic table*, in which the 82 elements then identified were arranged according to their chemical properties. The elements seemed to fit together in the table rather like the pieces in a jig-saw puzzle, although certain pieces appeared to be missing. From these gaps in the table, chemists were led to suggest that certain elements had not yet been discovered, and from the position of the gaps in the table they predicted certain properties of these unknown elements. In time, their predictions were to be proved correct, for all the missing elements have since been discovered (such as the metals scandium, gallium and germanium); the table, however, had been founded on some inspired guesswork, for although the chemists had found that the elements fitted neatly together according to their chemical properties, they had no idea why they should.

Advanced physics was eventually to give sound reasons for the existence of the periodic table; it also presented some simple evidence for the existence of elements. One such piece of evidence depends on the different coloured lights given out by hot vapours.

Later in this book we shall discuss how white light, such as sunlight, can be split up into its component colours to give a continuous spectrum; such a continuous spectrum is illustrated in the coloured Frontispiece to this book. Coloured light can similarly be split up into its component colours. You will no doubt be familiar with the intense yellow light given out by sodium vapour street lights and with the glaring bluish light of mercury vapour street lights. If the light from these lamps is split up into its component colours, line spectra are formed (see Frontispiece): compare these line spectra with the continuous spectrum of white light. Physicists have found that the same line spectrum as that shown for sodium is obtained every time that a vapour containing sodium is used; if you hold some cooking salt (sodium chloride) in a gas flame, you will see the same yellow light as that from a sodium vapour street lamp. Physicists have found that every element discovered by the chemists has a different and distinctive line spectrum. One element, helium, was in fact discovered in the sun's atmosphere by this method before it was discovered on Earth. Such agreements as those found

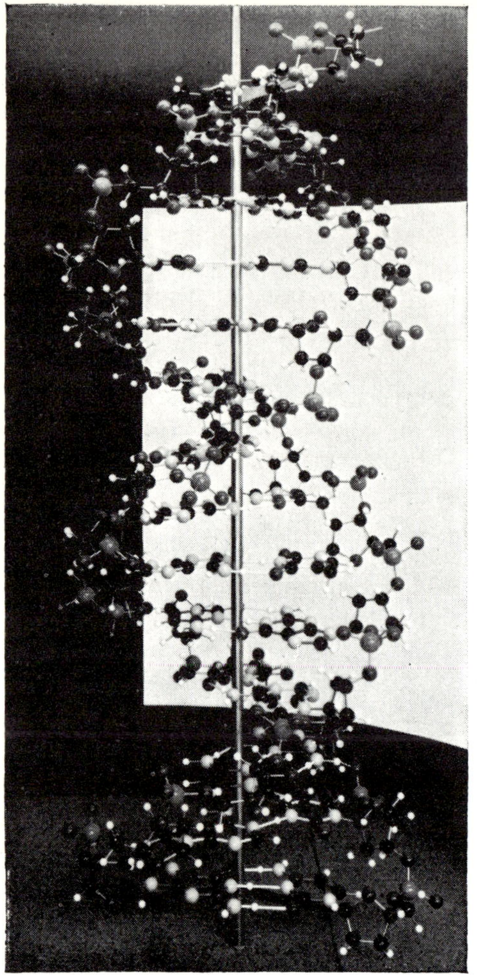

FIG. 26. *The photograph shows a model of a molecule of the nucleic acid DNA is formed from a complex arrangement of atoms of five elements: hydrogen; oxygen; carbon; phosphorus; nitrogen.*

between spectra and chemical behaviour greatly strengthens the evidence for elements.

Evidence for Atoms and Molecules

If we accept the evidence for elements, there are two possible models that we could imagine for matter. On one of these models, matter would be continuous; this would mean that if you cut up a piece of an element, then the bit you finished with, however small it might be, would still be the element. On the other model, matter would be made up of particles; the particles of any particular element would all be similar but they would be different from those of any other element. If you cut the particles up, however, you would no longer have your element.

You all probably know that the second model, the atomic model, is the one used by scientists, but you may not know why this has come to be accepted. Three pointers in this direction are given by elementary chemistry: the law of constant chemical composition; the law of multiple proportions; Avogadro's hypothesis.

The name of the *law of constant chemical composition* speaks for itself. It is found, for example, that water always consists of 1 g hydrogen to every 8 g oxygen (or, by volume, 2 volumes of hydrogen to every 1 volume of oxygen). It does not matter where the water comes from nor how it is split up nor how it is made, if it is pure then this is its composition.

On the continuous model, we would expect to be able to make the elements combine in any proportions, just as we can mix concentrated orange juice and water to make a drink as strong or as weak as we like. With the atomic model, on the other hand, we could well imagine a certain number of particles of one element combining with a certain number of particles of the other element, these numbers always being the same for a given compound. The name *atom* has been given to one particle of an element and the name *molecule* has been given to the group of atoms that is the smallest unit of a compound.

The *law of multiple proportions* states that in cases where elements combine together in different proportions, these proportions are related by simple whole numbers. For example, nitrogen and oxygen can combine in the following proportions:

1 g nitrogen and $\frac{4}{7}$ g oxygen $(1 : 1 \times \frac{4}{7})$
1 g nitrogen and $1\frac{1}{7}$ g oxygen $(1 : 2 \times \frac{4}{7})$
1 g nitrogen and $1\frac{5}{7}$ g oxygen $(1 : 3 \times \frac{4}{7})$
1 g nitrogen and $2\frac{2}{7}$ g oxygen $(1 : 4 \times \frac{4}{7})$
1 g nitrogen and $2\frac{6}{7}$ g oxygen $(1 : 5 \times \frac{4}{7})$

This suggests that atoms may, under some circumstances, combine together in different numbers to form different molecules. Thus an explanation is possible on the atomic model but it would be very difficult to find an explanation for this regularity using the continuous model of matter.

In 1808 Gay Lussac noted that when gases combined, the ratio of the volumes combining was always a simple ratio such as 1 : 1, 1 : 2, 2 : 3, etc. In volumes the nitrogen-oxygen compounds would look like this:

2 volumes of nitrogen and 1 volume of oxygen give 2 volumes of compound
2 volumes of nitrogen and 2 volumes of oxygen give 2 volumes of compound
2 volumes of nitrogen and 3 volumes of oxygen give 2 volumes of compound
2 volumes of nitrogen and 4 volumes of oxygen give 4 volumes of compound
2 volumes of nitrogen and 5 volumes of oxygen give 2 volumes of compound

and with water we have:

2 volumes of hydrogen and 1 volume of oxygen give 2 volumes of water vapour.

These volumes, of course, would be measured at the same temperature and pressure.

Three years later, Avogadro suggested that this would be the result of an atomic theory of matter, *if a given volume of any gas, at the same temperature and pressure, contained the same number of molecules.* A set of experiments was performed by Cannizzaro in 1858 to check one of the predictions of "*Avogadro's hypothesis*," as Avogadro's suggestion was called. Cannizzaro took equal volumes of a number of different gases each containing hydrogen, and found out the mass of hydrogen in each sample. Avogadro's hypothesis says that any sample of a gas with one hydrogen atom per molecule should have half the hydrogen, by mass, of all the samples with two hydrogen atoms per molecule, and so on. According to Avogadro's hypothesis, a sample of gas such as hydrogen chloride (which contains one atom of hydrogen per molecule) would contain half the amount of hydrogen of an equal sample of a gas such as water vapour (which contains two atoms of hydrogen per molecule). Cannizzaro took equal samples by volume of six different gases containing hydrogen, and obtained the following results:

Pure hydrogen	2 units by mass of hydrogen
Hydrogen chloride	1 unit by mass of hydrogen
Water vapour	2 units by mass of hydrogen
Ammonia	3 units by mass of hydrogen
Propane	8 units by mass of hydrogen
Methane	4 units by mass of hydrogen

Let us consider Cannizzaro's results. They suggest that there may be only one atom of hydrogen in a molecule of hydrogen chloride; two atoms of hydrogen in a molecule of water vapour; three atoms of hydrogen in a molecule of ammonia, etc. The possibility must also be considered that there are two, four and six atoms in the respective molecules, or three, six and nine; yet in the many experiments carried out, no gas was ever found containing less hydrogen than hydrogen chloride. We may accept the first set of figures and state that there is only one atom of hydrogen in a molecule of hydrogen chloride. This having been done, note how Cannizzaro's results indicate that pure hydrogen exists in a molecular form containing two atoms.

The chemist can now go on to compare the masses of the atoms of different elements and, the most conclusive evidence of all, can measure the number of molecules in a given volume of gas. The standard volume used is that occupied by 32 g of oxygen at 0° C and standard atmospheric pressure i.e. 22 400 cm^3. The number of molecules in this volume is about 6×10^{23}: this is called *Avogadro's number*. It is always found to be the same, however and wherever it is calculated, with any gas. The fact that we can calculate the number of molecules is very strong evidence indeed for their existence.

Physical evidence to support the existence of atoms and molecules occurred as early as 1827. In that year the biologist, Robert Brown, observed through a microscope the curious behaviour of grains of pollen suspended in water; they were neither still nor did they move steadily in any particular direction, but moved jerkily in all directions. Further investigation showed that any very fine particles suspended in air behaved in a similar way and an apparatus, which we shall discuss later (page 51), was designed to study the phenomenon more carefully. This curious movement of the grains of pollen was readily explained in terms of atoms and molecules; the molecules of the air, being in constant and random motion, continually collide with the grains of pollen. The observed fact that larger particles suspended in the air do not exhibit this movement was explained by the large number of air molecules that would be continually hitting such a particle on all sides, the effect of these many collisions cancelling each other out.

The discovery of the radioactive element radium in 1898, as we have already noted, led to the founding of an entirely new branch of physics devoted exclusively to a study of the atom. With this birth of atomic physics, matter was soon to yield up many of its secrets. Not only was evidence for the existence of atoms and molecules discovered, but also evidence for the existence of such particles as *electrons*, *protons* and *neutrons*, from which atoms are built up. Experiments were carried out to demonstrate the manner in which these atomic particles combined together to form the many different atoms while the particles themselves were accurately investigated and measured.

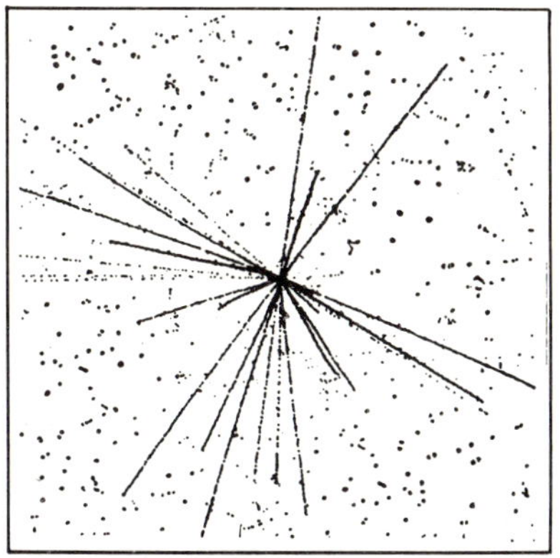

FIG. 27. *One of the first photographs of the disintegration of an atomic nucleus by bombardment with a high energy proton.*

The evidence presented by the physicists was mostly in agreement with the atomic theory of matter, and yet in very advanced physics some disagreements were noticed.

Already we have seen how Newton's theory of gravitation was replaced by Einstein's, and yet how Newton's theory is still perfectly adequate for our understanding of the facts considered in this book: we learnt that theories are not facts, but are a way of looking at the facts in order to understand them. And so it is with the atomic theory of matter; theory gave rise to observed facts, which in their turn gave rise to a new theory. The new theory does not concern us here, for when you come to study atomic physics in the last section of this book, you will find that the atomic theory of matter is all that you need for understanding the facts considered.

PHYSICAL PHENOMENA AND THE ATOMIC THEORY

So far in this chapter we have considered evidence for the existence of atoms and molecules. None of you, however, has ever seen an atom by itself, nor has anyone else. A table, for example, does not look as if it is a collection of particles; it looks and feels solid. Now it is time to see how the theory of atoms fits in with the relatively large-scale things we see happening around us in everyday life. While you study the main part of this book you should be aware of the millions upon millions of small-scale happenings that make up each event which is investigated.

Solids, Liquids, and Gases

Let us first see how the atomic theory explains the differences between solids, liquids, and gases. In solids the atoms do not move about, for if they did nothing would stay the same shape or pattern. It would appear that in some solids (see photograph, Fig. 2) the atoms or molecules are in a pattern, while in others there is no evidence of an orderly arrangement. How are the atoms kept in their places? Presumably the cause of this must be some attractive force between the atoms strong enough to hold each one in position.

In liquids the molecules obviously can move about. There are though, some limitations to this movement, for rain falls in drops and water in a bucket stays at the bottom of the bucket. This suggests that there are forces between the molecules but that these forces, although sufficient to keep all the molecules in a group, are not strong enough to keep individual particles in one particular place. In fact, the molecules are not even kept in their groups indefinitely; rainwater on a pavement, for example, will soon evaporate in summer sunshine.

In gases it is equally obvious that the molecules can move about and the speed of movement is evidently much greater than in liquids. Think of what happens when a smelly gas, such as coal gas, hydrogen sulphide or ether, is released on one side of a room. On the other side of the room you will smell it after only a very short time interval. You also know that you can walk through a gas very easily (air for example), so the molecules must be able to move out of the way easily. These facts, combined with the very low density of gases, suggest that the molecules are now far apart and moving independently. For this to be so the forces between the molecules in a gas must be negligible.

The pressure of a gas, on this theory, must be caused by the bombardment of the containing walls by the gas molecules. Atmospheric pressure, the pressure of about 10 newtons on every square centimetre of your body and on everything else, is caused by just this effect. If more molecules are put into a given space this theory predicts that the pressure should increase. You will know that this does happen from pumping up a bicycle tyre, for example.

In a liquid or in a solid the weight of all the molecules is transmitted directly to the support, since the molecules are effectively in contact.

Heat

Heat, as we have already said, is a form of energy. On the atomic theory of matter this energy can be the energy of movement of the atoms or molecules. The hotter a substance is, the faster the atoms or molecules are moving. In solids this movement can only be vibration, whilst in liquids and gases movement from place to place and rotation are also possible. Now we can see a reason for heat being needed to turn a solid into a liquid or a liquid into a gas (you know that heat is needed to boil away water). The heat supplied is the energy that the molecules need in order that they may escape from the pull of all the other molecules around them.

This theory also explains why the pressure of a gas goes up when we heat it. All the molecules are now moving faster and therefore collisions with the walls of the container are more violent and more frequent and the pressure increases.

Light and Sound

Some things you will meet in this book need something else apart from atoms to explain them. Light, radiant heat, and radio waves appear to have no matter attached to them. Since they can reach us from the sun and from distant stars, they appear also not to need matter for their transmission. Here there is a transfer of energy from one place to another with little or nothing happening to anything in between.

Other things, like sound and waves on the sea, have no atoms of their own but are the movement of atoms. Sound is the movement of air molecules backwards and forwards whilst water waves are the movement of water molecules up and down. Here energy is transmitted from one place to another without matter being transmitted as well, although matter is involved in the process. Sound, unlike light, cannot pass through a vacuum for sound is a function of matter.

Gravity

We will also meet forces, due to various causes, which act upon atoms. We have already mentioned the attractive forces which must exist if liquids and solids are to stay together. In addition there are gravitational, magnetic and electric forces. Gravity is the long-range attraction between pieces of matter. It affects all matter and so everything attracts everything else. This is the reason why you remain on the earth. There is a gravitational force between you and the earth which is quite strong enough to pull you down again if you jump upwards. We say that a "field of force" exists at any point where a force would be exerted on a suitable object placed there. It is the gravitational field of force due to the earth which causes all solids to fall to the lowest possible level, all liquids to find their lowest level, and the gases of the atmosphere to stay around the earth instead of drifting off into space. It is also why people in Australia and Britain both think they are standing upright though one is upside down in respect to the other. Gravitational forces, too, are the cause of the moon going round the earth and of the earth going round the sun.

Magnetism and Electricity

You will have met magnetic fields, it is certain, and you will know that magnets can exert forces on other magnets and on pieces of iron and steel even if they are not touching. Note that in this case the forces are not exerted equally on all matter; for example, a magnet will not pick up wood or copper. You may not have met electric fields yet, but you will do so later in this book, and you will find that they share many of the characteristics of magnetic fields.

It should also be mentioned, although it cannot be checked at this stage, that there can be waves (travelling disturbances) in fields of force. What we call light is just such a wave in electric and magnetic fields. This is the mechanism by which the energy is transferred from one place to the other. This is why light is called an electromagnetic wave. Similarly, radiant heat, ultra-violet light, radio waves and X-rays are also forms of electromagnetic radiation. Any atom or molecule which is moving may cause an electromagnetic wave to start out. When this happens some of the energy of motion of the atom or molecule is lost by the particle and it is this energy which is transferred elsewhere by the wave. The reverse process can also occur. The energy of a wave can be absorbed by a particle, the wave disappears and the particle has increased energy of movement. These processes are happening all the time, everything is giving out radiation and absorbing other radiation.

Finally, electricity: later in this book you will find evidence for the idea that an electric current in a wire can be explained as a flow of electrons (one of the three main constituents of atoms) through the wire. Electricity provides another means of transferring energy from one place to another.

Limitations of Science

A scientist looks at the world around him and attempts to explain what he sees.

In the first chapter of this book we saw that he achieves this only when experiment and theory go hand-in-hand. We saw also how sometimes theory becomes confused with fact; how theories must be recognized as only attempts to explain observed facts; how theories change.

In the second chapter we saw how, in order to explain what he sees, the scientist formed basic ideas of time, space, mass and temperature, and then devised accurate means of measuring these quantities. With clear ideas of

FIG. 28. *Energy released by the sun as heat and light; the transmission of this energy, through space, to the earth; the reflection, absorption and conversion of this energy. These are but a few of the problems of physics that this picture brings to mind.*

space and time, the scientist was able to form ideas of speed velocity and acceleration; similarly, from the idea of mass the ideas of force, energy, and power grew.

Turning to the matter from which the things around him are built up, we saw in this chapter how the scientist formed ideas of elements, compounds and mixtures, and how, in turn, he explained these ideas in terms of another idea, that of atoms. With the atomic theory the scientist was able to explain the differences he had already observed between solids, liquids and gases. He explained how water evaporates, the nature of heat and much else, yet, turning to the behaviour of radiant heat or light for example, the atomic theory could no longer offer an explanation for what was known to happen. A theory had once again shown its limitations.

As science advances and theory succeeds theory, perhaps the day will come when a scientist by means of a single theory will be able to explain *how* any part of the known universe works, be it the evaporation of water, the travel of light from some distant galaxy or the magnetic field of the earth. But *why* the universe works, why it or any part of it exists, are questions that neither experiment nor theory, in physics or in any other branch of science, will ever answer.

CHAPTER 4

QUESTIONS YOU MIGHT BE ASKING

THERE are many questions which inquisitive people might ask about the things and happenings they see around them. Here we shall consider some of the questions that can be answered by the physicist. A systematic study and thorough understanding of all that follows in this book will provide you with the answers.

Forces, Motion and Energy

What is gravity?
Is anything immune from gravity?
Does your weight change when travelling in a lift?
Would your weight and mass be different if you were on the moon?
How high is the atmosphere?
How does a barometer measure the pressure of the atmosphere?
Why does a soap bubble not collapse under such a pressure?
Why can people float in water but not in air?
What is the purpose of a three-speed gear on a bicycle?
Why can you lift a car with the help of only a car jack?
Why is it easier to pull a heavy garden roller rather than to push it?
Why do projectiles travel in curved paths?
Why do cyclists lean inwards when rounding a corner sharply?
Why do guns recoil?

FIG. 29. *Energy from food has been converted by the footballer into mechanical energy, some of which is now transferred to the ball.*

FIG. 30. *Balloons being filled with hydrogen. Hydrogen is much lighter than air; a balloon will rise when its weight is less than the weight of air it displaces. The balloons would also rise if filled with hot air, hot air being less dense than cool air.*

Heat

What is heat?
How does the heat of the sun reach the earth?
How could you measure the temperature of a furnace?
Why are regular gaps left between railway lines?
Why does a hot glass usually crack when touched with a cold knife?
What makes a balloon filled with hot air rise?
Why does the human body tolerate hot air when water at the same temperature would seriously blister the skin?
How does a refrigerator work?
Why does salt thrown on an icy pavement melt the ice?
What is the lowest temperature possible?
Why is the water at the bottom of a waterfall warmer than that at the top?
Why is a saw hot after use?
How does a car engine work?

FIG. 31. *Diesel-electric locomotive. Heat energy from burning fuel is here first converted into electrical and then mechanical energy.*

FIG. 32. *Materials through which light can pass, such as glass or water, also reflect light under certain conditions.*

Light

What is light?
How fast does light travel?
Can light travel round corners?
Is it possible to have light without heat?
What is a mirage?
How does a mirror work?
Why are many things coloured?
Why is it cooler to wear white clothes on a hot sunny day?
How is white light split up into coloured light?
Can light be stored?
How does a camera work?
How do microscopes and telescopes work?
How do our eyes see things?
Why do some people wear spectacles?

Sound

What is sound?
How is sound produced?
How fast does sound travel?
How does the ear work?

FIG. 33. *Each instrument in the orchestra can be recognized by the human ear even while the others are all playing.*

How do we measure sound?
Can there be sound which we cannot hear?
How does a bee in flight make a humming sound?
Why do sounds in an empty room sound louder than when the room is furnished?
Why are sounds at night more audible than during the day?
Why does the same note played on different instruments sound differently?
What is the real difference between notes of different pitch?
Why are there harmonies and discords?
How do musical instruments work?
How does a siren work?

Magnetism and Electricity

What is magnetism?
How are magnets made?
Why can some substances be magnetized but not others?
What is electricity and how is it connected with magnetism?
How fast does electricity travel?
Why do some things conduct electricity whilst others do not?
How does a dry battery produce electricity?
For what purpose are fuses used in the electric wiring of a house?

What is the difference between d.c. and a.c. electricity?
How is chromium plating carried out?
How does a telephone work?
How is sound recorded onto a film?
How do dynamos and electric motors work?
What causes a flash of lightning?

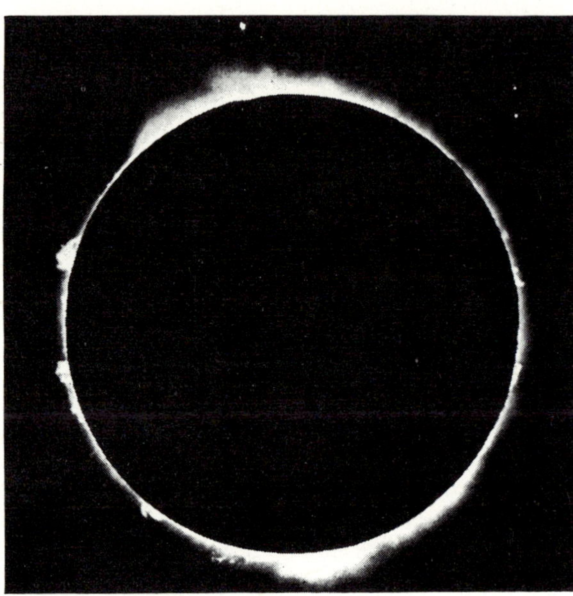

FIG. 34. *Total eclipse of the sun: the energy produced by the sun as heat and light is thought to be the result of nuclear fusion. Although nuclear fusion has been achieved by scientists in the hydrogen bomb, it cannot yet be controlled for peaceful uses.*

Nuclear Energy

What is an atom?
How big is an atom?
What are atoms made of?
How are atoms held together?
What produces the picture on a television screen?
What is radioactivity and why is it dangerous?
How is radiation detected?
How long does radioactivity last?
What are alpha, beta and gamma rays?
What are X-rays?
What are isotopes?
From where does the energy come when a nuclear explosion occurs?
How is the energy controlled in a nuclear reactor?
What is the difference between nuclear fission and nuclear fusion?

EFFECTS OF FORCE

CHAPTER 5

MOLECULAR THEORY OF MATTER

ALL gases have certain simple properties in common. They fill their container and exert a uniform pressure over all its surface. In this respect they differ from liquids and solids. The *pressure* of a gas is caused by the tiny moving particles called *molecules*, which are continually colliding with the molecules of the walls of the containing vessel and with each other. The effects of these continual collisions may be shown by using an apparatus similar to that shown in Fig. 35.

BROWNIAN MOTION

In the apparatus shown in Fig. 35, the two clips are opened and a puff of cigarette smoke is admitted to the metal box. The clips are then closed.

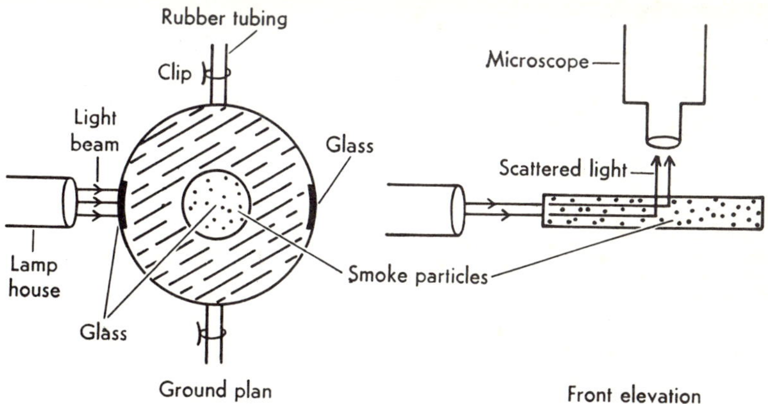

FIG. 35. *Apparatus for studying Brownian motion.*

A strong beam of light is directed through a glass window in the side of the box, so that it shines right through the box. Some of the light is scattered by the tiny smoke particles, upwards into a powerful microscope which is focused on some of the smoke particles. These particles of smoke are large enough to be seen through the microscope. The molecules of air in the box are much too small to be seen through the microscope. But if the smoke particles are observed for several minutes, it may be seen that they are continually moving in a jerky manner, through small distances, in all directions. What causes these small movements? The smoke particles are continually bombarded by the fast-moving molecules of air in the box. Thus, although the air molecules cannot be seen colliding with one another, the effects of their collisions with the smoke particles may be seen.

This phenomenon was first observed in the random motion of small particles suspended in water, caused by the bombardment of the suspended particles by the water molecules; the motion never ceased. The Scottish botanist Robert Brown (1773-1858) made this discovery and the phenomenon is named after him—the Brownian motion. Here is an experiment which will enable you to understand this motion.

Place a number of small marbles in a tray, together with two or three larger marbles. Continually shake the tray in a horizontal plane and notice that the movements of the small marbles are more rapid than those of the larger marbles, which are being jostled by the small ones. What do the small marbles represent? What do the large marbles represent?

Gases move easily, diffuse among one another and travel slowly through porous walls. Heating a gas makes its molecules move more quickly. It is assumed by scientists that the molecules of a gas exert no appreciable forces on one another except during collisions. When they come very close together, they exert strong repulsive forces for a very short time; this is a collision.

The molecules of a solid are in an orderly crystalline arrangement and do not wander about, as do the molecules of a liquid or a gas. They keep more or less to the same positions, but vibrate more and more, as the solid is made hotter. If heat energy is supplied to them, this orderly arrangement is broken down as the solid melts to liquid state, and the molecules are able to wander about in the space enclosed by the containing vessel and the free surface of the liquid. The molecules of a liquid are still very close together. When a liquid evaporates, the faster moving molecules escape, taking away more than an average share of heat energy and thus cooling the remaining liquid.

Molecules in gases are much further apart than they are in liquids and solids and move very quickly. The higher the temperature, the more quickly they move.

MOLECULAR ATTRACTION AND REPULSION

There is much evidence for the belief that matter consists of very small particles called *molecules*. These molecules are too small to see with an ordinary microscope, but we believe them to be in a state of continuous, irregular and rapid motion. Many of the properties of matter, whether in the form of solid, liquid or gas, may be interpreted in terms of the motion of molecules.

Liquids are very difficult to compress, that is, to crowd into a smaller space.

On the other hand, water in a tumbler remains there; it does not expand or fly apart. The fact that liquids are difficult to compress suggests that liquid molecules repel each other when they are very close together. But since the molecules of a liquid do not fly apart, they must also attract each other. There are thus two kinds of mutual forces between molecules: *repulsions* when the molecules come very close together and *attractions* at greater distances. The repulsive forces make it difficult to compress a liquid or to compress most solids.

If a solid is stretched, the attractive forces become more effective than the repulsive forces and the solid is left in *tension*, that is, trying to recover its original length. But you should understand that these attractive forces only act over short distances, too. The repulsive forces between molecules account for the collisions between the molecules of a liquid with each other, and the collisions between the molecules of the liquid and the molecules forming the walls of its containing vessel. These collisions produce the pressure of the liquid on the walls of the vessel.

The attractive force that causes like molecules to cling together is called *cohesion*. Cohesion binds the elementary particles of copper together to form a rivet; those of carbon together to form a diamond; and those of water together to form a drop of water or a lump of ice. The attractive force that causes molecules of one substance to cling to the molecules of a different substance, however, is called *adhesion*.

Capillarity

Place a very narrow bore glass tube, called a *capillary tube*, open at both ends, into water (Fig. 36). Note how the water rises to a greater height in the capillary than it stands in the vessel containing the water. Try the effect of using capillary tubes with wider and narrower bores, and you will find that the water rises highest in the narrowest tube. Place three clean capillary tubes of equal diameters, one into water, one into water to which a small quantity of detergent has been added and one into mercury. You will observe that water rises higher than the detergent solution. It seems that the adhesive forces between glass and water molecules are greater than the cohesive forces between water molecules. The

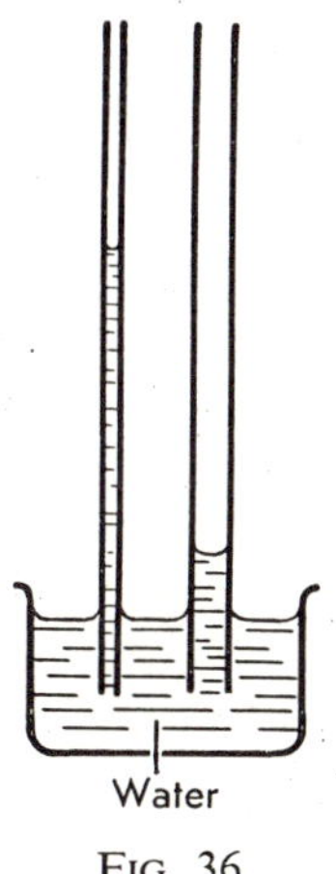

FIG. 36.

water rises up the tube until the upward pull of adhesion is balanced by the weight of the water in the tube. Capillary action causes liquids to be drawn into blotting paper, filter paper, soils, towels and lumps of sugar, to name but a few of the countless examples of this phenomenon. But observe what has happened to the capillary in the mercury (Fig. 37): the mercury is at a lower level than that outside. Furthermore, the surface of the mercury in the capillary is convex and not concave. This contrasting behaviour of the mercury seems to be caused by the cohesion between mercury molecules being greater than the adhesion between mercury and glass. In general, liquids are raised in open capillaries that they wet or to which they adhere. Liquids are depressed in capillaries that they do not wet or to which they do not adhere.

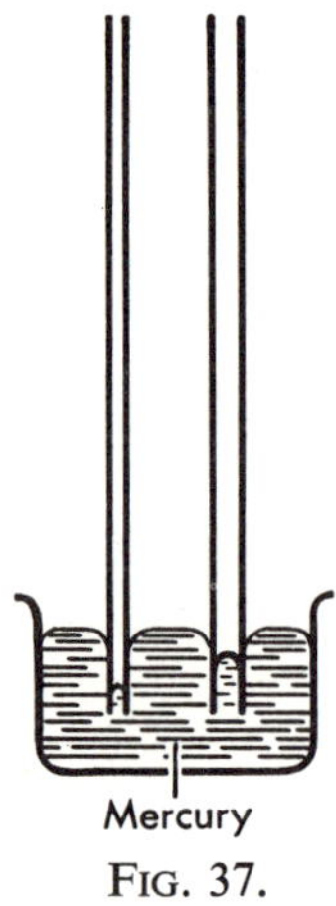

FIG. 37.

Surface Phenomena in Liquids

Try the following experiments:

1. Gently drop a double-edged, thin razor blade, with its large surface horizontal, on to the surface of water in a dish.

2. Gently drop a small needle (not a pin with a head) on to the surface of water in a shallow dish; or, if you find this difficult to do, float the needle on a small piece of blotting paper on the surface of water: after a while, the blotting paper will sink, leaving the needle floating. Now add some detergent or alcohol to the water and try to "float" the needle. In each of these experiments examine the surface of the water near to the edge of the blade and around the needle, with a hand-lens or magnifying glass. Notice that the surface is slightly depressed (Fig. 38).

3. Float three drinking straws on the surface of water in a bowl, to form an equilateral triangle *ABC*. Using an "eye-dropper" or small pipette gently drop one or two drops of a detergent at the point *D* inside the triangle, and note what happens (Fig. 39).

These interesting experiments indicate that there is a tension in the surface of a liquid, that tends to oppose an object trying to break through the surface, and also tends to prevent the molecules of the liquid itself breaking away through the surface. This force at the free surface of the liquid is called *surface tension.* It may be explained in terms of the molecular theory: each molecule of liquid attracts every other molecule of liquid (cohesion), but the

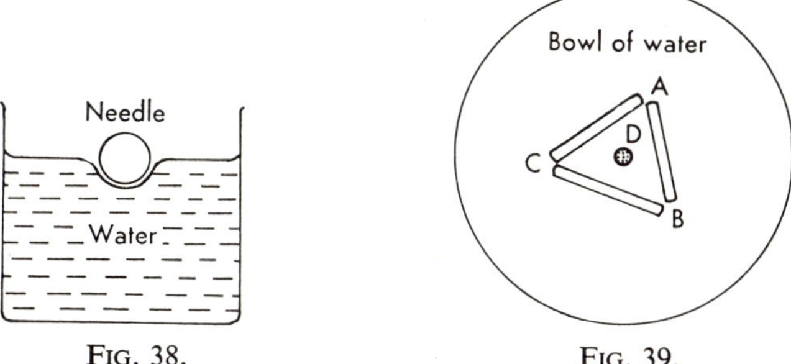

FIG. 38.

FIG. 39.

force decreases as the distance between the molecules increases. So when two molecules are separated by a certain distance, the mutual force of attraction becomes negligibly small. This distance is called *the radius of the sphere of molecular attraction.*

The molecules of a liquid are close together, and each molecule has an attracting force on a few molecules close to itself. But this attracting force decreases rapidly as the distance between molecules increases. For a molecule well within the body of a liquid, the average force on it during an interval of time will be zero (Fig. 40*a*). But for a molecule near to the free surface of the liquid (Fig. 40*b*), there will be more molecules pulling it into the body of the liquid than trying to pull it out of the liquid; there will be an average resultant force pulling the molecule into the body of the liquid. This force

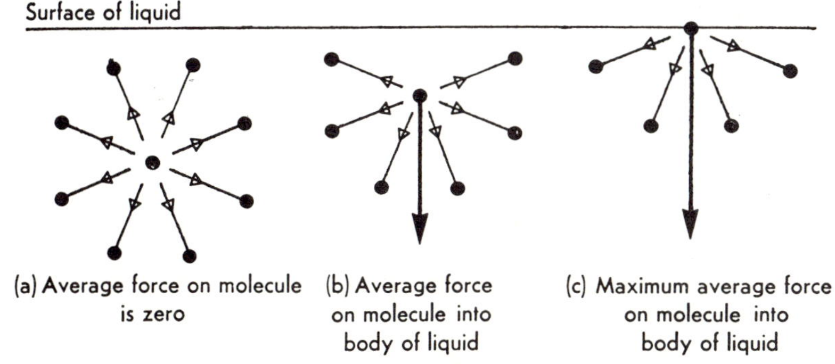

FIG. 40. *Forces acting on a molecule in a liquid.*

on any one molecule increases, the nearer the molecule gets to the free surface, reaching a maximum at the surface (Fig. 40*c*). It should be remembered that this surface effect on any molecule extends only a very short

FIG. 41. *Raindrops on a window-pane; a surface tension phenomenon.*

distance below the free surface (a few molecular diameters), but it makes it difficult for a molecule to escape from the body of the liquid.

Surface tension causes a falling drop of liquid to tend to assume a spherical shape. The sphere is the shape that has the minimum surface area for a given volume. Lead shot is produced in Bristol by allowing molten lead to fall from a sieve near the top of a tower into a pool of water many feet below. The liquid drops of lead tend to take the shape having the smallest possible area for a given volume, a spherical shape.

Notice what happens when water is sprinkled on to a cabbage leaf: it forms into drops, which easily roll off the leaf. Because of this, some form of soap, detergent or "wetting solution" is added to the sprays used for killing green fly and black fly on roses and vegetables. Since the wetting solution has a much lower surface tension than water, the solution sticks to the leaf, thus giving time to kill the pests. A wetting agent is also used in developing in photography to ensure that the developer spreads quickly over the film or print to prevent irregular and patchy development.

Viscosity

You will be aware that it takes much longer to pour treacle (or glycerin or lubricating oil) from one vessel to another, than it does to pour water. We say that treacle is more *viscous* than water. The viscous drag of a liquid depends upon the temperature of the liquid. Thus, the housewife will warm the golden syrup before serving it for lunch and it will flow from the sauce-boat more rapidly; its viscosity has been reduced.

Frictional resistance between solid surfaces is reduced if the surfaces are wetted. Some liquids, especially hydrocarbons, are very effective for this purpose. The solid surfaces are kept slightly apart by a layer of the lubricant and the "dry" friction is replaced by internal viscous friction, which is the lesser of the two.

It is probably the cohesion between molecules of the same liquid that causes resistance to flow, tending to restrict the moving of one layer of molecules relative to a neighbouring layer. This dragging effect on the faster moving molecules by the slower moving molecules varies considerably from one kind of molecule to another. When the temperature of a liquid is raised, the molecules move about more violently and are more able to overcome the cohesion between molecules and so the viscous drag is reduced.

Here is a simple but interesting experiment which you can carry out for yourself to illustrate the phenomenon of viscosity.

Compare the times required to pour, say, 50 cm³ of water, 50 cm³ of treacle at room temperature, and 50 cm³ of treacle at a temperature of about 50° C from one vessel to another.

DIFFUSION OF FLUIDS

Nearly fill a gas jar with water. Slowly pour a concentrated solution of copper sulphate (or potassium permanganate) through a long thistle funnel which reaches to the bottom of the jar. Since this solution is denser than water, a fairly sharp line of separation between water and the solution will result (Fig. 42*a*). Very carefully remove the thistle funnel; if you find that you are disturbing the copper sulphate, the funnel may be left in the jar. After several days, the coloured solution will spread upwards and, in time, the two liquids will mix to give a uniformly coloured solution (Fig. 42*b*). This movement of the heavier liquid up into the lighter one is known as *diffusion*. Both

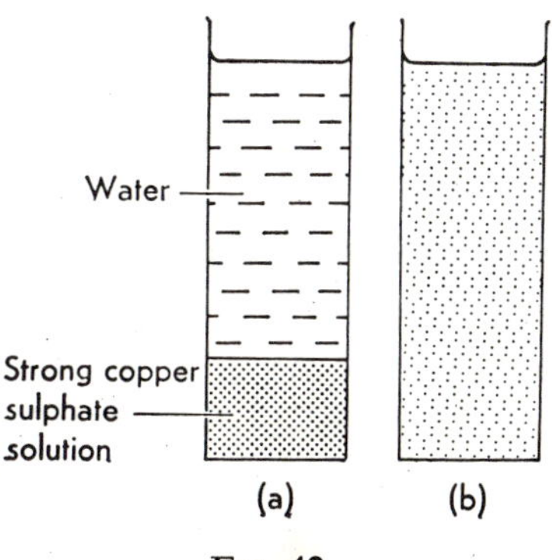

FIG. 42.

the particles containing the copper and the water molecules are moving all the time; the collisions that result cause some of the copper particles to spread throughout the whole body of the liquid. But the copper particles which give the blue colour to the solution are heavier than the water molecules. They move only slowly as a result of collisions.

Diffusion of Gases

If a gas jar of hydrogen is placed above a gas jar filled with air (Fig. 43) and allowed to stand for a short time, diffusion will take place. The contents of the lower jar will burn with a slight explosion when tested with a burning taper, showing that some of the lighter hydrogen molecules have diffused downwards.

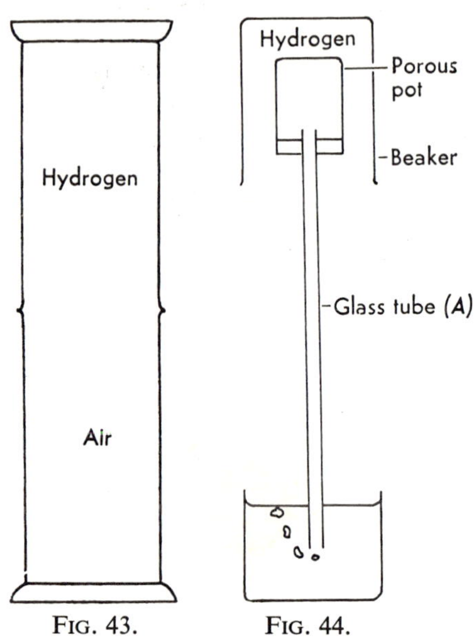

FIG. 43. FIG. 44.

The atmosphere we live in consists mainly of two gases: oxygen and nitrogen. Oxygen is heavier than nitrogen, so we might expect that the two gases would form two separate layers. But this is not the case. All gases interpenetrate each other in the same manner as the copper particles interpenetrated the water molecules. As a consequence of this *gaseous diffusion*, the atmosphere does not consist of two separate layers but of a homogeneous mixture.

Diffusion takes place more quickly in a gas than in a liquid because there are fewer molecules per $cm.^3$ But diffusion is still not a very quick process since even in a gas the molecules experience a large number of collisions in a short time. Furthermore, the rate of diffusion is not the same for all gases as can be shown by the following experiment.

Fig. 44 shows an unglazed earthenware pot, surrounded by an inverted beaker of hydrogen, coal gas, or any other gas lighter than air. The pot is full of air and fitted with a rubber bung through which passes a glass tube, *A*, dipping into water. A stream of bubbles will be seen to issue from the lower end of the tube, indicating that the hydrogen is diffusing into the pot at a faster rate than the air is diffusing out of it. When the beaker is removed, water will rise in the glass tube *A*, showing that the hydrogen diffuses outwards at a greater rate than air diffuses inwards.

A gas heavier than air (such as carbon dioxide) may be used for this test, by inverting the porous pot and placing the beaker of carbon dioxide upright. Water will rise in *A*, showing that the carbon dioxide diffuses more slowly than the lighter air. It has been shown experimentally that, for a gas, the following relationship between rate of diffusion and density holds true:

"*The rate of diffusion of a gas varies inversely as the square root of the density of that gas.*"

Osmosis

If wrinkled peas, raisins or dried prunes are left to soak in water, they begin to swell. The water passes inwards through the cell-wall of the fruit at a much greater rate than the juices pass from the inside of the fruit. In fact, if you taste the water after the fruit has swollen, you will detect very little flavour. The passage of the *solvent* through the cell-wall, but not of the *solute*, produces an effect called *osmosis*. The cell-wall of the fruit contains a layer through which this osmosis takes place and this is known as a *semi-permeable membrane.*

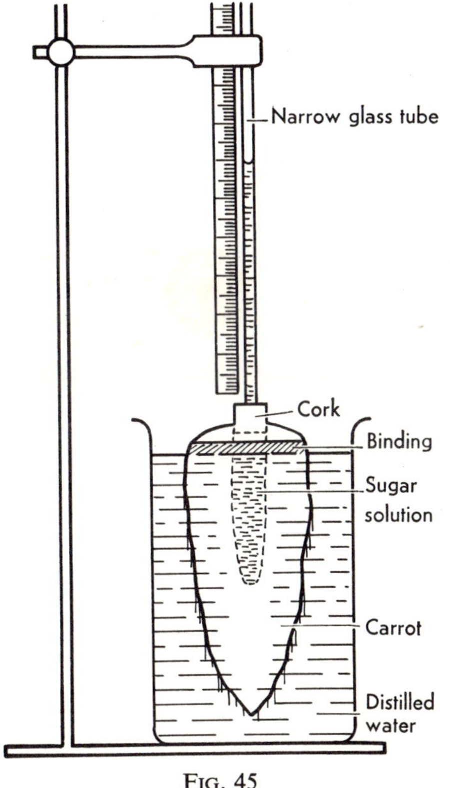

FIG. 45

Here is an interesting experiment that you might do at home. Cut off the wide end of a medium-sized carrot. Bind some tape around the carrot near to the top to prevent it from splitting. Bore a hole in the carrot 5-8 cm deep and about 1 cm wide. Fill the hole with a concentrated sugar solution, then tightly fit a cork through which passes a narrow glass tube about 1 m long. Place the carrot in a tall beaker containing distilled water (Fig. 45). You will notice that after a short time, the water begins to climb up the glass tube and might move $2\frac{1}{2}$ cm in 10 minutes at first. After many hours it will probably have climbed about 50 cm.

So your experiment shows that the distilled water passes inwards through the walls of the carrot more quickly than the water from the strong sugar solution passes outwards; the outer layers of the carrot evidently contain a

FIG. 46. *Osmosis has caused the 50 dried peas* (left) *to swell considerably* (right) *after only a few hours in water.*

semi-permeable membrane. But taste the water in the beaker; you will find that it is slightly sweetened. A few sugar molecules have passed through the cell-wall of the carrot showing, in this case, that the semi-permeable membrane is not a perfect one.

Osmosis plays an important part in the passage of liquids through plants, and in the absorption of food and disposal of waste by cells in plants and animals. It explains, in part, how sap solutions are forced to the tops of trees.

Can you now explain the wrinkled appearance and unpleasant "dry" sensation which results when you have had your hands for some time in water containing washing soda?

QUESTIONS

1. What is meant by (*a*) cohesion, (*b*) adhesion? Give examples.

2. Explain why the fabric of an umbrella, although porous, does not allow the rain to pass through it.

3. A light sieve made of a fine mesh gauze is coated with wax or resin and the sieve is found to float on water. Explain why.

4. Water does not pass through the canvas of a tent unless the inside of the canvas is touched by a finger. Explain this.

5. Why does water rise in a capillary but mercury is at a lower level in the capillary than in the vessel outside?

6. Close the bowl of a porous clay pipe with a bung, connect the stem by rubber tubing to a glass tube about 30 cm long and dip the other end of the glass tube into a bowl of water. Fill a jar with coal gas and place it inverted over the bowl of the clay pipe. Observe the results. Remove the jar of coal gas and again observe the results.

CHAPTER 6

SOLIDS IN MOTION

SPEED is the rate at which distance is covered, that is, the distance per unit time, found by dividing the distance by the time. It may therefore be expressed in terms of metres per second, kilometres per hour, etc. If a body travels equal distances in equal intervals of time, no matter how short those intervals of time may be, then the body is said to have *uniform speed.* Frequently a moving body does not have a uniform speed. It is still helpful, however, to know the *average speed* of a car, a speed-boat or an athlete. This is found by dividing the total distance travelled by the time taken.

$$v_{average} = \frac{s}{t}$$

For example, a car travelling the 176 km from London to Birmingham in $2\frac{3}{4}$ hours, travels at an average speed of $176/2\frac{3}{4}$ or 64 km/h. But the car does not move at 64 km/h for the whole of the 176 km. Similarly a very good sprinter covers 100 m in 10 seconds. His average speed is 10 m/s. He will be travelling faster than that as he crosses the finishing line. Why?

The distance that a moving body travels, is found by multiplying the average speed of the body, by the time taken to cover the distance.

$$s = v_{av}\, t$$

Velocity

Velocity strictly depends on three factors and not on two. These factors are distance, time, and direction. So if we want to know the velocity of an aircraft, we must state not only its speed, but also the direction in which it is travelling. Thus, an aeroplane might have a speed of 500 km/h, while its velocity is given as 500 km/h in a N.W. direction. Any quantity which requires a statement of two characteristics, magnitude (size) and direction, is called a *vector quantity* and may be represented by an arrow whose length drawn to scale represents the magnitude, and whose direction is in the direction of the vector. Thus, the velocity of the aeroplane mentioned above may be represented as in Fig. 47. Other vector quantities besides velocity are acceleration and force.

Quantities that are not affected by direction are called *scalar quantities* such as mass, volume, energy, quantity of heat. Scalar quantities may be added arithmetically, vector forces may not.

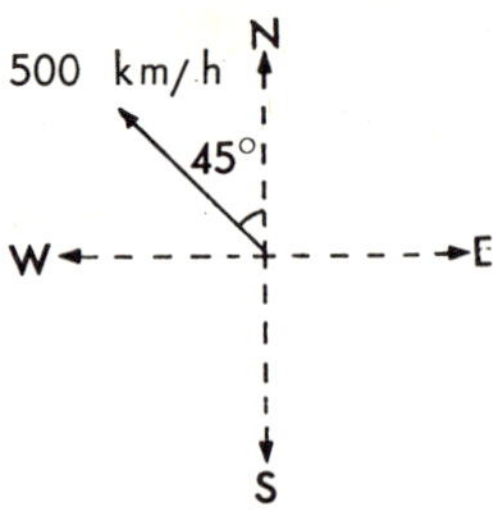

FIG. 47.

The motion of a body may be conveniently represented in graphical form by plotting distance travelled along the vertical axis and the time taken to travel that distance along the horizontal axis. Such a representation is known as a *space-time graph.*

The motion of a car travelling from rest at a uniform speed of 60 km/h for 4 min could be represented as in Fig. 48. The graph is a straight line.

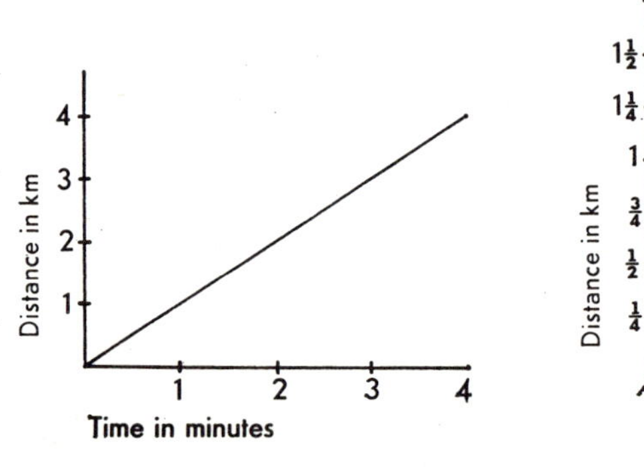

FIG. 48.

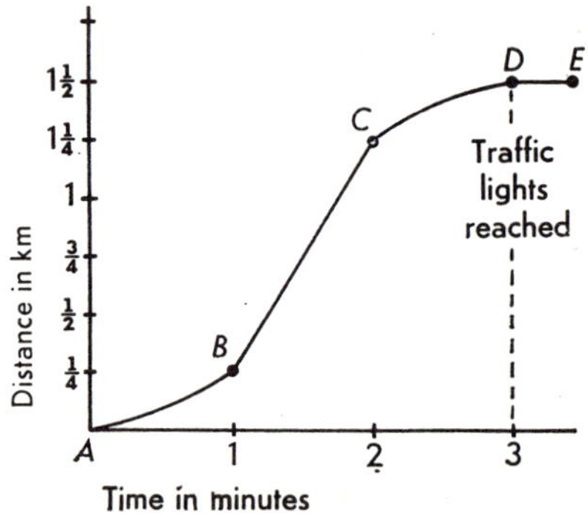

FIG. 49.

But a car cannot cover an actual journey in this way. Its journey is more like to be represented by a space-time graph something like that shown in Fig. 49. At *A*, the car starts moving from rest and its speed increases as the driver gradually changes into top gear. From *B* to *C* the car travels at uniform speed so *BC* is a straight line, and then from *C* to *D* the speed decreases as the driver applies the brakes as he approaches, say, some traffic lights. From *D* to *E*, the car is stationary, waiting for the lights to change to "GO."

The maximum and uniform speed is from B to C when 1 km is covered in 1 minute, that is 60 km/h. The average speed from start at A to stop at D is $v_{av} = s/t = 1\frac{1}{2}/3 = \frac{1}{2}$ km per min, or 30 km/h. The car's average speed while reaching the top speed of 60 km/h from A to B is $v_{av} = \frac{1}{4}/1$ km per min, or 15 km/h.

Acceleration

Whenever the velocity of a body is changing, the body is said to have *acceleration*, which is defined as the rate of change of velocity, or the change in velocity per unit time. The average acceleration is found by dividing the change in velocity by the time taken.

$$Acceleration = \frac{Increase\ in\ velocity}{Time}$$

Acceleration is measured in units of metres/second² (m/s²).

A body is said to have a uniform acceleration if it changes its velocity by equal increments in equal intervals of time, no matter how small the intervals of time may be. If the velocity of a body is increasing, the body is said to have a positive acceleration; if the velocity is decreasing, the body is said to have a negative acceleration or retardation or deceleration. Since velocity is a vector quantity, a body could have acceleration either because

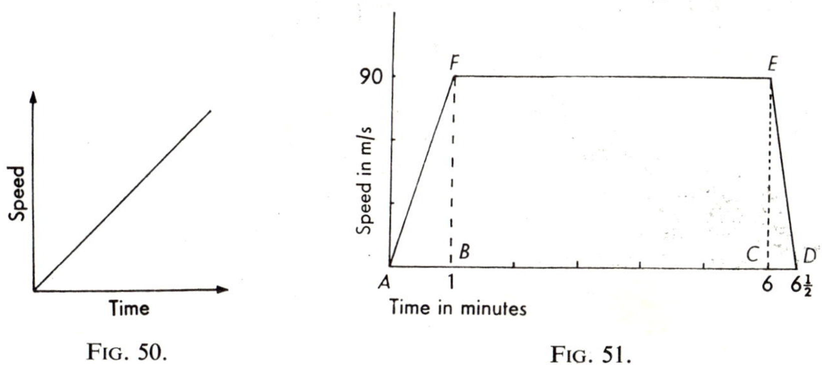

FIG. 50.

FIG. 51.

its speed is changing, or because its direction is changing, or because both are changing. Thus a train, moving at a steady speed of 50 km/h around a curve in the track, will have acceleration because its direction is changing all the time.

Fig. 50 represents a speed-time graph for a body travelling with a uniform acceleration. If a car moves from rest so that its speed increases uniformly from zero to 90 m/s in 1 minute, then its speed remains uniform for another 5 minutes, and then its speed reduces to zero in another $\frac{1}{2}$ minute when the brakes are applied, its journey might be represented by Fig. 51. As long as the car's acceleration is uniform from 0 to 90 m/s in 1 min (and this is important), the car's average velocity during the first minute is 45 m/s, so the distance it covers is $45 \times 60 = 2700$ m. During the next 5 minutes at a steady speed of 90 m/s the car will travel 90×300 m = 27 000 m. During the next and last $\frac{1}{2}$ minute the car's average speed is 45 m/s; so the distance it covers is $45 \times 30 = 1350$ m. Total distance covered is thus 31 050 m (or 31·05 km). We could obtain this answer from a speed-time graph by finding the area of the trapezium *ADEF* (Fig. 51), in terms of the scales used (not in cm² of graph paper), thus:

$$\begin{aligned} \text{Area of trapezium } ADEF &= \text{height} \times \tfrac{1}{2} \text{ the sum of parallel sides} \\ &= 90 \times \tfrac{1}{2}(300 + 390) \\ &= 90 \times 345 \\ &= 31\,050 \text{ m} \end{aligned}$$

Using this method we cannot obtain an answer accurate to 1 part in 1000. We therefore omit the 50 m, giving the answer as 31 000 m.

The graph for the actual journey of a car is more likely to look like Fig. 52. In this case the speed is varying all the time; the distance covered is still represented by the area *ABCD* under the curve. For example, suppose that paper, ruled in cm and mm, is used. Let the speed scale be 1 cm = 25 m/s and time scale be 1 cm = 10 s. Then 1 cm² (or large square) represents 25 m/s × 10 s = 250 m and 1 mm² (or small square) = 2·5 m. Suppose the area counted under the curve is 5 cm² (or large squares) + 306 mm² (or small squares), then total area of 5 + 3·06 = 8·06 cm² which represents 8·06 × 250 = 2015 m covered by the car during its test.

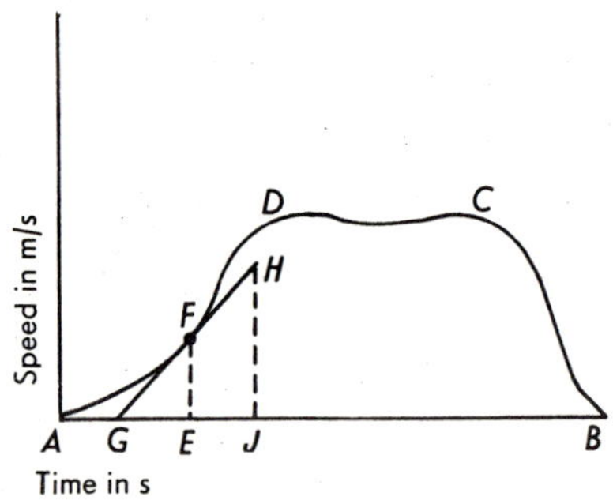

Fig. 52.

If the actual instantaneous acceleration of the car is required at any instant, say at time *E*, then a tangent should be drawn to the curve at *F* and the acceleration is then given by the slope of the curve at that point, namely, the slope of the tangent drawn to the curve at that point or *HJ*/*GJ*. Since velocity was plotted on the graph in units of m/s and time in seconds, the slope of the curve will give acceleration in units of m/s².

FORCE

Our present ideas about the causes of motion were first given by Galileo (1564-1642) and Isaac Newton (1642-1727). What do we associate with the cause of motion? Frequently we think of muscular pushes or pulls, such as those required to kick a football, or to pull a sledge. Such pushes or pulls we call *forces.* In physics, force is imagined to be the cause of all motion. The pull of a magnet on a nail is a force; it can change the motion of a nail.

If we wish to move a settee across a room we must apply a force all the time to keep it moving. Everyday experience seems to indicate that a force must constantly be applied to maintain a steady motion. We shall see that this is, in fact, not true.

Newton's Laws of Motion

In the days of the Greek philosopher Aristotle (384-322 B.C.), it was thought that a constantly maintained force was required to produce a constant velocity, and that, in the absence of force, bodies would come to rest. What, then, kept the heavenly bodies, such as the planets, moving in their orbits? Aristotle supposed that celestial matter was different in its nature from terrestrial matter and obeyed different laws. Many hundreds of years later, Galileo thought that the motions of heavenly bodies and earthly bodies were governed by the same, not different, laws. He stated that: any velocity once given to a body will be strictly maintained as long as there are no causes of acceleration or retardation. In other words, when no resultant force acts on a body, the body remains at rest, or continues moving in a straight line at constant speed. This is the *law of inertia,* and is usually referred to as *Newton's first law of motion.*

Newton's second law of motion states that: when a body is acted upon by a force, the rate of change of momentum is directly proportional to the impressed force and takes place in the direction of the force. The momentum of a body is the product of its mass and its velocity.

Thus, force is proportional to the rate of change of momentum. If the mass of the body remains constant, force is proportional to mass × rate of change of velocity, which is acceleration. So we can now say that force is proportional to mass × acceleration.

$$P \propto ma$$

Force can therefore be measured in terms of the change in motion which it produces and the name of the unit of force is the *newton* (symbol N), which will be defined later in this chapter. A force producing twice as great a change in motion would be 2 newtons (2 N), and so forth.

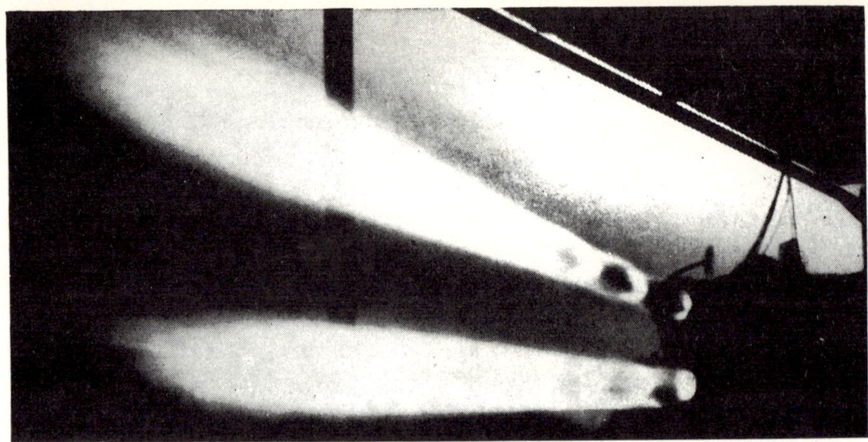

FIG. 53. *Rocket engines firing under test. The hot gases seen escaping at high speed from the back of the engines were found to be exerting a forward thrust of 72 000 N on the engines.*

Newton's third law of motion states that to every action there is an equal and opposite reaction. In other words, the mutual reactions of two bodies upon each other are always equal and oppositely directed.

Applying this law, we may say for example that the gravitational pull of the sun on the earth is equal and opposite to the pull of the earth on the sun.

In the jet engine, air used in the combustion of the fuel is compressed and mixed with the fuel, and introduced to the combustion chamber. Here it burns quickly.

The resulting hot gases expand rapidly and escape at high speed through a small turbine into the exhaust. When they escape, they exert a very great forward thrust on the aircraft, the hot gases experiencing an equal thrust in the backward direction. This, then, is a modern example of Newton's third law.

A simple supporting experiment may be carried out by yourself. Procure two spring balances, each reading up to about 200 newtons. Anchor one to a

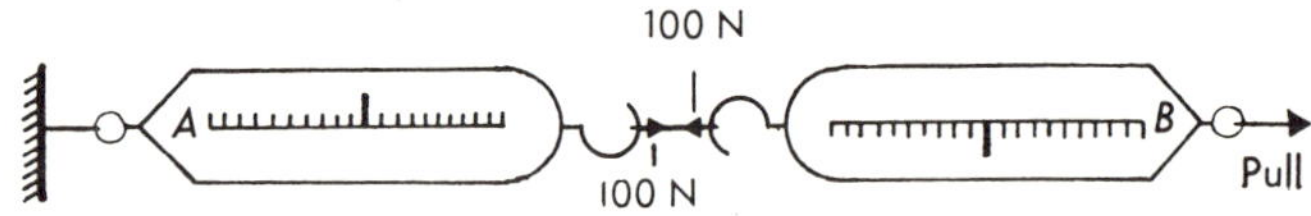

FIG. 54. *Apparatus for investigating Newton's third law of motion.*

hook on a bench or table, link the two together and pull on the free end of the other. You will note that each spring balance shows the same reading. Balance *A* pulls on balance *B* with a pull of say 100 N from *B* to *A*. Balance *B* pulls on balance *A* with a pull of 100 N from *A* to *B* (Fig. 54).

Does the Horse Pull the Cart or the Cart Pull the Horse?

In the light of Newton's third law of motion let us consider a horse pulling a cart. If the horse pulls on the shafts of a cart, and the shafts pull back on the horse with an equal force, how does the horse move? Newton himself answered this question in considerable detail.

Let us consider the horizontal forces acting on the horse (Fig. 55). He wants to move forward but the shafts pull back on him with a force P. He thrusts back on the ground with a force F_1 and the ground exerts an equal forward force on him. He will move forward if F_1 is greater than P ($F_1 > P$).

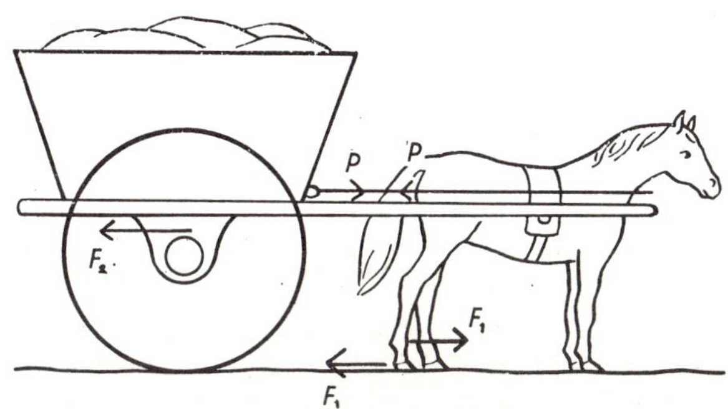

FIG. 55. *Newton's third law of motion applied to a horse and cart.*

F_1 depends on the friction between the horse's hooves and the ground. If the road is slippery, sometimes F_1 cannot become greater than P, and the horse cannot move the cart until sand is sprinkled on the road and F_1 is increased to a value greater than P.

Now let us consider the horizontal forces acting on the cart. There is a forward pull P on the shafts due to the horse. There is also a backward pull F_2, due mainly to the friction between the wheels and the axles. The cart will move forward when P is greater than F_2 ($P > F_2$).

Addition of Forces

If two or more forces act at the same time on a body, it is often possible to find one force which, in size and direction, will have just the same effect on the body as the two or more forces. Such a force is called the *resultant* of the two or more forces.

If two boys haul a boat from the water by pulling on the same rope in the same direction and one boy exerts a pull equal to 200 N while the other exerts a pull of 300 N, then the resultant force on the boat will be a pull of 500 N.

Let us consider this point further with the aid of a simple experiment. Take a light strong bar (Fig. 56) and support it by two spring balances, *P* and *Q*, in the manner shown. Attach to the bar *AB* a load of 200 N. You will notice that each of the two spring balances indicates a pull of 100 N. The resultant of the two spring balance pulls is 200 N in an upward direction. Notice that the bar *AB* does not move, however. This is because it is being pulled downwards by a force of 200 N; the bar *AB* is in equilibrium. This downward force which balances the upward pulls of the two spring balances *P* and *Q*, is called the equilibrant of these two forces. Thus, resultant and equilibrant are equal in magnitude, but opposite in direction.

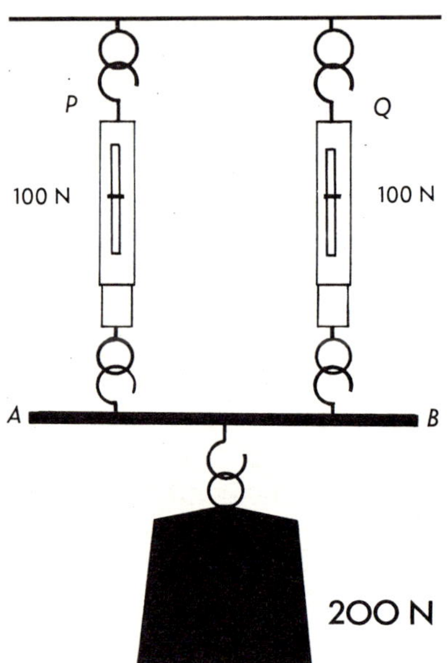

FIG. 56. *Considering the equilibrant and resultant of two forces.*

If two forces act in exactly opposite directions on a body, their resultant is equal to the difference between the two forces, acting in the direction of the greater force.

Resultant of Inclined Forces

We have seen that if two forces act in the same direction their resultant may be obtained by simple addition and if in opposite directions, by subtraction. But if they act in different directions, how then do we add their effects to find their resultant? We may use the apparatus shown in Fig. 57. The steel ring *A* is in equilibrium under the action of three forces when it is in the position shown. If the ring is moved to one side or upwards or downwards and released, it will come back to the position shown, the only position where the three forces are in equilibrium. The sum of the two readings indicated

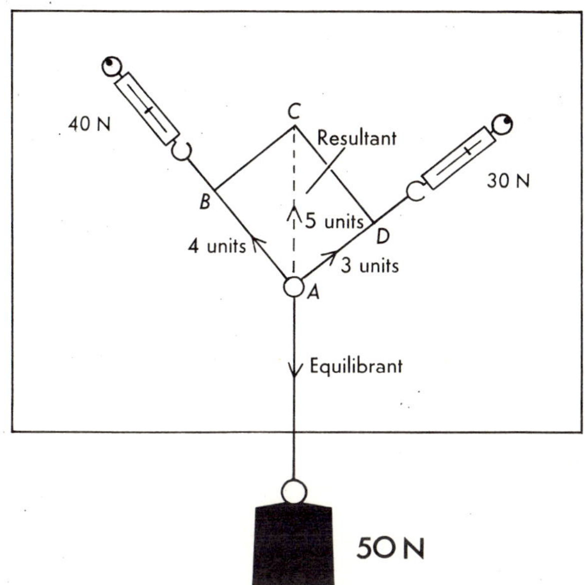

FIG. 57. *Parallelogram of Forces.*

by the spring balances is 70 N which is more than enough to balance the weight 50 N of the iron ring. So how is the resultant found? If a suitable scale is chosen, and AB marked to represent 40 N in size and direction and AD marked to represent 30 N in size and direction, and the parallelogram $ABCD$ is completed, it will be found that the diagonal AC will represent 50 N to scale, and will be opposite in direction to the weight 50 N on the iron ring. Many tests of a similar kind have led to the theorem known as *the parallelogram of forces* for finding the resultant of two non-parallel forces:

"*If two forces acting at a point be represented in size and direction by the adjacent sides of a parallelogram, then the diagonal which passes through their point of intersection will represent their resultant in size and direction.*"

This theorem may be used for adding or compounding any two vector quantities such as two velocities, or two accelerations, as well as forces.

For example, suppose that an aircraft is flying at 100 km/h relative to the air in a due easterly direction, and that a wind of 50 km/h is blowing from the south-west. Then the aircraft's velocity relative to the ground may be found by constructing a parallelogram of velocities *ABCD*, similar to the parallelogram of forces already considered. As before, a suitable scale is chosen, *AB* is marked to represent the velocity of the wind in magnitude and direction and *AC* is marked to represent the velocity of the aircraft relative to the air in magnitude and direction. When the parallelogram is completed, the diagonal *AC* will represent the resultant of the two velocities which is the velocity of the aircraft relative to the ground, in magnitude and direction.

Resolution of Forces

It sometimes happens that a force is most conveniently applied in one direction, while its effect is required in another direction. How can we find how much effect our force will have in the desired direction? This can be found by *resolving* our force.

For example, suppose you are mowing the lawn and can exert a force of 150 N along the shaft of the mower. How effective is this force in pushing the mower forwards over a horizontal lawn? Clearly you would not wish to get down on your knees in order to apply a force in the direction the mower

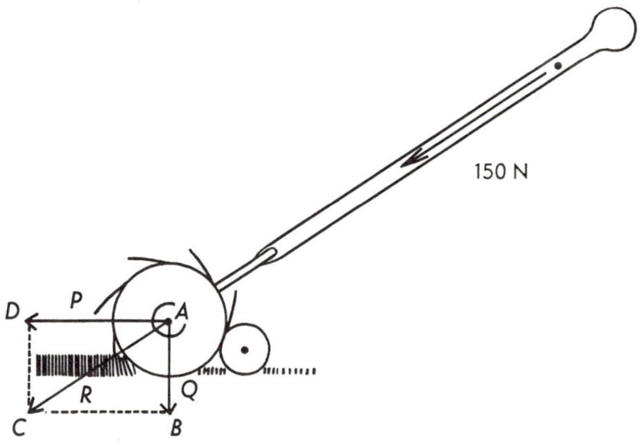

FIG. 58. *Resolution of forces applied to a lawn mower.*

is to travel. If we regard the 150 N as the resultant *R*, of two forces, one, *P*, in the direction the mower will travel, and one, *Q*, at right angles to this direction, then it will be the force *P* which will be effective in pushing the mower (Fig. 58). You may find the value of *P* by drawing the rectangle *ABCD* to scale, so that *AC* represents 150 N. *AD* will then represent the effective force which moves the mower.

The more upright you stand while pushing the mower (the greater the angle made between the shaft of the mower and the lawn), then the less

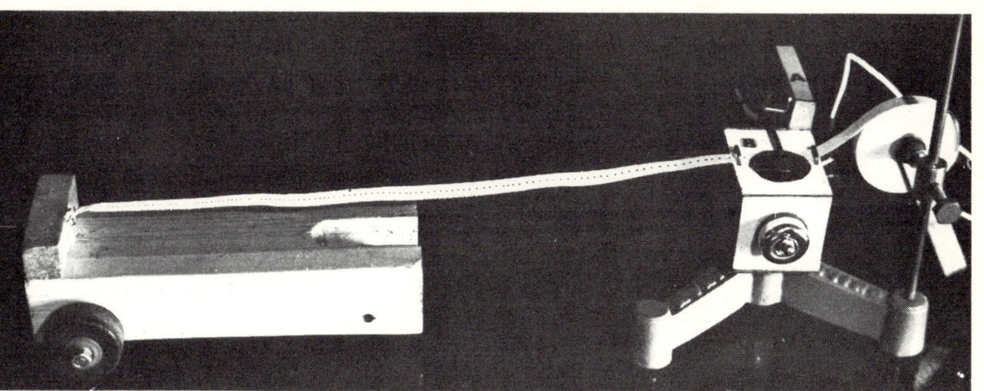

FIG. 59. *Apparatus for investigating force and motion. The trolley is attached to a roll of ticker tape which passes through a timing device.*

effective will be your push, and the mowing of the lawn will be more tiring. The forces P and Q are known as components of the applied force of 150 N. P is the effective component which moves the mower across the lawn, while Q is the ineffective component which tends to force the mower into the lawn and should be kept as low as possible.

To Investigate Force and Motion Experimentally

This may conveniently be done using a trolley similar to that shown in the photograph (Fig. 59). This trolley runs on three wheels, each having ball-bearings, so that there is little frictional force. If a force is applied to this trolley, it will accelerate. What must be measured to find the acceleration acquired when a given force acts on the trolley? Distance and time must be measured. Measurement of distance is not difficult. But the time intervals to be measured are very short and such devices as seconds clocks and stop watches are not sufficiently accurate. Fig. 60 shows a suitable timing device *A*, made from the make-and-break system of an electric bell. The hammer is replaced by an arm which vibrates above a rotating disc of carbon paper, so that it punches a carbon dot on to ticker tape which runs beneath the carbon paper.

B is a wooden block holding the ticker tape, guided by staples to run under the carbon disc. The other end of the ticker tape is attached to the trolley and as the latter moves along, the tape is pulled. As it flows under the carbon disc, it rotates the disc, so that one spot of the carbon is not quickly worn out. The vibrator is energized by a 2-volt accumulator. To find the frequency of vibration, the trolley is pulled along at a steady speed; at a signal, the switch is closed to start the arm vibrating and a stop watch is started. After 10 seconds the switch is opened and the vibrations cease. With the home-made model (Fig. 60), 603 vibrations were counted for a 10-second interval. So

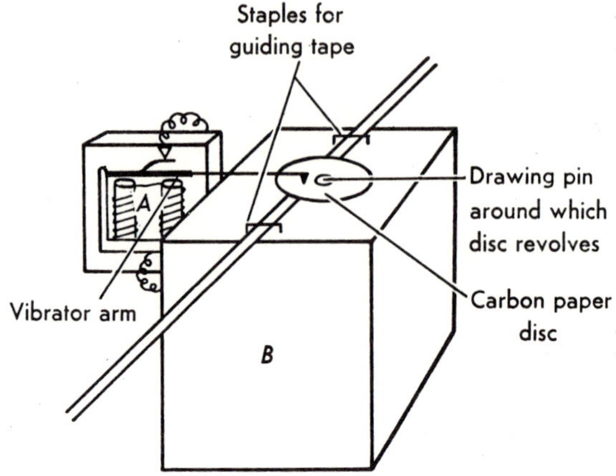

FIG. 60. *Timing device for investigating force and motion experimentally.*

the frequency of the vibrator was taken to be 60 vibrations each second; i.e. the interval between successive vibrations is $\frac{1}{60}$ second.

Experiment 1. *To find whether an accelerated motion is uniform and to estimate the acceleration.*

The trolley is placed on a long, flat, wooden track (Fig. 61), and the track is inclined slightly to the horizontal until it is found, by trial and error, that the trolley, when given a slight push, will roll down the track at uniform speed. Then the "diluted" effect of gravity on the trolley is sufficient to balance friction. A thread is now fastened to the front of the trolley, passed over a pulley and tied on to a weight carrier, weight 0·3 N, on to which slotted pieces of metal, each with weight 0·2 N, may be placed. When an accelerating force 0·9 N is applied and the trolley pulled down the track, with the vibrator operating, a trace is obtained on the ticker tape which shows that the dots are getting wider and wider apart as the trolley accelerates. The distances covered for successive intervals of 10 vibrations are

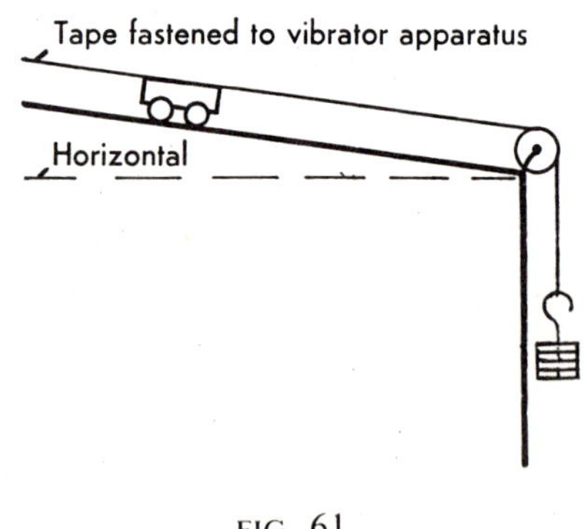

FIG. 61

measured and a table drawn up similar to Table 2, to find whether the acceleration is uniform.

We see that in the first 10 vibrations the trolley travels 1·0 cm, so its average

Table 2
Measurement of Uniform Acceleration

Distance d (cm)	Time t (s)	Average velocity v (cm/s)	Increase in v in $\frac{1}{6}$ s	Increase in v in 1 s
1·0	$\frac{1}{6}$	6·0	6·0	36·0
2·0	$\frac{1}{6}$	12·0	6·0	36·0
3·0	$\frac{1}{6}$	18·0	6·0	36·0
3·9	$\frac{1}{6}$	23·4	5·4	32·4
4·8	$\frac{1}{6}$	28·8	5·4	32·4
5·6	$\frac{1}{6}$	33·6	4·8	28·8
6·4	$\frac{1}{6}$	38·4	4·8	28·8
7·2	$\frac{1}{6}$	43·2	4·8	28·8
8·0	$\frac{1}{6}$	48·0	4·8	28·8
8·8	$\frac{1}{6}$	52·8	4·8	28·8

Average acceleration over first 51 cm = 32 cm/s².

speed for the first $\frac{1}{6}$ s is 6·0 cm/s. In the next 10 vibrations the trolley travels 2·0 cm, so its average speed for the second $\frac{1}{6}$ s is 12·0 cm/s. Thus in $\frac{1}{6}$ s the trolley's speed increases by 6·0 cm/s. Therefore in 1 second the speed increases by 36·0 cm/s. That is, the trolley's acceleration is 36 cm/s^2. Average acceleration for the first 51 cm is 32 cm/s^2.

Experiment 2. *To investigate the relationship between force and acceleration for a constant mass.*

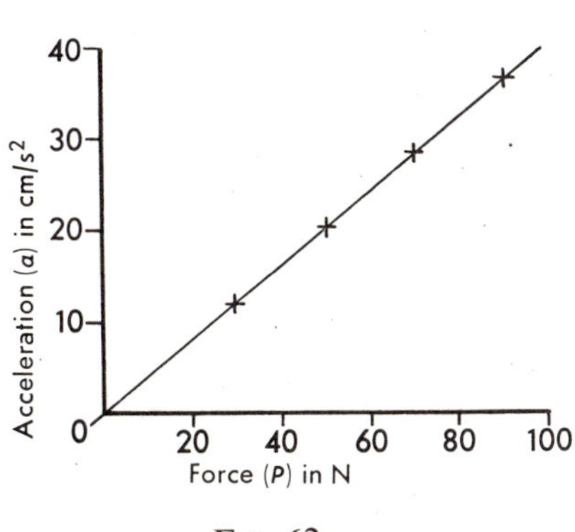

FIG. 62.

If Experiment 1 is repeated after transferring one of the slotted pieces of metal from the weight carrier to the top of the trolley then the accelerating force is 0·7 N instead of 0·9 N, but the total moving mass (trolley + weight carrier + slotted pieces) is the same as before. Another trace is recorded and the acceleration measured as in Experiment 1. Another piece is transferred from the weight carrier to the top of the trolley and another trace recorded, and so on. Thus the acceleration of the trolley for forces of 0·9, 0·7, 0·5, 0·3 N is measured (Table 3). A graph of acceleration a against force P (Fig. 62) shows a straight line passing through the origin. Thus:

$$a \propto P \text{ (mass constant)}$$

Table 3

Relationship Between Force and Acceleration When Mass is Constant

Force P (N)	Distance s (cm)	Time t (vib.)	t (s)	Acceleration $= 2s/t^2$* (cm/s²)
0·30	50	170	$\frac{17}{6}$	$100 \times \frac{36}{289} = 12$
0·50	54	140	$\frac{7}{3}$	$108 \times \frac{9}{49} = 20$
0·70	48	110	$\frac{11}{6}$	$96 \times \frac{36}{121} = 29$
0·90	51	100	$\frac{5}{3}$	$102 \times \frac{9}{25} = 37$

**N.B. The formula, acceleration* $= 2s/t^2$ *is discussed in Chapter 10.*

Experiment 3. *Test of the relation between moving mass and the force required for a constant acceleration.*

If three trolleys of equal mass are available, then this test may be performed. Measure the acceleration given to one trolley of mass M by a certain force P. Place a second trolley of mass M on top of the first trolley, total mass $2M$, and test whether a force of $2P$ will produce an acceleration equal to that obtained for one trolley of mass M. Then add a third trolley (total mass $3M$) and increase the force to $3P$ and see whether the acceleration is the same as that obtained before. In all three cases you will find that the acceleration is the same, indicating that when acceleration is constant, force is proportional to mass.

From Experiment 2, force $\propto$ acceleration (mass constant)
From Experiment 3, force $\propto$ mass (acceleration constant)
Combining these: force $\propto$ mass $\times$ acceleration

$$\text{or, } P = kma \text{ (where } k = \text{constant)}$$

It is possible to choose the unit of force to make mass equal to P/a. *We measure force by the acceleration it gives to unit mass.* In fact, *we define the unit of force as that force which will give to unit mass unit acceleration.*

A long time ago in London, the *pound* was chosen as a standard unit quantity of matter, or unit of mass. In Paris, the *kilogramme* was chosen as the unit of mass. The standard kilogramme is the mass of a cylinder of platinum-iridium alloy, carefully preserved at Sèvres, near to Paris. The standard pound is the mass of a platinum cylinder, carefully preserved at the Standards Office in London, but now seldom used.

Unit of Force

The unit of force is defined as that force which will give unit acceleration to unit mass.

In the Système International d'Unités (S.I.) in which the unit of length is the metre, the unit of mass the kilogramme, and the unit of time the second, the unit of force is the newton (symbol N). One newton will produce an acceleration of 1 m/s^2 in a mass of 1 kg.

1 newton will give an acceleration of 1 m/s^2 to a mass of 1 kg.
1 newton will give an acceleration of 1000 m/s^2 to a mass of 1 g.
1 newton will give an acceleration of 1 000 000 mm/s^2 to a mass of 1g.

WORK

Work, power, and energy are terms with a wide variety of meanings in literature and conversation. In physics we define them in a more restricted sense. Work is done when a force moves, and we define the work done as the product of the force and the distance moved by the point of application of the force in the line of action of the force.

$$W = F \times s$$

It is a measure of the energy transferred from one body to another.

Unit of Work

In S.I. the joule (symbol J) is the work done (energy transferred) when a force of 1 newton acts through a distance of 1 metre.

Power

Frequently we are interested in not only how much work is done, but also in how long it takes a machine to do a certain amount of work. We define power as the rate of doing work:

$$Power = \frac{Work}{Time} = \frac{Force \times Distance\ moved}{Time}$$

Unit of Power

In S.I. the unit of power is the watt (symbol W). It is 1 joule/second or alternatively 1 newton-metre/second.

1 watt = rate of working of 1 joule/second
1 kilowatt = rate of working of 1000 joule/second

Historically the unit of power was the horse power (H.P.). By experiment it can be shown that 1 H.P. = 746 watts.

Brake Horse Power Test

The power of machines is sometimes rated in brake horse power (B.H.P.), and sometimes it is calculated from the work required to operate the machines, which of course is greater than the B.H.P. because of work lost as heat in overcoming friction. You could test the B.H.P. of one of your electric motors, or model steam engines, in this way: fit a pulley to the shaft of the motor (Fig. 63). Use a flexible leather belt or a piece of cord as a friction band around the pulley, keeping it taut by using a suspended weight W. Allow

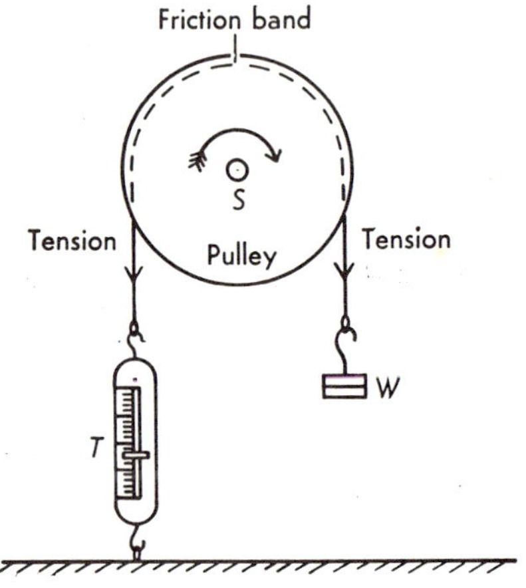

FIG. 63.

the motor to rotate the pulley against a suitable spring balance which will record the tension T. Then for every revolution of the pulley, the motor will do work equal to $(T - W) \times$ (circumference of pulley). Once the motor has attained a steady speed, the spring balance will read a fairly constant value of the tension in the left side of the belt and an average value may be obtained. The circumference of the pulley may easily be measured by wrapping a piece of string around it. Now comes the more difficult problem of measuring the time involved. If a tachometer (revolution counter) is available, its friction drive may be applied to the shaft of the motor at S and it will record the number of revolutions/minute made by the motor while under test. Here is

The power of a racing car's engine is often measured in brake horse power.

a record of an experiment made for such a test:

$$W = 2\text{ N}$$

$$\text{Average value of } T = 5{\cdot}3\text{ N}$$

$$\therefore \text{Force of friction} = 3{\cdot}3\text{ N}$$

$$\text{Circumference of pulley} = 0{\cdot}75\text{ m}$$

$$\text{Speed of motor} = 1400\text{ rev/min}$$

Work done by motor against frictional forces in one revolution $= 3{\cdot}3 \times 0{\cdot}75$ newton-metre or joule

$$\therefore \text{Work done by motor in 1 s} = \frac{1400}{60} \times 3{\cdot}3 \times 0{\cdot}75\text{ J}$$

$$\therefore \text{Brake horse power developed by motor} = \frac{140}{6} \times 3{\cdot}3 \times 0{\cdot}75\text{ J/s or W}$$

$$= 58\text{ W}$$

ACCELERATION OF A FREELY FALLING BODY

Any object released above the surface of the earth and allowed to fall freely will be pulled towards the centre of the earth and will therefore experience an acceleration. Such acceleration may be measured using the vibrator and ticker tape apparatus illustrated in the photograph (Fig. 59). A length of the tape is stuck on a wooden strip about 30 cm long and a strip of carbon paper is placed over the ticker tape and fastened to the wooden strip. The latter hangs vertically from a hook in a wall (Fig. 64). The vibrator arm is placed near to the bottom of the wooden strip, so that the arm vibrates to and fro in a horizontal plane. The current in the vibrator coils is switched on and the thread burned through at *A*. The strip falls vertically past the firmly-supported vibrator and a trace showing acceleration is recorded on the ticker tape. If the distance fallen for a counted number of vibrations is measured, the acceleration of a freely falling wooden strip may be calculated.

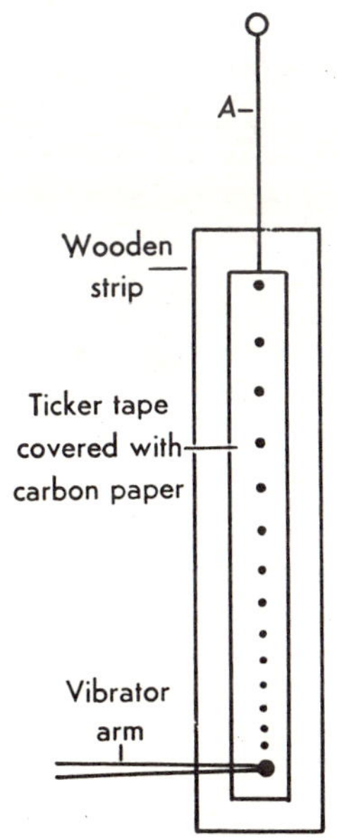

FIG. 64.

The Leaning Tower of Pisa: it was from the top of this tower that Galileo, while professor of mathematics at the university of Pisa, dropped bodies of varying mass to demonstrate that all bodies fall to earth with equal velocity.

Thus, if 14 vibrations are counted in a fall of 25·5 cm, let:

g = Acceleration of the falling wood,
t = Time to fall 25·5 cm from rest.

Then speed at end of time $t = g\,t$ cm/s
Average speed for time $t = \frac{1}{2}g\,t$ cm/s
Distance fallen in time t = Average speed × Time.

$$s = \tfrac{1}{2}g\,t^2$$

(Substituting for the measurements quoted above.)

$$25{\cdot}5 = \tfrac{1}{2}g(\tfrac{14}{60})^2$$

(Remember the vibrator makes 60 vibrations/second or hertz.)

$$g = 51 \times \frac{3600}{196} = 940 \text{ cm/s}^2$$

In this experiment, the fall is not strictly "free" because the wood might touch the wall slightly when falling and the air's inertia also slightly retards its fall. Both of these errors however are small, if you are careful with your adjustments of the apparatus. Tests with such an apparatus show that the acceleration of a freely falling body is close to 9·8 m/s^2 (or 980 cm/s^2). Now we call the force that the earth exerts on a body the weight of the body (symbol W). If the pull of the earth on a lump of iron of mass 1 kg gives it an acceleration of 9·8 m/s^2 and a mass of 1 kg is given an acceleration of 1 m/s^2 by a force of 1 N, then it follows that the weight on a mass of 1 kg is 9·8 N. Approximately this is taken as 10 N.

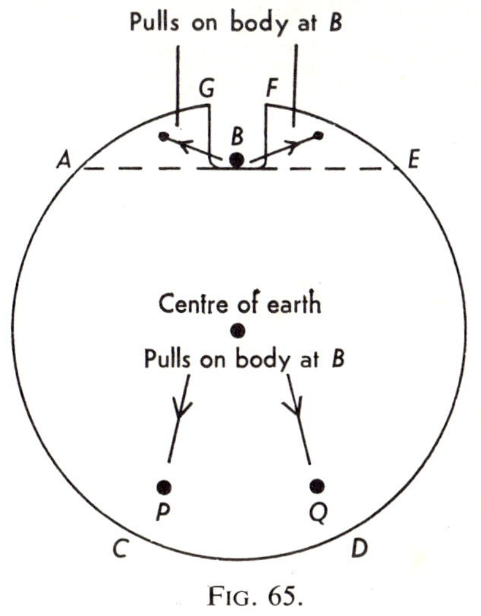

FIG. 65.

Is this always true? Not quite, because the pull of the earth on a body is not a constant pull. The amount of matter in a body, which we call its *mass*, remains constant (unless the body travels at enormously high speeds, approaching the speed of light), wherever we take the body (even to the moon). But the weight of the body, that is, the gravitational pull on it, depends on where the body is situated.

Since the earth is not quite a true sphere, but is flattened a little at the poles, a body is slightly nearer to the centre of the earth when at the poles, than it is when at the equator. Thus the weight of a body is slightly greater at the poles than at the equator, by about $\frac{1}{2}$ per cent. The weight of a body in Britain would be between these two values. If the body were taken above the surface of the earth, for example in a rocket, its weight would decrease, since it would be farther from the centre of the earth. If a body were taken down a very deep mine, its weight would also decrease. This statement probably surprises you. The reason is this: if the body is at *B* (Fig. 65), all particles of matter in the earth, in the space *ACDE*, pull *B* towards them. Particle *P* pulls on *B* along the direction *BP*. Particle *Q* pulls on *B* along the direction *BQ*, and so on. All these forces can be added together to give the equivalent of one force on *B* acting towards the centre of the earth. But all the particles comprising the earth in the space *AEFG* pull on *B* with a resultant force away from the centre of the earth. Thus the weight of the body is less at *B* than when the body is at the surface of the earth. You will probably be able to reason for yourself that the body will still have its original mass at the centre of the earth, but its weight at the centre of the earth will be zero. This does not mean that there would be no force acting on the body. There would be pulls in all directions, but the resultant pull would be zero.

Projectile Motion

A body will move in a curved path whenever the force that moves it has a component perpendicular to the direction of motion. The component of force along the path of the moving body changes its speed, but does not change its direction. A force perpendicular to the motion pushes the projectile sideways so that its direction changes and its path curves, giving it an acceleration perpendicular to its path.

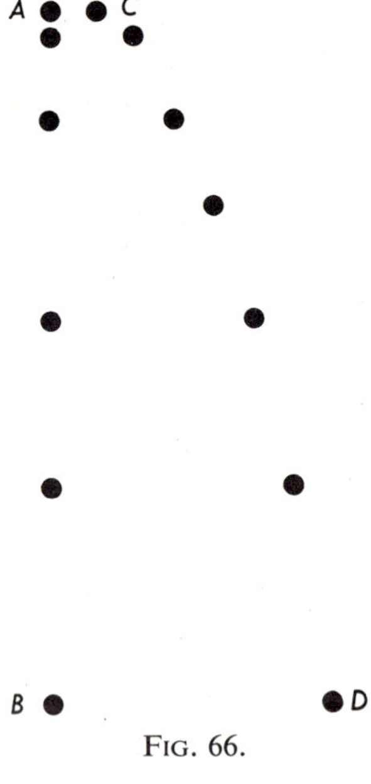

FIG. 66.

High speed flash photography of two ball-bearings painted black, against a white background, can give some interesting pictures. If you are keen on photography, you should try some of these experiments, using intervals between flashes of about $\frac{1}{20}$ or $\frac{1}{30}$ s. One ball is released from *A* (Fig. 66) so that it falls freely. The second ball *C* is given a small horizontal velocity to the right, being released at the same instant as *A*. *AB* shows the path of the first ball at short time intervals. *CD* shows the path of the second ball at equal time intervals. The force of gravity continually acts on the first ball giving it a uniform acceleration along *AB*. The vertical motion of the second ball is similar to that of the first, despite its horizontal motion. The horizontal motion of the second ball is at constant horizontal velocity, since no horizontal force continues to act on it. The presence of the downward force does not change the horizontal motion. The horizontal motion does not alter the effect of the downward force on the vertical motion. The shape of the path of the second ball is known as a *parabola*.

Motion in a Circle

There are many examples of uniform motion in a circle when the speed remains constant but the direction of motion continuously changes: particles, for instance, on your record-player, on the hands of a watch, or on the rotating aerial used in radar navigation all move with this type of motion.

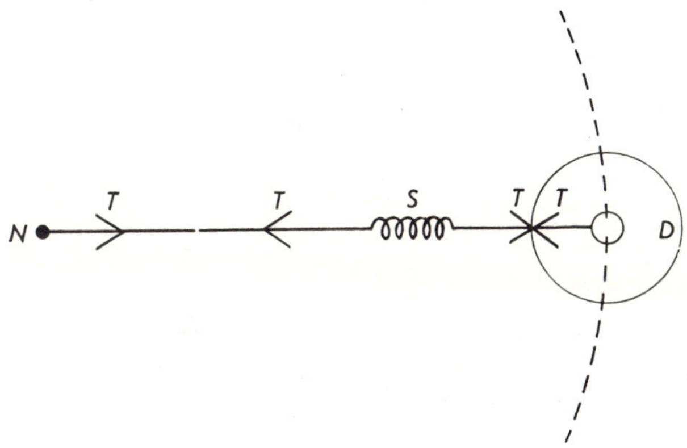

FIG. 67. *Apparatus for investigating motion in a circle.*

Consider the apparatus shown in Fig. 67. *N* is a nail in a table to which a spring balance *S* is connected by means of a string. The other end of the balance is connected by another string to the centre of a disc of solid carbon dioxide *D*. Since the solid carbon dioxide steadily vaporizes at room temperature, the disc is in effect sitting not on the table but on a layer of carbon dioxide gas. When *D* moves there will be very little friction acting on *D* as long as the table is smooth. If *D* is pulled outwards and then given a push at right angles to *DN* it will move at nearly uniform speed along the circumference of a circle, centre *N*. A series of photographs taken at equal short intervals through the slots of a stroboscope will show that *D* moves equal distances in equal intervals of time. Similar observations of the spring balance will show that the tension *T* in the spring is constant as long as *D* is moving at a constant speed.

FIG. 68. *In order to turn a corner at speed, motor-cyclists must lean inwards steeply to achieve the large centripetal force required.*

The extended spring exerts a force T, on D (Fig. 67), and it is obvious that T is always directed towards the centre of the circle N. It follows that the *acceleration* of the body is also towards the centre of the circle, although the *velocity* is at right angles to the radius. To make a body move in a circle with uniform speed there must be a force, called the *centripetal force*, at right angles to the motion and directed towards the centre of the circle. This force increases with the speed (in fact with the square of the speed, if the radius is constant). We will now discuss a few examples of centripetal force.

Throwing the Hammer

As the hammer is whirled round and round, the force making it move in a circle is the tension in the wire or handle. An equal and opposite force is exerted by the wire or handle on the thrower and if he is to remain where he is, he must so place his feet that the ground exerts an equal and opposite force on him. When the thrower lets go of the hammer there is no force making it move in a circle and so it flies off at a tangent.

Cyclist Turning a Corner

When a cyclist turns a corner he leans inwards. The force of the road on the cycle, *the reaction*, acts through the centre of gravity of the cycle and the rider. When the cycle and rider lean, the reaction also leans. The vertical component of the reaction (parallelogram of velocities, page 69) balances the weight and the horizontal component is the necessary centripetal force. This horizontal component must be provided by the friction between the tyres and the road surface. If the road is muddy or icy there is little friction and so, insufficient centripetal force. The cyclist fails to round the corner; he travels forward in a straight line; that is he skids. If the road is banked the reaction remains perpendicular to the surface of the road and no sideways friction is necessary.

FIG. 69. *The astronaut's weight (the attraction between his mass and the mass of the earth) keeps him in orbit around the earth. Without weight there would be no centripetal force to hold him in orbit and he would fly off into space.*

Car Turning a Corner

The centripetal force on a car turning a corner on a flat road is provided by friction between the tyres and the road. A passenger in the car tends to go on moving in a straight line but the sides of the car or the friction between him and the seat forces him to move in the same circle. He feels that he is being "flung outwards" and he talks about a "*centrifugal force*" making him do this. But for an observer outside the car the only force acting on the passenger is that exerted by the side of the car or by the seat towards the centre of the circle.

Aeroplane Changing Direction

The pilot banks the aeroplane so that the resultant force of the air on the wings acting perpendicular to them has a horizontal component which is the centripetal force. If the plane flies quickly in a sharp curve the acceleration produced may be many times bigger than the acceleration of a freely falling body.

Artificial Satellites

If a satellite is put into a circular orbit round the earth then the centripetal force necessary is provided by its weight, i.e. the attraction of the earth for the satellite. Inside such a satellite all objects are similarly moving under the centripetal forces of their weights and so appear weightless. This is a different condition from that inside a body moving in outer space. There, the objects would really be weightless as there would be no attraction towards the earth.

If the speed of the satellite decreased it would gradually get closer to the earth's surface; if the speed increased the radius of the orbit would increase and the satellite would finally leave the influence of the earth and fly off into outer space.

QUESTIONS

1. Which of the following quantities are scalar and which are vector: mass, displacement, force, momentum, speed, work, energy, acceleration, power?

2. Which of the following statements is correct for bodies at the surface of the earth?

(*a*) The force of gravity on a mass of 1 kilogramme is one newton.

(*b*) The force of gravity on a mass of one gramme is one newton.

(*c*) The force of gravity on a mass of one kilogramme is ten newtons.

(*d*) The force of gravity on a mass of ten kilogrammes is one newton.

3. Does a body in Australia fall in the opposite direction from a body in England? If so, how do you account for this?

4. Could a propeller-type of aircraft be used to fly to the moon? Give reasons for your answer.

5. Which requires the stronger ropes to support a given weight, a swing or a hammock? Use a diagram to explain your answer.

6. If you stand on a compression spring balance in a lift, and the lift suddenly starts to move upwards, what happens to the balance reading? What happens to the reading when the lift suddenly starts to move downwards? What happens to the balance reading when the lift is moving at a uniform speed, either upwards or downwards? If you suddenly bend your knees when standing on a bathroom scales, will the reading increase or decrease? Try it and see.

7. A sledge is pulled by a rope making 30° with the ground. The tension in the rope is 300 N. Find the horizontal and vertical components of the pull. How are both components effective in pulling the sledge?

8. The speedometer of a car starting from rest, showed the speed at the end of six successive seconds to be 4, 8, 12, 16, 20, 24 m/s. What was the average velocity in m/s; the distance travelled by the car in the time interval of six seconds; the acceleration of the car? How far did this car travel during the last second of its journey?

9. A sculler is able to row at 8 km/h in still water. He wishes to row a boat across a river 1 km wide, which flows at 5 km/h. In what direction must he row to arrive at a point on the far bank exactly opposite to his starting point on the near bank, and how long will it take him to cross?

10. If the acceleration of a freely falling body is 10 m/s^2 what is the velocity of a body released from rest, after $\frac{1}{4}$ s; $\frac{1}{2}$ s; 1 s; 2 s? What is the average velocity during the first $\frac{1}{4}$ s; during the first $\frac{1}{2}$ s; during the first 2 s? How far does the body fall in $\frac{1}{4}$ s; $\frac{1}{2}$ s; 1 s; 2 s?

CHAPTER 7

DIMENSIONAL CHANGES DUE TO APPLICATION OF FORCES

IF A FORCE is applied to a body, what kind of changes might be brought about?

1. The body might become longer or shorter.
2. The body might increase or decrease its volume.
3. The body might change its shape.
4. The body might change its temperature.
5. The body might break or crumble.
6. The body might change its state from solid to liquid, from liquid to vapour, or in the reverse order.
7. The body might change its colour, owing to the strains set up in it.
8. The body might change its properties, e.g. become demagnetized.
9. The pressure exerted by the body might change.

Perhaps you can think of other changes that might result from the application of a force to a body. Usually several of these changes take place at the same time when a force is applied. Let us study some of these changes.

Effect of Force on Length

What is force? If you stoop to pick up a purse lying on the pavement and it suddenly jerks away from you as you bend, you know that something is pulling the purse. On closer examination you find that the pull is provided by the hand of the small brother of your friend, holding a piece of string attached to the purse. If you wish to prevent the young rascal from running away, you will have to use a strong pull on his jacket. If he escapes, and in his excitement dashes towards an oncoming bicycle, you will have to pull quickly in a sideways direction to make the runner change his direction to avoid an accident. We may say that a force is a push or a pull. But to be more exact, we say that *a force is that which tends to make a body move when it is at rest; or to make it go more quickly or more slowly if it is moving; or to make it change its direction; or to make it change its shape.*

We have already defined force in terms of the amount of motion it can cause. What other effects can force produce? Let us investigate experimentally.

Hooke's Law

Let us find out how the length of a spring varies as the applied force stretching it is changed.

Support the upper end of a coil spring A by a split cork held in a clamp (Fig. 70). To the lower end of the spring attach a weight carrier B, whose weight is 0·5 N. Record the reading at C on a scale D held vertically by the wooden block E. Add a slotted weight of weight 0·5 N to B and record the new reading on D. Continue to take such readings, until the load on B reaches about 4 N. Remove the slotted weights one at a time, repeating the scale readings, until only the weight carrier is left again. Tabulate your results as shown in Table 4.

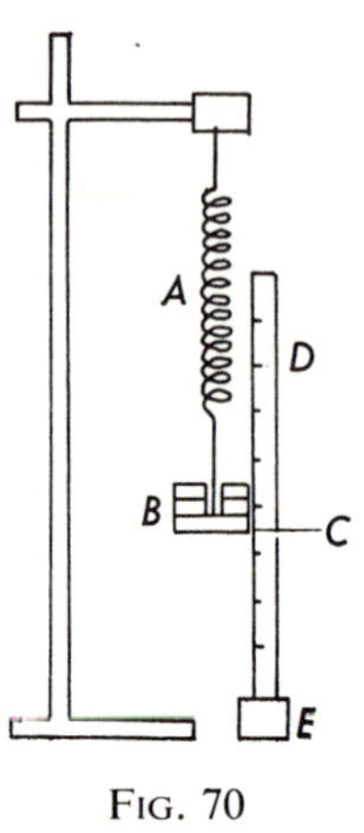

FIG. 70

Now plot a graph: load along x axis; scale reading along y axis (Fig. 71). The result is very nearly a straight line. This means that the ratio BA/BC will have a constant value for any two points on the graph. That is, the increase in load BC is directly proportional to the increase in stretch of the spring BA. If the load is increased by 2 N giving a stretch of 20 cm, then when the load is increased by 0·1 N the stretch will be 1 cm. We can now use the spring to measure the unknown weight of, say, a stone. If the stone is placed on the weight carrier and the reading level with C (Fig. 70) is 63·5 cm on the scale, then the weight of the stone + the weight of the carrier is 2·15 N. So the unknown weight of the stone is 1·65 N.

Since one newton is the unit in which our forces are measured, we may conclude from our experiment that the stretch of the spring is directly proportional to the force applied.

It was Robert Hooke (1635-1703), an English scientist, who first stated this relationship between force and stretch (*Hooke's law*):

"*The amount of change in shape of an elastic body is directly proportional to the force applied.*"

The spring balance is based on Hooke's law and is widely used to measure force—in newtons, etc. The stretch of the balance for a force of 2 N is twice that for 1 N.

In the coil spring experiment, the spiral coils of the spring were gradually opened wider, as the load increased. This is not the same as stretching the steel wire itself. Does it actually stretch when a force is applied? Since the

Table 4
Relationship Between Load and Extension for a Coil Spring

Load (N)	Scale reading (cm)			Extension (cm)
	Increasing load	Decreasing load	Mean	
0·5	47·5	47·5	47·5	—
1·0	51·7	51·7	51·7	4·2
1·5	56·6	56·6	56·6	4·9
2·0	61·5	61·6	61·6	5·0
2·5	66·7	66·8	66·8	5·2
3·0	71·8	71·9	71·9	5·1
3·5	76·8	76·8	76·8	4·9
4·0	81·9	81·9	81·9	5·1

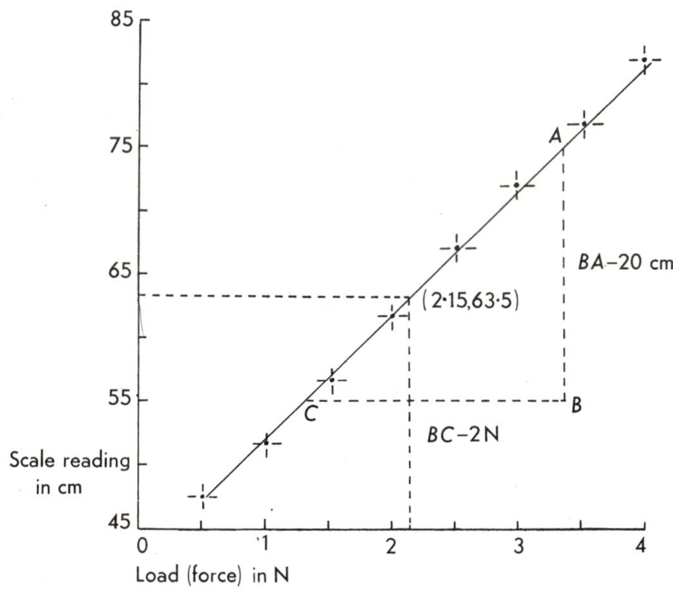

FIG. 71. *Graph illustrating Hooke's Law.*

stretch, if any, is much smaller than that of a coil spring, we shall need to use a more accurate method of measuring the small stretch.

Let us see whether Hooke's Law applies to the stretching of a nylon line.

This is an experiment you may perform yourself, using apparatus similar to that shown in Fig. 72. *AB* is a nylon clothes line, about 5 m long, about

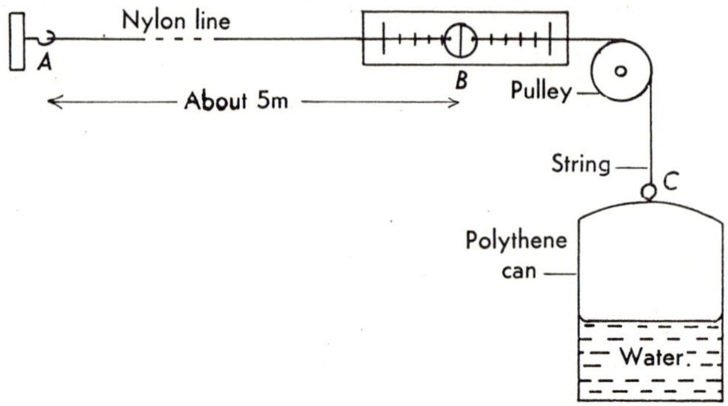

FIG. 72. *Apparatus to investigate the stretching of a nylon line.*

2 mm diameter, with a plastic ring at each end, and having about 500 N breaking strain. A piece of wire stretched across the plastic ring *B* will act as an indicator for measuring the stretch of the nylon against the fixed scale. Start with the polythene can (or a bucket will do) empty. Observe the reading on the scale below the wire across the ring *B*. Add 500 cm³ of water to the can. You will observe the ring *B* move a small distance as you add the water; record the new scale reading. Continue such observations until you have added about 5000 cm³ of water. The nylon line will have stretched about 7 cm altogether. Plot a graph similar to that in Fig. 71, using as your unit of load (force) 500 cm³ of water. See whether the nylon line used in your experiment has agreed with Hooke's Law. Will the string *BC* have stretched? If so, how will this have affected your results for the stretch of the nylon thread?

Nature of Stretch

When metals are examined under a microscope of high magnification they are seen to be made of very tiny crystals, each crystal being formed of a large number of atoms with a definite geometric arrangement. These crystals may be distorted by the application of a force, but recover their original pattern when the force is removed, provided the force used is not too great. If too great a force is used, slipping inside the crystals occurs, so that the specimen does not return to its original shape and size when the force is removed. The specimen is said to have exceeded its *elastic limit*. Usually a crystalline body like a metal can only be extended by about 1 per cent of its length before

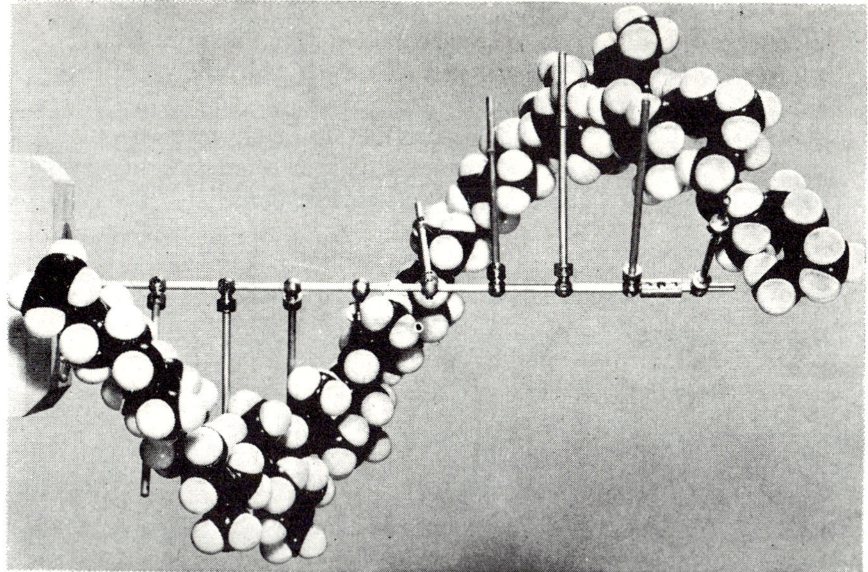

FIG. 73. *Model of the natural rubber molecule. This long molecule can take up many different shapes, giving to rubber its property of stretch.*

reaching its elastic limit, whereas rubber can be stretched to several times its original length. This is because the rubber molecule is built of many units, making it very long and thin. Each molecule is about $\frac{1}{400}$ mm long. Rubber in bulk may be likened to a mass of intertwined wriggling worms. When a stretching force is applied, the molecules tend to uncoil and when the force is removed, they coil up again. Therefore, rubber has great *extensibility*.

Stored Energy in a Stretched Material

When a coil spring or a length of rubber cord or a length of steel wire is stretched, work has to be done in each case and most of this work is stored in the strained specimen as energy. This energy can be converted into work. For example: the energy stored in the wound spring of a watch or clock may be used as work over a lengthy period of time in turning the wheels of the watch or clock; the energy stored in a stretched cord of elastic may be used to turn the propeller of a model boat; the energy stored in a stretched catapult band may be quickly converted into work, in propelling a stone from the catapult; the energy stored in the compressed spring of an air rifle may be used as work to eject the pellet from the rifle; the energy stored in a bent bow may be used to shoot an arrow. We say that an object that is in this kind of strained condition, i.e. able to do work, possesses *potential energy*.

Pressure in Liquids

The total force or thrust on the horizontal bottom of an open vessel with vertical sides and containing liquid (Fig. 74), is equal to the weight of the liquid. The mass of this liquid is equal to the product of its volume and density ($m = vd$), while its volume is equal to the product of its base area and its vertical height ($v = Ah$). It is now possible to express the force or thrust (F) on the bottom of the vessel in the simple equation:

$$F = Ahdg$$

where g is the force of gravity on unit mass.

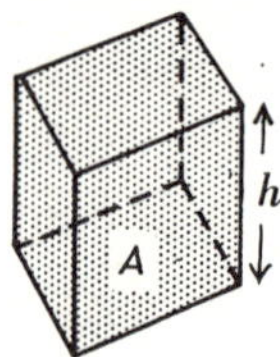

FIG. 74

Frequently it is more useful to know the force on each unit area of the base, rather than the total force on the base as a whole. This force per unit area is called *pressure* (p) and may be expressed as:

$$p = \frac{F}{A} = hdg$$

For example, if the vessel shown in Fig. 74 contains brine with a density of 1200 kg/m³ to a depth of 0·2 m, the pressure on the base of the vessel will be $0{\cdot}2 \times 1200 \times 10\ \text{N/m}^2 = 2400\ \text{N/m}^2$.

Measurement of Fluid Pressure

One type of pressure gauge is the *open-tube manometer*. This consists of a U-tube about half filled with a liquid such as water, oil or mercury, according

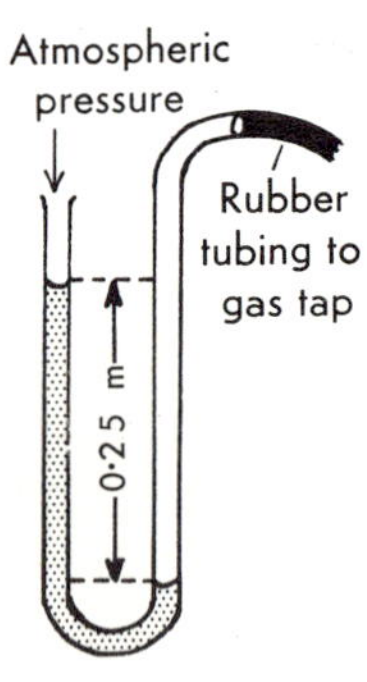

FIG. 75. *Open-tube manometer.*

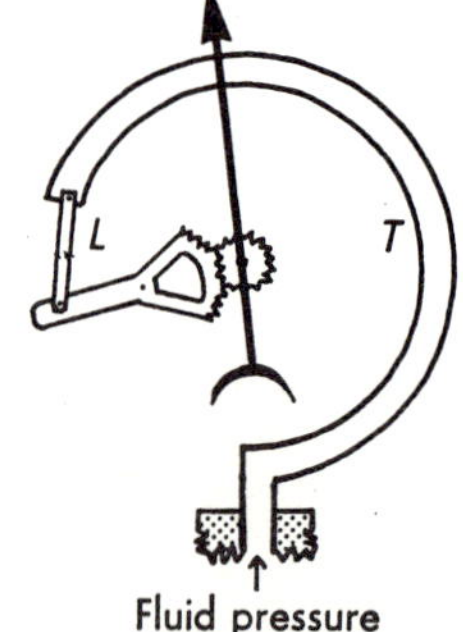

FIG. 76. *Bourdon gauge.*

to the degree of pressure to be measured. For example, pour some water into such a U-tube and connect one side of the U-tube by a length of rubber tubing to a coal gas tap (Fig. 75). Turn on the tap and the water in the open limb will rise to a height of about 0·25 m above the water in the limb connected to the gas supply. This means that the pressure of the gas exceeds the pressure of the atmosphere by the pressure of 0·25 m of water. Since the density of water is 1000 kg/m^3, the pressure of the gas exceeds that of atmospheric air by $0{\cdot}25 \times 1000 \times 10$ N/m^2 (remember that $p = hdg$).

If the water in the manometer is replaced by mercury, the manometer may be connected to the water tap and the pressure of the water supply measured. Why is the manometer liquid changed from water to mercury?

Another widely used device for measuring fluid pressure is the *Bourdon Gauge* (Fig. 76). It consists of a curved metal tube *T*, anchored to the inlet nozzle at one end, but connected at the other end by a lever *L* to cogged wheels, in such a way that if the coil uncoils or coils it rotates a pointer over a scale. When fluid (liquid or gas) pressure is applied inside the tube *T*, it straightens to an extent proportional to the pressure of the fluid. The greater the pressure the greater the angle through which the needle is rotated. Usually the gauge is calibrated to read in newtons per metre squared (N/m^2).

Such gauges are used to measure the pressure of steam in boilers, and of gas in steel cylinders.

Laws of Fluid Pressure

The French mathematician and writer Blaise Pascal (1623-62) stated the laws of pressure for fluids at rest. Remember, a fluid is something which flows and includes liquids and gases.

1. *The pressure is the same at all points which are at the same level in one fluid.*

2. *Fluid pressure on any surface is perpendicular to the surface.*

3. *At any place in a fluid, pressure acts equally in all directions.*

4. *Pressure is transmitted uniformly from one place to another throughout a fluid. (For example, if a piston is pushed at one place in a hydraulic system, the pressure exerted there is transmitted to every wall and to any other pistons in the system.)*

5. *The difference in pressure between any two places in the same fluid is found by multiplying the vertical difference of level h by the density of the fluid d by the force of gravity g on unit mass.*

$$\text{Pressure difference} = hdg$$

Let us consider the fifth law. Imagine a rectangular box drawn in the fluid, with base of area A m² and height h m from level Q to level P (Fig. 77). Let the density of the fluid be d kg/m³. The fluid in this box is in equilibrium (the fluid is at rest), so the resultant of all vertical forces on it must be zero.

Mass of fluid in box $= A\,h\,d$
Pull of earth on fluid (weight of fluid) $= A\,h\,d\,g$
Downward push of surrounding fluid on top of box $= p_P A$
($p_P =$ pressure of fluid at level p)
Upward push on bottom of box $= p_Q A$
BUT upward push = downward push as fluid is at rest

$$\therefore p_Q A = p_P A + A\,h\,d\,g$$

$$\therefore p_Q - p_P = h\,d\,g$$

i.e. pressure difference = vertical height × density × gravitational force on unit mass.

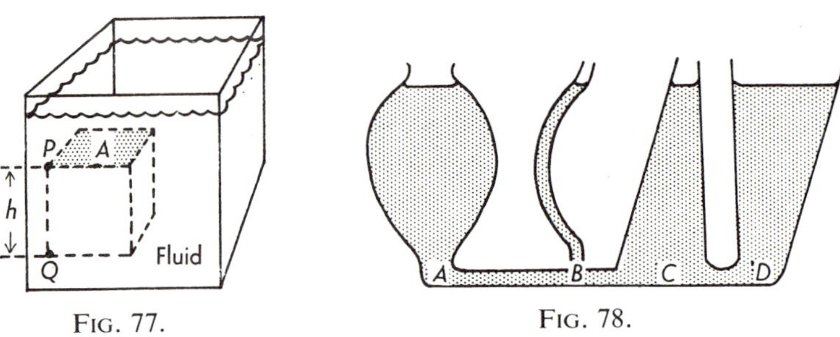

FIG. 77. FIG. 78.

Some of these laws of fluid pressure may be illustrated by a number of simple experiments.

1. *A liquid always finds its own level.*
When a set of open vessels of different shapes and sizes are interconnected, and a liquid is poured into one of them, it rises to the same vertical height in all the vessels (Fig. 78), provided none of the inter-connecting tubes are very narrow. Why this proviso? The shape and size of the vessels have no effect on the pressures at the points A, B, C, D, which are all at the same horizontal level, provided that no tube is so narrow that it causes a rise in level due to surface tension.

2. *Pressure depends upon the depth below the surface of a fluid.*
If holes are bored in the side of a cai. of water at different levels, and the

level of the water in the can is kept constant by regulating the tap, then the water from the bottom of the hole will eventually travel a much greater horizontal distance, showing that the pressure at the bottom hole must be greatest (Fig. 79).

Thus, the greater the depth below the surface, the greater will be the pressure exerted by the liquid.

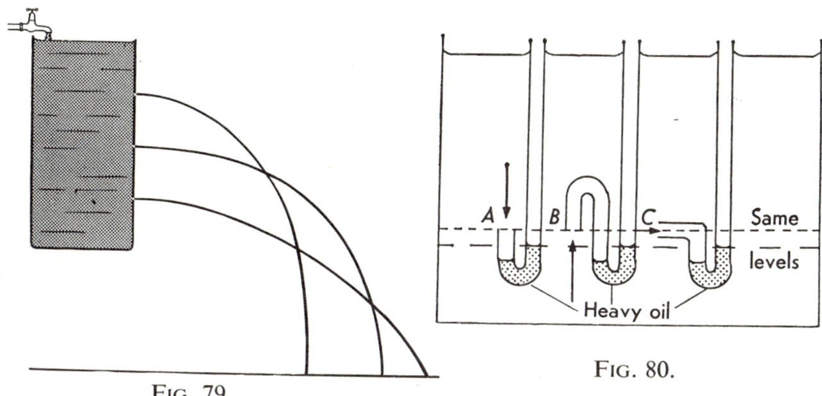

FIG. 79.

FIG. 80.

3. *The pressure at a given depth in a liquid acts equally in all directions.* This may be tested using an apparatus like the one shown in Fig. 80. Heavy oil is poured into the three open-tube manometers, until it stands at the same height in all three. The manometers are lowered into water; all the three open ends *A*, *B*, *C*, are at the same depth. It is found that the oil is pushed around all three manometers, until it stands at the same height in each, showing that pressures in the downward, upward and sideway directions, at a given depth, are all equal to one another.

Another simple experiment to support this statement may be done at home. Puncture an old rubber ball in several places, place it under water, force out the air and allow it to fill with water. Remove the ball from the water container and press the ball inwards with your thumb. The water shoots out of the holes in all directions, not merely in the direction in which it is pressed. The apparatus in Fig. 81 also illustrates the same fact. The cork plunger is removed, the glass sphere filled with water and the plunger replaced. When the plunger is pressed downwards, water issues from the tiny holes in streams at right angles to the surface of the glass sphere. The streams soon become curved, as shown, because of the pull of gravity.

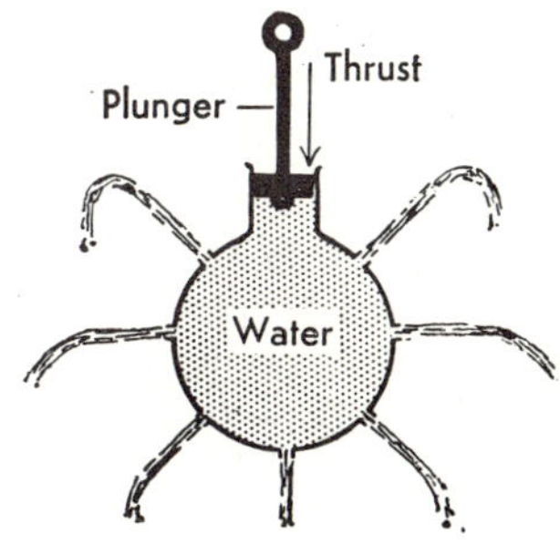

FIG. 81.

An interesting experiment to show how effective liquid pressure can be is the following: what column of water in a glass tube connected to a football bladder would be necessary to balance a piece of iron of weight 100 N placed on top of the bladder? You will probably be surprised at the result when you try it. Bind the teat of a football bladder tightly around one end of a glass tube about 1 m long. Place the bladder over a large sheet of squared paper. Pour water through the tube into the bladder, place a flat tin over the bladder and gently lower the piece of iron on to the tin. Water will rise up the tube to a level *A* and will then balance the iron weight resting on the tin (Fig. 82). The water will not all be forced out of the bladder, as you probably imagined it would. Your knowledge of liquid pressure should now enable you to calculate why the short column of water can balance the 100 N. Using a long pencil with a long tapering point, trace around the bladder in contact with the squared paper. Count the square centimetres of area in contact with the paper. Suppose this amounts to 400 cm² or 0·04 m². Then there is a thrust of 100 N on an area of 0·04 m², that is, a pressure of 2500 N/m². The height of a water column to balance this pressure, no matter what the diameter of the tube may be, would be found from the equation $p = hdg$. Then $h = p/dg = \frac{2500}{1000 \times 10} = 0{\cdot}25$ m (density of water is 1000 kg/m³). Now measure the height of *A* above the middle of the bladder and see how your answer agrees with the calculated result.

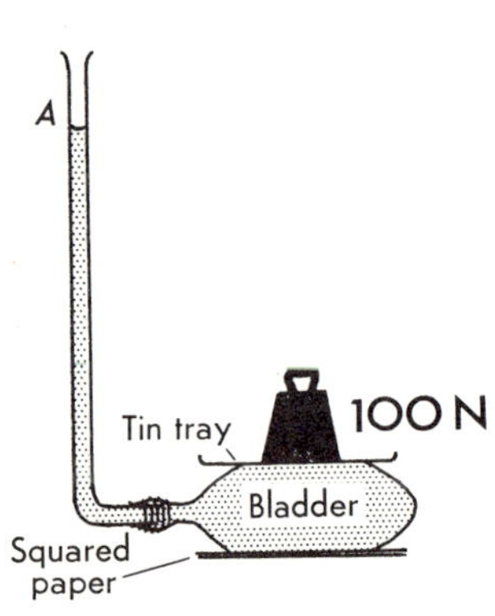

FIG. 82.

Transmission of Pressure

When an external force is applied to fluids, additional pressure is developed within them, and this is transmitted in all directions. Thus a small force applied over a small area at one place might result in a large force spread over a large area at another place, provided that the ratio force/area or pressure is the same at the two places. This is the principle of *the hydraulic press* which is an apparatus for producing a large force from a small one. It consists in its simplest form of a narrow cylinder connected to a wide cylinder fitted with pistons *A* and *B* (Fig. 83).

A mass of 1 kg placed on *A* where the area of cross-section of the tube is 0·01 m² exerts a force of 10 N on the liquid. The pressure will therefore be 1000 N/m² just under *A*. This will also be the pressure just under *B*. If the cross-section area of the tube is 1 m², the total thrust upwards will be 1000 N. However, if it is desired to raise a mass of 100 kg (weight = 1000 N) at

B through a distance of 0·1 m, then the effort at *A* must be moved through 10 m. In other words the work done by the effort is 10×10 J, and the work done on the load is $1000 \times 0{\cdot}1$ J.

The hydraulic press was invented in 1795 by Joseph Bramah, an English engineer from Yorkshire.

This principle is used in such devices as the hydraulic brake system of a motor-car, and in the hydraulic system for raising a car wheel to change a tyre, and the hydraulic system used in garages to raise the whole car for servicing. The great advantage of the hydraulic brake system for cars is that it results in equal pressure being applied to all the brakes. This was very difficult to achieve on the old direct lever system. As a result of the unequal pressures the car would tend to pull to one side or to skid.

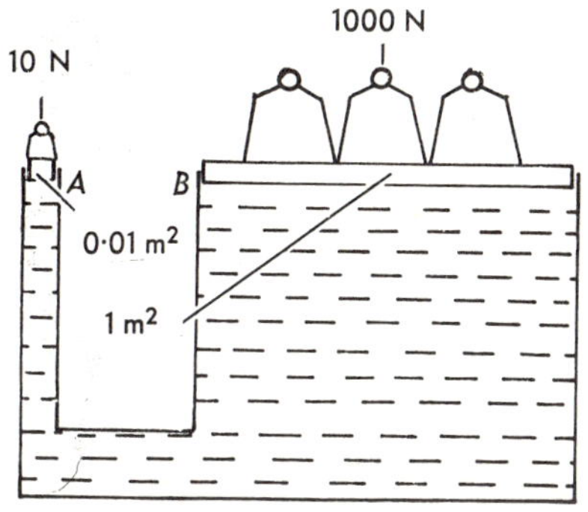

FIG. 83. *Hydraulic press.*

A new hydraulic device introduced in 1962 was the hydrolastic suspension as used in the Morris 1100 motor-car (Fig. 84). This is a device for moving liquid by means of a sudden increase in pressure in one part of the system. When the front wheel of the car hits a bump, a mixture of alcohol and water is forced through the pipes into the back springs, instantaneously raising the back of the car to balance with the front. When the rear wheel hits a bump, the reverse happens. The result is a smoother ride with no fore-and-aft see-sawing. The alcohol/water system is used since pure water would freeze in very cold weather.

FIG. 84. *Hydrolastic suspension.*

ATMOSPHERIC PRESSURE

Forces in gases have many effects, such as inflating car tyres, turning windmills, propelling rockets, keeping aircraft in flight, blowing down buildings, carrying away topsoil, moving clouds which carry the rain. Air is one form of matter and possesses the properties of all forms of matter, namely: air

FIG. 85. *Gases possess inertia. Wreckage left after winds of 150 km/h had blown the roofs from 50 houses.*

has weight; air occupies space; and air possesses inertia.

If a flask fitted with a stopcock is weighed on a balance sensitive to 0·01 g, and then the flask is connected to an efficient air pump and as much air pumped out as possible, the stopcock closed, and the flask re-weighed, it will be found that the flask weighs less than it did before the air was pumped out. If the volume of the flask is measured, it will be found that each 1000 cm^3 of the air it contains weighs about 1·29 g.

You are probably aware what happens when an inverted beaker is thrust into a deep glass tank of water. The air fills the beaker when the latter is near to the surface of the water. The lower the beaker is thrust, the less space does the air occupy, but the water only rises a little way into the beaker and, however much pressure you exerted on the air, the air would always occupy some space.

The property of inertia will be better understood when we have continued our study of forces at a later stage. But a simple experiment will help you here. Stand on an old table and hold an open sheet of paper as high as you are able. Release it, and count the seconds it takes to reach the floor. Now fold the paper many times so that it has a small surface area and repeat the fall from the same height. Notice that it now takes only a small fraction of the time it took before to reach the ground. The piece of paper had the same weight in each test, but in the first test it had to push aside much more air than in the second test, when it had a small area.

The air seemed "reluctant" to be pushed aside. We say that the air possesses *inertia.*

We have already seen how pressure in gases may be measured, using either an open U-tube manometer or a Bourdon pressure gauge. Simple experiments may be performed to show that atmospheric air exerts pressure:

1. Lower a tumbler into water, so that it fills with water. Invert the tumbler and raise it, without lifting the rim clear of the water. You notice that the tumbler remains full of water. The pressure of the atmosphere is sufficient to do this.

2. Completely fill a smooth-rimmed beaker with water and place a piece of thick paper or a postcard across the rim of the tumbler so that no air is trapped underneath (Fig. 86). Hold the paper on the surface of the tumbler and invert the tumbler. When the support is removed from the paper, the water does not run out. Why not? The beaker may be turned in any direction and the water will not run out. This shows that the pressure of the atmosphere is acting in all directions.

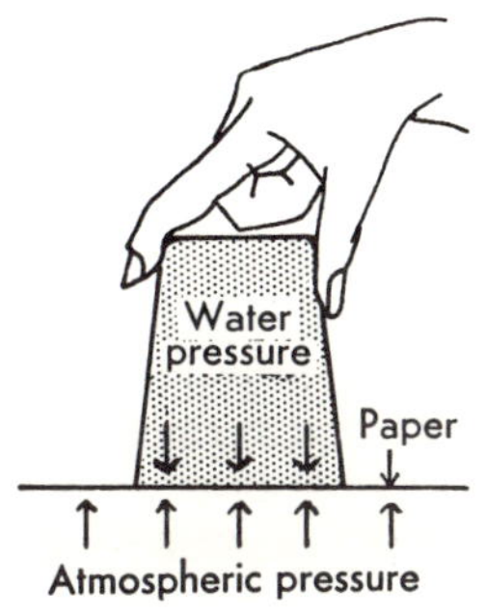

FIG. 86.

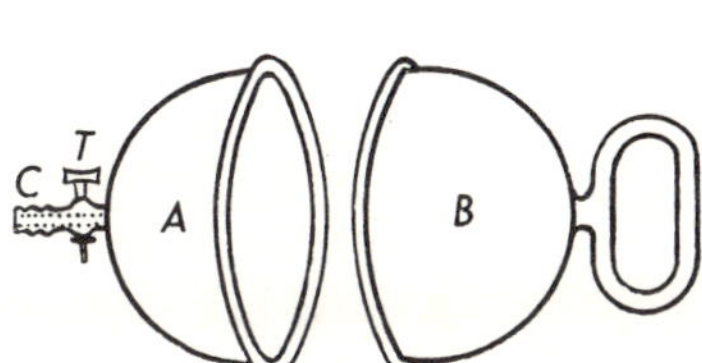

FIG. 87. *Magdeburg Hemispheres.*

3. *Magdeburg Hemispheres: A* and *B* (Fig. 87) are two metal hemispheres, with well-fitting flanges which prevent air from entering or leaving when *A* and *B* are placed together to form a sphere. *A* is also fitted with a tap *T* and a thread *C* so that a handle, similar to that on *B*, may be fitted. A hole passes through the axis of *CT*. *A* and *B* are placed in contact at their flanges, a thick rubber tube connects *C* to an efficient air pump, the tap *T* is opened and as much air as possible is removed from the metal sphere. The tap *T* is closed and the handle screwed on at *C*. It is usually found that two boys pulling on the handles as hard as they can, do not succeed in separating the hemispheres. This demonstrates how great the thrust of air is. Let us calculate why. Suppose the diameter of the sphere is 0·1 m. Atmospheric pressure is a little less than 110 000 N/m^2. Suppose the pump removes enough air to leave a difference of pressure between the outside and the inside of the sphere of 100 000 N/m^2. Then the thrust of the atmosphere on the outside of the sphere $=$ pressure difference $\times$ area $= 100\,000 \times \pi \times 0{\cdot}05^2 = 800$ N, nearly. It is little wonder that two boys find it difficult to separate the two hemispheres.

4. *Torricelli's method of measuring atmospheric pressure:* in 1643, the Italian Torricelli performed an experiment which you might repeat. A thick-walled glass tube, about 1 m long, which is closed at one end, and is clean and dry, is filled with mercury except for a length of about 10 mm of tube. Close the

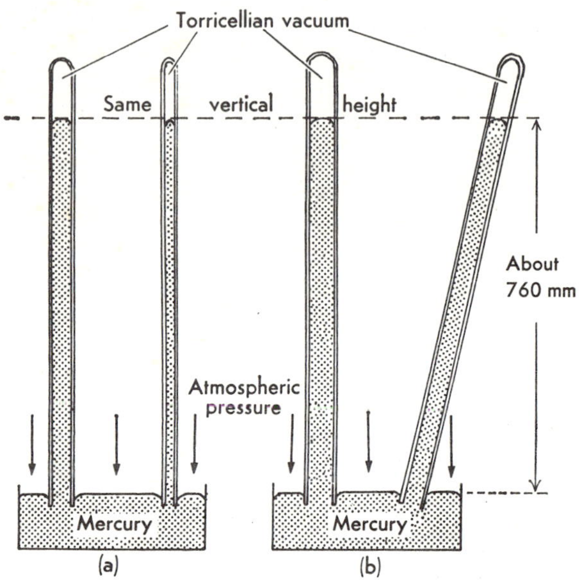

FIG. 88. *Torricelli's vacuum tubes.*

end with a finger and slowly invert the tube, so that the bubble of air runs to the closed end and back. This bubble of air should collect the small air bubbles that are likely to stick to the sides of the tube between the glass and mercury. Now completely fill the tube with mercury, place a finger over the open end and invert the tube so that the finger is below the surface of some mercury in a bowl. Remove the finger and the mercury falls a little way and then rests with a vertical height of about 760 mm between the surfaces of mercury inside and outside the tube (Fig. 88*a*). Repeat the experiment with another tube of smaller diameter than the first tube. You will find that the height of the mercury is equal to that obtained with the first tube, unless the second tube is much narrower and then you will find that the level is lower than in the wider tube, due to capillary action.

Now lean one of the barometer tubes away from the vertical (Fig. 88*b*). The mercury will be seen to move along the tube, so that the *vertical* height of the mercury remains the same as before, until the mercury eventually fills the tube. This latter fact confirms that there is a vacuum above the mercury in the barometer tube and such a vacuum is still called a Torricellian vacuum. Before Torricelli performed his experiment it was thought to be impossible to create a vacuum. (N.B.—You will learn, in a later section, that what exists above the mercury in the barometer tube is not quite a true vacuum, as there is some mercury in the form of vapour in the space.)

Fortin's Standard Barometer

Accurate barometers are required in scientific work; for instance in meteorological stations. One such barometer is the Fortin standard barometer which incorporates a means of setting the zero of the scale before the reading is taken (Fig. 89). The mercury reservoir is contained in a leather bag supported at the lower end of the barometer on a regulating screw *S*. If the atmospheric pressure changes, the level of the mercury in the reservoir rises or falls. The zero of the fixed scale in a Fortin barometer is the tip of an ivory pointer.

To take a barometric reading, *S* is adjusted to bring the surface of the mercury, seen through a glass window, just to touch the tip of the ivory pointer. The vernier is then moved by the screw *F* so that the vernier zero is level with the meniscus of the mercury in the upper end of the barometer tube. Then the scale and vernier reading gives the pressure of the atmosphere in cm of mercury. In the Kew barometer, the change in level of the mercury reservoir is allowed for when the fixed scale is made so that no adjustment of the screw *S* is required before a reading is taken.

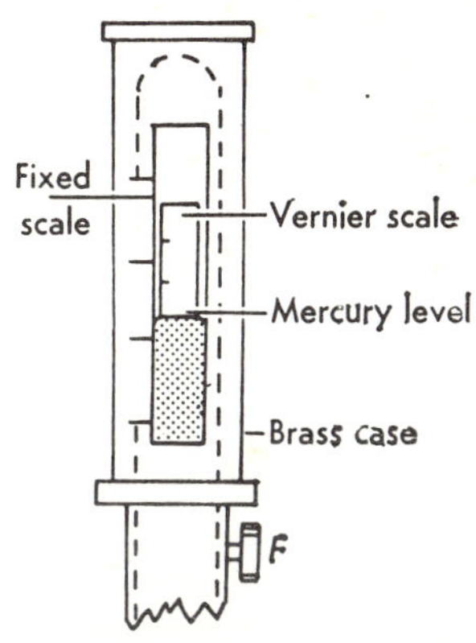

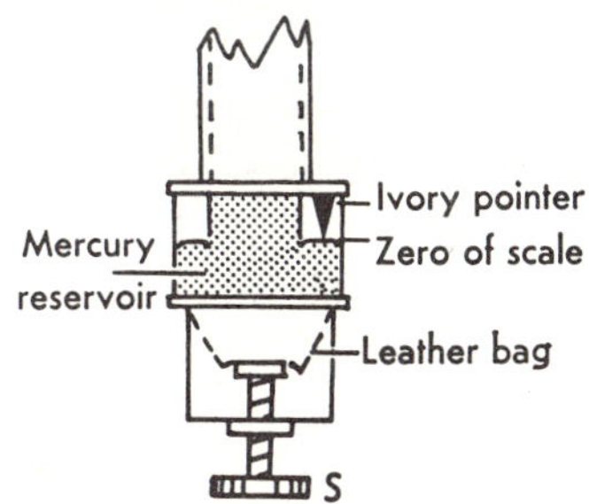

FIG 89. *Fortin's standard barometer.*

Most laboratories will possess a barometer of this type and you should certainly examine one and take a reading when the opportunity arises.

Aneroid Barometers

These barometers are so called because they contain no liquid. They are constructed as in Fig. 90. Unlike the standard barometer they may be carried about. The partially exhausted box is an airtight chamber made of thin metal with corrugated sides. Increase in atmospheric pressure pushes in the sides of the box and when the pressure decreases, the spring pulls the sides outwards. This motion is very slight but its effect is magnified many times by a system of levers. The small chain passing over the pulley moves the pointer over the dial on the face of the barometer. The dial is calibrated to read in inches or cm of mercury.

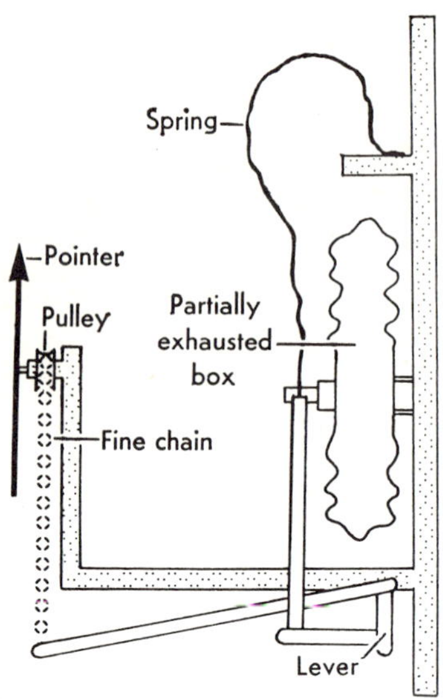

FIG. 90. *Aneroid barometer.*

Blaise Pascal, the French writer and mathematician, who carried out much of the early work on fluid pressure, discovered in 1646 that when a mercury barometer was taken up a mountain, the column of mercury decreased in height by about 2·5 cm for a climb of 270 m. If this decrease is known accurately, then an aneroid barometer may be calibrated so that it records height of ascent above the starting point. It is then called an *altimeter* and is used by mountaineers and by aircraft pilots.

How High is the Atmosphere?

Firstly let us calculate how high the atmosphere would be if it were all at the same density as it is at the surface of the earth. Since it can exert the same pressure as a column of mercury 76 cm high, and mercury is 13·6 times as dense as water, it can balance a column of water $0{\cdot}76 \times 13{\cdot}6$ m high, which is about 10 m. This would be the height of a water barometer. The pressure of the atmosphere $= hdg = 10 \times 1000 \times 10 = 10^5$ N/m².

The density of air at the surface of the earth is 1·29 kg/m³. If all the air had this density, the height of the atmosphere h would be given by

$$h = \frac{10^5}{1{\cdot}29 \times 10}$$
$$= 7750 \text{ m or } 7{\cdot}75 \text{ km}$$

But the higher above the surface of the earth the less dense is the atmosphere, so that the air extends to a height far in excess of 7·75 km, getting rarer as the height increases. This decrease in the density of the atmosphere with increasing height above the earth is in fact very rapid: 1·3 kg/m³ at ground level; 0·1 kg/m^3 at 20 km; 0·0004 kg/m^3 at 60 km; 0·000 000 8 kg/m^3 at 100 km. This explains one of the chief difficulties facing climbers of high mountains. At the summit of Mount Everest (8850 m) the air's density is only 0·3 kg/m^3 and the climber must take four breaths to obtain the oxygen obtained by one breath at sea level.

Balloon ascents have provided much information about the upper atmosphere, firstly using manned balloons and secondly using sounding balloons carrying pressure gauges and thermometers. Sounding balloons have reached heights of 30 km and rockets have carried instruments to greater heights. The study of the motion of meteors shows that the atmosphere reaches to a height of at least 160 km.

Measurements have shown that the temperature of the atmosphere decreases fairly uniformly to $-53°$ C at a height of about 13 km, above which the temperature remains almost constant for a great range of height. At very great heights the temperature of the atmosphere rises again. The region of fall in temperature is called the troposphere and the region above is called the stratosphere. The tropopause is the name given to the level at which the fall in temperature ceases. Since the stratosphere is above the region in which weather phenomena occur, travel by aircraft through the stratosphere is more comfortable. Air resistance is lower, noise is less, and jet engines work better. The cabins of aircraft must of course be kept at a pressure similar to that at ground level, else there would be too little air to breathe.

EFFECT OF FORCE ON THE VOLUME OF A GAS

You already know that a gas may be compressed into a smaller volume, and that when this happens, the pressure of the gas increases. The molecules of the gas are themselves incompressible. There are very many molecules of a gas in 1 cubic centimetre at normal pressure—about 3×10^{19}.

The size of one molecule is very small so that much of the space in which the molecules are moving is unoccupied by the molecules themselves. The molecules move at very high speeds at normal room temperatures, speeds of several miles per second, colliding with one another and with the molecules of the walls of the containing vessel or room. These collisions or impacts cause the gas to exert pressure. If the gas were compressed in a vessel closed by some kind of gas-tight piston into half of its volume, then there would be about twice as many collisions per second with the walls of the vessel, provided that the temperature of the gas has not changed. (A change in the temperature of the gas alters the speed of the molecules.) This would mean that the gas would exert twice the pressure it did before it was compressed. Whether this is so may be tested with an apparatus which we will now consider.

Boyle's Law

AB is a thick-walled glass tube containing dry air *AC*, enclosed by a column of mercury *CD* (Fig. 91). An air cylinder has sufficient volume for pressure

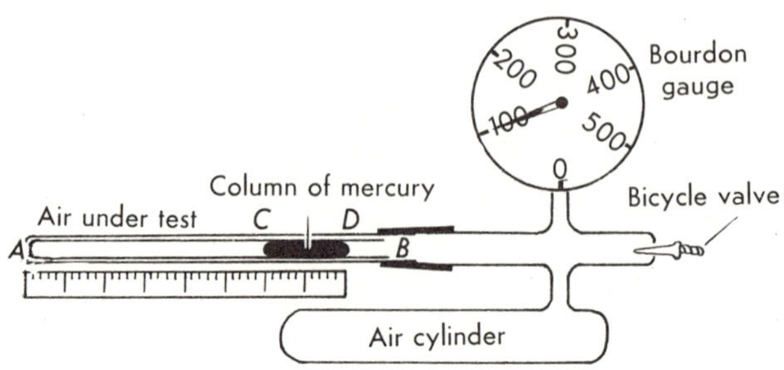

FIG. 91. *Apparatus for investigating Boyle's Law.*

to be controlled smoothly. The Bourdon gauge registers the pressure of the air in *AC*. When a bicycle pump is attached to the valve holder, air may be pumped into the apparatus, forcing the mercury column to the left, thus compressing the air in *AC*. The reading of the Bourdon gauge gives the resulting pressure of the air in *AC*.

The tube *AB* has a uniform cross-section, so the volume of tube occupied by air is proportional to the length *AC*. It can quickly be observed whether *AC* is halved when the pressure is doubled. A series of readings of pressure and volume of air may be taken. Tabulate your results (Table 5).

Table 5
Relationship Between the Pressure and Volume of a Gas

Pressure of air P kN/m²	Length of air column V m	$P \times V$	$\frac{1}{V}$
110	0·28	31	3·6
154	0·20	31	5·0
198	0·16	32	6·3
250	0·12	30	8·3
307	0·10	31	10·0
337	0·09	30	11·1

Be careful not to pump the air in quickly during this experiment, or else the compressed air will be heated, and the mercury column will move slightly to the right, as cooling takes place. (You will have noticed how the barrel of your bicycle pump gets very warm if you inflate a tyre quickly.)

It is important also to keep the mercury column *CD* horizontal in your experiment, else the gauge will not give the true pressure of air in *AC*.

You will notice that the product *PV* in column 3 is nearly constant. A useful exercise is to plot two graphs (Fig. 92):

1. *P* against *V*, which does not give a straight line.
2. *P* against $\frac{1}{V}$, which does give a straight line. When this line is produced it passes through the origin.

You will be able to verify from your results what Robert Boyle discovered from his experiments in 1662 and now known as *Boyle's Law:*

The volume of a given mass of gas is inversely proportional to its pressure, provided its temperature does not change."

Other gas laws are considered later (pages 172-76).

Use of Boyle's Law

Given a tube like the one shown in Fig. 93, it is possible to find the value of the atmospheric pressure at the time of the experiment by assuming Boyle's Law to be true for the column of air *AB* in the tube. It is important that the air is dry; a necessity that will be explained later in the book, in the section on Heat.

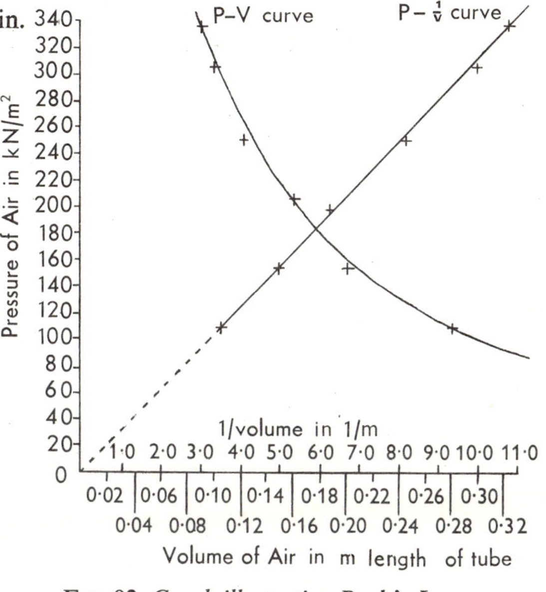

FIG. 92. *Graph illustrating Boyle's Law.*

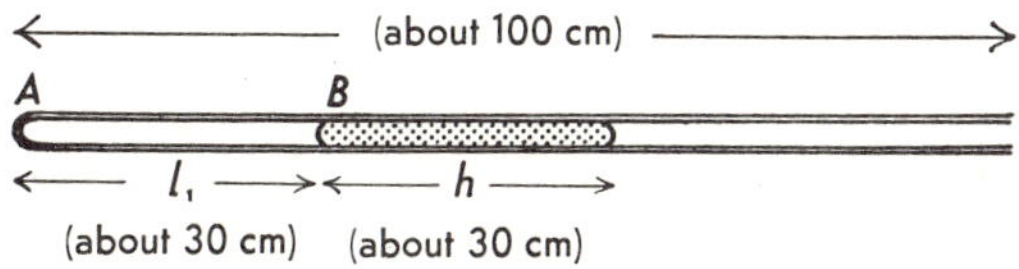

FIG. 93 *Apparatus to measure atmospheric pressure, assuming Boyle's Law.*

When the tube is in the position shown, the mercury column is in equilibrium under the forces acting upon it. The air column *AB* exerts a pressure equal to that of the atmosphere. Measure the length of this column l_1 and gently rotate the tube to a vertical position, with the open end upwards.

(Fig. 94*a*). The air now occupies less space than it did before because it is exerting a pressure equal to that of the atmosphere plus the pressure of a column of mercury h high. Measure the length of the air column, l_2. Now gently turn the tube upside down so that the closed end is uppermost (Fig. 94*b*). If you turn it slowly the mercury column will neither flow out of the tube nor break up into short columns.

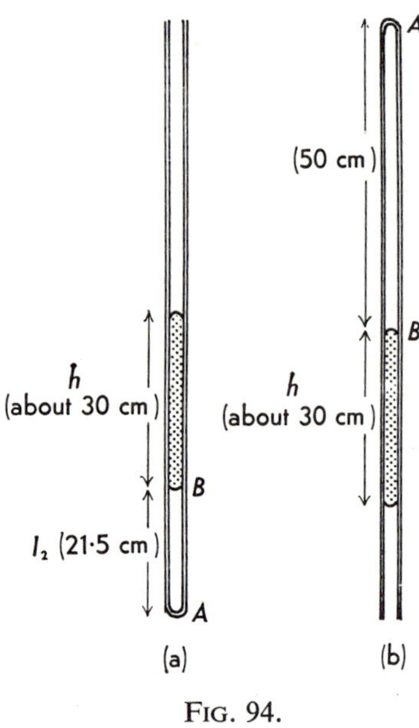

FIG. 94.

Now the air in AB is exerting a pressure equal to that of the atmosphere, less the pressure of a column of mercury h high. Measure the length of the air column, l_3. If it is assumed that Boyle's Law is true for the air in AB, then the product of pressure and volume should have a constant value for all three positions of the tube.

If we now let $l_1 = 30$ cm, $l_2 = 21{\cdot}5$ cm, and $l_3 = 50$ cm, call the cross-sectional area of the tube A, and the pressure of the atmosphere in cm of mercury p, then:

$$pv = \text{constant (Boyle's Law)}$$

Applying this to the air column AB in Fig. 93 and 94*a*:

$$\begin{aligned} p \times 30 \times A &= (p + 30) \times 21{\cdot}5 \times A \\ 30p &= 21{\cdot}5p + 645 \\ p &= 75{\cdot}9 \simeq 76 \end{aligned}$$

Applying Boyle's Law to the air column AB in Fig. 93 and Fig. 94*b*:

$$\begin{aligned} p \times 30 \times A &= (p - 30) \times 50 \times A \\ 3p &= 5p - 150 \\ 2p &= 150 \\ p &= 75 \end{aligned}$$

This gives an average value for the atmospheric pressure from the two sets of results as 75·5 cm (755 mm) of mercury.

QUESTIONS

(N.B. You may assume in numerical problems that the force of gravity on a mass of 1 kg is 10 N—actually it is nearer 9·8 N.)

1. Describe how you would make a simple barometer. Give two reasons why mercury is a better liquid than water for use in a barometer.

2. How do each of the following factors affect the pressure at the bottom of a quantity of liquid in a container?
(*a*) The shape of the container and the area of its base.
(*b*) The depth of the liquid and its density.
(*c*) The viscosity of the liquid.

3. What happens to a partially inflated hydrogen balloon when it is released? Why, after a time, does it cease to rise any higher?

4. Sometimes liquid will not flow freely from a sealed can when a small hole is pierced in the base, but if a second hole is pierced, it will flow freely. Explain this.

5. What is the pressure in a horizontal direction 10 m below the surface of the sea, if the density of sea-water is 1030 kg/m^3? What is the pressure in a vertical direction at the same place?

6. What is the weight of water in a swimming pool 30 m long, 10 m wide and 2 m deep? What is the pressure on the bottom in N/m^2? Density of water is 1000 kg/m^3.

7. If a water barometer reads 10·3 m and water has a density of 1000 kg/m^3, calculate the pressure of the atmosphere in N/m^2.

8. A car tyre is inflated to a total pressure of 180 kN/m^2 and its volume is 8000 cm^3. What volume would this air occupy if the valve of the tyre were opened so that all the air escaped? Assume that the atmospheric pressure is 100 kN/m^2 and that the temperature does not change.

9. A spring 75 cm long measures 100 cm when a force of 200 N is applied to it. What extra force must be added to make the spring double its original length?

10. Here is a practical exercise for you: use a tyre gauge to measure the pressure of the air in each of the four tyres of a car (remember that the gauge will measure the difference in pressure between the inside and the outside of the tyre). Measure the approximate area of contact of each tyre with the ground and then find the thrust of each tyre on the ground (thrust = pressure × area). Add together the four thrusts and from your result calculate the total weight of the car. Check your answer with the weight given in the car instruction book.

CHAPTER 8

SOLIDS MOVING ON SOLIDS

IF ONE solid object is placed upon another solid object, motion of the one relative to the other is always opposed by a force known as the *force of friction*, which varies greatly when the two kinds of surface in contact are changed. For example, it is easier to pull a sledge over ice or wet grass than over a concrete or stone pavement.

A block of wood B is placed on a horizontal table and a thread attached to it, the other end of which is fastened to a scale pan S, after passing over a pulley (Fig. 95). If a small weight is gently placed in the scale pan, the block will not move. It follows that a force of friction F between the lower surface of B and the table, equal in magnitude to the pull P in the string, acts in the opposite direction to P. If P is slightly increased by adding small weights to S, F increases to become equal to P. This cannot continue indefinitely, however, and if you try this experiment, you will find there is a limiting force of friction between the two surfaces and, if P exceeds this limiting force, the block B will move over the table. If B is given a slight push along the table, it will be found that B can be kept moving over the table at a constant speed, using a pull P less than that required to start B moving. We say that: *the force of static friction is greater than the force of sliding (or kinetic) friction.*

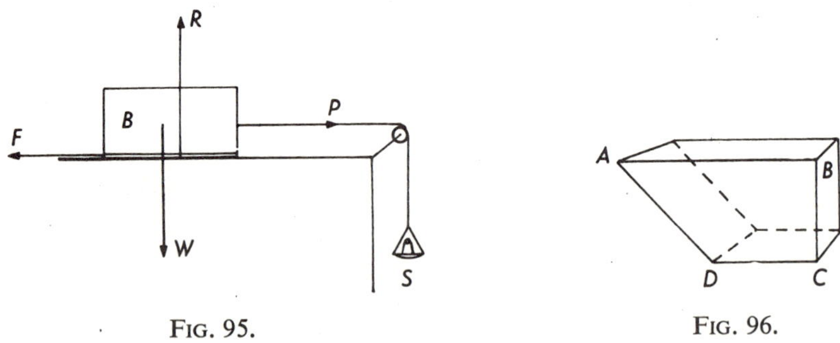

FIG. 95.

FIG. 96.

You may obtain useful information about the force of friction by carrying out some experiments.

In the apparatus of Fig. 95, use a block of wood shaped as Fig. 96 of known weight; in this case 1 N. Place the block with surface *CD* in contact with the table. Find the least mass that, placed in *S*, will keep the block moving at a steady speed over the table after the block has been given an initial slight push. Do not forget to include the weight of the scale pan in the total pull *P*. Place a mass of weight 1 N on the top of the wood block and again find the least force *P* required to keep the block and the mass moving at a uniform speed. Continue the tests adding a further equal mass each time. Repeat the experiment with surface *AB* in contact with the table. Notice (Table 6) how little difference there is between the limiting force of friction for the small area and for the larger one, which is twice as big, for the same normal reaction. Plot a graph of *F* against *R* (Fig. 97). The graph is a straight line which does not

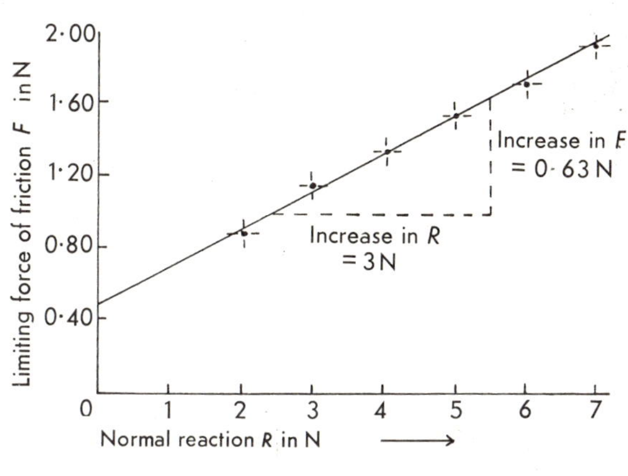

FIG. 97.

Table 6

Relationship Between Load and the Limiting Force of Friction

R (normal reaction) in N	*F* (limiting force of friction) in N	
Total load + weight of block	Large area	Small area (half-size)
2	0·86	0·91
3	1·10	1·16
4	1·31	1·31
5	1·51	1·56
6	1·71	1·71
7	1·91	1·95

pass through the origin, as it is found that the pulley's groove and bearings are causing their own force of friction of about 0·46 N.

The gradient of the graph $\left(\frac{\text{Increase in } F}{\text{Increase in } R} = \frac{0{\cdot}63}{3} = 0{\cdot}21\right)$ gives the *coefficient of sliding friction* between the two surfaces, symbolized by the Greek letter μ (mu).

$$\mu = \frac{F}{R}$$

Replace the block of wood by a block of different wood, or a block of metal or glass or plastic, and repeat the experiment. You will find that F/R is still a constant, but the value of the constant will be different from that for the first two surfaces used. Let us summarize the results.

Laws of Friction

1. When the two surfaces are at rest, the force of friction can take any value up to a certain maximum value called the limiting friction.

2. When motion occurs, the force of sliding friction F is proportional to the normal (at right angles to) force R between the surfaces. F/R is called the coefficient of sliding friction.

3. The coefficient of sliding friction is independent of the area of contact and also of the relative speed, within fairly wide limits.

The coefficient of sliding friction of wood on wood is 0·2 to 0·5 depending on the kind of wood; for steel on ice it is 0·02. Suppose that a block of wood weighing 40 N rests on a table. The coefficient of sliding friction between the two is 0·4. The pull required to move the block at a uniform speed over the table must balance the force of friction.

$$F = \mu R = 0{\cdot}4 \times 40 = 16 \text{ N}$$

Similarly, to pull a block of steel weighing 50 N at a uniform speed over ice will require a force of $0{\cdot}02 \times 50 = 1$ N.

Friction is sometimes a hindrance, but at other times it is essential for forward progress. So in some cases, engineers are trying to lower the effect of friction as far as possible, but in others, to keep it at a reasonably high value. Friction is essential between the soles of our shoes and the ground if we are to start walking or running and again when we stop. Friction between the shoe and the ground is necessary if we wish to change direction. Motors can only drive machinery when there is sufficient friction between the belt drive and the pulley.

Without friction, your plate would not remain on the table if the latter were inclined very slightly to the horizontal. On the other hand, if the drawer of your sideboard sticks you rub soap on it since the limiting force of friction between soap and wood is less than that between wood and wood.

Rolling Friction

To give your friend a ride over a rough pathway, it is easier to pull him in a small hand-cart running on wheels, than it is to drag him over the same surface on a sledge.

The effects of friction are much reduced by using wheels or rollers. When a wheel starts to roll, the point of contact with the ground is stationary for a very short time and acts as a pivot about which the wheel turns. If *B* (Fig. 98) is a shaft attached to the bearing of a wheel and axle, then a pull *P* on the shaft, forces the bearings against the axle and tends to drag the wheel along the ground. The opposing force of friction *F*, which is equal and opposite to *P*, prevents the wheel from sliding or skidding and produces a turning effect, called a *couple*, which turns the wheel about *C*. The wheel moves forward long before *P* reaches the limiting value of friction between the wheel and the ground.

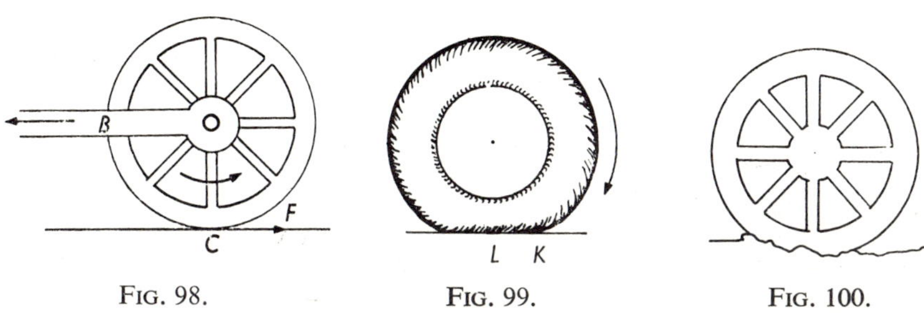

FIG. 98. FIG. 99. FIG. 100.

In actual practice, however, there is always a certain amount of sliding when a wheel rolls across the ground because wheels tend to become flattened where they make contact with the ground (Fig. 99). In this case the point *L* is momentarily at rest while at the point *K* the tyre is moving over the ground against the force of friction. The more the wheel is flattened the more important this sliding will be, for as every motorist knows, if he keeps his tyres soft, they will wear out far more quickly than if he had kept them inflated at the correct pressure.

Without rolling friction, motion by wheel, one of man's far-reaching inventions, would not be possible; the wheel would merely spin round, as when a car tries to move away from a wet, muddy parking field. A factor which affects rolling friction is the depression which the wheel causes on the surface below it (Fig. 100). On a hard surface, the depression is slight, but on a soft surface it can be considerable and this means that the wheel has to roll up a slight hill, causing more work to be done. The harder the surface the better, and car tyres with high pressures cause less friction than the same tyres with low pressures. Again, the undesirable depression is reduced (as in farm tractors) by using wide tyres and having wheels of large diameter.

Because rolling friction is very small, engineers use ball-bearings and roller-bearings to hold rotating parts of machinery.

How do tyres of large diameter and width and with a heavy tread help a tractor to move across soft and muddy ground?

Lubrication

When two dry sticks are rubbed together, the work done against friction produces heat and a fire can be started. If the pistons and cylinders of an engine were perfectly dry, there would be so much friction, and so much heat produced in working against this friction, that the metals would become extremely hot, and piston and cylinder would "seize," or weld together. But if the surfaces are properly lubricated, loss of power due to friction is reduced considerably and so is the wearing down of the surfaces. Lubricated surfaces are not in contact with one another; a layer of liquid separates them. In this way sliding friction is replaced by liquid friction which is very much less. Another important point about lubricated surfaces is that the friction changes very little as extra load is put on the rotating shaft, whereas, with dry surfaces, the friction would increase proportionately to the load.

Causes of Friction

Friction has been studied for a long time, but scientists do not agree on one simple cause of friction. It is probable that there are many causes. Among them is certainly the intermeshing of surface irregularities. Microphotographs show that the harder projections of surfaces in contact plough through the softer portions, snapping some of them off no matter how well the surfaces have been polished. When two flat steel surfaces are placed together, the actual area of contact is only a very small fraction of the apparent area, so the pressures over these areas of contact are very great and might be of the order of 10^4 N/cm^2. Thus, when one metal slides over another, the heat produced locally might be enough to melt the two metals at the area of contact and fuse them together.

Another cause of friction is the adhesion between the two kinds of molecules brought into contact and these molecular forces could be the most important cause of friction. Another theory says that friction is partly caused by electrostatic forces brought into being when one surface is rubbed on another. You will understand this idea better when you have read the section dealing with electrostatics.

QUESTIONS

1. How is the force of sliding friction affected by the area of the rubbing surfaces and by their relative speeds?

What is meant by the coefficient of sliding friction?

2. Explain the use of cylindrical rollers in dragging along a heavy crate over a floor with a rough surface.

3. What force will be required to pull a block of iron of weight 10 N up a slope inclined at 45° to the horizontal, if the coefficient of friction between iron and the slope is 0·2?

4. Slide a block of wood over the surface of a horizontal table and note the distance it travels. Now sprinkle a few small polystyrene beads over the surface of the table and push the block of wood with a force similar to that used before and note how far the block travels. What observation can you make about the movement of the polystyrene?

5. Place a 5p piece on the turn-table of your record player, close to the centre and rotate the table. Then move the coin gradually farther outwards and note what happens. Explain your observation. Place a sheet of shiny paper on the turn-table and the coin on top of that and compare results with those obtained with the turn-table alone.

6. A horizontal force of 8 N is required to pull a box of weight 40 N at a uniform speed along a horizontal floor. What is the coefficient of sliding friction? What force acts perpendicular to the floor? What is it called?

CHAPTER 9

SOLIDS IN FLUIDS

FIRSTLY, let us remind ourselves that a fluid is something that can flow from one place to another. Fluids include all liquids and gases. We have two aspects to examine: the effect of placing a solid in a fluid; the effect of moving a solid through a fluid.

You will already be aware that an object suspended in water appears to weigh less than when it is suspended in air. You know that if you are lifting a bucket which you have filled with water from a tank, the bucket appears to weigh more and more as you gradually lift it clear of the water. Also, if you are sitting on the side of a swimming bath, it is more difficult to move your hand through the water than to move your hand through the air.

Archimedes' Principle

Try this experiment: weigh a stone by suspending it from a spring balance. Weigh the same stone when it is completely immersed in water in a beaker, or a bucket. It appears to weigh less than before. By how much is its weight reduced? This may be answered by use of a device known as a displacement can (Fig. 101*a*). Fill the displacement can with water until it overflows from the spout. Weigh an empty beaker *B*. Carefully lower the stone into the displacement can, catching the water which overflows. Weigh the beaker and displaced water and hence you will know the weight of the displaced water. You will find that it is equal to the apparent loss in weight of the stone when suspended in water.

Now to convince yourself that this is not just true for water, repeat your experiment with brine and with paraffin. You will find different upthrusts on the same stone, but in all cases you will find that the apparent loss in weight of the stone is equal to the weight of liquid it displaces. These results are summarized in the principle named after Archimedes (287 to 212 B.C.).

Whenever a solid is totally or partially immersed in a fluid, it experiences an upthrust equal to the weight of fluid it displaces.

Now do you expect that a solid can really lose some of its weight while remaining in the same place? If weight is not lost, then what has happened to it?

A lump of iron in air weighs 4 N. This lump of iron in water weighs 3·5 N.

A beaker of water placed on a compression balance weighs 2·5 N. When the lump of iron is lowered into the beaker of water so that all the iron is immersed in the water, without touching the bottom or the sides of the beaker, the compression balance shows a reading of 3 N (Fig. 101*b*). We

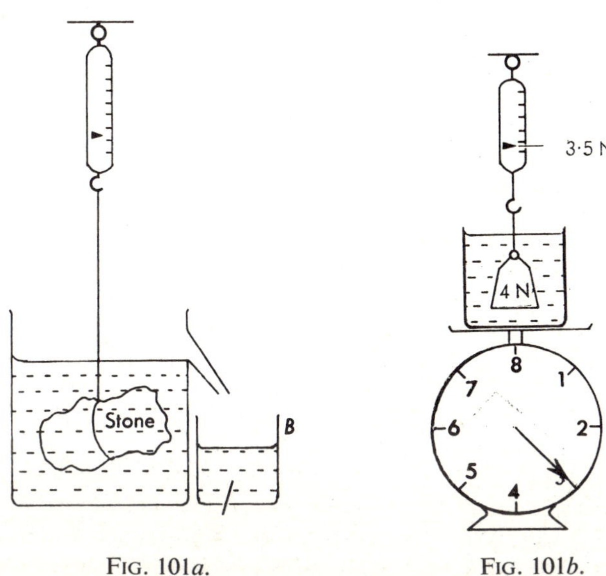

FIG. 101*a*. FIG. 101*b*.

find that the lump of iron appears to lose 0·5 N of weight, while the beaker of water gains 0·5 N of weight. 3·5 N of the iron's weight is supported by the spring balance and 0·5 N of its weight by the beaker of water. Another way of expressing this is to say that: the upthrust of the water on the iron is equal to the downthrust of the iron on the water. To help you to understand this, refer again to the theory concerning pressure difference in a fluid on page 91.

If the iron weight is gently lowered so that it rests on the bottom of the beaker, then the pull on the spring balance will be reduced to zero and the reading of the compression balance will be 6·5 N. The compression balance is supporting all of the iron's weight.

Relative Density

The relative density of a substance is the ratio of the weight of a specimen of the substance to the weight of an equal volume of water. To find the relative density of copper you weigh a lump of copper first in air and then when it is completely immersed in water. Let the weights, in newtons, be respectively a and b. The upthrust of the water on the copper is $(a - b)$.

But this is the weight of the volume of water displaced by the copper, and this volume is the same as the volume of copper.

$$\text{Relative density of copper} = \frac{\text{Wt. of copper}}{\text{Wt. of water displaced}} = \frac{a}{a-b}$$

This number tells you how many times the copper is as dense as water. If the same lump of copper is now weighed in another liquid and its weight, in newtons, is c, then the upthrust of this liquid on the copper is $(a - c)$. Thus the volume of liquid that has a weight of $(a - c)$ is equal to the volume of water that has a weight of $(a - b)$ and both volumes are the same as the volume of copper.

$$\text{Relative density of liquid} = \frac{a-c}{a-b}$$

For example, if you examine the results for Fig. 103, you will see that

$$\text{R.D. of iron} = \frac{\text{Wt. of Iron}}{\text{Upthrust of water on iron}} = \frac{4\text{ N}}{\frac{1}{2}\text{ N}} = 8$$

Floating Bodies

If a body floats on water, then its whole weight is supported by the water and the upthrust or weight of water displaced should equal the weight of the floating body. This may be tested experimentally by weighing a test tube and then floating it in a displacement can similar to that used in Fig. 101*a*, catching the displaced water and weighing the latter. Add some lead shot to the test tube, reweigh it and again float it in the displacement can, catch the displaced water and weigh it.

Repeat the experiment for a number of different weights by adding more lead shot and then repeat the experiment when the test tube with lead shot is floated in another liquid, such as brine.

Your results will indicate the *law of a floating body: a floating body displaces its own weight of fluid.*

When high altitude balloons are used for meteorological or cosmic ray research, the balloon is partially inflated with hydrogen and then released, with the necessary instruments and photographic plates attached to the balloon. The upthrust on the balloon is equal to the weight of air displaced by the balloon and its attachments. This upthrust is greater than the weight of the balloon + hydrogen contained + attachments. Therefore the balloon is carried upwards.

As the balloon rises, the pressure of the atmosphere on it decreases, so the pressure of the hydrogen inside causes the balloon to swell and when it has reached a certain altitude, the balloon bursts, and the instrument cases fall. Parachutes attached to the cases open and bring the instruments safely to the ground for scientific examination. A balloon, with a more rigid envelope

Partially filled with hydrogen this balloon rose to a height of 24 km, carrying a 3000 kg telescope with camera attached, to study the planet Mars.

that cannot swell much, rises when it is released and, as it rises, the upthrust upon it decreases as the density of the atmosphere gets less at the higher altitudes. Eventually the balloon floats at the height where its weight is equal to the weight of air it displaces.

The liner *Queen Elizabeth* 2 has a weight of about 66×10^7 newtons, so it displaces when afloat a volume of water (about 66×10^3 m^3) that has a weight of 66×10^7 newtons. When a submarine is floating in water, its weight is equal to the weight of water it displaces. In order to submerge, water is admitted to ballast tanks built into the hull of the submarine so that the weight of the submarine is increased, without increasing its volume, until the weight of the submarine becomes a little greater than that of the water it displaces. When the submarine is to surface again, compressed air is driven into the ballast tanks, expelling some of the water from them, until the weight of the submarine again becomes equal to the weight of water displaced.

The Hydrometer

The relative density of a liquid may be found by using a simple hydrometer, which you can make for yourself.

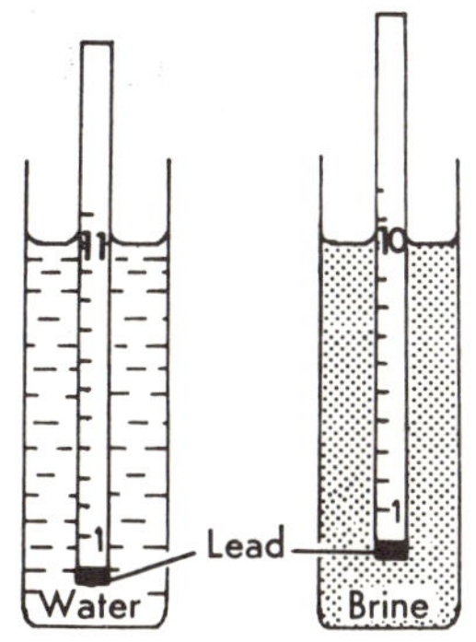

FIG. 102.

Weight a strip of wood with a small piece of lead at its lower end, so that the wood will float with its length vertical. Varnish the surface of the wood so that it will not soak up water. Attach a simple uniform scale to the strip of wood. Float the strip of wood in water and measure the length of it immersed (Fig. 102). Wipe off the water and float the hydrometer stick in brine and measure the length immersed. From your measurements you may obtain the relative density of brine. Thus, if your simple hydrometer floats with 11 cm of length immersed in water and with 10 cm of length immersed in brine, the relative density of brine $= \frac{11}{10} = 1{\cdot}1$.

This is the reason:

$$\begin{aligned}
\text{Let the weight of the hydrometer} &= w \text{ in newtons}\\
\text{Cross-sectional area of hydrometer} &= A \text{ in m}^2\\
\text{Density of water} &= d_1 \text{ in kg/m}^3\\
\text{Density of brine} &= d_2 \text{ in kg/m}^3\\
\text{Then volume of water displaced by hydrometer} &= 0{\cdot}11\ A \text{ in m}^3\\
\text{Weight of water displaced by hydrometer} &= 0{\cdot}11\ Agd_1 \text{ in newtons}\\
\text{Volume of brine displaced by hydrometer} &= 0{\cdot}10\ A \text{ in m}^3\\
\text{Weight of brine displaced by hydrometer} &= 0{\cdot}10\ Agd_2 \text{ in newtons}
\end{aligned}$$

By the law of flotation, the weights of the water and the brine are both equal to w, the weight of the hydrometer:

$$0{\cdot}11\ Agd_1 = 0{\cdot}10\ Agd_2$$

$$\therefore \frac{d_2}{d_1} = \frac{11}{10}$$

It follows that the relative density of the brine is 1·1. The experiment may be repeated for any liquid and in general:

$$\textit{Relative density of liquid} = \frac{\textit{Length of hydrometer immersed in water}}{\textit{Length of hydrometer immersed in liquid}}$$

Hydrometers are usually constructed on the basic pattern of the one shown here. The centres of gravity are kept low by the lead shot or mercury in the bottom so that they float upright in the liquid. The stems of hydrometers are made narrow so that small changes in relative density will show an appreciable change in level.

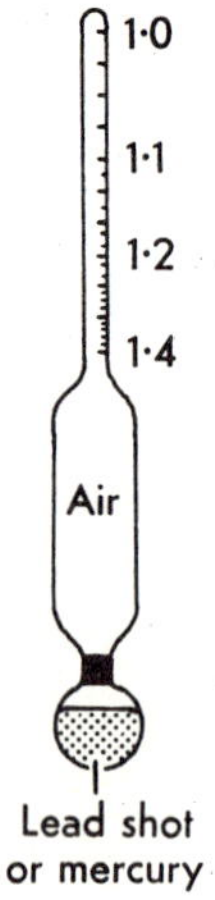

They are calibrated by floating them in liquids of known densities and marking the appropriate scale on the stem. Such hydrometers are used for testing the R.D. of the acid in lead accumulators like those used in motor cars; for testing the R.D. of milk or of wines and spirits; the percentage of sugar in sugar solutions; and the purities of many substances used in the chemical industry.

1·0
1·1
1·2
1·4
Air
Lead shot or mercury

SOLIDS MOVING THROUGH FLUIDS

The forces acting on solids moving at various speeds through fluids are very important today: in high speed flight; fast motoring; and fast water travel. Such forces are also important to the successful athlete, if he is to be just a little better than his rivals. Here are a few simple experiments to illustrate the effects of these forces.

Place a sheet of paper across two supports (Fig. 103). Blow underneath and parallel to the paper (NOT upwards), gently at first, and then more

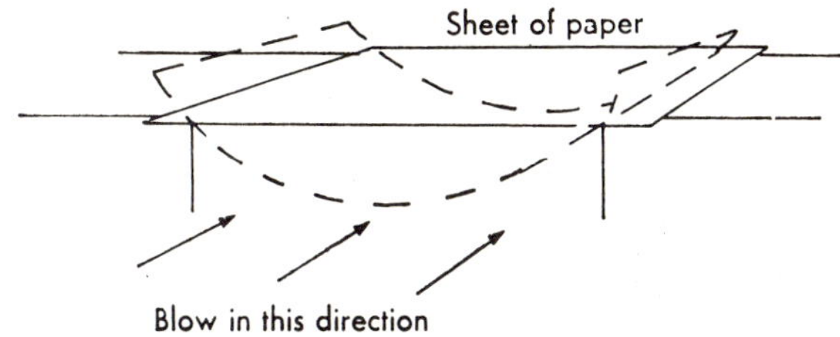

FIG. 103.

strongly. You will notice that the sheet of paper curves downwards (dotted lines) and not upwards, as you might suppose.

Support a sheet of paper by your two hands at *A* and *B* and blow across the drooping paper in the direction shown (Fig. 104*a*). You find that the paper rises towards the stream of moving air (dotted lines).

Bernoulli's Principle

Both experiments indicate that the stream of moving air near to the surface of the paper causes a decrease in pressure in that region, and the faster the air moves the greater the decrease in pressure.

This principle was discovered by a Swiss experimenter Bernoulli (1700-82)

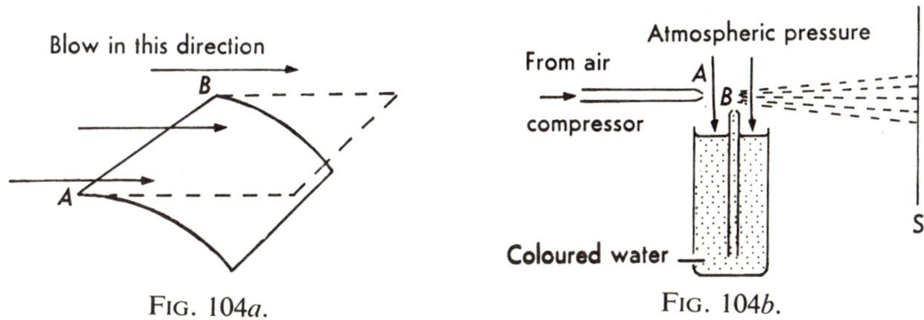

FIG. 104*a*. FIG. 104*b*.

and it now bears his name. Expressed simply it states: *as the speed of a fluid increases, the decrease in pressure becomes greater.*

Again, if a glass tube *A* (Fig. 104*b*) be connected to an air compressor (such as that used for inflating car tyres) and the stream of air molecules be directed over a glass tube *B*, placed in a vessel of coloured water, a fine spray of coloured water may be received on a screen at *S*. The fast moving mass of air above *B* will reduce the pressure in this region below the atmospheric pressure on the surface of the water in the vessel. The water will be forced to the top of the tube *B*, and some of it mixing with the moving air molecules will be carried as a spray towards *S*. This is the principle of the spray gun used for oil and paint.

A Bernouilli tube may be used to demonstrate this principle, as in Fig. 105*a*. If a wide tube *BC* has a narrow neck at *A* and side tubes fitted to act as pressure gauges, then when water flows through the tube, it will flow faster at *A* than at *B* or *C*. This will give a lower pressure at *A* than at *B* or *C*, as the rise in the side tubes will indicate.

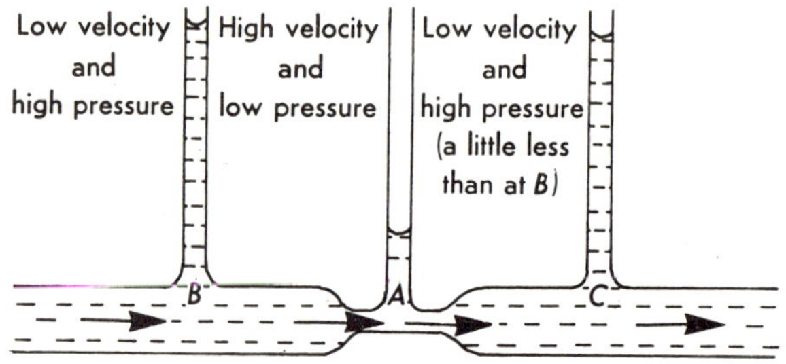

FIG. 105*a*. *Apparatus demonstrating Bernouilli's principle.*

Newton's Second Law will help to explain Bernouilli's principle. Consider a small mass of water flowing along the tube of Fig. 105*a*. When it reaches *A* it is moving faster than it was at *B* and therefore has more momentum. Between *B* and *A* the water must have accelerated. This requires a forward force which is provided by the fluid pressure of the surrounding water. This suggests that the pressure at *B* must be greater than the pressure at *A* to produce acceleration in the direction *B*→*A*.

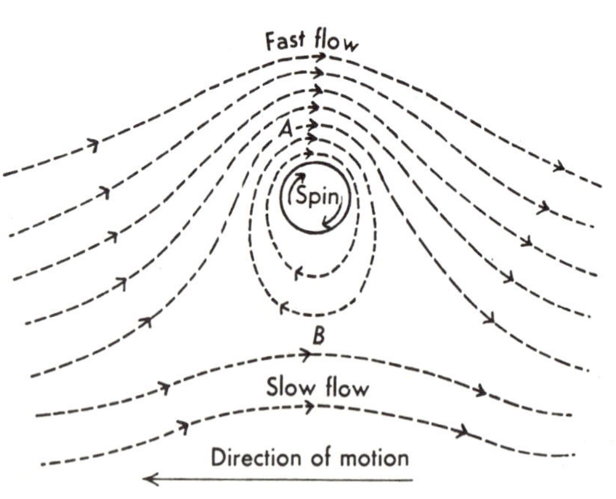

FIG. 105*b*. *Motion of a cricket ball.*

We should now be able to understand why a bowler when spinning a cricket ball can make it swerve as it moves through the air. If we look at Fig. 105*b* we can see that the spin on the ball increases the relative air speed at *A*, while decreasing it at *B*. Hence there will be slightly less sideways pressure on the ball from the side *A* than there will be from side *B*. So the ball will tend to swing slightly upwards in its flight. This swerve is most easily generated when the ball is new and has enough seam for the bowler to get a firm grip for spinning the ball.

What Lifts an Aeroplane?

When air flows past a flat plate *AB* inclined at an angle θ to the direction of flow, the pressure of the air on the top surface is less than that underneath (Fig. 106). There is a force on the plate in an upward direction. The force of

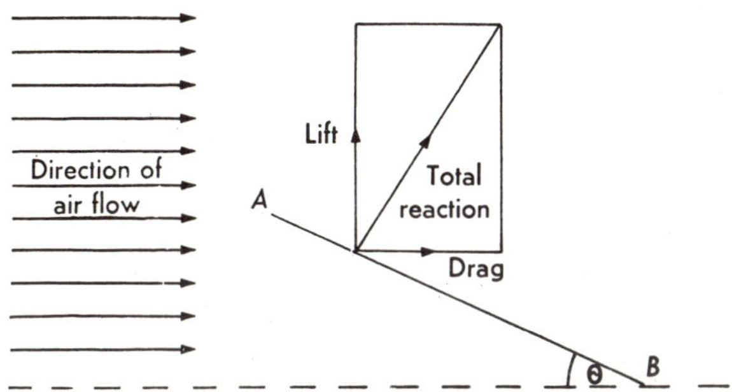

FIG. 106. *Forces acting on an aeroplane.*

total reaction at right angles to the plate may be resolved into two component forces; one vertically upwards called the *lift* and the other horizontally backwards called the *drag*. A smaller angle of inclination will give a lift which is large compared with the drag, but the total reaction will decrease, so it is useless to try to get lift with a very small angle of inclination.

Turbulence and Streamline Flow

We saw in Chapter 5 that there is always a *viscous drag* whenever the molecules of a fluid are set in motion. For example, when oil is poured from a can, it can be seen (especially when small air bubbles are trapped in the oil) that adhesion between oil and steel molecules causes the oil in contact with the can to remain stationary, while the oil above flows over it, the fastest moving oil molecules being those at the surface. At slow speeds of flow, the particles of liquid follow definite paths or stream lines, but above a certain speed called the *critical velocity*, the liquid is thrown into turbulence.

A wide glass tube is submerged in a tank filled with water. The tank is kept gently overflowing, maintaining a constant head of water above the tube. Water is siphoned from the tank, through the tube, at a speed con-

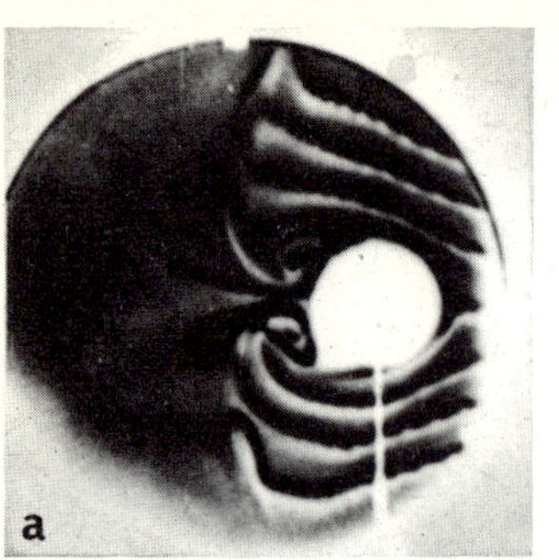

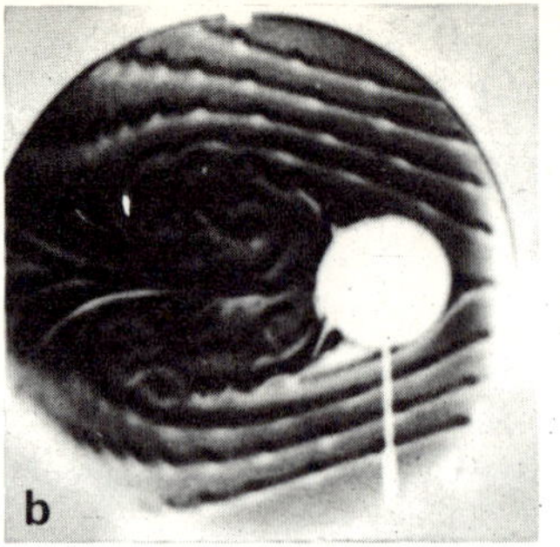

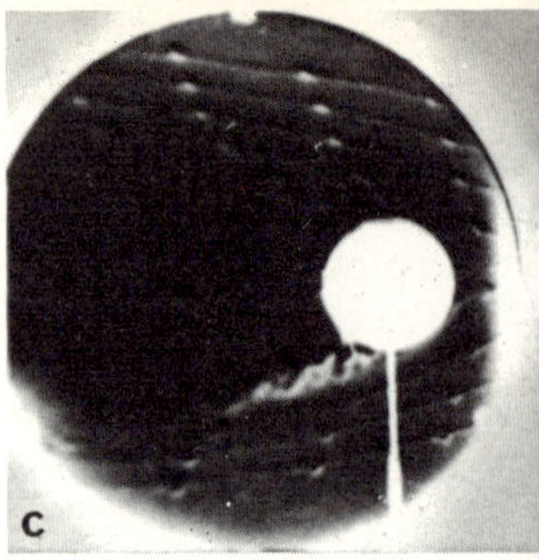

As rate of flow increases past a cylinder (a), *the wake behind breaks down into turbulence* (b), *becoming completely turbulent* (c).

trolled by the tap *C* (Fig. 107). Coloured water is fed from a funnel *B*, through a fine capillary jet at *A*, into the mouth of the wide glass tube. The tap *C* is slightly opened and water siphons from the tank through the tube. When the water is moving slowly, the flow is streamline; the liquid flowing in the manner of concentric tubes of which the outer one is close to the wall of the glass tube. The innermost stream within the tube may be observed by means of the jet of coloured water.

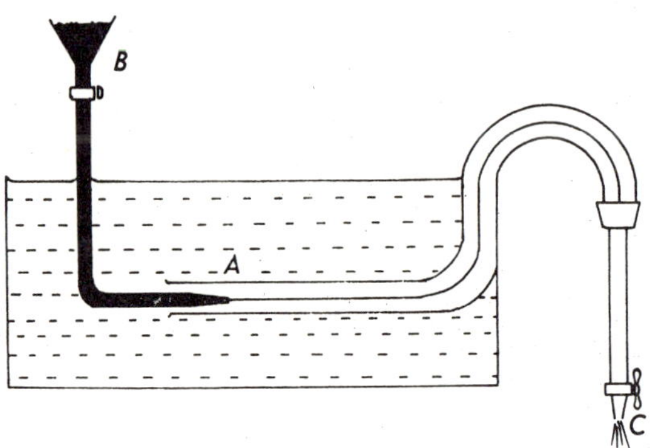

FIG. 107. *Apparatus for investigating turbulence.*

When water is being siphoned from the tank slowly, the coloured jet is seen not to mix with the surrounding liquid. When the tap *C* is opened further, the velocity of the water siphoning from the tube becomes large enough to cause turbulence. When this occurs the coloured jet starts to mix with the surrounding liquid.

Similarly, when the velocity of air increases beyond a certain critical value, turbulence is produced; a fact which causes problems in the design of aeroplanes, since turbulence can result in a considerable drag on the aeroplane. In an attempt to minimize this turbulence, experiments with differently-shaped wings (aerofoils) are carried out in a wind tunnel. Suitable light crystals are injected into the air stream flowing round the wing and photographs are taken. These photographs show a streamline flow at low air speeds and turbulence at high speeds (top of page). They also show that turbulence varies from one shape to another for the same air speed; the aerofoil shown in (d) on page 121 creates far less turbulence than a sphere.

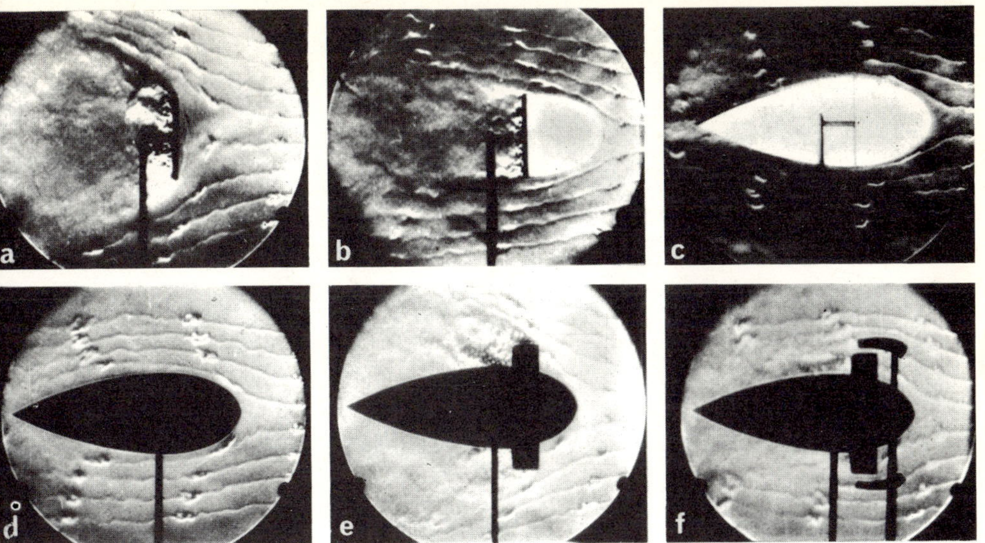

Complete turbulent wake (a) *is reduced* (b) *and further eliminated as the body's shape is streamlined* (c) *and* (d). *Introduction of an engine causes turbulence again* (e), *reduced by modifications of shape* (f).

QUESTIONS

1. If two metal spheres having equal diameters but different weights are released in water at the same moment, will they fall at the same rate? Test your answer experimentally if you can.

2. If water is added to milk, how is the relative density of the milk affected? Describe the construction and mode of use of an instrument for measuring the relative density of milk.

3. If you are sitting in a rowing boat and haul someone out of the water into the boat, why does the person's weight seem to increase as you lift him farther out of the water?

4. A fresh egg will usually sink in a bowl of water. If salt is slowly dissolved in the water, the egg can be made to float just breaking the surface of the water. Explain this. What can you say about the relative density of the egg?

5. A piece of metal weighs 100 N in air, 60 N in water and 70 N when immersed in petrol. Find (*a*) the relative density of the metal, (*b*) the relative density of the petrol.

6. A balloon which holds 1000 m^3 of gas is filled with hydrogen at atmospheric pressure. The fabric has a mass of 100 kg. What load will the balloon lift? (Density of hydrogen = 0·09 kg/m^3. Density of air = 1·29 kg/m^3. Gravitational constant = 10 N/kg.)

7. Suspend two table tennis balls by threads about 1 m long, so that the balls are separated by about 3 cm. Direct a fine jet of air between the balls and notice the result. What principle does your experiment illustrate?

CHAPTER 10

MOTION AT A UNIFORM ACCELERATION IN ALGEBRAIC FORM

LET us remind ourselves what we mean by a uniformly accelerated motion. It is one in which the velocity increases by equal increments in equal intervals of time, no matter how short the time interval may be. For example, if the acceleration is 30 m/s^2, then the velocity will increase by 30 m/s in 1 second or by 3 m/s in 0·1 second or by 300 m/s in 10 seconds, and so on. In other words, if we plot a velocity/time graph we shall have a straight line with a positive slope.

Equations of Linear Uniformly Accelerated Motion

Let u = initial vel. of the body, at some instant of time;
v = final vel. of the body after the lamps of time t;
s = distance travelled by body in time t;
a = uniform acceleration of the body;
Then $v = u +$ increase in vel. in time t;
a = increase in vel. in 1 s, at = increase in vel. in t s

$$\therefore v = u + at \ldots\ldots\ldots\ldots\ldots\ldots\ldots (1)$$

As long as the acceleration is uniform (and this is important) then distance = (average vel.) × time.

$$s = \frac{(u + v)}{2} \times t$$

$$= \frac{(u + u + at)}{2} \times t$$

$$\therefore s = ut + \tfrac{1}{2}at^2 \ldots\ldots\ldots\ldots\ldots (2)$$

Or, we might express s in terms of u, v and a, instead of in terms of u, t and a.

$$\text{From equation (1), } t = \frac{v - u}{a}$$

$$\text{but } s = \text{(av. vel.)} \times \text{time,}$$

$$\therefore s = \frac{(v + u)}{2} \times \frac{(v - u)}{a} = \frac{v^2 - u^2}{2a}$$

$$\therefore 2as = v^2 - u^2$$

$$\text{or, } v^2 = u^2 + 2as \ldots\ldots\ldots\ldots\ldots\ldots (3)$$

Equations (1), (2) and (3) are called the equations of a linear uniformly accelerated motion and may be used in the solution of problems. Frequently, however, the problem may be solved from first principles, using the fact that distance = (average vel.) × time.

For example: The road test of a new sports car said that it reached 80 km/h from standstill in a time interval of 7 s and 120 km/h in a time interval of 19 s. Top gear acceleration increased the speed from 80 km/h to 100 km/h in 5 s. Is the acceleration the same for each of these tests?

In everyday life the speed of a car is measured in km/h but for calculation it is better to use the standard of metre/second. It is easy to show that

$$1 \text{ km/h} = \frac{1}{3{\cdot}6} \text{ m/s}$$

$$\text{First test increase in velocity} = \frac{(80 - 0)}{3{\cdot}6} \text{ m/s}$$

$$\text{Time interval} = 7 \text{ s}$$

$$\text{Acceleration} = \frac{80}{3{\cdot}6 \times 7} = 3{\cdot}1 \text{ m/s}^2$$

$$\text{Second test increase in velocity} = \frac{(120 - 0)}{3{\cdot}6} \text{ m/s}$$

$$\text{Time interval} = 19 \text{ s}$$

$$\text{Acceleration} = \frac{120}{3{\cdot}6 \times 19} = 1{\cdot}8 \text{ m/s}^2$$

$$\text{Third test increase in velocity} = \frac{(100 - 80)}{3{\cdot}6} \text{ m/s}$$

$$\text{Time interval} = 5 \text{ s}$$

$$\text{Acceleration} = \frac{20}{3{\cdot}6 \times 5} = 1{\cdot}1 \text{ m/s}^2$$

Thus the acceleration during the first test was the greatest.

Now let us find the distance travelled during each test.

$$\text{First test average velocity} = \tfrac{1}{2}\left(\frac{80}{3{\cdot}6} + 0\right) \text{ m/s}$$

$$\text{Distance travelled} = \text{average velocity} \times \text{time}$$

$$= \tfrac{1}{2} \times \frac{80}{3{\cdot}6} \times 7 = 78 \text{ m}$$

$$\text{Second test average velocity} = \tfrac{1}{2}\left(\frac{120}{3{\cdot}6} + 0\right) = \frac{100}{6}\ \text{m/s}$$

$$\text{Distance travelled} = \frac{100}{6} \times 19 = 320\ \text{m}$$

$$\text{Third test average velocity} = \tfrac{1}{2}\left(\frac{100}{3{\cdot}6} + \frac{80}{3{\cdot}6}\right) = 25\ \text{m/s}$$

$$\text{Distance travelled} = 25 \times 5 = 125\ \text{m}$$

Using the equations of motion given above, check all these results.

Momentum Changes When Two Bodies Interact

If a boy and a man are standing close to one another on ice and the boy pushes the man, they both move apart in opposite directions, but the boy will move faster than the man.

If we measure the velocities of the man and the boy, we find that they are in inverse ratio to their masses. For example, if the boy has a mass of 50 kg and the man 75 kg, then if the man is pushed at a speed of 1 m/s, we find that the boy is pushed in the opposite direction at a speed of $1\frac{1}{2}$ m/s. In other words, the product of the mass and velocity ($m \times v$), in each case is the same (75 kg m/s). Their *momenta* are equal in size but opposite in direction.

Conservation of Momentum

This result may be tested more accurately using apparatus similar to Fig. 108. A and B are two similar wooden trucks on wheels having very little friction. They are separated by a spring S of suitable tension and tied together

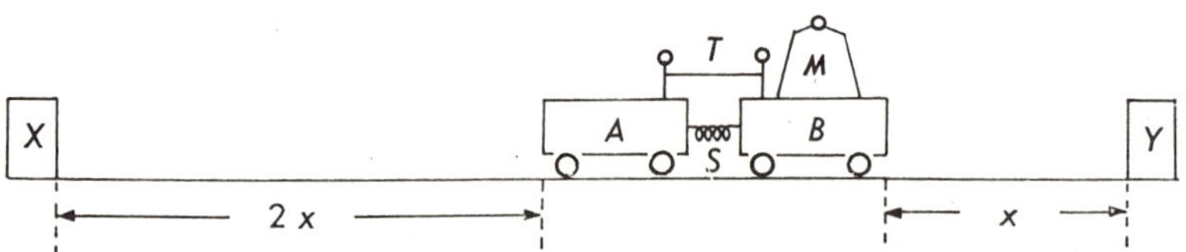

FIG. 108. *Apparatus for investigating conservation of momentum.*

by thread at T. The trucks are placed on a smooth, horizontal table, the thread is burned through at T and the spring S thrusts the trucks apart, exerting equal forces in opposite directions. It is found that A and B

move through equal distances in opposite directions before coming to rest.

A mass M, equal to the mass of truck B, is now placed on top of B and the experiment is repeated. Wooden blocks X and Y are placed in such a position, found by testing, that A hits X and B hits Y at the same instant after the thread has been burned. It is found that the distance A has travelled is twice the distance B has travelled in the same time; that is, the average velocity of A is twice that of B.

By adding other masses to trolleys A and B and carrying out similar tests, it will be found that the experiments show that the (mass of A) $\times$ (vel. of A) is always equal to the (mass of B) $\times$ (vel. of B), after the two have separated. This illustrates a very important principle of physics, known as *the law of conservation of momentum:*

"*when two or more particles collide, or separate, the total momentum in any given straight line remains constant, provided no external force acts in that direction.*"

Thus if m_1, m_2 are two masses colliding and separating; u_1, u_2 are their velocities before collision; v_1, v_2 are their velocities after collision, then:

$$m_1u_1 + m_2u_2 = m_1v_1 + m_2v_2$$

Momentum Equation

Newton's second law of motion states that: "when a body is acted upon by a force, the rate of change of momentum is directly proportional to the impressed force and takes place in the direction of the force." Suppose that a body of mass m is moving at velocity u and is acted upon by a force P for a time t, changing its velocity to v. Then by Newton's second law,

$$P = \frac{\text{change in momentum}}{\text{time}}$$

$$= \frac{mv - mu}{t}$$

$$\therefore Pt = mv - mu$$

This is known as the momentum equation and the product Pt is called the impulse of the force.

$$\textit{Impulse of force} = \textit{Change of momentum}$$

It follows that if a body moving with a momentum $m\,u$ is brought to rest in a very short time, a large force is necessary to make this change of momentum. This is the principle involved in the use of a hammer to drive a nail into wood, or of a pile driver to drive concrete pillars into the ground or the bed of a river for bridge supports.

Good cricketers follow their strokes through. The longer the bat and ball are in contact, the greater the ball's velocity.

Because of this principle, the experienced cricketer relaxes his hands and arms, drawing them towards his body as he catches a ball, so that the ball's momentum is reduced to zero comparatively slowly and so the force on the cricketer's hands is reduced. The inexperienced cricketer tends to stiffen his hands and arms, decreasing the time that the ball is in contact with him while the momentum is being reduced to zero and so the ball hits his hands with great force, often knocking back the hands and the ball drops from them. You can apply this principle to the catching of a rugby ball and to the relaxing rather than the stiffening of one's body, during a fall. "Follow-through" is important to the effectiveness of a cricketer's or tennis player's stroke, for the longer that the bat or racquet is in contact with the ball, the greater is the impulse of the force and so the greater is the velocity of the ball.

The magnitude of the force needed to cause the continuous change in momentum can be calculated by calculating the rate of change of momentum.

For example:

A jet of water is projected perpendicularly against a wall with a velocity of 50 m/s. The area of the jet is 0·5 m^2 and a volume of water 0·04 m^3 is projected each second. Assuming that there is no splashing back (i.e. the water after striking the wall just falls to the ground), what is the force and pressure on the wall?

The water loses all its momentum on striking the wall. As a volume of 1 m³ of water has a mass of 1000 kg—

$$\text{In 1 second the mass of water hitting the wall} = 0{\cdot}04 \times 1000$$
$$= 40 \text{ kg}$$
$$\text{Momentum of this water} = 40 \times 50$$
$$= 2000 \text{ kg m/s}$$

This is lost in 1 second.

$$\therefore \text{Rate of change of momentum} = 2000 \text{ kg m/s}^2$$
$$\therefore \text{Force} = 2000 \text{ kg m/s}^2 \text{ or N}$$

This acts over an area of $0{\cdot}5$ m².

$$\therefore \text{Pressure} = \frac{2000}{0{\cdot}5}$$
$$= 4000 \text{ N/m}^2$$

In this example what gains the momentum that the water has lost?

Here is an experiment which uses the principle of conservation of momentum.

To find the muzzle velocity of an air rifle pellet: a block of wood *A* (Fig. 109) is supported by four strings, each about 2 metres long. A scale *S* is placed behind *A*. An air rifle pellet is weighed on an accurate balance and loaded into the rifle. The rifle is then placed on a support and pointed in a horizontal direction at the centre of the face of *A* that is opposite the rifle.

An observer looking through a telescope at a safe distance, takes a reading on the scale *S*, opposite the left edge of the block *A*. The operator holding the rifle, fires the pellet into the middle of the wooden block, and the observer notes how far the left edge of *A* swings along the scale *S*.

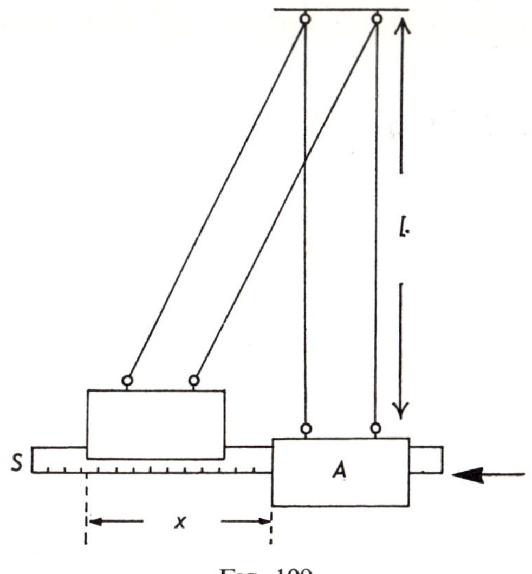

FIG. 109.

The movement of the block is circular movement and may be represented by Fig. 110. The centre of the circle shown represents the point from which the four strings are suspended; the radius of the circle is equal to the length of one of the strings; x is equal to the sideways displacement of the block as measured on S; while h is equal to the vertical rise of the block when it swings after having been struck by the pellet from the air rifle.

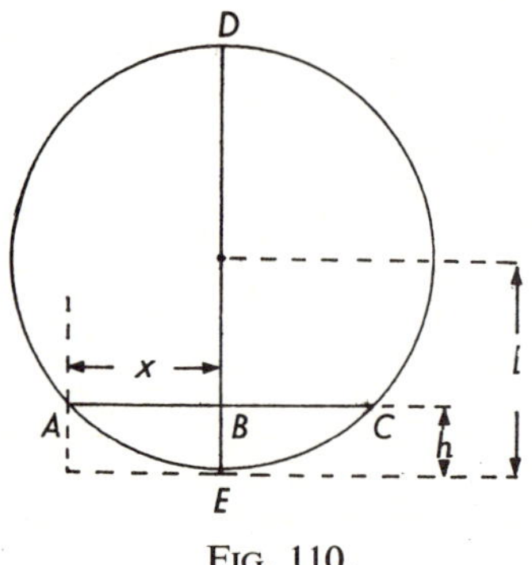

FIG. 110.

We can now calculate h:

$$AB.BC = DB.BE$$

(Theorem of intersecting chords)

$$x^2 = h(2l - h)$$

Since h is the order of a few cm, while $2l$ is about 400 cm, h is negligibly small compared with $2l$.

$$\therefore x^2 = 2lh$$

$$\therefore h = \frac{x^2}{2l}$$

Assuming that the potential energy of the wooden block at the top of its swing = kinetic energy at moment of impact:

$$mgh = \tfrac{1}{2}mv^2$$

$$\therefore v^2 = 2gh$$

$$\therefore v^2 = \frac{x^2}{l}g$$

Let m = mass of pellet
V = velocity of pellet leaving rifle
M = mass of wooden block
v = velocity of wooden block and pellet after collision, the pellet remaining embedded in the wooden block.

Then momentum before collision = mV

Momentum after collision = $(m + M)v$

As momentum must be conserved, $mV = (m + M)v$

But m is negligibly small compared with M:

$$\therefore V = \frac{M}{m}v$$

$$= \frac{M}{m}x\sqrt{\frac{g}{l}}$$

WORK AND ENERGY

Energy is defined as the ability of a body to do work. Much of the energy we use comes from the burning of some kind of fuel, such as food in our bodies, coal or petrol or diesel oil in various forms of engines. In order to do work, energy is frequently changed from one form into another. For example, heat from burning coal is used to heat water, which gives up much of its heat energy to warm a room. Or, the energy stored in oil is changed to the energy of motion of a diesel locomotive. Or, the energy from a battery or from a dynamo producing an electric current by rotation in a magnetic field is used to revolve an electric motor. In such changes, energy is never created nor destroyed; it is only changed from one form to another or transferred from one place to another, sometimes by some kind of wave motion. This is one of the important principles of science and is known as *the principle of the conservation of energy*.

Kinetic Energy and Potential Energy

The ability of a body to do work because of its motion is called *kinetic energy* (*K.E.*). The ability of a body to do work because of its position or state is called *potential energy* (*P.E.*). The water behind the dam of a reservoir possesses gravitational potential energy, and the force of gravity will do work on it, as it falls through the conduit pipes towards the power house. Its P.E. changes into K.E. and this energy of motion is used in turning the turbine which drives the generator. The wound-up spring of a watch or clock has strain P.E. and the spring can do work as it uncoils.

Work and Kinetic Energy

We have defined work as the product of force and the distance moved in the direction of the force. While the force acts, we are transferring energy of motion (K.E.), from the producer of the force to the body. Let a force P pull the body from rest for a distance s. The work done on the body is Ps. Then if m = mass of body and a is the acceleration it acquires through the action of the force P, then:

$$P = ma$$
$$\therefore Ps = mas$$

If P is a constant force, then a is constant and the velocity the body reaches in a distance s is given by the equation

$$v^2 = u^2 + 2as$$
$$\text{But } u = 0$$
$$\therefore v^2 = 2as$$
$$\therefore Ps = mas$$
$$= \tfrac{1}{2}mv^2$$

The energy transferred to the body in giving it motion, what we call its kinetic energy, is equal to $\frac{1}{2}mv^2$. In other words, if a body of mass m is moving at speed v, then it has to get rid of an amount of energy equal to $\frac{1}{2}mv^2$ in coming to rest. In the case of a moving motor car, most of this energy is converted into heat produced from the friction between the brake shoes and the drum of the wheel when the car is brought to rest. When one moving body collides with another body at rest or with another moving body, then energy is transferred from one body to another. Thus if a bullet fired from a rifle enters a block of wood suspended by strings, the block gains kinetic energy. This energy is converted into gravitational potential energy as it swings upwards in the arc of a circle.

If, after a collision, two bodies have a total K.E. equal to their total K.E. before collision, then this is called an elastic collision. The interactions between moving bodies that we see around us are never perfectly elastic. Frequently some of the original K.E. is turned into heat and sound energy. Some interactions are nearly perfectly elastic, such as collisions between billiard balls and the collisions between molecules in a gas.

To remind you then: *in an elastic collision between bodies, momentum and K.E. are conserved; in an inelastic collision, momentum is conserved, but K.E. is not conserved.*

QUESTIONS

1. Do you consider physics to be a study of matter and energy and the transformation of energy from one form into another?

2. One engine raises a volume of water 0·5 m³ through a height of 80 m in 4 minutes; another raises a volume of 0·4 m³ through a height of 60 m in 2·4 minutes. Which engine has the greater power?

3. A perpetual motion machine is a machine that continuously produces more power than is supplied to it. Why is it not possible to produce such a machine?

4. How long will it take a motor of power 0·5 kW to raise a mass of gravel of 500 kg through a height of 15 m?

5. A bricklayer of mass 60 kg carries a load of bricks of mass 25 kg up steps 20 m high. What fraction of the gravitational potential energy he produces is useful?

6. A 110 g cricket ball moving at 50 m/s is caught by a fielder who brings the ball to rest in a time 0·1 second. What is the average force experienced by the fielder's hands?

7. If a 14 g bullet is fired at a velocity of 360 m/s from a 4·2 kg rifle, what will be the velocity of the rifle, as the bullet is ejected?

8. If a rocket burns fuel at the rate of 2 kg/s and the gas is emitted at the rear end of the rocket with a velocity of 1800 m/s, what force is exerted on the rocket and in which direction?

MECHANICAL AND HEAT ENERGY

CHAPTER 11

WORK AND ENERGY

HAVE you ever watched a large crane lifting up heavy girders during building construction? It exerts a tremendous force in lifting the heavy lengths of steel, and must use up a large quantity of energy in doing this work. Where does this energy come from?

Before attempting to investigate this problem, we should first make sure we know the exact meanings of the terms work, power, and energy.

We have already learnt Newton's law which states that an object will remain in a state of rest or of uniform motion in a straight line unless acted on by a force. Also, such a force causes motion and acceleration in the direction in which the force acts.

If we pick up a book from the floor and lift it vertically upwards to place it on a table, we must exert a force at least equal and opposite to that of gravitational attraction. Furthermore, we are conscious of doing physical work in this process and say we have used up some energy. When a force acting on a body succeeds in moving the body we say work is done. It is important to note that work is not done if the force does not succeed in moving its point of application. If we push hard against a locked door without opening it, we do no work on the door. Only when we move the door open is work done.

The amount of work done is measured by the product of the force and the distance moved in the direction of the force. In the case of lifting up a book of mass m through a vertical height of h, we have:

$$\text{Force on book} = mg$$
$$\text{Distance moved} = h$$
$$\text{Work done in lifting book} = hmg$$

(m is measured in kilogrammes, h in metres, g in newton/kilogramme, and the work done in joules).

In doing this amount of work, an amount of energy must be expended. *We define energy as a capacity for doing work.* In this particular case we are considering mechanical energy, but of course there are other forms of energy, e.g. electrical, heat, atomic, etc., which will be discussed later. If something has energy it is capable of exerting a force and doing work.

Potential energy stored in a bent bow is about 5 joules: this, transferred to the arrow as kinetic energy, is able to send it a distance of about 300 m.

Potential Energy and Kinetic Energy

In lifting the book on to the table, work has been done and energy expended. What then has happened to this energy?

When work is done on something an amount of energy is given to it, and this energy is equal to the amount of work done. This follows a fundamental law of nature that states that *energy can neither be created nor destroyed. It can only change from one form to another.*

The energy that was used up in raising the book has in fact been stored up in it by virtue of its increased height above the earth's surface. This energy of position is termed *potential energy*, and is equal in this case to the work done in lifting up the book, or *mgh*. Strictly it is an increase in potential energy relative to its original position lying on the ground, but we can take this original potential energy lying on the ground as zero.

To make this point clearer, let us consider the staircase shown in Fig. 111, with a weight *mg* on the top step *D*. We may choose the level of the steps *A*, *B* or *C* as indicating zero potential, and so can define the potential energy of the weight *D* in three ways:

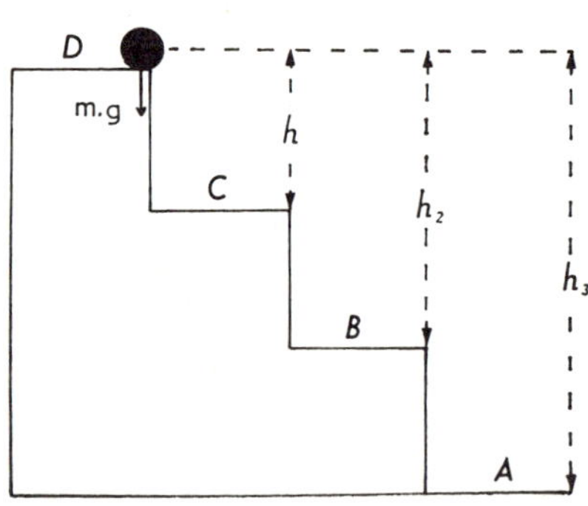

FIG. 111.

1. mgh_1 above step C.
2. mgh_2 above step *B*.
3. mgh_3 above step *A*.

The normal way of defining potential energy however, is by considering the surface of the earth or ground level as zero. The main point to remember is that it is the change in potential energy of a body which is important.

There are many other forms of potential energy, for the concept of energy of position is a general one. For example, the pellet on the end of a catapult's stretched elastic has potential energy, and will fly off when released, or a small iron screw near a magnet has potential energy and will move to the magnet if free to do so.

If the book is allowed to fall from the table, being acted on by a gravitational force, it will accelerate downwards at the rate of 9·8 m/s².

Now obviously as the book falls it loses its energy of position. Its potential energy decreases but, at the same time, its velocity increases. A body which is in motion is capable of doing work. A moving rifle bullet can punch a hole through a piece of wood before it comes to rest. This energy of motion, which enables the body to do work, is called *kinetic energy*.

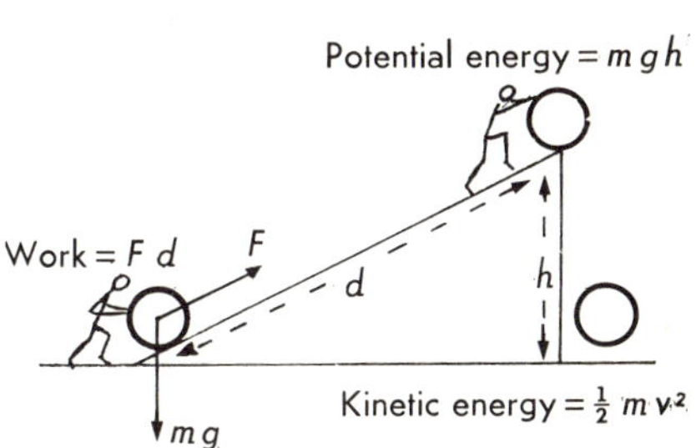

Fig. 112.

What is the relation between potential energy and kinetic energy?

Consider a stone of mass m falling from rest under gravity through a height h (Fig. 112). We know it accelerates from rest at g m/s².

$$\text{Let its velocity on striking the ground} = v$$

$$\text{Initial velocity, at rest} = 0$$

$$\text{Average velocity in falling} = \frac{(0 + v)}{2}$$

$$\text{The time taken to fall through height } h \text{ m} = \frac{\text{Distance}}{\text{Average velocity}}$$

$$= \frac{h}{v/2}$$

$$= \frac{2h}{v}$$

Throughout this time of $2h/v$ the stone accelerates at g and thus reaches a final velocity on reaching the ground of $g2h/v$. This velocity we have already called v.

$$\therefore\ v = \frac{2hg}{v}$$

$$\therefore\ \frac{v^2}{2} = hg$$

But the initial potential energy of the stone which equals the work needed to raise the stone from the ground up to height $h = mgh$.

$$\therefore\ mgh = \frac{mv^2}{2}$$

At the moment of striking the ground, all the potential energy has been converted into kinetic energy and this kinetic energy $= \frac{1}{2}mv^2$. To summarize:

$$\text{Potential energy} = mgh$$
$$\text{Kinetic energy} = \frac{1}{2}mv^2$$

where m is the mass of the body in kg, h the height of the body above the earth in m, v the velocity of the body in m/s, and g the gravitational force on 1 kg or the acceleration due to gravity (i.e. 9·8 N/kg or 9·8 m/s²).

$$F = dK \text{ (constant)}$$

Simple Pendulum

We can learn a lot about potential and kinetic energy by experimenting with a simple pendulum, a metal bob suspended on the end of a string from a fixed support (Fig. 113).

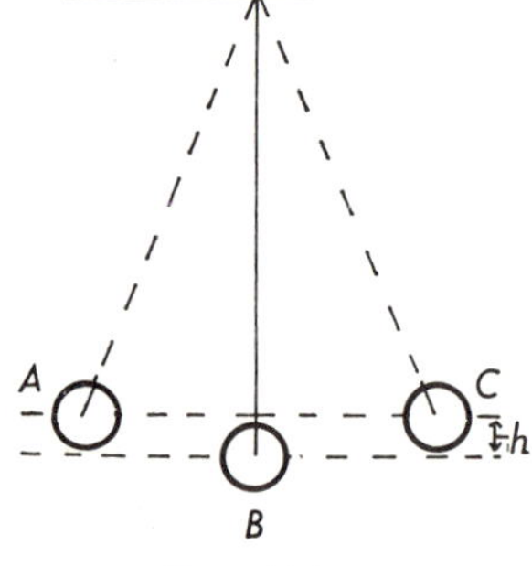

FIG. 113.

First the bob is pulled to one side, to position A. It has been raised through a height h and work has been done in doing it. This work is stored as potential energy.

On releasing the bob it swings down and accelerates, gaining kinetic energy and at the same time losing potential energy, until at B all its original potential energy has been changed into kinetic energy. From B to C kinetic energy decreases and there is a corresponding increase in potential energy until point C is reached. Thus the energy of the pendulum bob alternates between potential and kinetic energy. Gradually the energy is lost because a small amount of work is done against the air resistance and the friction of the support and the amplitude of the swing of the pendulum decreases.

Energy Stored in a Spring

Energy can be stored in springs by either stretching them or compressing them and can be used later to drive watches, toy trains, or, in the case of a pogo stick for example, to lift up a child into the air.

We know according to Hooke's law that, when a spring is stretched, the greater the extension the greater is the tension in the spring. As we stretch the spring we do work, and energy is stored in the spring as potential energy.

Let us repeat the experiment of progressively loading a spring with weights and recording the extension resulting (page 86).

We find that the force F used to stretch the spring is not constant, but increases in proportion to the elongation d of the spring.

$$F = dK \text{ (constant)}$$

FIG. 114. *The motive force in a watch, supplied by energy stored in a compressed spring, is controlled by the hair spring shown here. Energy from the main spring alternately compresses and stretches the hair spring which, through an escapement mechanism, replaces the need for a pendulum by giving to the watch a regular beat.*

The average force applied when the spring is stretched from its normal length to some final extension, *d*, may be found:

$$F_{\text{average}} = \frac{\text{Initial tension} + \text{Final tension}}{2}$$

$$= \frac{0 + Kd}{2} = \tfrac{1}{2}Kd$$

Since the distance moved is *d*,

$$\text{Work done, } W = F_{av}d = \tfrac{1}{2}Kd^2$$

If we now compare this result with a straight line graph obtained by plotting load against extension we find that the area under the graph $= \frac{1}{2}Kd^2$, which is proportional to the work done in stretching the spring.

Consider a spring that requires a mass of 0·4 kg in its pan to give it an extension of 0·2 m, the mass having been added gradually.

Average force during extension	$= \frac{1}{2}(0{\cdot}4 \times 9{\cdot}8)$N
	$= 1{\cdot}96$ N
Distance extended	$= 0{\cdot}2$ m
Work done	$= 1{\cdot}96 \times 0{\cdot}2$ J
	$= 0{\cdot}392$ J

This is the strain potential energy stored in the spring.

Now let us by experiment find the mass which fastened to the spring (Fig. 115*a*) and suddenly released just stretches the spring through 0·2 m before being jerked upwards again. This mass is found to be 0·2 kg. Since it has fallen through a vertical height of 0·2 m it has lost gravitational potential energy equal to $(0{\cdot}2 \times 9{\cdot}8) \times 0{\cdot}2$ J. This value (0·392 J) is exactly the same as the work done in stretching the spring by adding the load of 0·4 kg gradually applied. And, of course, so it should be since in both cases the same amount of strain energy is stored in the spring as it is extended the same amount.

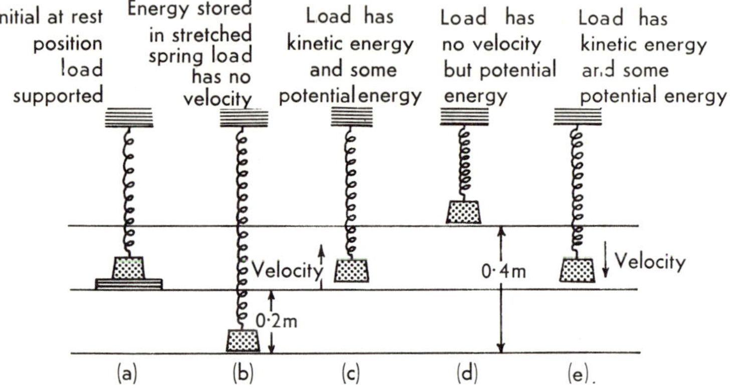

FIG. 115. *Energy changes in an oscillating spring.*

We notice that the 0·2 kg load in the second experiment is not sufficient to hold the spring extended an amount of 0·2 m. There is an upward resultant tension of $(0{\cdot}4 - 0{\cdot}2) \times 9{\cdot}8$ N $= 1{\cdot}96$ N. This causes the spring to begin contracting again and lifts up the 0·2 kg load so that it bobs up and down. At the top of its oscillations it possesses only potential energy due to its position (Fig. 115*d*). At the bottom of its oscillation it has lost this potential energy, which is then stored up in the stretched spring (Fig 115*b*). Half-way up (Fig. 115*c*), or half-way down (Fig. 115*e*), at the point where the spring is in its state of equilibrium, the 0·2 kg load is moving with considerable velocity and possesses kinetic energy in addition to some potential energy due to its height above the bottom point. At the top and the bottom of the oscillations the load is momentarily at rest.

To help us understand these ideas of work and energy better, let us calculate the work done in raising a metal ball of mass 4 kg vertically through a height of 3 m above ground level and also the potential energy of the ball at a height of 1 m above ground. When the ball is dropped freely from a height of 3 m calculate its kinetic energy when it has fallen to within 1 m of the ground, and hence find its velocity at this point:

$$
\begin{aligned}
\text{Work done in raising the ball 3 m} &= mgh \\
&= 4 \times 9{\cdot}8 \times 3 \\
&= 117{\cdot}6 \text{ J}
\end{aligned}
$$

$$
\begin{aligned}
\therefore \text{ The initial potential energy of the ball} & \\
\text{when 3 m above ground} &= 117{\cdot}6 \text{ J} \\
\text{Potential energy at 1 m above ground} &= 4 \times 9{\cdot}8 \times 1 \\
&= 39{\cdot}2 \text{ J} \\
\text{Loss in P.E. in falling through 2 m} &= 117{\cdot}6 - 39{\cdot}2 \\
&= 78{\cdot}4 \text{ J} \\
\text{Thus gain in kinetic energy} &= 78{\cdot}4 \text{ J} \\
\text{But gain in kinetic energy} &= \tfrac{1}{2}mv^2
\end{aligned}
$$

$$
\begin{aligned}
\therefore v^2 &= \frac{2}{m} \times 78{\cdot}4 \\
&= \frac{156{\cdot}8}{4} \\
&= 39{\cdot}2 \\
\therefore \text{velocity } v &= 6{\cdot}3 \text{ m/s}
\end{aligned}
$$

Power

When energy is used and work done, we are often interested in the time taken. Have you noticed how much easier it is to climb a hill by walking round an easy gradient path rather than going straight up the side? The height to the top is the same in each case, and so is the gain in potential energy or work done. In one case, however, the work is done more gradually. It takes longer. We call the rate of working the power, and it can be calculated by dividing the work done by the time taken to do it.

In our original example of lifting up a book of mass m through a vertical height h, the work done was mgh. If this work had been completed in a time interval t, the power used would have been mgh/t. This is measured in watts.

The rate of working or power of machines is of great interest in everyday life. Commercial electric motors, for instance, generally have their power indicated upon them, e.g. 373 watts. The horse power (H.P.) and the watt are units of power which we have already discussed (see page 75).

James Watt, the developer of the steam engine, introduced the idea of the horse power as a unit to measure rates of working. He calculated by experiment the rate at which a normal horse could work, and having increased his estimate by 50 per cent (to be on the safe side), called the rate one horse power. It represents a rate of working of 746 watts.

FIG. 116. *With the aid of a powerful crane, a fallen lintel at Stonehenge weighing 18 tons was easily and quickly replaced. Consider how many men were required, and for how long, to carry out the same work when Stonehenge was originally built more than 3000 years ago.*

Calculate your own horse power in climbing up a flight of stairs:

$$\begin{aligned}
\text{Vertical height of stairs ascended} &= 6 \text{ m} \\
\text{Weight of person climbing} &= 500 \text{ N} \\
\therefore \text{ Work done in climbing} &= 500 \times 6 \\
&= 3000 \text{ J} \\
\text{Time taken in climbing} &= 8 \text{ s} \\
\therefore \text{ Power (rate of working)} &= \frac{3000}{8} \\
&= 375 \text{ W} \\
\text{or Horse power} &= \frac{375}{746} \\
&= 0{\cdot}5 \text{ H.P.}
\end{aligned}$$

MACHINES

Until comparatively recently in history the only power available for doing most work was man power, supplemented by the power of certain beasts of burden like the horse and elephant. Man sought methods to enable him to do work more easily and conveniently. He soon discovered that he could raise heavy loads by dragging them up a slope or inclined plane rather than attempting to haul them vertically upwards. This must have been the method employed in lifting up the stone cross-pieces in Stonehenge (Fig. 117). Even if distance x is much farther than height h, the actual work done in raising the load to the top is the same. The work can be done slowly and less power used. Today, by means of modern machines man is able to lift great loads easily and quickly.

FIG. 117. *Building Stonehenge.*

Theory of the Lever

One of the simplest and commonest machines in use is the lever. A tight-fitting tin lid can easily be levered open by using a steel rod or even a penny. A see-saw is also an example of a lever. How does it work?

When two boys balance on a see-saw, they must arrange themselves one on either side of the central balance point. Furthermore, it is soon discovered that if one boy is much heavier than the other, he will weigh down his end unless he sits closer to the centre than his lighter companion.

Let us investigate the working of such a see-saw by using a light wooden bar, about a metre long, marked off in centimetres, and pivoted by means of

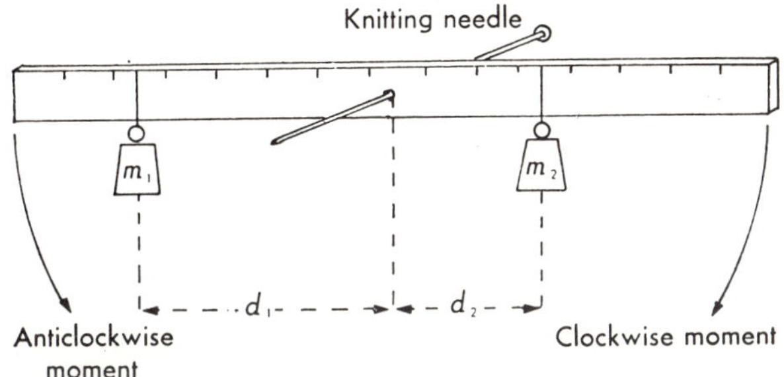

FIG. 118. *Theory behind a see-saw.*

a knitting needle through a hole in the centre (Fig. 118). Initial balancing of the bar can be achieved by adding a little plasticine to one end if necessary. Suitable 100 g or 200 g masses can be hung from the bar.

On investigation, using equal masses on opposite sides of the pivot or fulcrum, it is found that the distances of the masses from the fulcrum are the same when balanced.

Various masses can be hung from each side of the bar and adjusted to balance the bar horizontally. The masses and their distances from the fulcrum should be tabulated (Table 7).

Table 7
Taking Moments About a Fulcrum

Clockwise Moments			Anti-clockwise Moments		
Mass	Distance from fulcrum	Weight × Distance	Mass	Distance from fulcrum	Weight × Distance
m_1	d_1	$m_1 \times d_1 \times g$	m_2	d_2	$m_2 \times d_2 \times g$

The turning effect of each weight is measured by the product of the weight and its distance from the fulcrum. This product is called the *moment* of the force.

With various combinations of weights on the bar it will be found that the sum of the products of weights and distances on the right hand side equals the sum of the products of the weights and distances on the left hand side.

This balance rule is known as the *principle of moments* and may be fully expressed as follows:

For a body in equilibrium under the action of a number of forces in one plane, the sum of the clockwise moments about any point equals the sum of the anti-clockwise moments about the same point.

For our bar to balance, or to be in equilibrium, the sum of the clockwise moments about the fulcrum must equal that of the anti-clockwise moments.

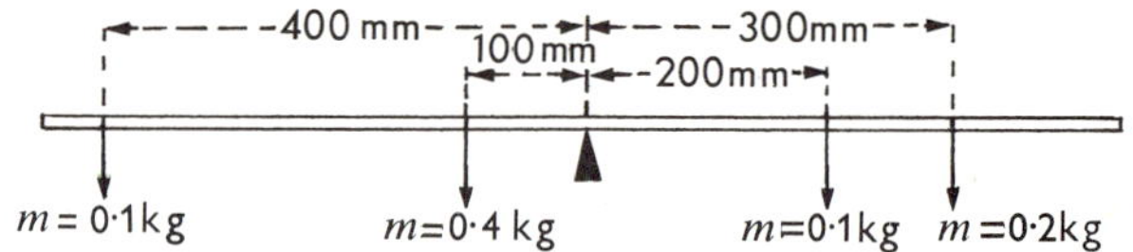

FIG. 119. *Bar in equilibrium about a fulcrum.*

The forces of gravity on the masses 0·1 kg, 0·4 kg, 0·1 kg, 0·2 kg, i.e. the weights, are respectively 1 N, 4 N, 1 N, 2 N (very closely taking $g = 10$ N/kg instead of 9·8 N/kg).

Now let us consider whether the bar shown in Fig. 119 is in equilibrium:

$$\begin{aligned} \text{Clockwise moments} &= 0{\cdot}1 \times 0{\cdot}2 + 0{\cdot}2 \times 0{\cdot}3 \text{ newton metres} \\ &= 0{\cdot}08 \text{ N m} \\ \text{Anticlockwise moments} &= 0{\cdot}4 \times 0{\cdot}1 + 0{\cdot}1 \times 0{\cdot}4 \text{ N m} = 0{\cdot}08 \text{ N m} \end{aligned}$$

Since the clockwise moments equal the anti-clockwise moments the bar is in equilibrium.

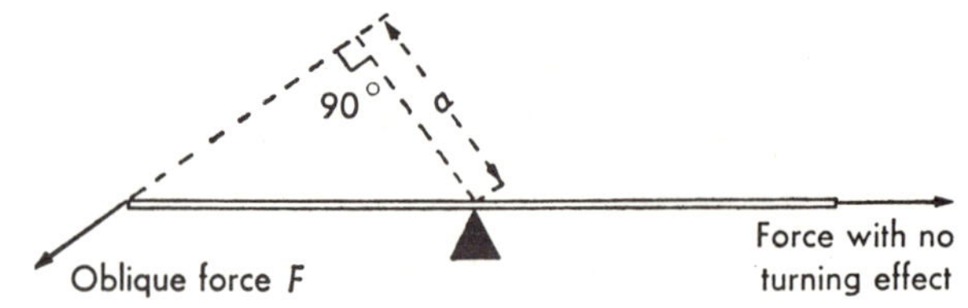

FIG. 120. *Moment of an oblique force about a fulcrum* = Fa.

It will be obvious that if a force is exerted pulling along the bar instead of perpendicularly to it, as in the case of the weights, there will be no turning effect, or moment.

When the turning force is oblique or at an angle to the bar it is found that the effective turning moment is the product of the force and the perpendicular distance from the fulcrum to the line of action of the force: $F \times a$ (Fig. 120).

Centre of Gravity

Let us try to balance a uniform rectangular card on a knife edge. It is found that the card balances only when the knife edge is vertically below the centre x of the card (Fig. 121).

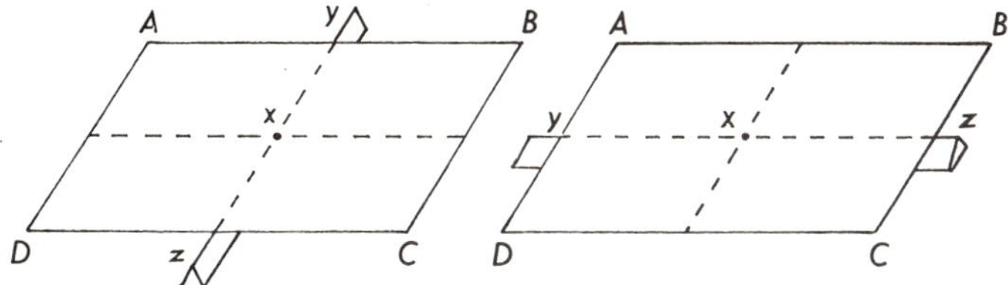

FIG. 121. *Balancing a uniform card on a knife edge.*

This is what one would expect since the two halves of the card on opposite sides of the knife edge exert equal and opposite moments about the fulcrum.

The card can be balanced in different positions by turning it, but always the knife edge is under the centre of the card x. With care the card may be balar.ced on a pin point placed at x.

Imagine the card made up of many minute particles of mass m_1, m_2, m_3,

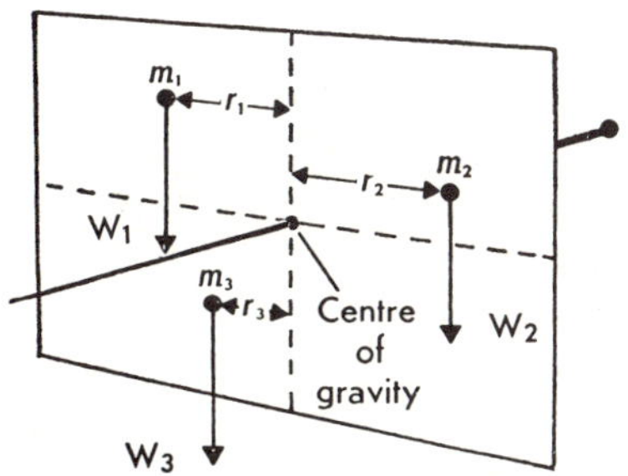

FIG. 122. *Centre of gravity of a uniform card.*

etc., at distances r_1, r_2, r_3, etc., from the centre (Fig 122). The force of gravity acting downwards on each particle will exert a turning moment about the central fulcrum, calculated by the product of the weight of the particle multiplied by its distance from the fulcrum r. Some of these moments will be clockwise and some will be anti-clockwise. For equilibrium the algebraic sum of these moments must equal zero:

$$W_1r_1 + W_2r_2 + W_3r_3 + \text{etc.} = \Sigma\, Wr = 0$$

There is one point only in a body about which this moment sum is zero irrespective of how the body is orientated: this point is called the centre of gravity. The whole of the weight of the body can be considered to act through this point.

For a uniform body of regular shape like the card or a metre rule, the centre of gravity is located at the geometrical centre and can be determined by measurement. The centre of gravity of an irregular card can be found by simple experiment (Fig. 123*a*).

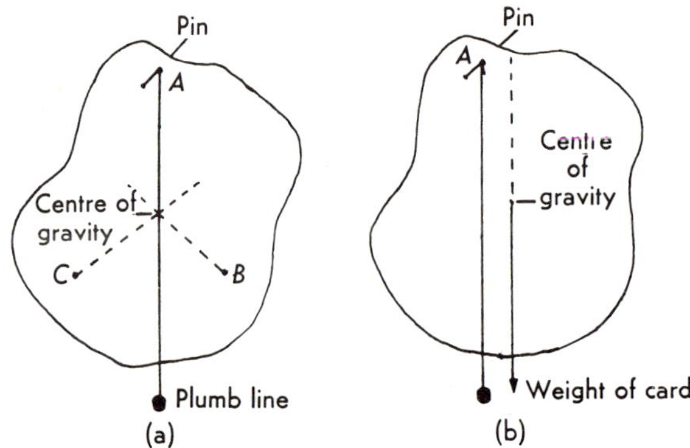

FIG. 123. *To find the centre of gravity of an irregular card.*

The card is suspended freely by means of a pin from a point *A* near the edge of the card. From the pin is hung a plumb line made from a weight attached to the end of a thread. A pencil line is marked carefully on the card immediately behind the thread to mark the vertical. This procedure is then repeated with the card suspended from two (or more) other points *B*, *C*, near the edge. The lines drawn intersect at the centre of gravity of the card. This may be verified by balancing the card on a pin point placed at this point.

When the suspended card is displaced a little to one side (Fig. 123*b*) the whole weight of the card, considered to be acting through the centre of gravity, exerts a turning moment about the suspension point *A*. This causes the card to swing back to its "at rest" position with the centre of gravity vertically below the point of suspension. In this position the centre of gravity descends to the lowest position possible. Considering all the weight acting at this point it can be seen that the potential energy is least in this position.

Equilibrium

Let us consider a solid wooden cone standing on its base (Fig. 124*a*). It is in equilibrium. Any slight tilt must raise the centre of gravity, which in this case is situated a quarter of the way up the central axis (Fig. 124*b*). On release the cone will fall back to its original stable position. The cone is said to be in *stable equilibrium* for after a slight displacement, on release, it returns to its original position. The weight of the cone acting down through the centre of gravity exerts a restoring moment about the fulcrum *F*.

When the cone is balanced on its apex it is in equilibrium and the centre of gravity is at its highest (Fig. 124*c*). Any slight displacement lowers the centre of gravity and the cone falls over (Fig. 124*d*). This is known as *unstable equilibrium*. The weight acting through the centre of gravity exerts an upsetting turning moment about *F*.

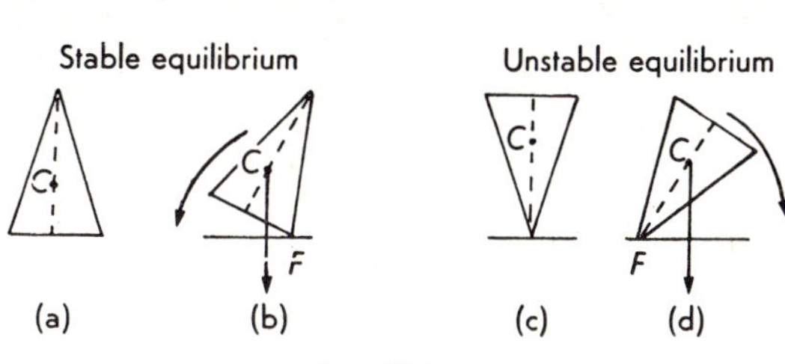

FIG. 124.

A cone lying on its side will rest in any position. It can roll without raising or lowering the centre of gravity. There is no turning moment caused by the weight of the cone. This type of equilibrium is known as *metastable* (*neutral*) *equilibrium*.

It will be seen that if the cone, standing upright on its base, is tilted through a large enough angle the centre of gravity will be vertically above the point on the edge *F* about which it is tilted. Any further tilt will cause the cone to fall over. If the centre of gravity is near the apex of the cone quite a small tilt will cause this upset. On the other hand if the centre of gravity is near the base a very large tilt will be needed to tip the cone so that the weight acting down through the centre of gravity falls outside the base of the cone to cause upsetting. By loading the base the centre of gravity is lowered and greater stability achieved. A popular balancing toy has its base so heavily loaded that when it is laid on its side it immediately rights itself (Fig. 125).

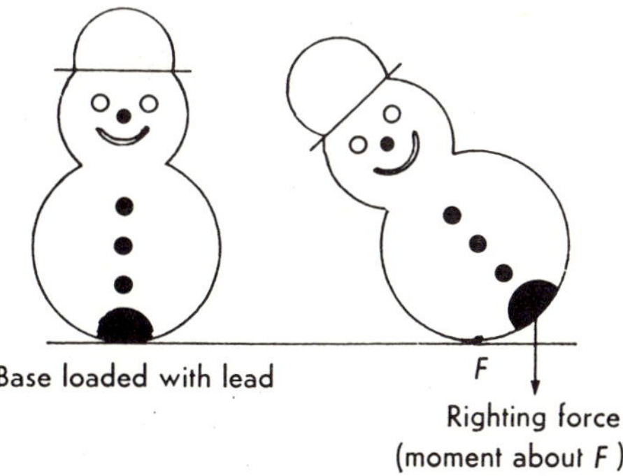

FIG. 125.

FIG. 126. *A double-decker bus must have a very low centre of gravity in order to be stable even when fully loaded on the top deck and cornering at speed. Before use, all such buses must undergo the "tilt test". The top deck is loaded with 30 sand-bags each of mass 70 kg. The bus is then tilted to a regulation 28°, or beyond, out of the vertical. This low centre of gravity ensures that the bus has great stability, particularly when cornering.*

Weighing a ruler. The weight of a ruler can be measured by using the principle of moments. The ruler is pivoted at a known distance from its centre of gravity and then balanced horizontally by adjusting a known mass m on the opposite side of the fulcrum. The distance y of this mass from the fulcrum is measured. The whole of the weight of the ruler may be considered to act through the centre of gravity (in this case this is also the centre of the ruler) and the moment of this weight is balanced by the weight on the other side.

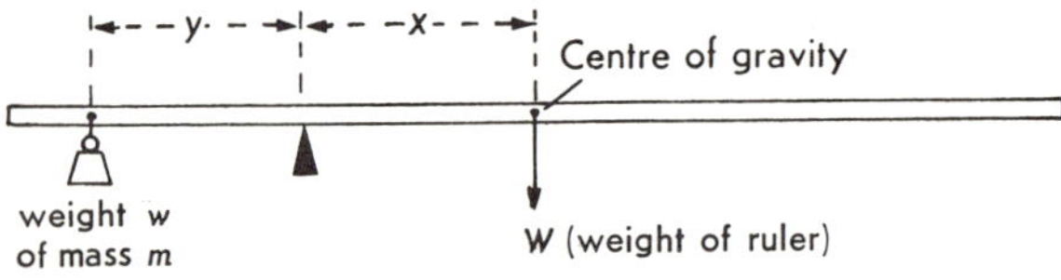

FIG. 127. *Weighing a ruler by pivoting it off-centre.*

If w $(= mg)$ is the weight of the mass and W is the weight of the ruler,

$$\text{then } wy = Wx$$

$$\text{and } W = \frac{wy}{x}$$

Mechanical Advantage and Velocity Ratio

The principle of moments is utilized in all lever applications. A common balance or pair of scales uses equal lever arms, so that to obtain equal moments on both sides, equal masses which necessarily must have equal weights must be placed in the two scale pans (Fig. 128). Many levers use arms of unequal length, as in the crowbar.

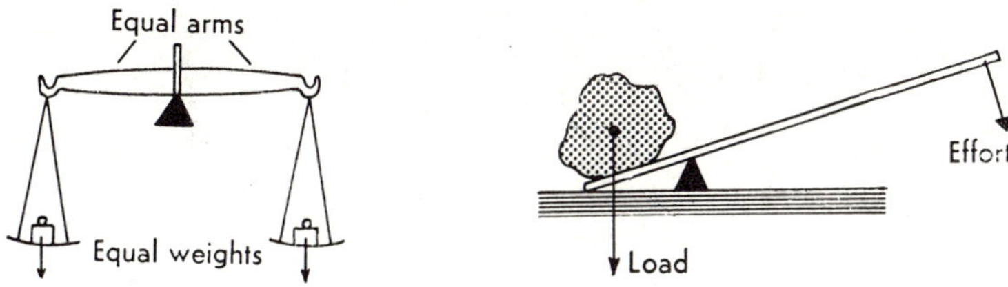

FIG. 128. *Pair of scales.* FIG. 129. *Crowbar.*

In talking about levers, the word *load* is sometimes used for the force that the lever exerts on a body, sometimes for the body itself, and sometimes for the weight of the body that is moved. In reading the following paragraphs you must be quite clear which meaning is being used.

In the case of a crowbar (Fig. 129), it is seen that the effort acts at a much greater distance from the fulcrum than the load, and hence to achieve equal moments, a large load can be balanced by a small effort. This "gain" in force is called the mechanical advantage.

$$\text{Small effort} \times \text{Large distance} = \text{Large load} \times \text{Small distance}$$

$$\textit{Mechanical advantage} = \frac{\textit{Load}}{\textit{Effort}}$$

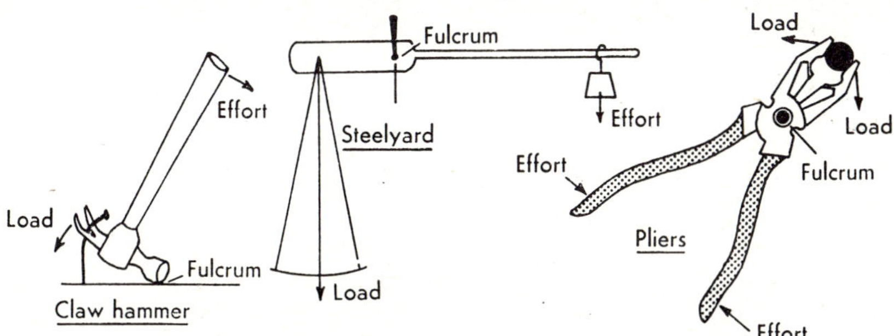

FIG. 130. *Three common applications of the lever.*

Similar mechanical advantages are obvious in other lever applications like pliers, steelyards, and the claw-hammer (Fig. 130). The effort arm in each case is considerably longer than the load arm.

At first sight one might think that less work is done in raising a large stone by using a long crowbar (Fig. 129), than by a direct lift. Consideration of the work done in each case however, by comparing the products of forces and distances moved, reveals this not to be the case.

Looking at Fig. 131, it can be seen that while the large load L moves a small distance a, the small effort E moves a large distance b. Moreover it is seen that the ratio of the distance b moved by the effort to the distance a moved

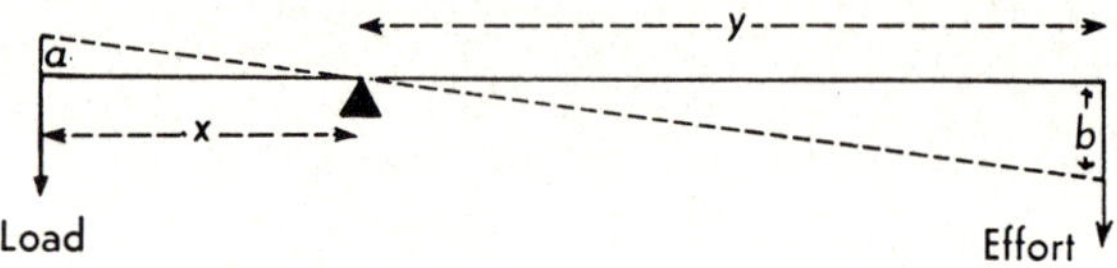

FIG. 131. *Velocity ratio of a simple lever.*

by the load in the same time, is equal to the ratio y/x, the distances of effort and load from the fulcrum. This ratio is known as the *velocity ratio.*

In an ideal lever obeying the principle of moments:

$$L \times x = E \times y$$

$$\therefore \frac{\text{Load}}{\text{Effort}} = \frac{y}{x}$$

$$\text{But from geometry (Fig. 131)} \frac{y}{x} = \frac{b}{a}$$

$$\text{and} \frac{\text{Load}}{\text{Effort}} = \frac{b}{a}$$

$$\therefore L \times a = E \times b$$

This shows that the work done on the load is equal to the work done by the effort. Ideally this is true, but in practice friction at the pivot introduces energy losses and more work must be done by the effort than is done on the load. The machine is not quite 100 per cent efficient, although it may be nearly so.

The *efficiency* of a machine is taken to be the ratio of the work done by the machine to the work put into the machine by the effort:

$$\text{Efficiency} = \frac{\text{Work output}}{\text{Work input}}$$

$$= \frac{L \times \text{Distance moved by load}}{E \times \text{Distance moved by effort}}$$

$$\text{but } \frac{L}{E} = \text{Mechanical Advantage}$$

$$\text{and } \frac{\text{Distance moved by effort}}{\text{Distance moved by load}} = \text{Velocity Ratio}$$

$$\therefore \textit{Efficiency} = \frac{\textit{Mechanical Advantage}}{\textit{Velocity Ratio}}$$

N.B.: Efficiency, since it is fractional, is generally multiplied by 100 and expressed as a percentage.

Simple Levers

Simple levers can be classified into three distinct types or orders:

Type 1 (Fig. 132*a*), where the fulcrum *F*, is situated between the effort *E*, and the load *L*. Examples of this type of lever are the see-saw or the common balance.

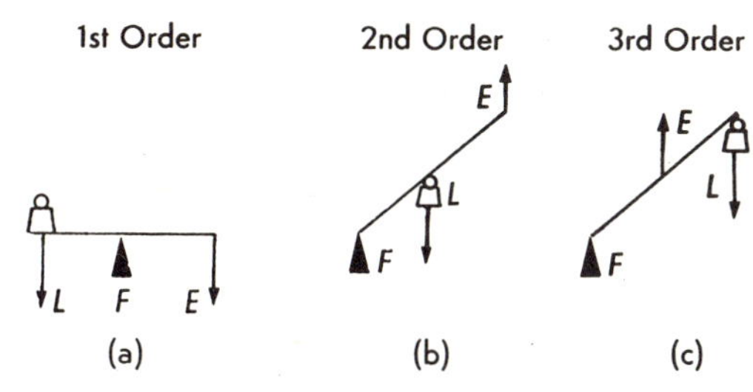

FIG. 132. *The three types of simple lever.*

Type 2 (Fig. 132*b*), where the load *L* is situated between the fulcrum *F* and the effort *E*. Examples of this type of lever are the wheel barrow or the action of opening the lid of a box (Fig. 133).

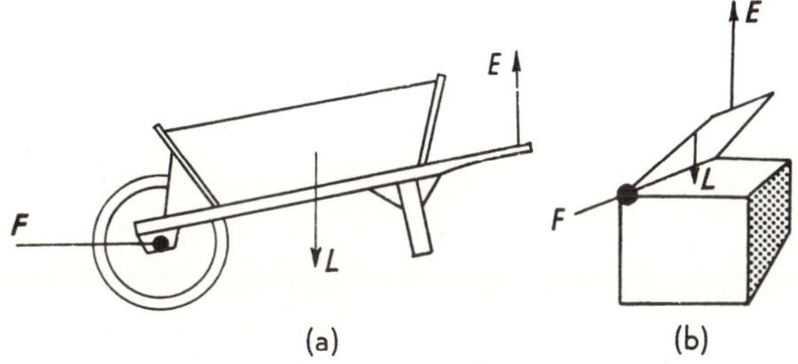

FIG. 133. *Two common examples of 2nd order lever.*

Type 3 (Fig. 132*c*), where the effort *E*, is situated between the fulcrum *F*; and the load *L*. Examples of this type of lever are your own forearm (Fig. 134) or a pair of sugar tongs.

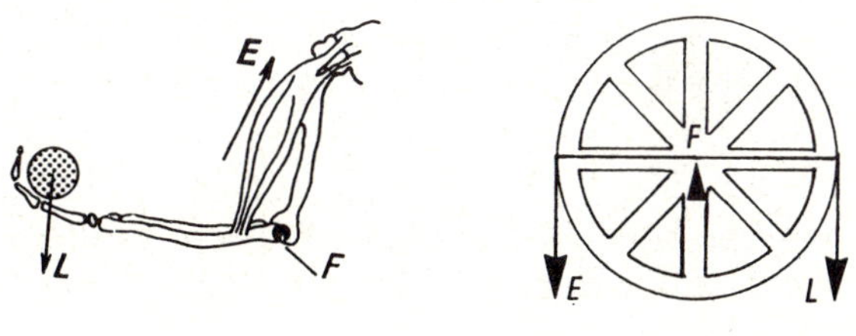

FIG. 134.

FIG. 135.

Wheel and Axle

A wheel is in many ways just a special case of a Type 1 lever, with the axle as the fulcrum, and the spokes of the wheel as equal lever arms (Fig. 135).

In the wheel and axle however, a large wheel is attached to the same axle as a small diameter drum. An effort is applied to a rope round the wheel, which raises a load suspended by a rope round the drum (Fig. 136).

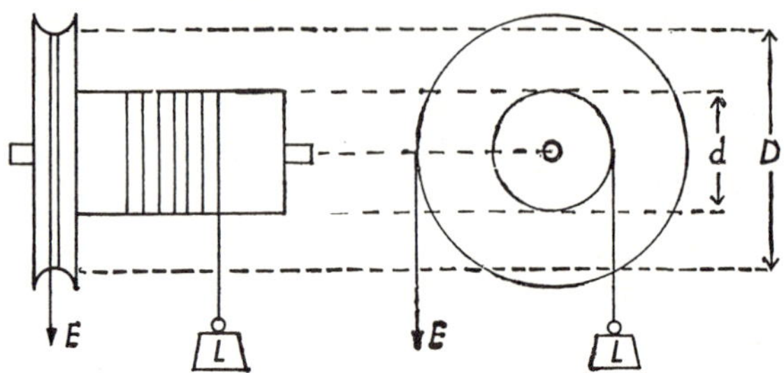

FIG. 136. *Wheel and axle.*

Consider what happens when the wheel and axle complete one rotation; the load is raised through a height equal to the circumference of the drum:

$$\text{Circumference of drum} = \pi \times \text{Diameter of drum } (d)$$
$$\text{Work done on the load} = L\pi d$$

The effort E, applied to the rope round the wheel, will move in the same time through a distance equal to the circumference of the wheel:

$$\text{Circumference of wheel} = \pi \times \text{Diameter of wheel } (D)$$
$$\text{Work done by the effort} = E\pi D$$

If there are no frictional losses and the work input equals the work output:

$$L\pi d = E\pi D$$
$$\therefore \frac{L}{E} = \frac{D}{d}$$

The velocity ratio D/d equals the mechanical advantage L/E in this perfect case. In practice, the mechanical advantage is a little less than the velocity ratio, and the efficiency is less than 100 per cent.

The Winch

Most frequently the wheel is replaced by a crank handle as in a simple winch or, in the case of the bicycle, by the pedals. In the bicycle the motive force is transferred by means of a chain to the driving wheel at the rear. Examination of a bicycle will show that the number of teeth on the pedal wheel is greater than on the rear axle wheel; often 48 to 18. It follows that for each rotation of the pedals the rear wheel turns $\frac{48}{18}$ times.

The introduction of gears into a machine enables greater velocity ratios to be obtained as in the geared bicycle and geared winch. It should be noted

FIG. 137. *If the gear on the right rotates in a clockwise direction, in which direction will the other two gear wheels rotate?*

however, that in a winch, one complete revolution of the effort crank causes only a fractional rotation of the load drum. The velocity ratio is greater than 1. In a bicycle, however, one rotation of the pedal crank causes the rear wheel to turn $\frac{48}{18}$ times; the velocity ratio is less than 1, since the wheel has a greater diameter than the pedal crank.

In the winch shown in Fig. 138, the small gear wheel turns twice for each rotation of the large wheel (ratio 20 : 10). If the teeth number N and n respectively, the small wheel will turn N/n times for each turn of the big wheel.

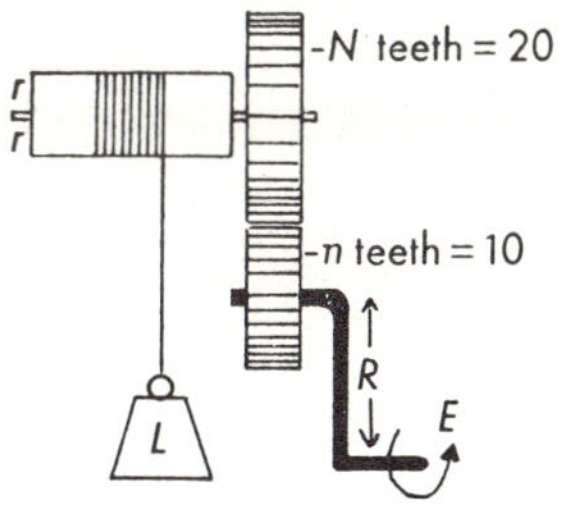

FIG. 138. *Winch.*

One rotation of the drum raises the load L a distance of $2\pi r$ and turns the large gear wheel once. In the same time the small wheel and the crank handle each turn N/n times.

For each revolution of the crank handle the effort moves a distance $2\pi R$.

Thus for each rotation of the drum the effort moves through a distance $2\pi R \times N/n$:

$$\text{Velocity ratio} = \frac{\text{Distance moved by effort}}{\text{Distance moved by load}}$$

$$= \frac{2\pi\ RN}{2\pi\ rn}$$

$$= \frac{RN}{rn}$$

It is interesting to note that the introduction of an extra idler gear wheel in the train of gears has no effect on the velocity ratio but merely preserves the original direction of rotation.

Screw Jack

The screw jack used in lifting motor cars (Fig. 139) is similar in some ways to the geared winch. A turning effort E is applied to a double-headed crank handle. This turns a bevel gear wheel meshed to a crown wheel, rotating a nut through which is threaded a vertical screw with a helical thread of small pitch. (The pitch of a screw is the distance apart of the threads.) Thus if a rotation of the effort handle turns the nut one revolution and drives up the screw a small distance equal to the pitch of the thread p, then the velocity ratio would be $2\pi r/p$ and may have a large value.

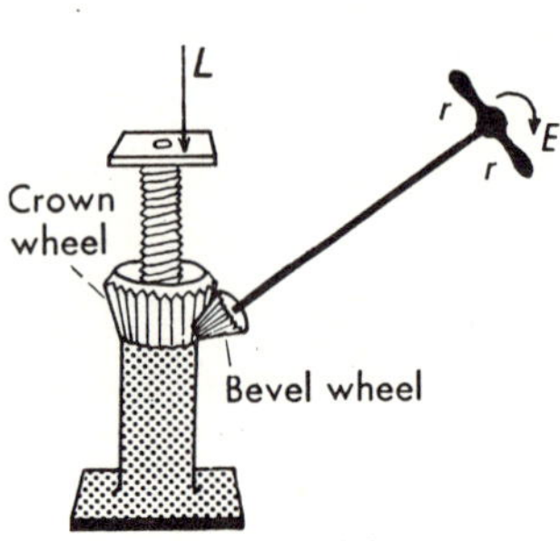

FIG. 139. *Screw jack.*

Calculate the velocity ratio of such a screw jack where the radius of the crank handle is 50 cm and the pitch p of the screw is 1 cm:

$$\text{Velocity ratio} = \frac{2\pi 50}{1} = 100\pi = 314 \text{ approx.}$$

If an effort of 25 N is applied to each end of the handle giving a total effort of 50 N, the mass that can be lifted assuming no frictional losses will have a weight of 50 × 314 N = 15 700 N. Of course, there will be quite large frictional losses reducing the efficiency to as little as 30 per cent and in this case the mass that can be lifted will only have a weight of 30 per cent of 15 700 N = 4710 N. This is advantageous in that the low efficiency prevents the car jack from running down again under action of the load L when the effort E is removed (Fig. 139).

Pulleys

Pulleys are machines used for lifting heavy loads. *Common sheaved pulleys* have two blocks, within which individual pulley wheels rotate independently. Though in practice these pulley wheels are on a common axle, they are frequently depicted one below the other (Fig. 140).

When effort E is applied sufficient to overcome friction and the load, rope is pulled out and the load L rises. It can be deduced that when L is raised 1 m, each of the suspending ropes must also shorten by 1 m resulting in this case in a total effort movement of 4 m as there are 4 suspension ropes. Thus the velocity ratio is 4. Similarly with a 3-sheaved pulley and 6 ropes the velocity ratio is 6.

Simply counting the ropes to find the velocity ratio does not hold good in the case of an *Archimedes-type pulley* (Fig. 141). Here a rise of 1 m by the load follows a shortening of the rope round pulley A by 2 m (1 m on each side). Now since the upper end is fixed, it follows that B is raised 2 m. Similarly pulley C is raised 4 m and finally from the fixed pulley D 8 m of rope is pulled out by the effort E. The velocity ratio is seen to be 8.

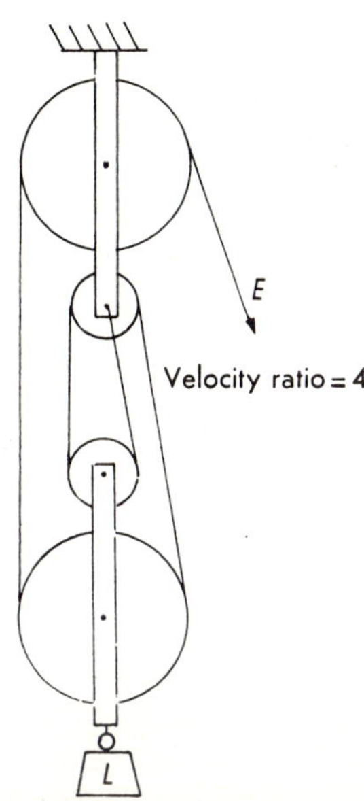

Two sheaved pulley blocks
FIG. 140.

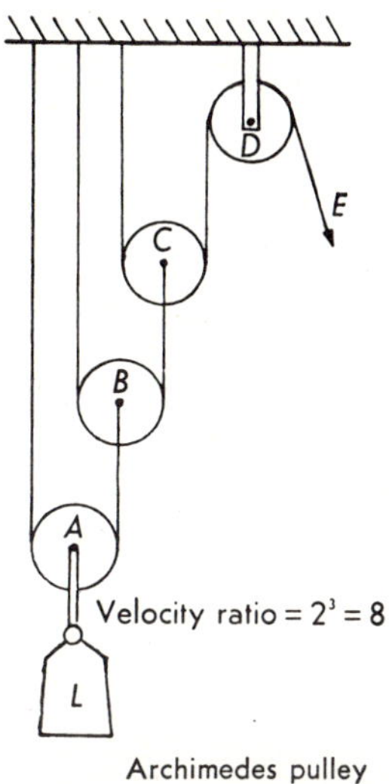

Archimedes pulley
FIG. 141.

It can be seen that with 4 such pulleys the velocity ratio is given by $2^3 = 8$. (There are 3 moving pulleys.) With an extra pulley, making 5 in all, of which 4 are moving, the velocity ratio is $2^4 = 16$. In general with n moving pulleys the velocity ratio is 2^n.

The Weston Differential Pulley (Fig. 142) has in its upper block two co-axial pulleys cast together as one unit, but having slightly different radii. The pulley wheels have teeth (or indents) and revolve together on one axle. A continuous chain passes round the larger wheel, under the load wheel, and over the smaller wheel at the top. Links of the chain fit over the teeth so that the chain is held securely. Let us consider a Weston pulley with N teeth in the larger wheel and n teeth in the smaller wheel. (The load wheel plays no significant part in this machine.)

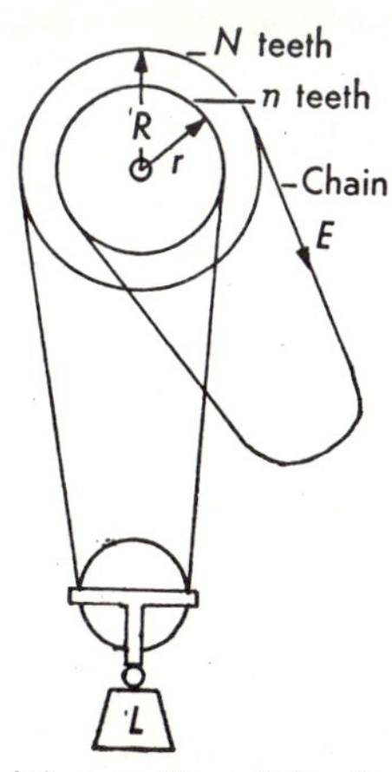

Weston differential pulley
FIG. 142.

FIG. 143. *A block of five common sheaved pulleys (velocity ratio = 10) is being used by this giant crane in the Port of London to lift a mass well in excess of 100 000 kg.*

One complete rotation of the two wheels allows N links of the chain to be pulled up on to the large wheel and at the same time n links to be pulled off the small wheel. The part of the chain supporting the load is thus shortened by an amount equal to $N - n$ links. This is equivalent to raising the load $(N - n)/2$ links since both sides of the chain must be equally shortened.

In the same time (one rotation), the effort has pulled out N links of chain into the slack side loop.

$$\text{V.R.} = \frac{\text{Distance moved by effort}}{\text{Distance moved by load}} = \frac{N}{\frac{N-n}{2}} = \frac{2N}{N-n}$$

It is immediately apparent that if n is just one link less than N a very large velocity ratio (V.R.) is achieved.

The efficiency is low since there is a lot of friction; more than the weight of the load itself. This proves most useful in that the load will not "run backwards" or "overhaul" when the effort is removed. The frictional resistance always opposes motion and when the effort is removed it opposes the load and is sufficient to prevent it falling.

To determine practically the efficiency of an actual pulley block system the following procedure may be used.

First the velocity ratio of the machine is determined. This may be calculated by inspecting the pulley system, counting supporting ropes and moving pulleys, etc., or by actual measurement. In the latter case the distance moved by the effort rope is actually measured using a rule while the load is raised by a measured amount.

A known load is then placed on the pulley and an increasing effort applied, by adding more and more weights to the effort rope, until the load is just on the point of being raised. The load divided by this effort equals the mechanical advantage (M.A.).

The efficiency is then calculated by dividing M.A. by V.R. and multiplying by 100 to convert the answer into a percentage.

For example, in the two-sheaved pulley block (Fig. 140) an effort of 1 N was found to be just enough to start to lift a load of weight 3 N.

$$\text{Mechanical Advantage} = 3/1$$

$$\text{Velocity Ratio} = 4$$

$$\therefore \text{ Efficiency} = \frac{\text{M.A.}}{\text{V.R.}} \times 100\% = \frac{3 \times 100}{1 \times 4}\% = 75\%$$

QUESTIONS

1. A force of 400 N is used to drag a load steadily along a horizontal road for 50 m. How much work is done?

2. A man who weighs 800 N climbs steadily up a mountain slope that rises at 30° to the horizontal. If he climbs from 1000 m up to 1500 m at an average walking speed of 1 m/s, calculate (*a*) how much work he does in the climb, (*b*) his rate of working.

3. State briefly what is meant by the moment of a force about a point. A uniform metre rule is pivoted on a needle through a hole drilled on the 40 cm mark. A 24 g mass hung from the 5 cm mark just balances the rule. What is the mass of the rule? If the volume of the 24 g mass is 3 cm^3 and it is suspended from the rule so that it is immersed in water, where must it be hung now to balance the rule?

4. Outline carefully the energy changes which occur when a child jumps up and down on a pogo stick.

5. What is meant by the efficiency of a machine?

Describe how you could determine by experiment the efficiency of a bicycle.

6. The wheel of a wheel and axle machine is 20 cm in diameter, and the axle is 4 cm in diameter. If the machine is 50% efficient, what effort is needed to lift a load of 250 N?

7. A Weston differential pulley has wheels with 15 and 14 cogs in the upper block. In an experiment it is found that an effort of 100 N just lifts a load of 1050 N. What is the efficiency of the machine?

CHAPTER 12

MEASURING THE QUANTITY OF HEAT

WHENEVER we examine a machine, as detailed in the last chapter, we find the energy supplied by the effort more than the work obtained from the machine. If indeed energy cannot be destroyed, what has become of this extra energy?

Let us consider what happens when a piece of steel is sawn through using a hack saw. A considerable amount of energy is used in moving the blade backwards and forwards, yet no obvious resultant increase in the kinetic or potential energy of the steel or blade occurs. If we feel either steel or blade, however, we immediately notice they have become hot; their temperatures have risen. This leads us to the solution of the problem: heat is a form of energy. Whenever a machine works, there is frictional resistance, and the energy used up in overcoming this is converted into heat energy:

Energy supplied by effort = Work done + Heat energy losses.

We know that heat can be liberated as a result of chemical and electrical processes without involving mechanical energy. It is possible also to convert heat energy into mechanical energy and do work as in the steam engine or internal combustion engine. This concept of heat as a form of energy was not clear before the middle of the nineteenth century.

Burning gas in a Bunsen burner is a common laboratory source of heat. We can use such a burner to investigate some of the effects of supplying heat energy to a body.

If a piece of iron wire is held with one end in a Bunsen flame it is noticed that the end in the flame becomes hot. The heat has raised the temperature of the iron. It will soon be noticed that the end of the wire remote from the flame also becomes hot, showing heat energy has passed along the wire. As more heat is supplied to the wire it glows red, and then at a higher temperature still appears almost white hot, bends and may melt, changing from a solid to a liquid form.

We can consider the temperature of a body as its relative hotness or coldness. In everyday life it is measured in degrees Celsius (also called degrees Centigrade). This way of measuring temperature depends on the temperature of melting ice and of steam from boiling water. Its symbol is °C. In science, temperatures are based on quite a different idea and are measured in units called kelvin (symbol K). However, the units of differences of temperatures are the same on both scales, so in this book a temperature is given on the Celsius scale and a difference of temperature is given on the Kelvin scale. Thus we say, for instance, that when the temperature of some water rises from 15° C to 20° C, its temperature rises by 5 K.

Heat generally flows from a region of high temperature to one of a lower temperature, e.g. from the red-hot end of a steel poker held in a fire to the colder end outside the fire.

Heat then is a form of energy that increases the temperature of a body to which it is added, provided that the body does not change state from solid to liquid or liquid to gas. It is necessary to add this last proviso because continued heating of a vessel of boiling water, for example, causes no rise of temperature above the boiling point. The energy is used in changing the water to steam. This point will be discussed later (Chapter 14).

To clarify this important distinction between heat and temperature, let us carry out a simple illustrative experiment. Place a beaker half full of water on a tripod and steadily supply it with heat by means of a Bunsen burner. The temperature of the water will be observed to rise slowly either by recording with a thermometer or merely by placing a finger in the water. After a minute or two hold a small tin tack (using tongs) in the same flame. It can be seen glowing red hot after just a few seconds showing it to be much hotter than the water, though it must have received considerably less heat. If now the tack is dropped into the water, a hiss will be heard as the red hot tack gives up some of its heat to the water and in so doing is cooled to the water's temperature. It is clear that heat and temperature are different.

Let us now further investigate the effect of heat in raising the temperature of bodies of different masses and different materials.

Three identical thin metal cans (Fig. 144*a*, *b*, *c*) are placed on tripods and filled respectively with 200 g water, 400 g water and 400 g of oil, all at room temperature. Three Bunsen burners are carefully adjusted by eye to give

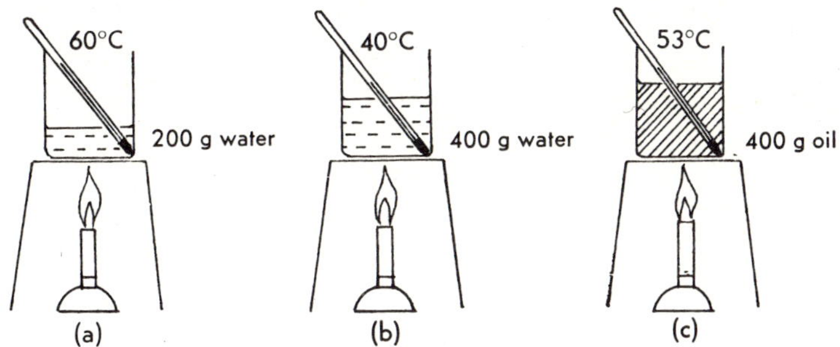

FIG. 144. *Temperature rise depends on mass and nature of material heated*

equal small flames and thus to supply heat at the same rate. The three cans are then heated simultaneously and recordings of the temperatures of their contents are made every minute.

Consider such an experiment in which the initial temperature of the three cans and their contents is 20° C. The following observations may be obtained, depending on the size of Bunsen flame used:

After 5 min temperature of 200 g water = 60° C
temperature of 400 g water = 40° C
temperature of 400 g oil = 53° C

Since the cans are being heated at the same rates, we can draw the conclusion that *the rise in temperature produced by a quantity of heat varies with the mass of the material and the nature of the material heated.* (N.B. If the cans are thin they have little effect on the rises in temperature.)

Units of Heat Energy

To examine the rise in temperature of a given mass of water when supplied by a measured quantity of heat energy, one can heat 400 grammes of water in a light container (expanded polystyrene is excellent), using an immersion heater. A 36-watt heater gives a steady rise in temperature.

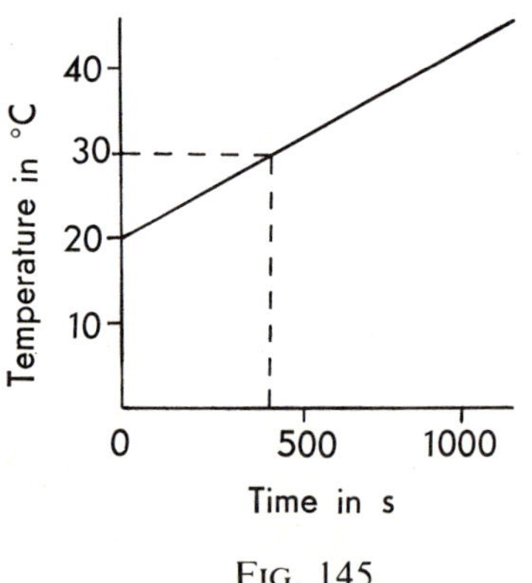

FIG. 145

In an experiment it took 470 seconds to heat up the water from 20° C to 30° C, that is, a rise in temperature of 10 K. Thus the rate of rise in temperature was 1 K in a time interval of 47 seconds. During that time 36 J of energy were supplied every second. Thus the energy required to heat up the water was 36 × 47 J = 1692 J. It follows that the energy needed to raise the temperature of 1 gramme of water through 1 K is 1692/400 J = 4·23 J. More accurate experiments show that 4·18 J are required.

Obviously 4·18 kJ will be needed to raise the temperature of 1 kg water of water through 1 K.

You will notice that the unit of heat energy—the joule—is the same as the unit of all other forms of energy. At one time another unit was commonly used—the *calorie.* This was the heat energy required to raise the temperature of one gramme of water through 1 K (or to be more precise from 14·5° C to 15·5° C). It was shown by experiment that 1 calorie was equivalent to 4·18 joules. This was known as the mechanical equivalent of heat. In biology, another unit is used—the *Calorie*, which is equivalent to 1000 calories and is therefore also called the *large calorie* or *kilocalorie.* It is often misleadingly printed with a small initial letter in place of a capital letter. Finally, the heating value of fuels (the calorific value) is quoted in *therms.* 1 therm = $1{\cdot}043 \times 10^8$ joules.

Heat Capacity

We use (Fig. 146) two identical containers (made of expanded polystyrene), containing respectively 1 kg of water and 1 kg of oil, and a cylindrical block of copper also of mass 1 kg standing on an insulating pad (cork or polystyrene). The initial temperatures of all three bodies are taken with mercury thermometers.

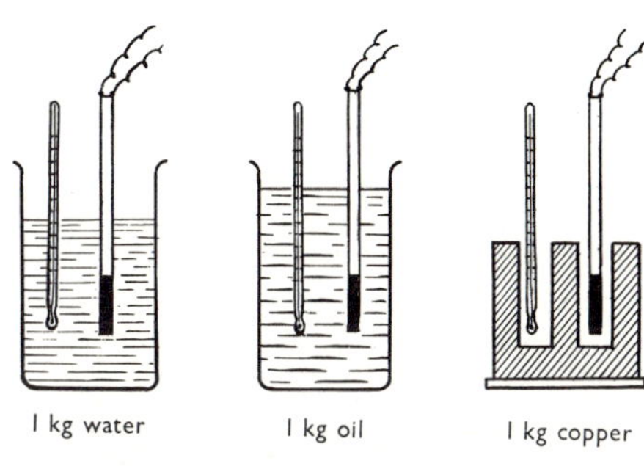

FIG. 146.

If they have been left out before the experiment is started, these three temperatures will be the same.

Identical immersion heaters are placed in each body and are switched on for the same period of time. After switching off, the temperatures of the three bodies are again taken. Suppose these results are obtained:

Power of each heater	= 40 W
Time switched on	= 300 s
Heat supplied to each body	= 12 000 J = 12 kJ
Rise in temperature of water	= 3 K
Rise in temperature of oil	= 5 K
Rise in temperature of copper	= 30 K

The heat required to raise the temperature of a body through 1 K is called the *heat capacity* of the body.

$$\text{Heat capacity of the water} = \frac{12}{3} = 4 \text{ kJ/K}$$

$$\text{Heat capacity of the oil} = \frac{12}{5} = 2{\cdot}4 \text{ kJ/K}$$

$$\text{Heat capacity of the copper} = \frac{12}{30} = 0{\cdot}4 \text{ kJ/K}$$

i.e. the copper, for instance, needs only 0·1 of the heat that the water needs to increase its temperature by the same amount.

$$\text{Relative heat capacity} = \frac{\text{Heat required to raise the temperature of a body}}{\text{Heat required to raise an equal mass of water through the same rise in temperature}}$$

$$\text{Relative heat capacity of copper} = \frac{0{\cdot}4}{4} = 0{\cdot}1$$

$$\text{Relative heat capacity of oil} = \frac{2{\cdot}4}{4} = 0{\cdot}6$$

Specific Heat Capacity

The *specific heat capacity* of a substance is defined as the amount of energy needed to raise the temperature of 1 kg of the substance through 1 K. In elementary work it is assumed that its value is the same at all temperatures, so that in experiments a considerable rise of temperature can be used. But it must be remembered that this assumption does not hold good at very low temperatures. The specific heat capacity of a substance can be found by direct heating, as in the experiment above.

Specific heat capacity of water = 4 kJ/kg K
oil = 2·4 kJ/kg K
copper = 0·4 kJ/kg K

Another way if finding the specific heat capacity is to use the *method of mixtures*. When a hot body is placed in contact with a cold body, heat will flow from the hot body to the cold body until the temperatures of the two bodies are the same. If this "mixing" is done in an insulated container, it can be assumed that the heat given out by the hot body equals the heat taken in the cold body. Knowing all the quantities except the specific heat capacity of the material of one of the bodies, this latter can be calculated.

Finding the Specific Heat Capacity of a Solid

A lagged calorimeter is used. This consists, in its simplest form, of a thin walled metal can A which can be placed inside another container B, often a larger vessel lined with wool or felt, which acts as a heat insulator. Sometimes air is used as an insulator between the inner calorimeter and the outer container, with the calorimeter on a cork insulating base.

If the material to be examined is a solid, like a piece of iron, its mass m, is first found by weighing, and then it is heated to about 100° C. This can be done by suspending the metal on a thread in the steam above a water bath or by using a more elaborate steam bath such as the Nicholson Heater (Fig. 147). This latter apparatus has the advantage of keeping the solid dry by insulating it from the steam itself. In both methods a thermometer is placed close to the heated metal to record its temperature θ°.

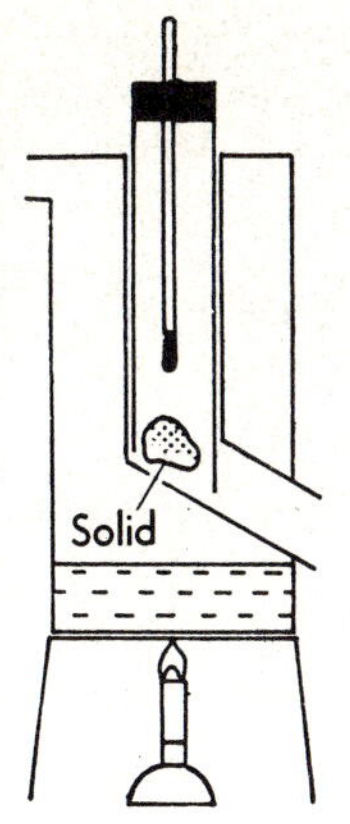

FIG. 147.

Meanwhile the calorimeter itself is weighed, first empty, together with its stirrer, and then about two-thirds full of cold water. This gives the mass m_c of the calorimeter and stirrer, and m_w the mass of the cold water. The calorimeter and stirrer must be made of a metal of known specific heat capacity s_c. They are often of either copper or aluminium.

The calorimeter with its water is placed inside the heat-insulating jacket and its steady temperature θ_1° C is recorded. After some considerable time, perhaps 15 minutes, when the metal specimen has fully heated up on its water bath, its temperature θ° is recorded.

Quickly, to avoid heat loss to the air, the hot metal is transferred to the cold water in the calorimeter and, after stirring the mixture, its final temperature is measured, θ_2° Results may be conveniently noted down as shown in Table 8 and the specific heat capacity of the metal found by equating the quantities of heat involved.

Table 8

Finding the Specific Heat Capacity of a Solid

	Calorimeter	Water	Hot solid
Mass	m_c	m_w	m
Initial temperature	θ_1	θ_1	θ
Final temperature	θ_2	θ_2	θ_2
Change in temperature	$\theta_2 - \theta_1$	$\theta_2 - \theta_1$	$\theta - \theta_2$
Specific heat capacity	s_c	4·2	s
Quantity of heat	$m_c\ (\theta_2 - \theta_1)\ s_c$	$m_w\ (\theta_2 - \theta_1) \times 4{\cdot}2 \times 10^3$	$m\ (\theta - \theta_2)\ s$

Assuming there has been no loss of heat during the experiment, we may assume that heat given out by the hot solid, equals the heat gained by the water and the calorimeter.

$$m(\theta - \theta_2)s = m_c(\theta_2 - \theta_1)s_c + m_w(\theta_2 - \theta_1) \times 4{\cdot}2 \times 10^3$$

$$\text{Specific heat capacity of metal}(s) = \frac{(\theta_2 - \theta_1)\ (m_c\, s_c + m_w \times 4{\cdot}2 \times 10^3)}{(\theta - \theta_2)\, m}$$

$$\text{Its units are } \frac{\text{joule}}{\text{kilogramme} \times \text{degree kelvin}} \text{ or J/kg K.}$$

In the above equations the product $m_c\, s_c$ is called the *heat capacity* of the calorimeter. It can be found by a separate experiment or it can be calculated knowing m_c and s_c.

Solids in finely divided form, copper turnings or lead shot, can be conveniently heated by placing in a test-tube heated on a water bath. Their masses can be determined at the end of the experiment by weighing again the calorimeter, water, and metal, and subtracting the mass of calorimeter and water already found.

Finding the Specific Heat Capacity of a Liquid

The specific heat capacity of a liquid can be found by placing a known mass of the liquid in the calorimeter instead of water, as in the last experiment. A hot solid made from material of known specific heat capacity is then added and the experiment carried out as before. In the final equation, this time, the heat gained by the liquid is given by $m_1 (\theta_2 - \theta_1) s_1$, where s_1, the specific heat capacity of the liquid, is the only unknown quantity and can be calculated. The heat given out by the hot solid is known because this time its specific heat capacity is known.

This method of mixtures can only be used directly with good-conducting solids because heat losses do occur during the period needed to complete heat transfer from the hot solid to the liquid. However, under normal conditions, these heat losses to the surroundings are so small that the specific heat of a poor conductor like a rubber stopper can be determined reasonably accurately by this simple method. More elaborate cooling corrections must be applied if a higher degree of accuracy is required, or if a large bad-conducting solid is used.

Specific Calorific Value

The amount of heat energy liberated when various fuels or foods are used up is of great importance in modern life, and the term *specific calorific value* (or heat of combustion) is often seen quoted, e.g. in Gas Board accounts.

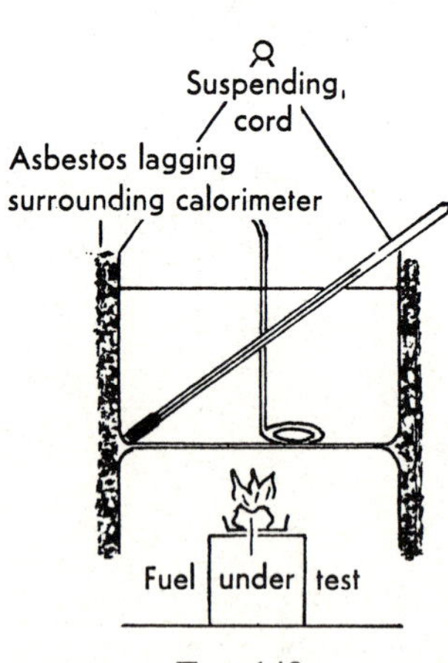

FIG. 148.

The specific calorific value of a substance may be defined as the quantity of heat liberated when unit mass of the substance is completely burned. The heat of combustion of domestic gas is often expressed in "therms per cubic foot." It can be taken that 1 therm per cubic foot is equivalent to 10^9 J/m^3.

A method of measuring the specific calorific values for some common substances involves a simple double calorimeter as shown in Fig. 148. The calorimeter consists of two thin copper cans soldered together and surrounded by a heat-insulating jacket of asbestos. The calorimeter is suspended by strings from a stand and contains a weighed quantity of cold water at an initial temperature several degrees below room temperature. A thermometer and a stirrer are also required.

A measured mass of the fuel to be tested, e.g. methylated spirits, is placed in a small container (a metal bottle cap with a little cotton wool will do). This fuel is ignited under the calorimeter and completely burned.

The water in the calorimeter is stirred and the final highest temperature recorded by a thermometer. The specific calorific value is then calculated as in the actual experimental results quoted below:

Mass of methylated spirits	= 4g
Mass of calorimeter	= 200 g
Mass of cold water in calorimeter	= 800 g
Initial temperature of water	= 17° C
Final temperature of water	= 51° C
Specific heat capacity of water	= 4·2 kJ/kg K
Specific heat capacity of material of calorimeter	= 420 J/kg K

Remember that in calculations masses must be expressed in kilogrammes:

Rise in temperature of water and calorimeter = 34 K

$$\text{Heat taken in by water} = 0{\cdot}8 \times 4200 \times 34 \text{ J}$$

$$\text{Heat taken in by calorimeter} = 0{\cdot}2 \times 420 \times 34 \text{ J}$$

$$\text{Total heat taken in} = 34(3360 + 84)\text{J}$$

This is given out by 0·004 kg of fuel burning.

$$\text{Heat given out by 1 kg burning} = \frac{34 \times 3444}{0{\cdot}004} = 29\,274\,000 \text{ J}$$

∴ The specific calorific value of methylated spirits is 29 000 000 J/kg or 29 000 kJ/kg or 29 MJ/kg

It is essential in this experiment that the calorimeter be large in diameter if reasonable results are to be obtained. The same method may be used to determine the specific calorific value of candle wax: a candle heats the water in the calorimeter. The candle is weighed before and after burning and blown out when a satisfactory temperature rise has been achieved.

QUESTIONS

1. Into a copper can weighing 600 g, is poured 100 g of water at 80° C. If the initial temperature of the can was 20° C, calculate the final temperature, assuming there are no heat losses. (Specific heat capacity of copper is 420 J/kg K.)

2. Calculate how much heat is needed to raise the temperature of an aluminium pan containing 1·5 kg of water from room temperature 20° C to boiling point 100° C, assuming no heat losses. The mass of the pan is 0·5 kg and the specific heat capacity of aluminium is 820 J/kg K.

3. Describe how you would determine the specific heat capacity of some oil by means of an experiment.

4. What is meant by the specific calorific value of a fuel?

Describe how you would discover approximately the specific calorific value of the wax of a candle by a laboratory experiment.

CHAPTER 13

MEASUREMENT OF TEMPERATURE

ALREADY we have seen that temperature is a measure of the hotness or coldness of a body relative to something else. It dictates the direction in which heat energy flows, for heat naturally flows from a region of high temperature to one of low temperature. Judging the temperature of luke-warm water in a bowl by placing one's finger in it proves most unreliable. If your finger is held in hot water for a minute before testing, the conclusion will be that the water in the bowl is cold (obviously with reference to the hot water). If on the other hand you hold your finger first in cold water before testing, your reaction will be that the water in the bowl is hot (with reference to the cold water). It is necessary to use a precise instrument to record temperatures and, of course, there must be a fixed agreed reference point and a scale for calibration. Such instruments are *thermometers and pyrometers.*

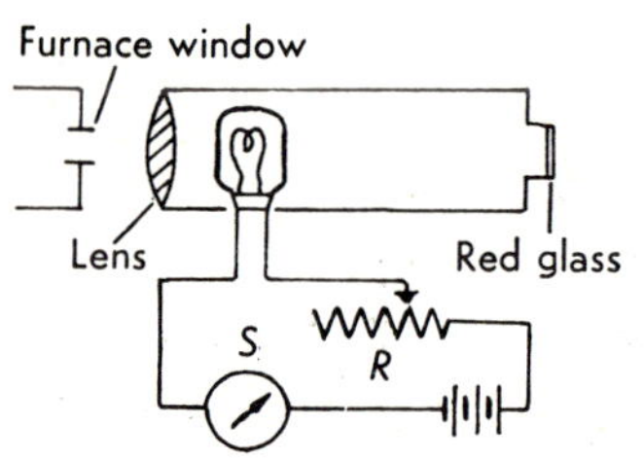

FIG. 149. *Optical pyrometer.*

Some physical property that varies with temperature and can be measured is used. What are such properties?

We know that the colour of a heated wire changes with temperature through dull red to bright red and then to white, but these changes are only obvious at high temperatures. However, they form the basis of the *optical pyrometer* used in measuring furnace temperatures (Fig. 149).

By means of a variable resistance *R* in the lamp circuit, the filament colour of the lamp is matched against that of the furnace and the ammeter scale *S* records the temperature directly, having been calibrated previously for known temperatures.

THERMOMETERS

A more commonly used physical property is that of the volume of a liquid. Let us fill a small round-bottomed flask (100 cm^3) full of water coloured with a little potassium permanganate. When a rubber stopper, through which passes a glass tube, is placed in the neck of the flask, some liquid rises up the tube. A little manipulation will result in the liquid occupying a level in the tube, as shown in Fig. 150. If the flask is now placed in warm water the level of the liquid will rise up the tube, showing expansion has taken place. On placing in cold water the level falls, indicating contraction. We have now a crude thermometer, but it is uncalibrated. The level of the liquid rises and falls with changes in temperature. Such a device is strictly called a thermoscope.

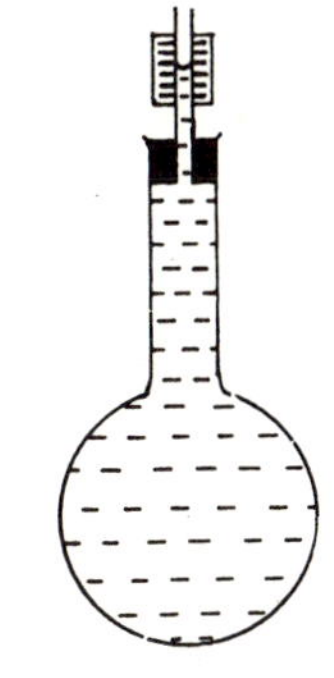

FIG. 150.

Celsius Scale

We need a reference point and a scale. It is found that scales, as defined below, differ according to the liquid used. When mercury and alcohol are used, however, the differences are very small and can usually be detected. If water is used the difference from other scales is very great and water is therefore not a suitable thermo-metric liquid. Both alcohol and mercury are used, being carefully sealed into a glass bulb blown on the end of a length of capillary tubing.

For a fixed reference point the melting point of pure ice (or freezing point of pure water) is normally used. Under normal conditions of pressure for pure water this point is found to be constant.

To fix a scale the difference between the temperature of melting ice (the lower fixed point) and a second upper fixed point, the temperature at which pure water boils under standard atmospheric pressure, may be taken as 100 degrees. This is the Celsius scale. The ice point is called 0° C and the water boiling point 100° C. It is a centigrade scale as the interval the two points is divided into one hundred equal parts. The scale can be extended above 100° C and below 0° C. Temperatures below 0° C are indicated by a minus sign, e.g. − 18° C.

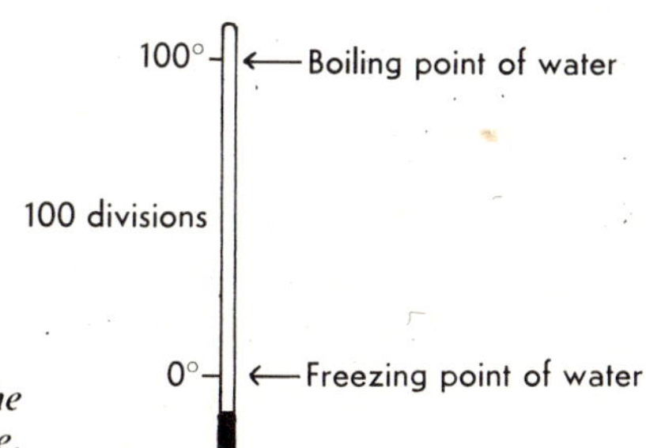

FIG. 151. *Calibration of the Celsius scale of temperature.*

The thermodynamic scale, sometimes called the Absolute or Kelvin scale, is defined in rather a different way but for elementary work it can be taken that the zero of the scale is absolute zero (0 K), the melting point of ice is 273 K and the boiling point of water 373 K. The size of the degree is the same on both scales and that is why a change in temperature is measured in kelvin (symbol K).

At one time the Fahrenheit scale was used in Britain. On this scale there were originally six degrees between the lowest temperature then obtainable (the temperature of a freezing mixture) and the temperature of the human body. Each degree was later divided into sixteen parts, so the body temperature became 96° F. Still later it was defined, using the same fixed points as the Celsius scale.

Although the scale now used is named after Celsius, a Swedish scientist, there is some evidence that Linnaeus, the Swedish botanist, really deserves the credit for inventing it.

Mercury freezes to a solid at — 39° C and boils at 359° C. These temperatures thus indicate the outside limits of use of mercury as a thermometric liquid. Since alcohol boils at 78° C it cannot be used to measure high temperatures, even that of boiling water. It does, however, remain liquid down to — 115° C, and is often used in thermometers for low-temperature measurement. A red dye is added to make the liquid more visible. Mercury itself can be easily seen because of its silvery colour.

The normal laboratory thermometer contains mercury and is scaled between — 10° C and 110° C. The bulb is made of very thin glass to enable heat to pass readily to the mercury from the outside. This results in rapid response to its surroundings. Also, the capillary tube is of fine bore so that even a small expansion of the mercury in the bulb pushes up the mercury in the stem a measurable distance. Other scale ranges are also used.

Maximum and Minimum Thermometers

It is often convenient to record the maximum and minimum temperatures reached during a day, as in meteorology. Special kinds of thermometers enable this to be done.

A maximum thermometer, which is of the mercury in glass type, contains a small index. When the temperature rises, the expanding mercury pushes this index along to record the highest temperature but, as the temperature falls, the mercury contracts and leaves the index behind (Fig. 152). The maximum temperature is thus indicated by the end of the index nearest to the mercury thread. The thermometer is normally used in a horizontal position and the index can be brought back to the mercury by tilting.

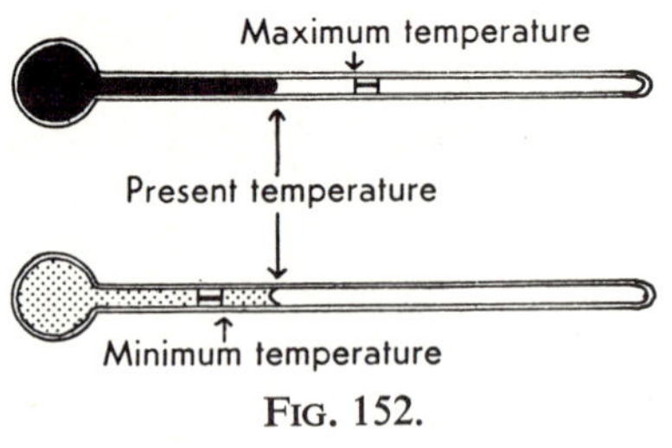

FIG. 152.

A minimum thermometer is similar but contains alcohol instead of mercury. In this an increase of temperature causes the alcohol to expand and flow past the index. On cooling, however, the contraction of the alcohol drags back the index because of the surface tension pull. The minimum temperature is recorded by the end of the index nearest the alcohol (Fig. 152).

In 1782 Six devised a combined maximum and minimum thermometer, constructed as shown in Fig. 153. The thermometric liquid is alcohol, while the mercury in the tube acts merely as a "pusher" for the two indices. Bulb *A* is full of alcohol which expands and contracts with rise and fall of temperature, and forces the mercury round the tube. Bulb *B* is only partly filled with alcohol and contains air and some alcohol vapour. The two indices, which are made of iron, are fitted with springs to hold them against the inner walls of the glass tubing. The mercury pushes the indices along, leaving the bottom end of index *X* registering the highest temperature and the bottom end of index *Y* registering the lowest temperature experienced. The indices can be reset in contact with the mercury column by means of a magnet. No attempt whatsoever must be made to shake the indices back, since this will ruin the thermometer.

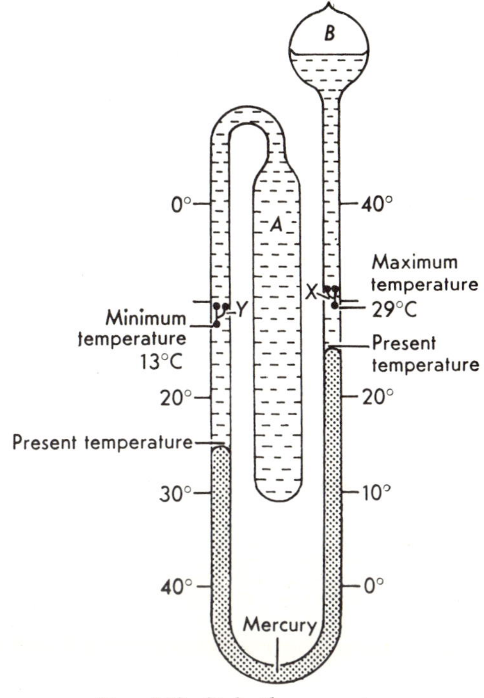

FIG. 153. *Six's thermometer.*

Clinical Thermometers

A special type of maximum thermometer is the clinical type used by doctors to measure patients' body temperatures (Fig. 154). It has a very fine bore with a kink or constriction above the bulb. Placed in a patient's mouth, under the tongue, the mercury in the bulb expands and moves up the stem past the kink to record the temperature. On removal from the mouth, however, contraction of the mercury causes the fine mercury thread to break at the kink, leaving the mercury above the kink still indicating

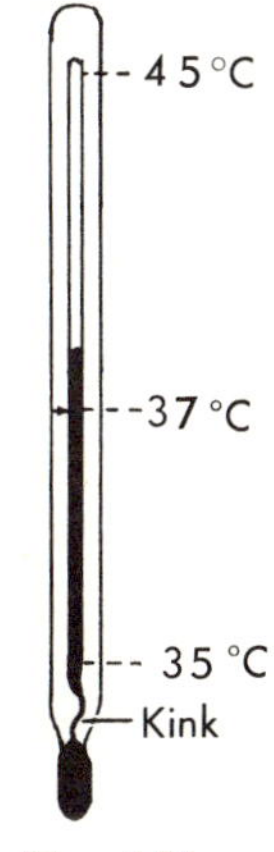

FIG. 154.

the maximum temperature reached. The mercury below the kink contracts down into the bulb. Shaking will make all the mercury run back into the bulb.

This thermometer is calibrated between 35° C and 45° C in fifths or even tenths of a degree. Thickening of the glass stem on the side opposite the graduations magnifies the alcohol thread.

Bi-metallic Thermometers

A bi-metallic thermometer uses the expansion of a compound coil of invar (a metal alloy of negligible expansion) and brass to record changes in temperature. When the temperature rises the spiral will curl clockwise and move the pointer over the scale. A simple home-made model can be made using a compound strip of invar and nickel steel (Fig. 155). The theory behind this thermometer will be discussed in the section that follows on the expansion of metals.

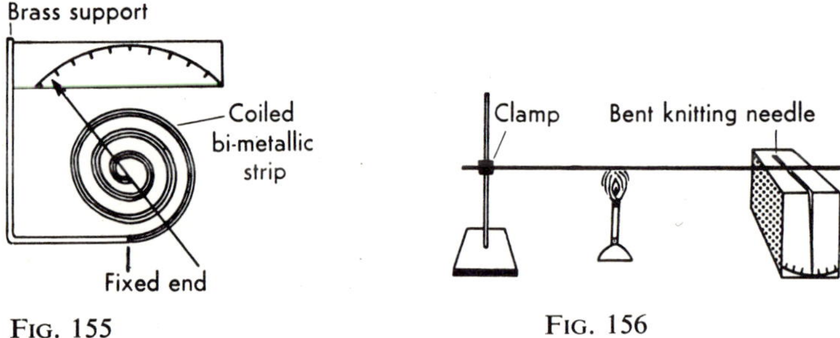

FIG. 155

FIG. 156

EXPANSION OF METALS

Do metals expand when heated? How much do they expand?

If a bar of metal, e.g. brass or copper, about a metre long, is heated strongly it does not seem to change much; but, if it is firmly clamped at one end (Fig. 156), and the free end is placed on a bent needle supported on a glass plate, the needle will be seen to move. The expansion of the rod can take place in one direction only and rolls the needle over the glass, causing quite a large movement of the pointer over the scale. The actual linear expansion of the metal, however, is very small compared with that of liquids and is often less than 0·01% for each degree rise in temperature.

This fractional increase in length of a solid per degree rise in temperature is known as the *Linear Expansivity*, α.

If the original length of a metal bar is l_0, the expansion per degree rise will be $l_0\ \alpha$, and the expansion due to a rise in temperature t will be $l_0\ \alpha\ t$. It follows that $\alpha = \dfrac{\text{expansion}}{l_0\ t}$.

If l_t is the length of the bar at $t°$ C and l_o the length at 0° C:

$$l_t = l_o + l_o \alpha t$$
$$\therefore l_t = l_o (1 + \alpha t)$$

Writing the units can be difficult. It can be seen from the expression that the unit is $\frac{1}{\text{kelvin}}$. When abbreviated this is written either as "per K" or as "K^{-1}."

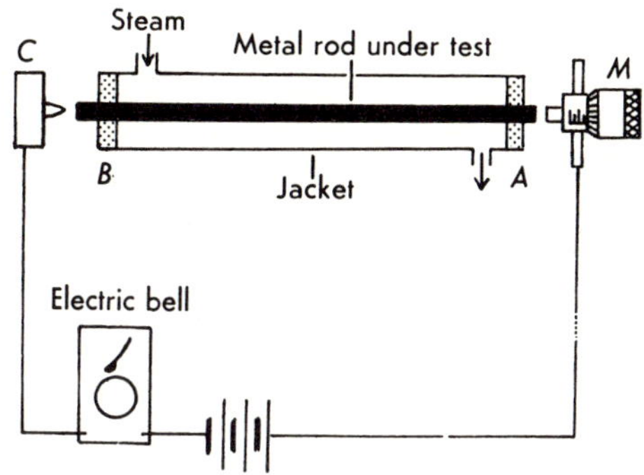

FIG. 157. *Apparatus to measure linear expansivity.*

Measuring Linear Expansivity

There are many different kinds of apparatus for measuring the linear expansivity of a metal. One such apparatus is similar to that shown in Fig. 157, where a bar of the metal to be examined is surrounded by a jacket through which steam can be passed. The bar is about 50 cm long and can slide horizontally when the screw of the micrometer gauge *M* is turned to press up against one end.

The initial length l_0 of the bar is measured at room temperature with a metre rule. The room temperature $t°$ C is noted. The micrometer is screwed up so that it just touches end *A* of the bar, and end *B* is just touching the terminal stop *C*. The reading of the micrometer gauge is read to the nearest thousandth of a centimetre.

In some types of this apparatus there is an auxiliary electrical circuit as shown, containing a battery and electric bell. When both ends of the test bar are touching the end screws the circuit is complete and the bell rings.

Steam at 100° C is passed through the jacket until the bar reaches 100° C. We shall see how this may be checked later. The micrometer gauge is then readjusted to touch the end of the bar so that contact is also made at end *B*. The electric bell will ring again. This new reading of the micrometer screw gauge is noted, and the difference between the two readings of the micrometer gives the amount of linear expansion *y* of the bar. It is wise to continue further steam heating of the bar and take a further reading of the micrometer to ensure full expansion has occurred. This third reading should be the same as the second reading if the bar was at 100° C:

$$\text{Expansion of the bar} = y$$
$$\text{Increase of temperature} = (100 - t)\text{ K}$$
$$\text{Original length of bar} = l_o$$
$$\therefore \text{ Linear Expansitivity, } \alpha = \frac{y}{l_o(100 - t)}$$

The unit will be K^{-1}.

Applications of Metal Expansion

Though the coefficients of linear expansion of metals are small, in the case of a great length of metal, as in a steel railway line or steel bridge, the total change in length between summer and winter temperatures may be considerable. Compensation for such changes in length can be made, e.g. gaps are often left between railway lines, and bridges may be mounted on rollers to permit changes of length or they may be fitted with some other kind of expansion joint.

Failure to compensatc for such expansions could result in very serious damage as the forces involved in expansion or contraction are very large. In riveting together metal plates this contraction force is used. A very hot metal rivet is placed through holes in the two metal plates and hammered to flatten over the rivet and fix it. On cooling, contraction of the rivet occurs and grips the two plates tightly together (Fig. 158).

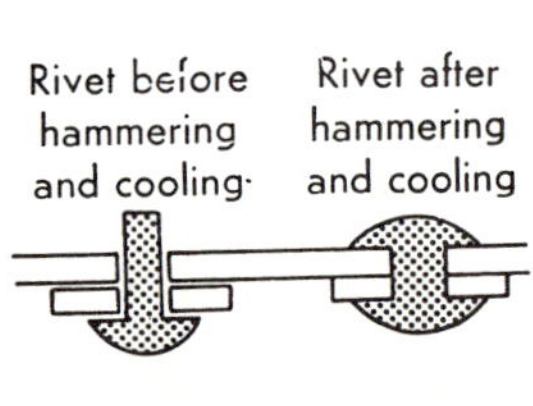

FIG. 158.

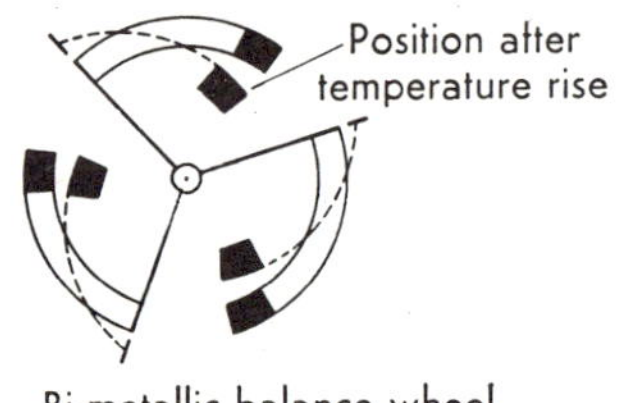

FIG. 159.

In sensitive mechanisms like the metal balance wheel of a watch or the pendulum of a clock, even small changes in length can introduce errors. Here the difference in expansion between two different metals is frequently utilised in a compensating arrangement.

If two dissimilar metal bars are closely riveted together and then heated, the metal with the greater coefficient of linear expansion will expand more, causing the originally straight bar to curve. Such a bar is called a bi-metallic bar.

The weights in the balance wheel of a watch are arranged on the circumference and the wheel, in sections, is made from a bi-metallic strip using brass and steel, for example. A rise in temperature causes the brass to expand more than the steel and the wheel curves bringing the weights inwards slightly (Fig. 159). This compensates for the normal outward expansion of the wheel spokes, and results in the weights remaining at a constant distance from the centre pivot of the wheel. The time kept by the watch is then almost independent of any temperature fluctuations.

Invar, a nickel-steel alloy, has a negligible coefficient of linear expansion and is often used in accurate machines and instruments.

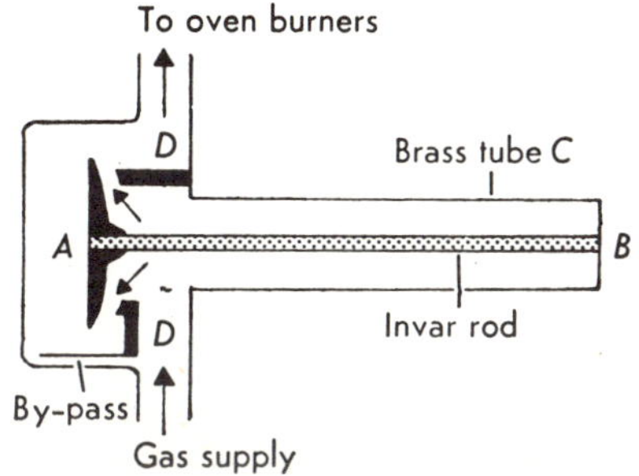

FIG. 160. *Gas oven thermostat.*

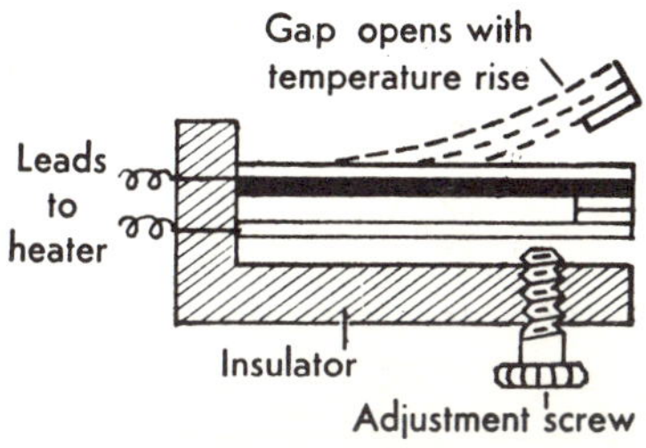

FIG. 161. *Electrical thermostat.*

In gas ovens, thermostats are used to maintain constant temperatures. They consist fundamentally of a bar of invar *AB*, fixed inside a brass tube *C* (Fig. 160). One end, *B*, of the invar is fixed to the oven, and the gas supply is fed in through the copper tube *D*, and past the valve at *A*. A rise in temperature causes negligible change in the length of the invar rod *AB*, while expansion of the brass tube reduces the gap *AD* restricting the gas supply. A fall in temperature has the reverse effect of increasing the gas supply. In practice a small by-pass enables a little gas to flow to the burners even when a high temperature has closed the main thermostat passage.

A bi-metallic strip is frequently used as a thermostat in conjunction with an electric circuit, in an electric iron for example (Fig. 161). A rise in temperature causes the strip to bend as a result of differential expansion of the two constituent metals. As a result the heating circuit is broken. On cooling the bi-metallic strip straightens again, and the heating circuit is restored.

LIQUID EXPANSION

We saw from our experiment with the flask of coloured water that a liquid normally expands on heating. In the case of liquids and gases we speak of cubical expansion or volume expansion. *The cubical expansivity* γ, is the fractional increase in volume for unit rise in temperature. Thus if we let V_0 be the volume at 0° C and V_t the volume at t° C:

$$V_t = V_o(1 + \gamma t)$$

The volume of a fixed mass of liquid increases as its temperature rises. Since density is the mass of unit volume of a substance, it follows that increasing the temperature decreases the density:

$$\text{Original density } (d_0) = \frac{\text{Mass of liquid}}{\text{Original volume}} \left(\frac{m}{V_o} \text{ at } 0° \text{ C}\right)$$

$$\text{New density } (d_t) = \frac{\text{Mass of liquid}}{\text{New volume}} \left(\frac{m}{V_t} \text{ at } t° \text{ C}\right)$$

$$\therefore d_t/d_o = V_o/V_t$$
$$= 1/(1 + \gamma t)$$

Let us now carry out an interesting experiment using Hope's apparatus (Fig. 162). The inner metal cylinder is filled with water at room temperature and surrounded by a freezing mixture of ice and salt. Readings of the two thermometers θ_1 and θ_2 are taken every minute and plotted on a graph (Fig. 163).

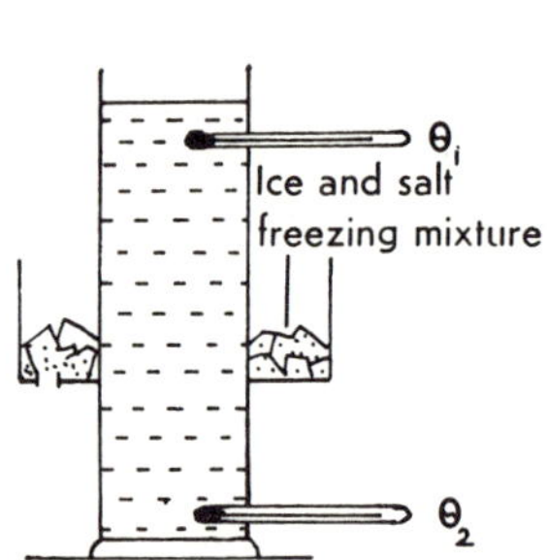

FIG. 162. *Hope's apparatus.*

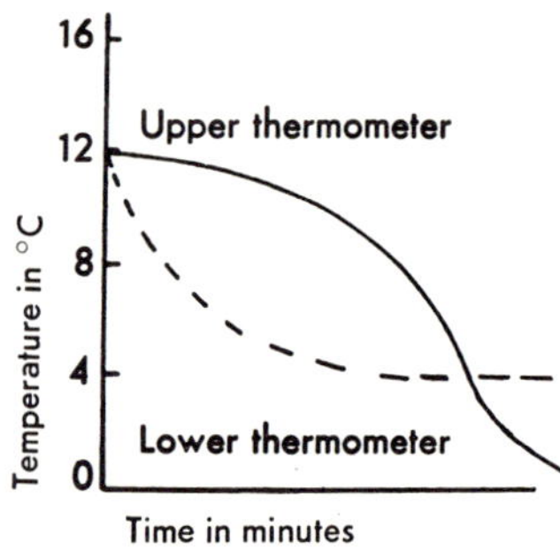

FIG. 163. *Graph for Hope's apparatus.*

What do we conclude from our results? The lower thermometer rapidly falls to 4° C. and then remains there until the upper thermometer falls to 0° C. and ice begins to form on the surface. This indicates that water at 4° C. must have a greater density than at any other temperature and, sinking to the bottom, stays there. Once all the water has been cooled down to 4° C., any further cooling causes a reduction in density and this colder less dense water then rises to the surface and ultimately freezes.

This phenomenon is of very great importance in nature, where the water at the bottom of a deep pond never cools below 4° C, even when a thick layer of ice forms on the surface.

Measuring Cubical Expansivity

Dulong and Petit first measured the real cubical expansivity of a liquid (mercury) in 1816. A simplified form of their apparatus (Fig. 164) consists of a U-tube containing the liquid under examination. One limb is surrounded by a glass steam jacket and the other limb by a water jacket. Initially when the liquid in both limbs is at the same temperature and has the same density, the levels of liquid in the tubes are the same. Steam passing through one jacket raises the temperature of liquid in column *A* to θ_1° C (approximately 100° C). Expansion occurs and causes a diminution of the liquid density to d_1. Ice cold water is passed through the other jacket and cools the liquid to θ_0° C causing a corresponding increase in density to d_0. When θ_1 and θ_0 are steady, the heights of the two liquid columns, h_1 and h_0, are carefully measured.

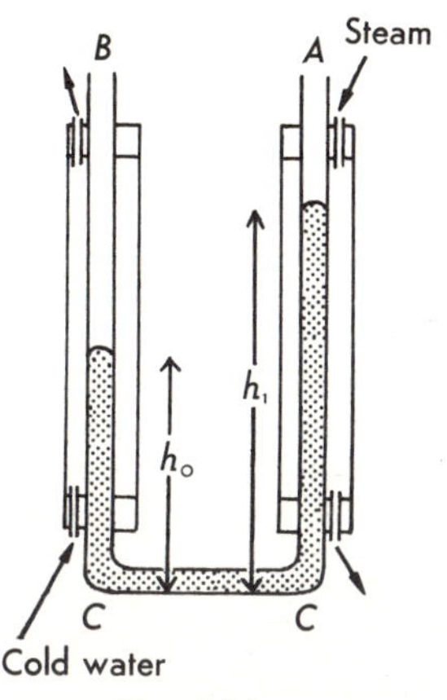

FIG. 164.

From our knowledge of hydrostatics we know that for equilibrium at the bottom of the liquid columns *CC*, the pressure due to both columns must be equal:

$$h_1 d_1 g = h_0 d_0 g$$

$$\therefore \frac{d_1}{d_0} = \frac{h_0}{h_1}$$

$$\text{But } \frac{d_1}{d_0} = \frac{1}{(1 + \gamma\theta)}$$

$$\text{Where } \theta = \text{change in temperature } (\theta_1 - \theta_0)$$

$$\therefore \frac{h_1}{h_0} = \frac{d_0}{d_1}$$

$$= 1 + \gamma\theta$$

$$\therefore \gamma\theta = \frac{h_1}{h_0} - 1$$

$$\gamma = \frac{h_1 - h_0}{h_0\theta}$$

This value of γ is the real cubical expansivity of the liquid. Modern modifications of this method make it more accurate.

If we completely fill a relative density bottle with a liquid at room temperature $\theta_1°$ and then heat the bottle to $\theta_2°$, both the bottle and the liquid contents expand. Because the liquid expands more than the glass, liquid is expelled from the bottle and weighing before and after heating determines the mass of liquid expelled from the bottle. Calculation gives reasonably accurately (for not too large temperature differences) the apparent cubical expansivity of the liquid:

$$\gamma_a = \frac{\text{Mass of liquid expelled}}{\text{Mass of liquid left} \times \text{Temperature increase}}$$

This apparent expansivity must be less than the true expansivity because the glass bottle itself expanded a little. It can be shown that:

$$\gamma = \gamma_a + \text{cubical expansivity of the glass}$$

GASEOUS EXPANSION

We have seen that liquids expand on heating. Do gases also expand? Let us fit a glass tube through a one-holed bung in the mouth of a 200 cm^3 flask. If we place the end of the tube under water and warm the flask with our hands bubbles of air will be seen to be forced out of the tube. The air in the flask expands readily and appears to have a large coefficient of expansion. A simple experiment will enable us to investigate this expansion in more detail.

Expansion at Constant Pressure

Into a clean glass capillary tube about 25 cm long and sealed at one end is introduced a trace of concentrated sulphuric acid to dry the air inside. The tube is warmed over a flame and the open end is then dipped into mercury in a dish. On cooling a thread of mercury about 1 cm long is drawn into the tube and acts as an index enclosing a fixed mass of dry air (Fig. 165). (An alternative method is to draw a thread of mercury 1 cm long into a clean dry capillary tube and then to seal one end of the tube with a cement such as Araldite.) If the tube is of uniform bore the volume of the air will be proportional to its length. This may be measured by attaching a rule to the tube by rubber bands.

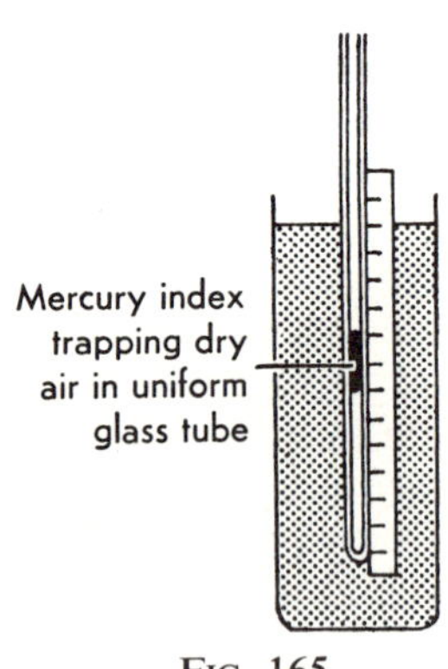

FIG. 165.

The tube is completely immersed in a water bath and warmed slowly. Periodically (at about 10 K intervals) the heating is discontinued and the water stirred to ensure that the enclosed air is at the same temperature as that of the surrounding water.

The temperature and the length of the air column are recorded and the results are plotted on a graph (Fig. 166). The straight line graph shows that the air expands regularly. If the volumes corresponding to temperatures

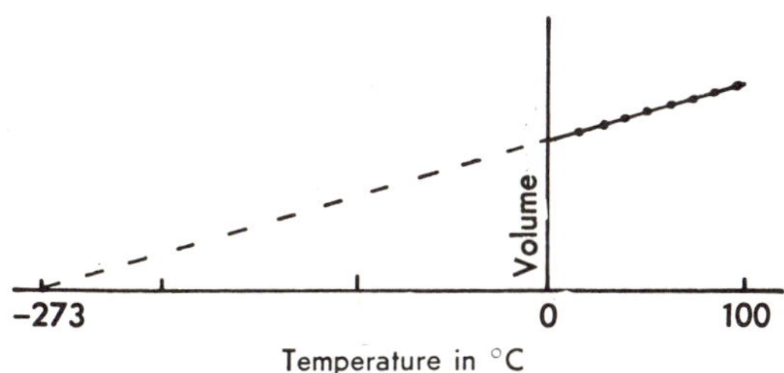

FIG. 166. *Temperature/volume graph for fixed mass of gas at constant pressure.*

of, say, 0° C and 50° C are read off then the cubical expansivity can be calculated:

$$\text{Cubical expansivity of a gas} = \frac{\text{Increase in volume of gas from } 0°\text{ C to } t°\text{ C}}{\text{Volume of gas at } 0°\text{ C} \times t}$$

It will be found to be about $\frac{1}{273}$. So if the temperature were to decrease and the volume were to decrease at the same rate it would be zero at −273° C. The graph (Fig. 166) when traced back cuts the temperature axis at about −273° C. (In practice the air would liquefy long before this temperature was reached.) This temperature −273° C is called the *absolute zero* and is believed to be the lowest temperature that can ever be attained. A temperature within 0·01° C of absolute zero has so far been achieved.

It must be noted that throughout this experiment the air in the tube was at constant pressure, namely the pressure of the atmosphere, as it was separated from the air outside by only a small mercury index.

This behaviour of dry air is typical of all gases and was first discovered by Charles in 1781. The results of many experiments can be stated in *Charles' law* (sometimes called Gay-Lussac's law).

The volume of a fixed mass of gas at constant pressure increases by $\frac{1}{273}$ *of its volume at* 0° *C for each kelvin rise in temperature.*

Increase of Pressure at Constant Volume

A mass of gas enclosed in a vessel of fixed volume cannot expand when it is heated. The heating increases the kinetic energy of the gas molecules and their increased motion increases the pressure on the sides of the vessel.

To investigate this behaviour, a glass vessel containing air and a little concentrated sulphuric acid to dry it is connected to a mercury manometer (Fig. 167).

The air bulb is totally immersed in a vessel of water that can be heated and stirred. Initially the temperature of the air in the bulb is taken to be the same as that of the surrounding water. The mercury manometer reservoir is adjusted to bring the level to a fixed mark X and the pressure head XS is then recorded. By adding this to the atmospheric pressure (obtained from a barometer) the total pressure of the air in the bulb is obtained.

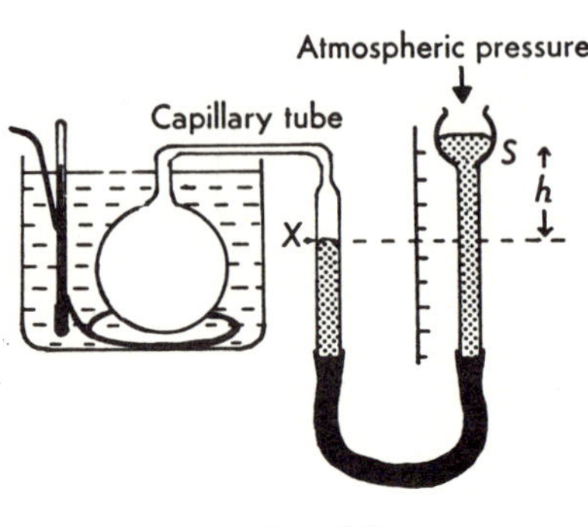

FIG. 167.

The water is heated until the temperature has risen about 10 K. Heating is then discontinued and, after stirring to ensure that the air has attained the water temperature, this temperature is recorded. As before the level of mercury is restored to X (in order to keep the volume constant) and the total air pressure is calculated.

This procedure is repeated over a range of temperatures up to 100° C and a temperature/pressure graph is drawn (Fig. 168). It is similar to the

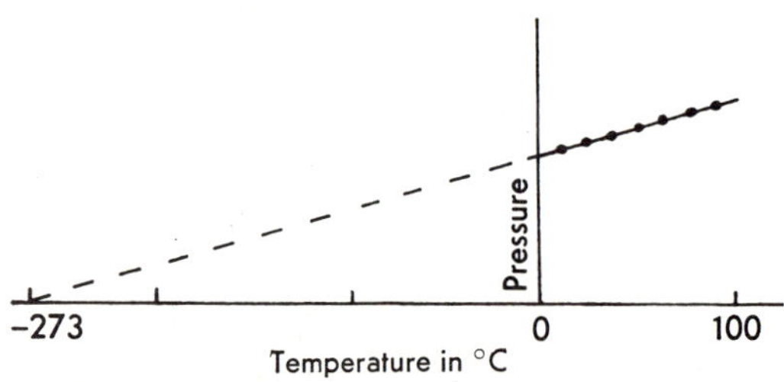

FIG. 168. *Temperature/pressure graph for fixed mass of gas at constant volume.*

graph for Charles' law and the coefficient of increase of pressure is found to be $\frac{1}{273}$, the same value as the cubical expansivity. We can say:

for a fixed mass of gas at constant volume the pressure increases by $\frac{1}{273}$ *of its value at* 0° *C for each kelvin rise of temperature.*

Absolute Temperature

To go back to Charles' law we have, if the volume at 0° C is V_0 and the volume at t° C is V_t:

$$(V_t - V_0) = \frac{t}{273} \times V_0$$

$$V_t = V_o \left(\frac{t + 273}{273}\right)$$

$$\frac{V_t}{273 + t} = \frac{V_0}{273}$$

Now, for a given mass of gas, $\frac{V_0}{273}$ is a constant:

$$\therefore\ V_t \text{ is proportional to } (t + 273)$$

This quantity, first introduced by Lord Kelvin, is called the *absolute temperature* and is expressed at T kelvin. The *absolute zero* is given by $T = 0$, that is:

$$t + 273 = 0$$

$$\therefore \text{ Absolute zero} = -273^\circ \text{ C} = 0 \text{ K}$$

as we found from the graph on previous page.

We can say therefore that *the volume of a fixed mass of gas at constant pressure is proportional to its absolute temperature.* (This is *not* Charles' law but is simply, at this stage, a definition of absolute temperature.)

It follows, from the experiment we did on measuring the pressure of a gas at different temperatures when its volume was kept constant, that *the pressure of a fixed mass of gas at constant volume is proportional to its absolute temperature.*

Gas Equation

We have already found (see Boyle's law, page 102) the connexion between the pressure and volume of a fixed mass of gas at constant temperature. If we use P, V, T, for the pressure, volume, absolute temperature of a gas we have:

(1) $PV =$ Constant (at constant temperature).

(2) $\frac{V}{T} =$ Constant (at constant pressure).

(3) $\frac{P}{T} =$ Constant (at constant volume).

Combining these we have the gas equation:

$$\frac{PV}{T} = \text{Constant (for a fixed mass of gas).}$$

When quoting a volume of a gas we must obviously state the pressure and temperature at which it was measured. It is useful to have a standard temperature and pressure (s.t.p.) and these have been chosen as 0° C (i.e. 273 K) and 760 mm of mercury (or 1 atmosphere). The gas equation is used to calculate the volume of gas at s.t.p. if its volume at another temperature and pressure is known. An example will make this clear.

A balloon containing air at 800 mm mercury pressure and at a temperature of 17° C has a volume of 580 cm³. What will be its volume at 760 mm mercury pressure and at a temperature of 0° C?

From the combined gas laws $\frac{P_0 \times V_0}{T_0} = \frac{P_1 \times V_1}{T_1}$ for constant mass.

In using this equation it does not matter what units each quantity is expressed in so long as the units are the same on both sides of the equation. Therefore we can use cm³ for volume and mm mercury for pressure. As the mass is constant:

$$\frac{800 \times 580}{273 + 17} = \frac{760 \times V_1}{273 + 0}$$

$$V_1 = \frac{273 \times 800 \times 580}{290 \times 760}$$

$$= 522$$

Therefore the volume of gas is now 522 cm³.

QUESTIONS

1. Compare the relative merits of mercury, water and alcohol as thermometric liquids in glass thermometers.

2. Describe the construction and operation of Six's maximum and minimum thermometer.

3. What is meant by the term linear expansivity? Describe how you would measure this quantity for copper.

4. Railway lines, made of steel, are 40 ft long and are laid with gaps of $\frac{1}{4}$ in between them at 0° C. If the linear expansivity of steel is 0·000 012 5 K^{-1}, at what temperature would the gaps just close? (In this calculation it does not matter what unit of length is used so long as the expansion and the original length are measured in the same one.)

5. Explain carefully how it is that a creature at the bottom of a deep pond can survive even when there is a thick layer of ice on the top of the pond.

6. State the law of Charles. Describe how you would verify it.

7. An air bubble of volume 1 cm³ is released at the bottom of a pond 3 metres deep. Calculate its volume when it has risen to the surface. Assume no change in temperature. (Atmospheric pressure = 760 mm mercury. Relative density of mercury = 13·6.)

8. In an electrolysis experiment 40 cm³ of hydrogen gas were collected at 15° C and under a pressure of 800 mm mercury. If the density of hydrogen is 0·09 kg/m³ at standard temperature and pressure calculate the mass of hydrogen collected. (Hint: calculate the volume of the hydrogen at s.t.p., then remember that 1 m³ = 1 000 000 cm³.)

CHAPTER 14

CHANGE OF STATE

HEAT causes substances to expand and sometimes to change colour, but in some circumstances heat will actually cause a change of state, from solid to liquid or from liquid to gas. If ice is heated, for example, it gradually melts and changes into water. If water is heated, it eventually boils and changes into steam.

Melting Points and Boiling Points

Let us begin heating steadily some ice in a can taken straight from a deep-freeze at, say, $-25°$ C. We shall carefully maintain a constant heating rate and record the temperature every minute. When plotted on a graph our results will be as shown in Fig. 169.

The temperature of the ice rises steadily from $-25°$ C as heat is supplied to it. Since its specific heat capacity is 2·1 kJ/kg K each joule of heat energy

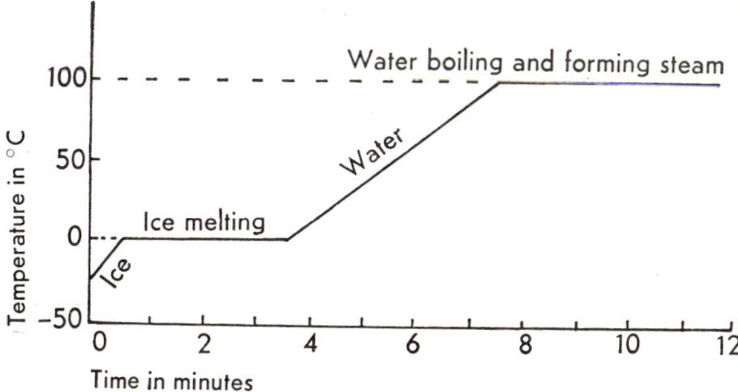

FIG. 169. *Temperature/time graph for ice-water-steam conversion.*

raises the temperature of 1 g of ice 0·48 K until all the ice reaches the temperature of 0° C, when the ice begins to melt. At this point the heat being supplied is used up in changing the ice from the solid state to water, and no further temperature rise occurs until all the ice has melted. Once all the ice has changed to water, the temperature again begins to rise steadily. Since the specific heat capacity of water is 4·2 kJ/kg K, each joule of heat energy now supplied raises the temperature of one gramme of water only 0·24 K.

The rise in temperature is therefore not so rapid; the temperature continues to rise to 100° C when it remains steady; the water boils and changes into steam. At this temperature all the heat being supplied is used in changing the state of the water from liquid to vapour.

Latent Heat and Specific Latent Heat

On examining the results it is found that considerable amounts of heat are needed to change the state from solid to liquid and from liquid to vapour. This heat is known as *latent heat*, or "hidden heat", and is used in changing the state of inter-relationship and energies of the molecules of the substance by increasing their separation and their potential energies.

Quantitatively, the specific latent heat of fusion of a substance is the amount of heat needed to change 1 kg of the substance from the solid to the liquid state without changing its temperature.

The specific latent heat of vaporization of a substance is the quantity of heat needed to change 1 kg of the substance from the liquid to the gaseous (or vapour) state without changing its temperature.

When steam condenses to water, this latent heat is released and warms the object on which condensation occurs. Latent heat is also released when water freezes to ice. Similar latent heat exchanges occur for other substances.

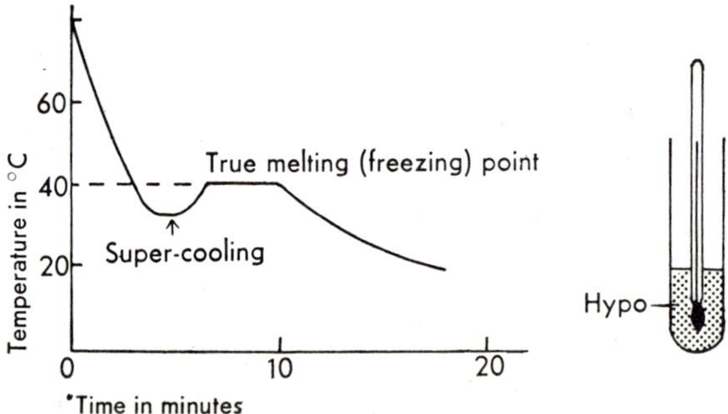

FIG. 170. *Temp./time graph for the cooling of molten sodium thiosulphate.*

An interesting illustration is found when a test-tube containing molten hypo (sodium thiosulphate) is allowed to cool. If the temperature of the hypo is recorded every minute and a graph of the results is plotted, a curve is obtained as in Fig. 170.

The temperature of the molten hypo falls quite rapidly to a temperature below its normal melting point. This is because it is losing heat to its cooler surroundings. Then quite dramatically the temperature begins to rise very quickly up to about 40° C, its normal melting point, and remains constant there until all the hypo has solidified. This sudden rise of temperature is due to latent heat being released as the hypo solidifies and is sufficient to more than compensate for the normal heat losses to the surroundings. Once all the hypo has solidified, a gradual fall of temperature is again observed as the hypo cools down towards the lower temperature of its surroundings. This phenomenon, of a substance remaining liquid below freezing point, is known as *super-cooling.*

More commonly, molten substances, such as naphthalene, begin to solidify as soon as their freezing point is reached, and the fall in temperature is merely arrested and stays constant until solidification is complete.

Effect of Impurities on Melting Points

Experiments reveal that pure substances have definite fixed melting points and boiling points and, in fact, knowledge of such temperatures is a valuable aid to identification. It is essential that the materials being examined are completely pure because it is found that impurities depress melting points and raise boiling points.

That common salt added to ice forms a freezing mixture is well-known, and temperatures down to −18° C can be obtained in such a mixture. This effect is of importance in promoting the melting of ice on roads during winter time. Salt (often mixed with sand) is sprinkled on the icy roads, which may be at a temperature of, say, −5° C. As the new melting point (or freezing point) of the ice-salt mixture is lower than this, perhaps −7° C, the ice melts since it is now above its depressed melting point. Other compounds, like calcium chloride, may also be used in conjunction with ice to form a freezing mixture. The salt forms an aqueous solution in traces of water which are present. More ice melts, diluting the solution, and drawing the latent heat required for the change of state from its surroundings. This causes cooling and a fall in temperature.

When salt is added to water it is found that the solution boils at a slightly higher temperature than normal, i.e. above 100° C.

Effect of Pressure on Melting Points

The melting point of a substance may also be changed by pressure. Substances that expand on melting, have their melting points raised by an increase in pressure.

Water expands when it freezes and contracts on melting. Thus an increased pressure causes a lowering of the melting point.

A wire supporting a piece of iron of mass 5 kg is placed over a block of ice. It is best to carry out this experiment in an outside temperature below 0° C,

FIG. 171. *Regelation largely explains the formation of glaciers from snow, and their flow down the mountain. Pressures exerted by surrounding snow and ice cause the succession of melting and freezing by which glaciers are able to flow at speeds of up to 2 metres per day.*

the melting point. Gradually the wire will be observed to cut its way down through the ice block, by melting the ice immediately underneath it where great pressure is being exerted (Fig. 172). The water so produced freezes again above the wire and after a considerable time the wire will pass right through the block, which is still left intact. This succession of melting and freezing is an example of *regelation.* The loss of potential energy of the 5 kg mass on moving down through the ice provides the heat necessary to melt the ice below the wire. Re-freezing occurs as this heat is slowly lost to the colder surroundings.

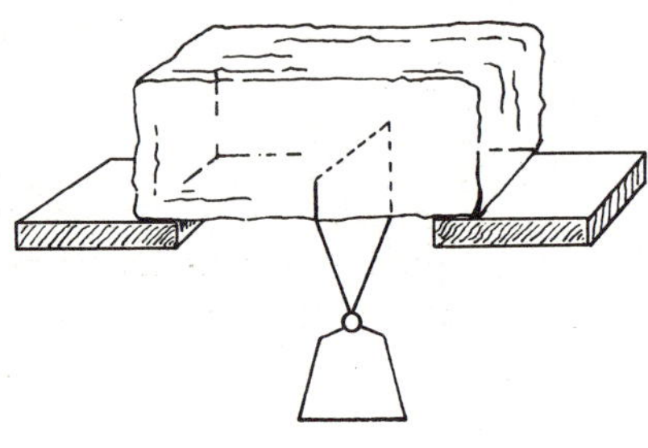

FIG. 172.

Effect of Pressure on Boiling Points

The boiling point of a liquid is greatly affected by pressure. An increase in pressure raises the boiling point, while a reduction in pressure lowers the boiling point. This may be easily demonstrated.

Partly fill a distillation flask with water and add a few pieces of broken pottery so that "bumping" is prevented when the water boils. The flask should be fitted with an extended outlet tube which dips under mercury contained in a gas jar (Fig. 173*a*). When the water is heated it eventually boils, and steam forces its way through the small amount of mercury in the jar. To do this, the pressure of the steam must equal atmospheric pressure plus the pressure due to the depth h mm of mercury: this total pressure equals $(P_a + h)$ mm mercury.

The water thus boils under increased pressure and it is found that the temperature recorded by the thermometer in the steam reads more than

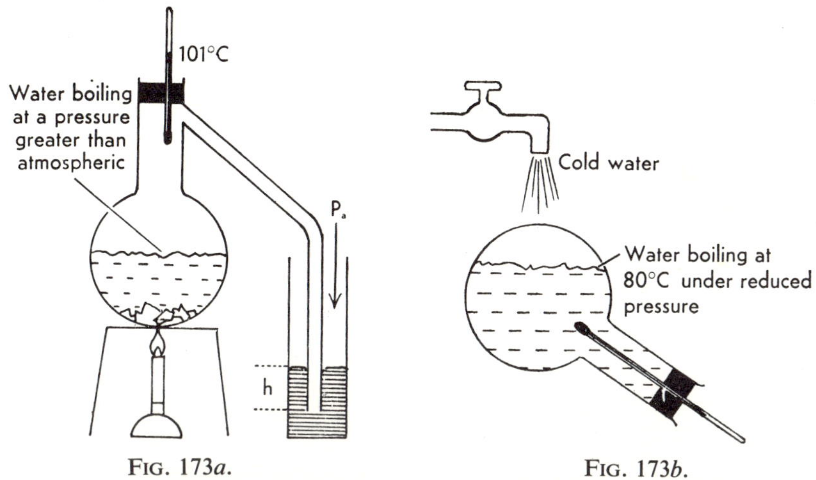

FIG. 173*a*. FIG. 173*b*.

100° C. By pouring more mercury into the jar (raising the height h), the pressure is increased and the boiling point raised several degrees above 100° C. In fact, an excess pressure of 27 mm mercury raises the boiling point 1 K.

Increase in boiling point is used in pressure cookers. The screwed lid increases the pressure above atmospheric, raises the boiling point of water, and since organic substances break down more rapidly at higher temperatures, cooking is quicker.

If a round-bottomed flask half-filled with water is heated to 100° C, the water boils normally at atmospheric pressure. Now, if the flask is quickly closed when still full of steam, with a rubber stopper and a thermometer, and placed under water from a cold tap, it will be noticed that the water continues to boil although the thermometer registers a temperature many degrees below 100° C (Fig. 173*b*). Temperatures of 60° C may be achieved with the water still boiling. The cold tap water cools the top of the flask and causes the steam to condense. This condensation causes a reduced pressure inside the flask. To release the rubber bung from the flask at the end of this experiment, one must heat the flask up to 100° C again.

At the top of a mountain where the pressure may be considerably less than normal, water boils at temperatures well below 100° C, and it is difficult to brew a good cup of tea.

Finding an Approximate Value for the Specific Latent Heat of Fusion of Ice

The specific latent heat of ice, i.e. the heat energy required to change 1 kg of ice into water at 0° C, can be determined experimentally by the following simple experiment.

Two large plastic funnels are arranged side by side in retort stands so that their spouts dip into measuring cylinders placed below. Both funnels are filled almost to the top with crushed ice at 0° C. The water formed from any ice that melts will run down into the measuring cylinders.

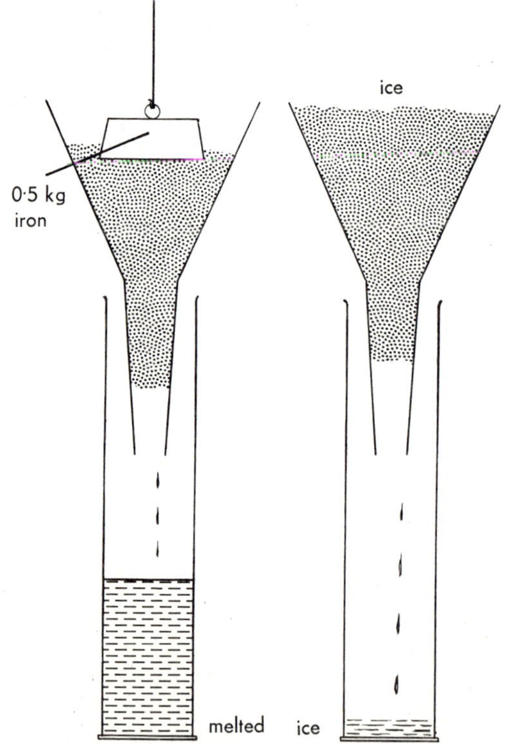

FIG. 174. *Finding the specific latent heat of fusion of ice.*

A piece of iron is suspended by a thread in boiling water until it attains a temperature of 100° C. It is then quickly transferred across and placed in one of the funnels. As the iron rapidly cools down, a good deal of the ice melts and the resulting water runs down into the cylinder and its volume is measured. However, some of the melting is due to heat going into the ice from the air surrounding the funnel. This amount is found by noting the volume of water running into the second cylinder during the same period of time as the iron takes to cool down the temperature of the ice.

The following are results from an actual experiment:

Difference of the two volumes of water collected	= 65 cm³
Mass of iron	= 0·5 kg
Specific heat capacity of iron	= 0·4 kJ/kg K

$\therefore$ Heat given out by iron in cooling = mass × fall in temperature × S.H.C.

$$= 0{\cdot}5 \times (100 - 0) \times 0{\cdot}4 \text{ kJ}$$

Mass of water formed = 0·065 kg

(Remember that in calculations all masses must be in kg: 65 cm³ of water has a mass of 65 g, i.e. a mass of 0·065 kg.)

$$\text{Heat to melt 1 kg of ice} = \frac{0{\cdot}5 \times 100 \times 0{\cdot}4}{0{\cdot}065} = 323 \text{ kJ}$$

$\therefore$ Specific latent heat of fusion of ice is 323 kJ/kg.

More accurate experiments give a value of 334 kJ/kg.

Finding the Specific Latent Heat of Vaporization of Water

The specific latent heat of vaporization of water, that is the heat needed to change each kg of water into steam at 100° C, can be found simply by boiling off water using an immersion heater.

A large, thin, lagged can of known weight, containing about 100 cm³ of water, is counterpoised on a household type balance (Fig. 175). An immersion heater is carefully suspended in the water without resting on the bottom of the can and upsetting the balance. The heater is switched on and the temperature is noted every 50 seconds until the water boils.

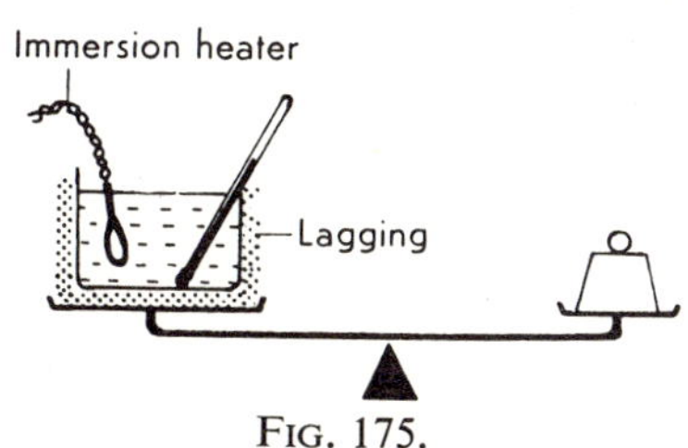

FIG. 175.

When the water is boiling freely, the masses on the balance are adjusted to counterpoise the can once more, and then a mass of 20 g is removed quickly from the right hand scale pan and a stop watch started. The time t that elapses before balance is restored is noted by the watch, and this is the time required to boil off 20 g of water. These observations may be repeated for 40 g, 60 g, etc., of water boiled off.

The following results are from an actual experiment:

Mass of can	= 150 g
Mass of water in can	= 800 g
Average rise of temperature in 50 s	= 7 K
Average t for 20 g	= 100 s
Specific heat capacity of water	= 4·2 kJ/kg K

In this simple experiment we neglect the heat taken in by the can in the first part of the experiment.

Heat taken in by the water in 50 s $= m \times s \times \theta$

(m = mass in kg, s the S.H.C. in J/kg K, and θ the rise in temperature in kelvin)

$$= 0{\cdot}8 \times 4{\cdot}2 \times 7$$
$$= 5{\cdot}6 \times 4{\cdot}2 \text{ kJ}$$

This is assumed to be the same as the heat energy given out by the immersion heater.

Heat given out by the immersion heater in 100 s $= \dfrac{100}{50} \times 5{\cdot}6 \times 4{\cdot}2$ kJ

This is used to boil away 20 g of water.

Heat to boil away 1 kg of water $= 2 \times 5{\cdot}6 \times 4{\cdot}2 \times \dfrac{1000}{20} = 2352$ kJ

∴ The specific latent heat of vaporization of steam is 2352 kJ/kg.

More accurate experiments give a value of 2260 kJ/kg.

Frequently in laboratories the specific latent heat of vaporization of water is determined by a method of mixtures. In this method steam from a steam generator passes through a water trap and is directed on to the surface of a weighed quantity of cold water or known temperature in a lagged calorimeter (the heat capacity of the calorimeter must be known). After a suitable rise of temperature of the water has been noted the steam is turned off and the calorimeter and contents weighed again to find the weight of steam that has been condensed to water. The heat lost by the steam in condensing and cooling is equated to the heat gained by the calorimeter and cold water. This enables the specific latent heat of steam to be calculated.

Cooling by Evaporation

Liquids, like water, when normally left for some time exposed to the surrounding air in an open vessel, evaporate at room temperature. The liquid gradually changes state into the vapour form and is carried away by air currents. Heat energy for the process is normally supplied by the surroundings. A current of air blown across the surface of the liquid will encourage evaporation by removing the vapour as soon as it is formed. This will cause cooling since latent heat needed for vaporization will have to be drawn from the liquid itself. If carbon tetrachloride (cleaning fluid) is spilled on one's hand it vaporizes quickly, drawing heat from the hand and causing a sensation of cold. Blowing across it intensifies the cooling. This principle of cooling by evaporation is used in butter coolers and in refrigerators.

Water may be cooled to freezing point by a similar method. A glass beaker, or metal can, containing a little ether is placed on a large drop of water on a wooden block. Air from a pump is bubbled through the ether (Fig. 176). This causes rapid evaporation of the ether and consequent cooling. The required latent heat of vaporization is taken from the ether itself, and by conduction, from the beaker and the water below it. The water rapidly freezes. Wood is a bad conductor of heat and does not allow heat to flow through it to the beaker.

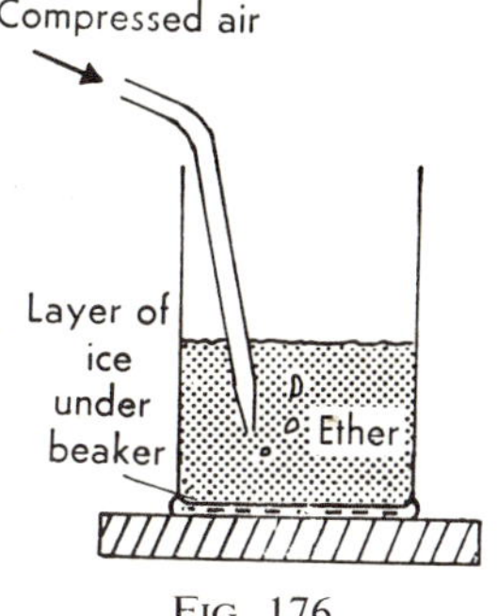

FIG. 176.

QUESTIONS

1. What is meant by latent heat of vaporization and latent heat of fusion? Trace the changes caused by heating a kettle of cold water to boiling.

2. What effects does pressure have on melting points and boiling points? Describe experiments to illustrate these effects for water.

3. Explain carefully why a mixture of salt and sand is often spread on icy roads in winter-time.

4. Describe how you determine practically the specific latent heat of fusion of ice. Point out any precautions to take to ensure an accurate result.

5. Calculate how much heat is needed to change 0·2 kg of ice at 0° C into water at 40° C (specific latent heat of fusion of ice = 334 kJ/kg).

6. An aluminium kettle of mass 300 g contains 940 g of water at 20° C. It is heated by an immersion heater that supplies energy at the rate of 800 W. Assuming no heat energy is lost to the surroundings, and no evaporation until the boiling point is reached, calculate:

(*a*) how long it will take for the water to reach boiling point;

(*b*) how long it will take to boil away 50 g of water.

(Specific heat capacity of water = 4·2 kJ/kg K; specific heat capacity of aluminium = 0·84 kJ/kg K; specific latent heat of vaporization of water = 2260 kJ/kg.)

7. Describe carefully what happens when a test tube containing hot molten naphthalene is allowed to cool rapidly in air.

CHAPTER 15

HEAT ENERGY AND MECHANICAL ENERGY

WE HAVE already discussed the wasted energy due to frictional heat losses when working a machine or sawing a piece of wood:

Work put into a machine = Output of useful work + Heat losses

Some mechanical energy is transformed into heat energy which is wasted, resulting in loss of efficiency. This idea was not clearly understood until the middle of the nineteenth century.

The Nature of Heat

Before the nineteenth century it was thought that heat was a mysterious fluid called "caloric," which could enter or leave a body and change its temperature or state. This fluid had to be weightless and invisible for no change of weight was detectable on heating a lump of metal, for example, and no visible liquid exuded from a hot metal bar as it cooled. It is interesting to note that a century earlier the "phlogiston theory" suggested that an invisible fluid was responsible for the oxidation of materials. When a substance burned in air it was thought it lost phlogiston. The fact that the substance increased in weight as a result of burning was explained by suggesting that phlogiston had negative weight!

Towards the end of the eighteenth century doubt began to fall on the "caloric theory." Count Rumford observed that when cannons were bored, the metallic chips the cannon and the borer all became hot, despite the fact that there was no apparent source of heat. In a later experiment he used a blunt borer driven by horses, surrounded the cannon with a box and filled the box with $2\frac{1}{2}$ gallons of water. After boring the cannon for 140 minutes, the water was seen to boil. Where had all this heat come from? Many explanations were given but Rumford was convinced that the work done by the horses had somehow been converted into heat.

Confirmation of the idea that work could be converted into heat was soon to follow from Sir Humphry Davy, who had noticed that if two pieces of ice were rubbed together with sufficient vigour, enough heat was produced to melt the ice; "the immediate cause of heat is motion," he wrote in 1812.

The concept that heat is a form of energy interchangeable with other forms of energy was finally becoming clear, and it followed that as, at that time,

FIG. 177. *Polishing a steel wheel gives rise to much heat and a shower of sparks. Mechanical energy of the rotating buffing wheel is being converted into heat energy: the possibility of such a conversion was not realized until 1812.*

heat and mechanical energy were measured in different units there must have been a relationship between these units. In the early 1840s Prescott Joule, a wealthy Manchester brewer with a private laboratory, found this relationship by many different experiments. He studied the heat produced in driving water through fine bore tubes and the mechanical work needed to do it. He investigated the rise in temperature of the water falling from the top to the bottom of a Swiss waterfall where the gravitational potential energy is changed into heat. (He is supposed to have carried out this experiment on his honeymoon.) He stirred water by a paddle wheel driven by falling weights (see below). In all these experiments he found that the relationship was constant, that is, when equal amounts of mechanical energy, however produced, disappeared, the same amount of heat always appeared.

First Law of Thermodynamics

This fact led to the formulation of the first law of thermodynamics: *when any kind of energy is transformed into heat energy or heat energy is transformed into another kind of energy, then the total amount of energy remains constant*. In other words, *energy can be neither created nor destroyed.* This is the principle of the conservation of energy. It is the most fundamental principle in physics and indeed in all science.

Joule's Experiment

Joule's most celebrated experiment is illustrated in Fig. 178.

The masses *W*, *W* could be raised by turning the handle *H* with the pin *P* out, so disconnecting the handle from the paddle shaft. The pin *P* was then made to connect the weight drum and paddles so that, when released, the weights fell through a known height *h* and turned the paddles.

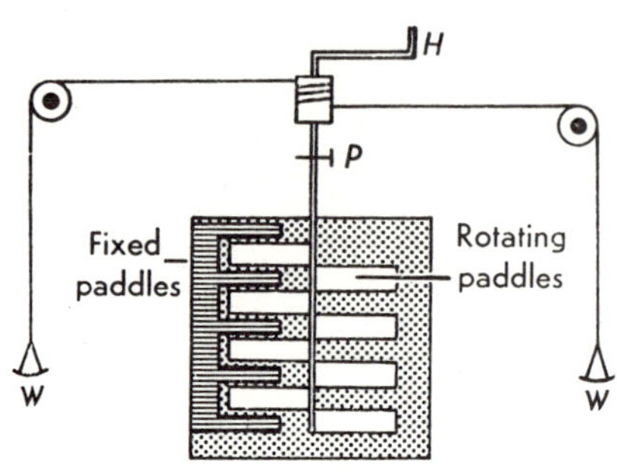

FIG. 178

The friction of the rotating paddles heated the water and caused a slight rise in temperature. This process was repeated a number of times, and the final temperature rise of the water was noted. The masses *W* had lost potential energy in falling and this energy was calculated, using the units of mechanical energy then in vogue. The heat produced was found in the heat units then being used, by knowing the mass of the water and the heat capacity of all the apparatus and measuring the rise in temperature. Joule found that however many times he did the experiment, using different amounts of energy (and even using mercury instead of water), the ratio of mechanical energy to heat produced was always the same. Since Joule's day we have come to realize that there is no fundamental difference between mechanical energy and heat, heat being the kinetic energy of the small particles that make up matter, so we measure them both in the same unit—the joule.

Finding the Specific Heat Capacity of Copper, Heat being Produced Mechanically

A simple laboratory experiment to determine the specific heat capacity of copper uses a solid copper cylinder of known weight attached to a non-conducting plastic axle (Fig. 179). Round this cylinder, which can be rotated by a handle, is wrapped a thin copper belt supporting a 5 kg mass as shown in the sketch. The copper cylinder is polished so that friction is not great and the copper belt slips over the cylinder as it is turned by the handle. The handle is turned a known number

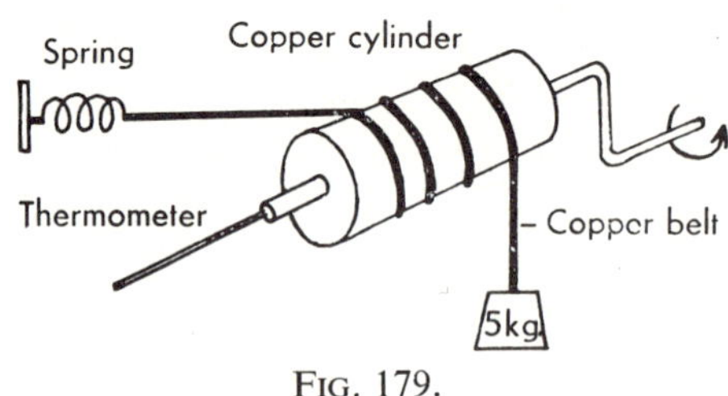

FIG. 179.

of times (about 100) at a rate just sufficient to support the 5 kg mass, so that the spring is not stretched or under tension.

The frictional resistance is the force of gravity on the 5 kg mass and the work done is this force × distance moved i.e. the force × circumference of the drum × number of revolutions. The mechanical work is transformed into heat, this heats up the copper and its rise in temperature is noted.

In an experiment these results were obtained:

Mass on end of belt	= 5 kg
Diameter of copper cylinder	= 4 cm
Number of revolutions	= 100
Mass of copper cylinder	= 460 g
Mass of copper belt	= 20 g
Rise in temperature	= 3 K
Gravitational constant	= 9·8 N/kg

Let the specific heat capacity of copper be s.

Heat taken in by cylinder and belt (both are heated through the same rise in temperature): $= (0{\cdot}46 + 0{\cdot}02) \times s \times 3$ J

Mechanical energy used in rotation $= \dfrac{5 \times 9{\cdot}8 \times (4 \times \pi) \times 100}{100}$

Assuming that all this turns to heat in the copper:

$$0{\cdot}48 \times s \times 3 = 5 \times 9{\cdot}8 \times 4 \times \pi$$

$$s = 428 \text{ J/kg K}$$

$\therefore$ Specific heat capacity of copper = 430 J/kg K

A commonly described experiment for finding the specific heat capacity of lead is to allow lead shot to fall down a cardboard tube repeatedly and to record the rise in temperature of the lead shot resulting from the heat produced. This experiment gives a fair result if the loss of kinetic energy (resulting from the rotation of the tube) is added to the loss of gravitational potential energy when the cardboard tube is rapidly inverted.

Finding the Specific Heat Capacity of a Liquid, Heat being Produced Electrically

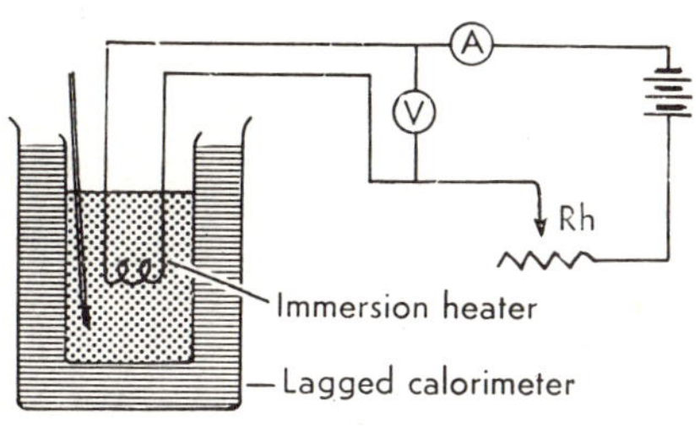

FIG. 180.

In Fig. 180 a measured current is passed through an immersion heating coil in a calorimeter containing a known mass of the cold liquid. The electrical energy used in a given time is readily calculated from the current, the P.D. and the time as detailed below. The heat produced is calculated from the rise in temperature and the masses and heat capacities of the calorimeter and the water it contains.

Mass of liquid $= M$
Mass of calorimeter $= m$
Specific heat capacity of liquid $= S$
Specific heat capacity of material of calorimeter $= s$
Rise in temperature $= \theta$
Time current flows $= t$
Current $= I$
Potential difference $= V$
Heat taken by calorimeter $= m \times s \times \theta$
Heat taken in by liquid $= M \times S \times \theta$
Energy given out by heater $= V \times I \times t$

$$MS\theta + ms\theta = VIt$$

$$S = \frac{VIt - ms\theta}{M\theta}$$

M and m are measured in kilogrammes, t in seconds, V in volts, I in amperes.

The rheostat (Rh) is included in the circuit to regulate and maintain a constant current flowing.

Second Law of Thermodynamics

In these experiments all the mechanical work or electrical energy is converted into heat. This is very often possible, but it is impossible to achieve complete conversion of heat energy into mechanical work partly due to frictional losses, but mainly because a fraction of the heat supplied is ejected from an exhaust. Man is, however, very interested in harnessing heat energy by converting it into mechanical energy to drive machines, so he is also interested in reducing heat losses and increasing efficiency.

The principle of converting heat energy into mechanical energy is summed up in the second law of thermodynamics.

Maximum mechanical energy obtainable from heat energy is independent of the machine used to carry out the conversion but is dependent on the temperatures of the bodies between which the transfer of heat is made.

Heat energy is converted into mechanical energy by heat engines. These absorb a quantity of heat from a source at a high temperature, convert some of this heat into mechanical energy and then gets rid of the rest of the heat to a sink at a low temperature. The fraction of the heat absorbed that is converted into mechanical energy depends on the difference of temperatures between the source and the sink.

FIG. 181. *Newcomen's engine of* 1717 *was one of the first heat engines to find a practical application. The caption to this contemporary drawing reads,* "*The engine for raising water* (*with a power made*) *by fire.*" *Remember that the concept of inter-converting heat energy and mechanical energy and mechanical energy was not to be realized for another* 100 *years. This engine is discussed on page* 214.

HEAT ENGINES

Most heat engines, such as steam engines and internal combustion engines, use heat energy to raise the pressure of a vapour or gas. In a steam engine, heat from burning coal or oil causes water to boil changing it to the vapour form. Since this takes place in an enclosed boiler, the steam pressure becomes much higher than normal atmospheric pressure and can be used to push a piston along a cylinder. Connected to the piston by a rod and crank arrangement is the driving wheel which is turned by the movement of the piston and performs useful work.

Steam Engine

The typical modern slide valve type of steam engine (Fig. 182) uses steam pressure to push the piston backwards and forwards, the inlet and outlet of steam being controlled through the ports *p*, *q* and *r* by a slide valve.

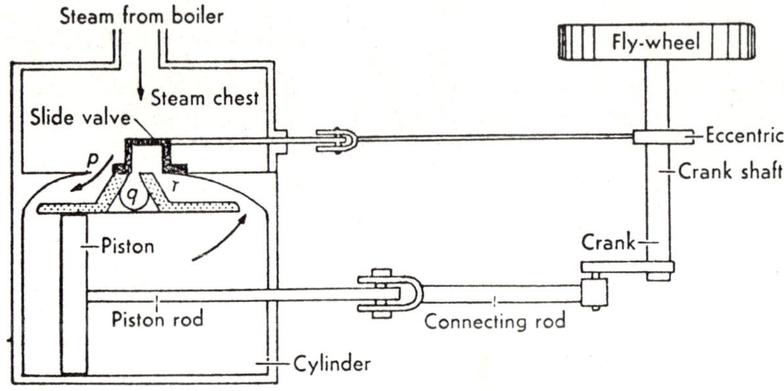

FIG. 182. *Slide valve type of steam engine.*

With the slide valve in the position shown, steam enters the cylinder through *p* forcing the piston to the right. As this happens, the slide valve moves to the left, opening the exhaust port *q* through which the steam leaves the cylinder, and then opening *r* through which fresh steam enters the cylinder, *p* and *q* having by this time closed. Fresh steam having entered through *r*, the piston is now forced to the left.

The piston rod sliding backwards and forwards, being connected to the crank shaft, causes the crank shaft to revolve. The crank shaft is connected to a heavy flywheel which is used to drive the engine during the two short periods when the piston can exert no turning power (when the connecting rod and crank are in a straight line).

The slide valve is connected to an eccentric attached to the crank shaft. This enables the slide valve to move to the left when the piston moves to the right.

Petrol Engine

In the petrol engine of a car (Fig. 183), an explosive mixture of petrol vapour and air is drawn by suction into each of the cylinders in turn, and then compressed by the upward movement of the piston. This upward motion and compression is helped by a heavy flywheel connected to the piston by a crank shaft. When the piston is near the top of its stroke, a spark from the sparking plug ignites the gases releasing a large quantity of heat. This heat

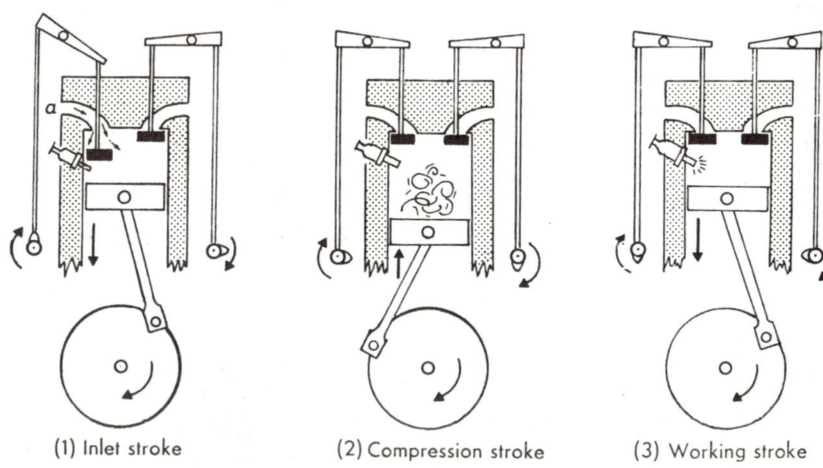

FIG. 183. *The four strokes of a petrol engine.*

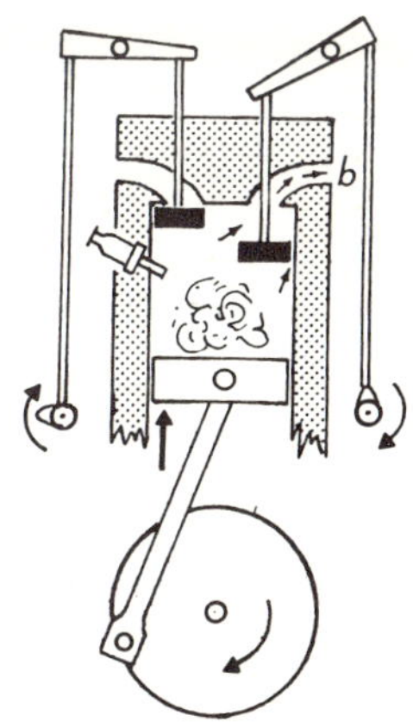

raises the pressure of the gases in the cylinder, forcing down the piston, turning the crank shaft and driving the car. At the same time energy is stored in the flywheel during this working stroke to drive the piston through the three non-working strokes which are to follow.

1. *Inlet stroke.* The piston moves down the cylinder, the intake valve *a* is opened by a cam and a mixture of gas and air is admitted.

2. *Compression stroke.* The inlet valve closes, the piston moves up the cylinder and compresses the gas mixture. At the top of this stroke, a spark passes between the points of the sparking plug and ignites the mixture.

3. *Working stroke.* The heat released expands the gases in the cylinder and drives the piston downwards again.

4. *Exhaust stroke.* A cam opens the exhaust valve *b*; the momentum of the flywheel drives the piston upwards again and the exhaust gases are driven from the cylinder.

Diesel Engine

In a diesel engine the upward motion of the piston compresses the air in the cylinders and thus raises the temperature (this is the same compression heating effect observed when a bicycle tyre is pumped up). At the top of the stroke a fine spray of fuel is injected into the hot air, in which it burns, giving out heat energy. This raises the pressure, causes the air and gases to expand, forces down the piston, and propels the motor.

In each case that we have discussed, heat energy from the burning fuel is changed into useful mechanical energy. A lot of energy is wasted however. A normal steam engine is only about 7% efficient, and a motor car is perhaps 15% efficient.

High-pressure steam, produced by burning fuels and heating water, can be made to do work in a turbine. Here the steam impinges on thousands of blades attached to a common axle, and its pressure and motion (kinetic energy) are used in turning the turbine and perhaps an electric generator connected to it.

Refrigerator

The refrigerator is a different kind of heat engine and really pumps heat from one place to another: it is a heat pump. Compression-type refrigerators

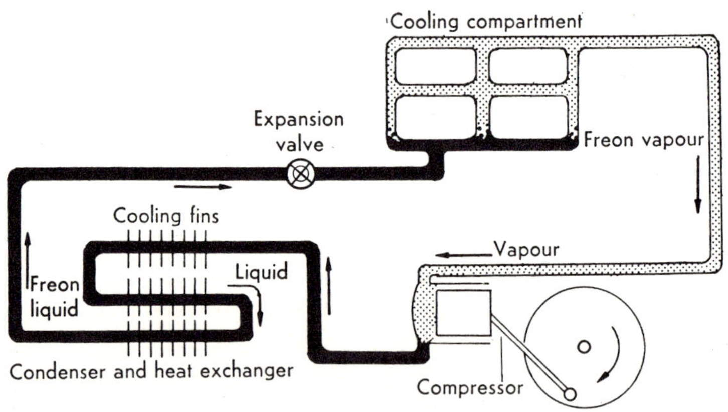

FIG. 184. *Compression-type refrigerator driven by electricity.*

are electrically driven and use electrical energy to cause the initial compression of Freon (a vapour at room temperature) by means of a cylinder and piston arrangement (Fig. 184).

The compression causes the Freon vapour to condense into a liquid which circulates to the freezing compartment. Here, between the expansion valve and the pump, there is a greatly reduced pressure and Freon changes back to a vapour again, expands and becomes cold. Heat is drawn from the freezing compartment to provide the necessary latent heat energy for the Freon to vaporize.

The compression of the Freon causes local heating at the back or top of the refrigerator and this heat must be carried away by cooling fins, draughts and convectional air currents.

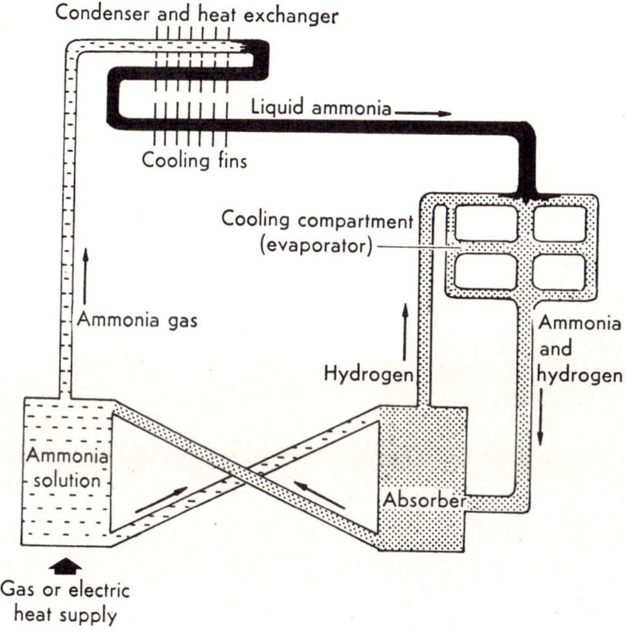

FIG. 185. *Absorption-type refrigerator driven by gas or electricity.*

In absorption-type refrigerators (Fig. 185), a small heater is used to boil ammonia gas out of solution. The gas develops a high pressure and is liquefied by cooling in a condenser. The liquid ammonia passes to the freezing compartment where it is allowed to evaporate rapidly, so cooling the compartment. This evaporation of the ammonia is assisted by feeding hydrogen into the evaporator. The ammonia gas dissolves in water in the absorber and returns to the boiler.

Jets and Rockets

In jet aircraft fuel is burnt to produce heat energy. This energy is used to impart a high velocity and hence momentum to the exhaust gases which stream out at the rear of the plane and drive the plane forward.

FIG. 186. *Launching of an American rocket into space. The gas stream, formed by burning fuel, issues with such enormous velocity that it is able to drive the rocket upwards.*

Rockets also use heat energy from burning fuels to produce the required mechanical energy to drive them forwards. They contain their own fuel and oxygen supplies, unlike a jet engine which needs a supply of air from outside for combustion. The heat energy released from burning the fuel (alcohol or kerosene) in oxygen (from liquid oxygen) makes the exhaust gases expand and issue backwards from the rocket at very high speeds. This backwards momentum of the gases imparts an equal and opposite forward momentum to the rocket by a reaction principle. (Blow up a balloon and then release it so that the air can rush out of the neck. The balloon will fly off in the opposite direction by the same reaction principle.)

QUESTIONS

1. Explain how you find the specific heat capacity of iron, using the work done against friction as source of energy.

2. In falling from the top to the bottom of a waterfall the water is found to rise in temperature by 0·5 K. Explain why this is so, and calculate the height of the waterfall, assuming the water at the bottom has lost all its kinetic energy. (Assume that acceleration due to gravity = 10 m/s^2 and specific heat capacity of water = 4 kJ/kg K. This is a very approximate calculation, so approximate values are sufficient.)

3. How would you attempt to find the specific heat capacity of aluminium using electrical energy as your source of heat?

4. What principle is expressed by the Second Law of Thermodynamics?

5. Describe the operation of a four-stroke petrol engine, and trace the conversion of heat energy into mechanical energy.

6. Write a brief description of the operation of an electrical refrigerator, and trace the energy changes occurring in the system.

7. A model steam engine can lift a 5-kilogramme mass through a vertical height of 86 cm in 60 seconds. Calculate the working power of the engine in watts. If the engine burns 0·25 grammes of methylated spirits in 60 seconds what fraction of the total heat energy available is used mechanically? (Specific calorific value of methylated spirits = 32 000 kJ/kg.)

CHAPTER 16

KINETIC THEORY

THE statements made by Sir Humphry Davy that "the immediate source of heat is motion" and that "heat is a form of energy" lead one to enquire where and how energy is stored when a substance is heated.

In the chapter on the molecular theory of matter (page 51) we considered gas as being made up from minute particles called molecules which were in a state of perpetual and random motion. Let us now recall what we learnt in that chapter and relate it to what we have learnt about kinetic energy.

Kinetic Theory of Gases

The moving molecules have kinetic energy, equal in each case to $\frac{1}{2}mv^2$, where m is the mass of the molecule and v is its velocity. The fact that the molecules do not slow down in time and finally come to rest, signifies that they are perfectly elastic and can collide with each other or the walls of their containing vessel without loss of energy or velocity. Here then is the store house of heat energy—in the molecules themselves. The motion of the molecules is quite random, but at any defined temperature the average kinetic energy of the molecules is constant. A rise in temperature, caused by supplying more heat energy, increases the velocity of the molecules, and their kinetic energy. Cooling has the reverse effect, and if the cooling continues to absolute zero (−273° C, 0 K), the kinetic energy of the molecules reduces to zero. It may be deduced that *the mean kinetic energy of the molecules varies directly with the absolute temperature.*

The explanation of the behaviour of gases in terms of molecular motion is known as the *Kinetic Theory of Gases.* The bombardment of the walls of a vessel containing a gas by the moving molecules is responsible for gas pressure. Similarly, a ball rebounding from a wall exerts a pressure on the wall, and this pressure depends on the rate of change of momentum of the ball. The change of momentum of a gas molecule on rebounding from the wall of the containing vessel depends on the molecule's speed, and hence also on the absolute temperature, since the speed depends on the absolute temperature. It follows that the pressure of a gas also depends on the absolute temperature.

This inter-relationship between the pressure, volume and absolute temperature of a fixed mass of gas is expressed in the general gas law:

$$\frac{PV}{T} = \text{constant}$$

This is a combination of both Boyle's Law and Charles' Law (page 175).

The increased energy of molecules on heating solids, liquids and gases, leads to expansion. The more rapidly moving molecules occupy more space, and being more widely separated, possess greater potential energy.

Surface Tension

Molecules at the surface of liquids occupy a rather special position, for it is assumed that molecules exert a force of attraction on one another and here all the attraction must pull back towards the body of the liquid. In the middle of the liquid other molecules are attracted equally in all directions by surrounding molecules. It is thus not easy for a molecule to break away from a liquid surface into the air above it. To do so it requires a considerable velocity and kinetic energy. This assumption of molecular attraction explains the phenomena of surface tension, and also why the surface tension of liquids decreases as their temperature rises: greater molecular velocity and expansion reduces the inter-attraction.

Surface molecules must have a potential energy with reference to the molecules within the main body of a liquid, since, for a molecule to reach the surface, work must be done against the resultant inward attraction of the rest of the molecules. This idea of energy held in the surface of a liquid helps to explain surface tension effects. For example, liquids tend to reduce their surface area and hence reduce their surface potential energy. They tend to form spherical drops with the least surface area consistent with a certain volume. There is a tensile stress in the surface of liquids. This tension in the surface of a liquid can be illustrated easily by the two following simple demonstrations.

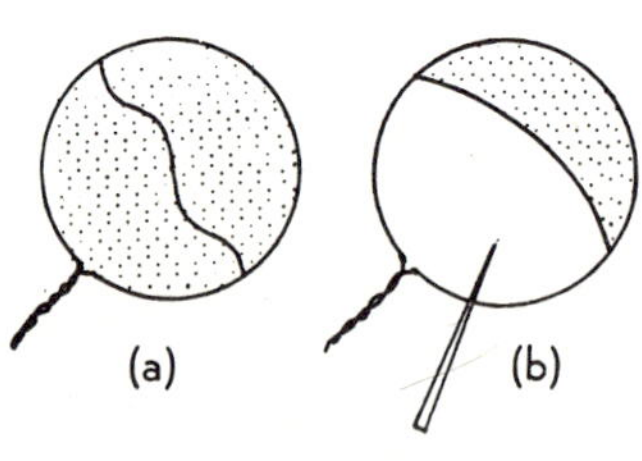

FIG. 187.

A wire loop with a thread tied loosely across its diameter is dipped into some strong detergent solution, and a film forms across the wire loop. The thread hangs loosely in the film (Fig. 187*a*), but if one part of the film is punctured by means of a hot needle the thread is seen to be pulled tight by the tension of the film remaining (Fig. 187*b*).

A clean glass funnel is dipped into a saucer of strong detergent and a film forms across the mouth of the funnel. It will be noticed that this film slowly contracts and climbs up the funnel, reducing its surface area as it climbs.

Evaporation and Vapour Pressure

As the temperature of a liquid rises, the reduced inter-attraction and increased kinetic energy of the molecules enables more and more to escape from the liquid into the air space above. Under normal room temperatures many liquids readily lose molecules in this way, and this is called evaporation. The rate of evaporation increases with rise of temperature and heating. Evaporation is also encouraged if the escaping vapour molecules are carried away (by air currents) as soon as they break free from the liquid surface.

Above a liquid surface, air molecules and vapour molecules move about rapidly at random and bombard the liquid surface. This continuous bombardment provides the atmospheric pressure acting on the surface. The actual *partial pressure* exerted by the vapour molecules themselves is known as the *vapour pressure.* A state of dynamic equilibrium may exist when just as many liquid molecules leave the surface as vapour molecules return. In this condition we say the space above the liquid is saturated. If more molecules are leaving the liquid than are returning from the vapour above, we say the space is unsaturated.

As the temperature rises more and more molecules change from the liquid to the vapour state. As this change takes place, the molecules move farther apart, becoming more widely separated from one another. Since molecules mutually attract one another, work must be done and energy used up to effect this change of state. The potential energy of the molecules has been increased even though, if their temperature remains constant, their average kinetic energy is unchanged. It is only the more energetic molecules (with above average kinetic energy) which escape; thus extra heat energy (latent heat) must be supplied to maintain temperature high enough to enable the evaporation to continue. The escaped vapour molecules exert greater and greater vapour pressure as the temperature increases, a fact that can be easily demonstrated.

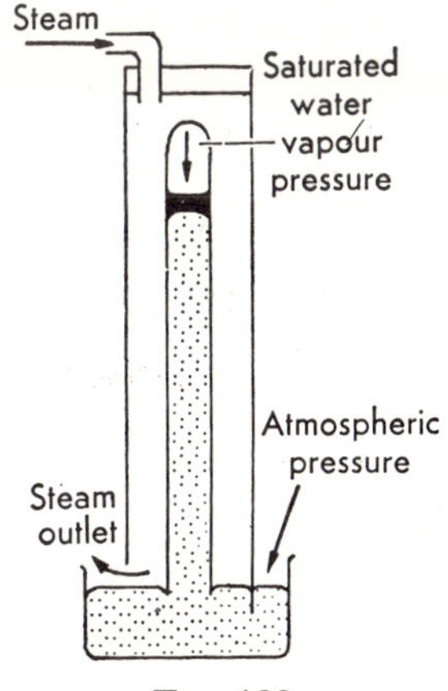

FIG. 188.

A simple mercury barometer is set up inside a glass tube (Fig. 188). A few cm³ of water are introduced into the barometer at the bottom and being less dense immediately rises to the top of the mercury. Here evaporation occurs and the space becomes saturated with water vapour. (There must still be some liquid water on top of the mercury if saturation is to be maintained.) It will be noticed that the mercury level falls slightly due to the partial pressure of the water vapour.

Now steam is passed into the outer surrounding glass tube and as the temperature rises, more water evaporates from the top of the mercury, and the water vapour pressure increases, forcing down the mercury level in the barometer tube. The fall in mercury level indicates that the saturated vapour pressure is increasing. It continues to increase until at 100° C. the mercury level has fallen right down to the level of the mercury in the reservoir. The water vapour pressure must then be equal to the atmospheric pressure. This in fact provides us with *a new concept of boiling point*, for a liquid boils when its saturated vapour pressure equals the external pressure.

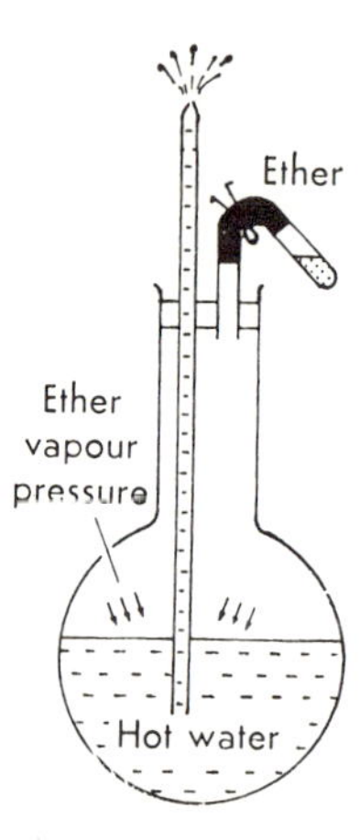

Fig. 189.

Different liquids exert different vapour pressures at the same temperature. If a few drops of ether, for example, are introduced into the barometer in the last experiment (instead of water) at room temperature, a great fall in the level of the mercury will occur, indicating that ether has a large saturated vapour pressure even at room temperature.

A round-bottomed flask is half filled with warm water and corked with a two-holed rubber bung (Fig. 189). Through the holes pass a tall glass tube drawn out to a jet at the top and reaching nearly down to the bottom of the flask, and a short glass tube connected by rubber tubing to a small sealed tube. Into the sealed tube is placed a few cm^3 of ether and the rubber tube is closed with a spring clip. On opening the clip and allowing the ether to run down into the flask, rapid evaporation of the ether takes place and builds up a large vapour pressure, for the temperature of the water in the flask may be near to the ether's boiling point. This pressure, acting on the surface of the water, forces water up out of the jet in a fountain.

QUESTIONS

1. Describe an experiment to illustrate surface tension, and explain in terms of the molecular theory why liquids exhibit surface tension.

2. What is meant by vapour pressure?

Describe an experiment to investigate how the vapour pressure of a liquid varies with temperature.

3. Explain carefully in terms of the molecular theory how the boiling point of a liquid depends on pressure and temperature.

4. A vessel of fixed volume contains 1 gramme of oxygen gas at 0° C. When another gramme of oxygen is introduced into the vessel, still at the same temperature, the pressure of gas in the vessel is doubled. On raising the temperature the pressure is increased further. Explain these facts in terms of the kinetic theory of gases.

CHAPTER 17

HEAT TRANSFER

HAVE you ever held a poker in a coal fire and watched the smoke go up the chimney? Your face becomes flushed with the heat of the fire. How does this happen? How does the end of the poker in your hand get hotter and hotter? Why does the smoke go up the chimney? All these problems involve the transfer of heat energy from the fire, which is at a high temperature, to another place at a lower temperature. Heat normally travels from a region of high temperature to a region of lower temperature. This is implied by the second law of thermodynamics. But how does the heat energy travel?

CONDUCTION

An experiment will help to explain what does happen in the case of a metal rod. Several rods of different metals are fixed into the side of a metal box containing water which can be heated to boiling point by an immersion heater (Fig. 190). The rods are coated with wax (by dipping them into a tall jar of hot water on which floats a layer of molten wax). Small ball bearings are then attached to the wax on the rods.

As the temperature of the water in the container is raised it will be noticed that wax begins to melt along the four rods. As the wax melts the ball bearings fall off. Heat energy is passing along the rods. The inner end of each rod is at a high temperature and the remote end at a lower temperature. We say a *temperature gradient* exists along the rods. Atoms in the rods at the hot end will have greater kinetic energy than those at the cold ends. This energy will be transferred along the bar by mutual collisions. This process of heat transfer is known as conduction.

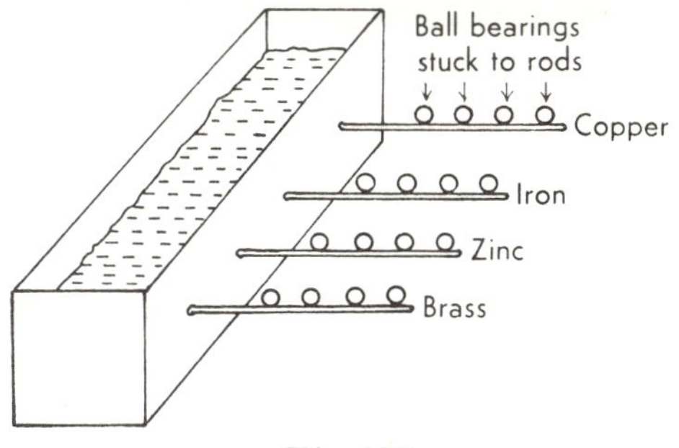

FIG. 190.

Modern theory shows that "free" electrons in a metal play a prominent part in conducting heat. The electrons possess kinetic energy and are in random motion. They can thus transfer heat energy from the hot end to the cold end of a metal bar. The presence of a cloud of free electrons in a metal gives it not only good heat-conducting properties, but also good electrical conductivity.

Heat Insulators

It will be seen that heat travels faster in some metals than in others, for the ball bearings at the end of the copper rod, for example, fall off before those at the end of the iron rod. Some materials, e.g. glass, are poor conductors of heat. One end of a glass rod may be held in the hand while the other is heated to red heat in a Bunsen flame. Very little heat is conducted along the bar. Very poor conductors of heat are called *heat-insulators*, and are frequently used in every-day life. Pan and kettle handles are often made of poor-conducting solids like Bakelite, and table mats are often made of cork. Asbestos is a poor heat-conductor (or good heat-insulator) and is extensively used for lagging pipes, as is expanded polystyrene.

Comparing the Conduction of Wood and Copper

That wood is a poor conductor and copper is a good conductor can be readily shown by a simple experiment. A wooden rod, with one end turned down in a lathe, just fits into a copper cylinder of the same diameter (Fig. 191). A piece of thin paper is then wrapped once round the two joined cylinders covering the join and making good contact with both the wood and the copper. When this paper is heated gently with a Bunsen flame by passing the cylinder through the flame several times, it is found that the paper covering the wood chars while that covering the copper remains white. This is because heat is conducted away from the paper in contact with the copper very quickly, preventing the temperature of the paper rising to the point of charring. The wood cannot conduct heat away so quickly and the paper there chars.

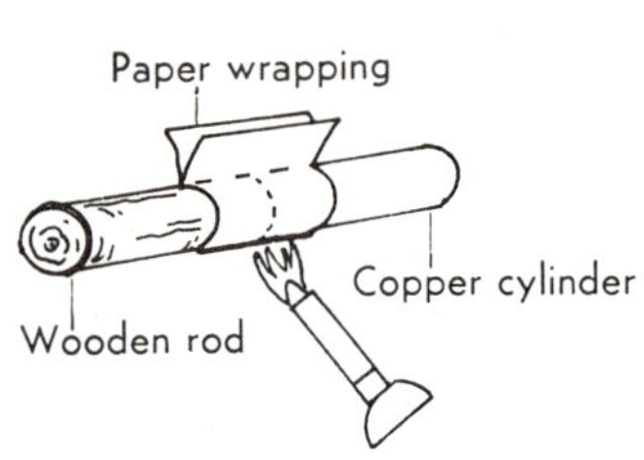

FIG. 191.

Miner's Safety Lamp

The excellent heat-conducting properties of copper make it a suitable material to use in the bits of soldering irons, where rapid heat transfer is required. Copper gauze is used also in the safety lamp which was invented by Sir Humphry Davy in 1813. His invention depended on the two facts that gases can only burn when they become hot, and that the metal copper is a good conductor of heat.

Two simple experiments will illustrate the principle on which the lamp works (Fig. 192).

If a fine copper gauze is held above a lighted Bunsen flame, the gas does not burn above the gauze, even when the gauze is lowered quite close to the Bunsen tube itself. Unburnt gas comes through the gauze and can be ignited by holding a match above the gauze.

If a cool copper gauze is held a little above the end of a cold Bunsen burner (half an inch) and a lighted match is applied above the gauze when the gas is turned on, a flame will appear above the gauze but not underneath it. In each of these experiments the copper gauze conducts heat away so quickly that the gas cannot reach a temperature high enough for it to ignite.

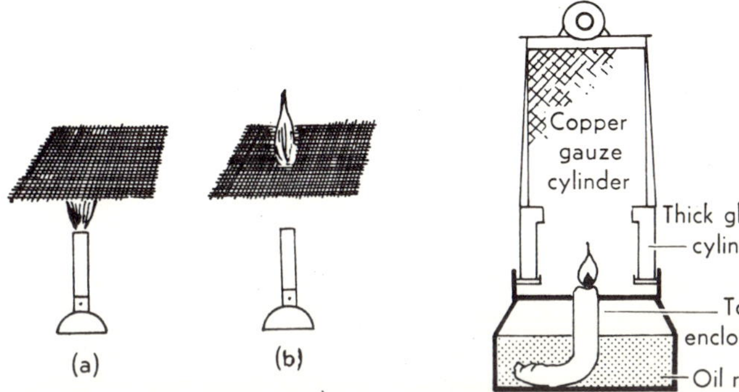

FIG. 192.

FIG. 193. *Miner's safety lamp.*

Sir Humphry Davy invented his lamp to reduce the number of fatal mine explosions caused by the inflammable gas methane, or fire-damp, which is often found in coal mines. Mixed with air, methane explodes when it comes into contact with a naked flame. In the Davy lamp a simple oil burner was surrounded by a cylinder of fine copper gauze mounted above another cylinder of thick glass (Fig. 193). Even when explosive gases came into contact with the outside of the gauze, the temperature was below their ignition point as a result of the good conducting properties of the copper. The methane mixture could, however, penetrate through the gauze and produced a bluish flame inside the glass cylinder. Thus the lamp served the dual purpose of protecting the miners from possible explosions and also of indicating the presence of explosive gases. Modern oil heating lamps used in garages during cold weather are similarly constructed.

Rate of Heat Flow

The rate at which heat energy is transferred through a solid plate depends on the temperature difference between the two end surfaces and the distance,

L, separating the ends (Fig. 194). This temperature difference $\frac{(t_1 - t_2)}{L}$ is known as the *temperature gradient.*

The rate of flow also depends on the area of the plate surface A and the characteristic thermal conductivity k of the material. An equation involving these variables enables the rate of heat flow to be calculated.

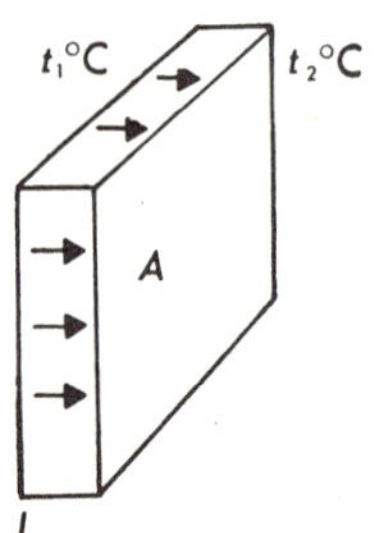

FIG. 194.

$$\text{Rate of heat energy flow} = k\,\frac{(t_1 - t_2)}{L}\,A$$

Example: The temperature inside a living room is 20° C while the temperature outside is 0° C. Calculate how rapidly heat is being lost through a glass window of area 2 m × 2 m if the glass is 3 mm thick. The thermal conductivity of glass k is 1 W/m K.

$$\text{Temperature gradient} = \frac{20 - 0}{0{\cdot}003}$$

$$\text{Area of conducting plate} = 2 \times 2\ \text{m}^2$$

$$\text{Rate of heat transfer} = k\,\frac{(t_1 - t_2)}{L}\,A$$

$$= \frac{1 \times 20}{0{\cdot}003} \times 4 = 26\ 700\ \text{W}$$

Heat is therefore conducted outwards at the rate of 30 kW.

(Note: the various values are given to only one figure and therefore the answer must be to only one figure.)

CONVECTION

With the exception of mercury, liquids are poor heat-conductors and gases are generally worse. The molecules are more widely separated and transfer of energy from molecule to molecule is not so easy. Also, with the exception of mercury, there are no "free" electrons available in liquids and gases to transport the heat energy.

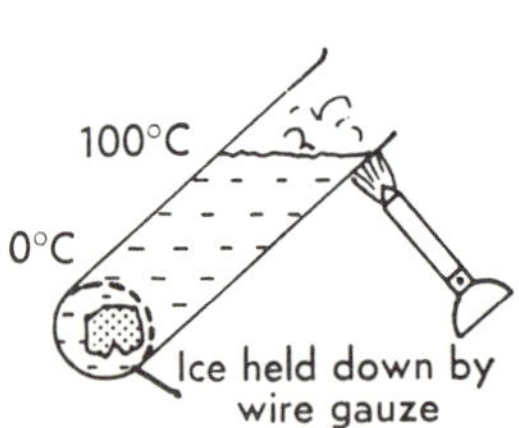

FIG. 195.

It is owing to the poor heat conducting properties of air that wool garments are effective, for between the wool fibres air is trapped, forming a heat-insulating layer. Liquids and gases do, however, transfer heat quite effectively. When heat is supplied to the bottom of a kettle of water the water at the top soon becomes hot. How does this transfer of heat in fluids occur? That it is not due to conduction can be seen from the following experiment.

A boiling tube is half filled with water and a pellet of ice is held at the bottom by means of some wire gauze (Fig. 195). The water at the surface can be boiled, at 100° C, by heating with a small Bunsen flame, while the ice at the bottom is still obviously at 0° C. Now the experiment is repeated with the ice floating on the water surface, and the Bunsen flame is applied at the bottom. Rapidly all the water is heated and the ice melts.

To investigate what happens in this second case, stir some sawdust in a beaker of water, and then apply a small Bunsen flame to the bottom to one side. Sawdust particles will be seen rising up above the heated area, circulating, and then descending at the farther side of the beaker (Fig. 196). Such a circulating fluid current is called a convection current, and this method of heat transfer is known as *convection:* the motion of the hot body itself carrying its heat with it.

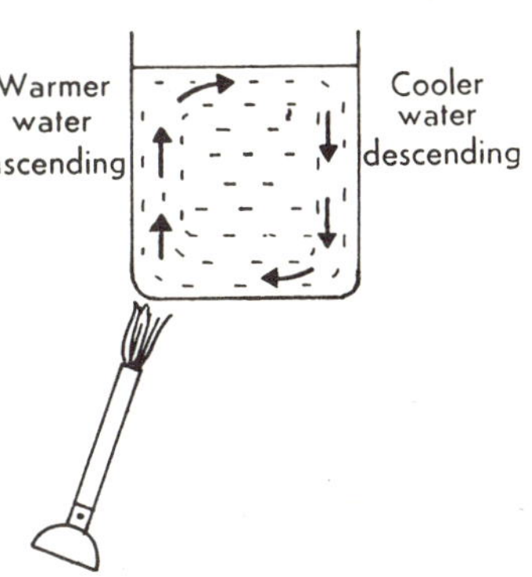

FIG. 196. *Convection current.*

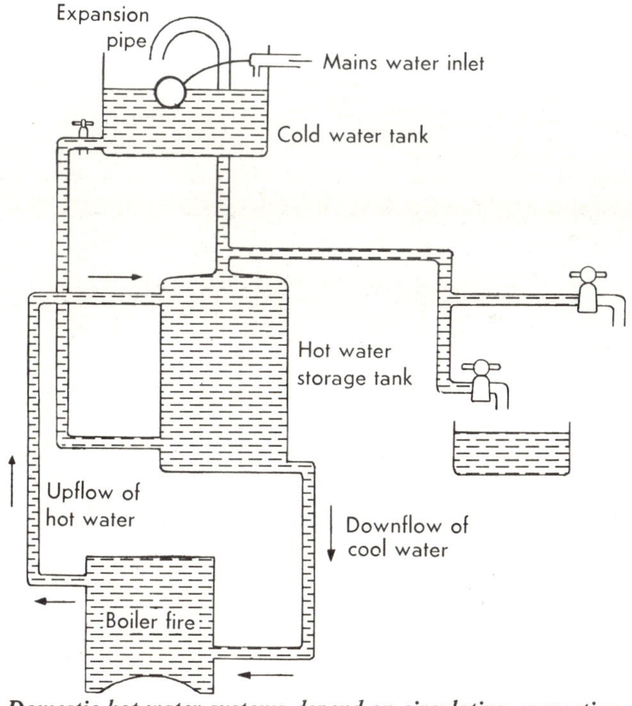

FIG. 197. *Domestic hot water systems depend on circulating convection currents.*

The circulating current is caused by changes in the density of the fluid resulting from expansion (page 170). The operation of a domestic hot water system (Fig. 197) depends on such circulating convection currents.

Like water and other liquids, gases also are fluids and show convectional heat transfer. A whirl turbine can be used to show this. A paper spiral can be cut from a sheet of paper and supported by a pin or wire at the centre. Such a spiral held above a heat source such as an electric light bulb acts like a turbine and spins round, driven by rising convectional air currents (Fig. 198).

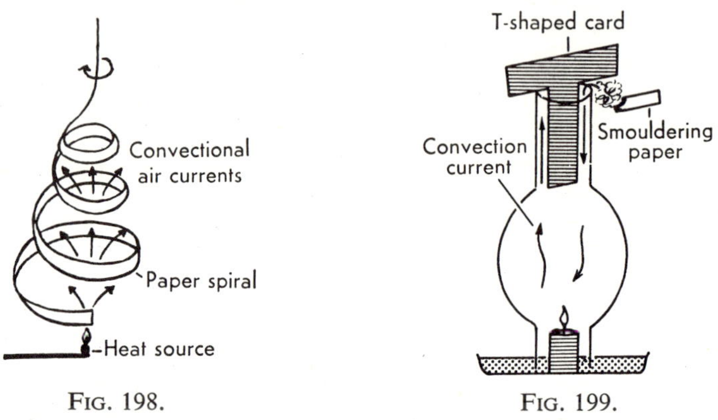

FIG. 198.

FIG. 199.

The reality of the air currents may be clearly shown by arranging a burning candle in a saucer of water with a tall lamp glass enclosing it (Fig. 199). Air cannot enter below because of the water trap and finds some difficulty

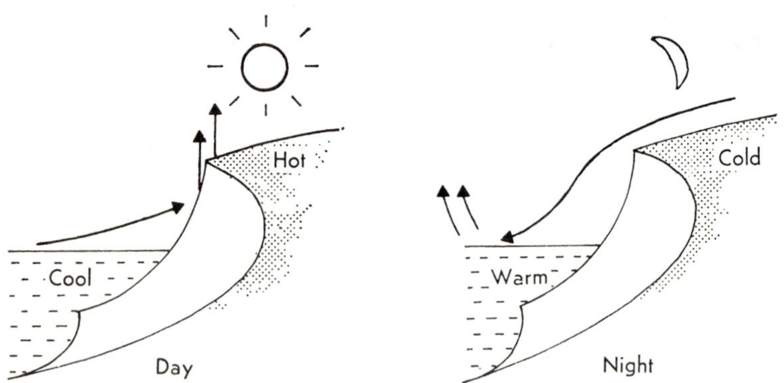

FIG. 200. *Formation of land and sea breezes by convection currents.*

in passing down the tube quickly enough to support the burning of the candle, which may go out. A T-piece of cardboard placed on the top of the lamp glass improves the burning of the candle and obviously a fresh supply of air must be passing down the tube. When a smouldering piece of blotting paper is held at the top it will be seen that smoke is drawn down one side of the T-piece and up the other. The hot candle flame has set up a circulatory convection current. In the original example of the coal fire, hot air rises, carrying smoke up the chimney, and cool air flows in at the bottom to replace it.

Such convectional currents are responsible for land and sea breezes (Fig. 200). During the day sunshine heats the land more quickly than the sea, largely because only the surface layer of earth is heated while a larger mass of water is heated due to wave movement. The air above the land becomes warm, rises, and sets up a convectional air current, forming a sea breeze because cool air from the sea rushes in to replace it. At night the converse happens. The land cools more quickly than the sea, which retains its heat better, and keeps the air in contact with it warmer. Thus a convectional air current is set up in the reverse direction giving rise to a land breeze.

RADIATION

Heat reaches us direct from the sun, having crossed some $1{\cdot}5 \times 10^9$ kilometres of space. This heat transfer is independent of solids and fluids and is entirely different in nature from conduction and convection. It is the transfer of heat energy by means of electro-magnetic waves, which travel in a similar manner to radio waves and light waves. Only the wave-lengths differ for the speed is the same in all three cases, namely 3×10^8 metres per second. This method of heat transfer is known as radiation and is the method by which heat is beamed from an electric fire, and by which heat in the fire example at the beginning of this chapter was thrown out into the room. Only when electro-magnetic radiation, in this case infra-red radiation, is absorbed by a material object, is heat energy produced. Space through which the radiation travels is not heated, unless occasional atoms present in the space absorb a little of the radiant energy.

A convex lens will focus the sun's radiant heat rays well enough to ignite a match. It is even more striking to fill a round-bottomed flask with a solution of iodine in carbon disulphide and hold the flask of black liquid in the path of the sun's rays. The invisible heat rays are focused by the flask and can ignite a match. It can easily be seen that heat rays are reflected and refracted in the same way as light rays.

Black and Polished Surfaces

Do bright surfaces radiate heat better than dull black ones? Do bright surfaces absorb heat better than dull black ones? Some simple experiments

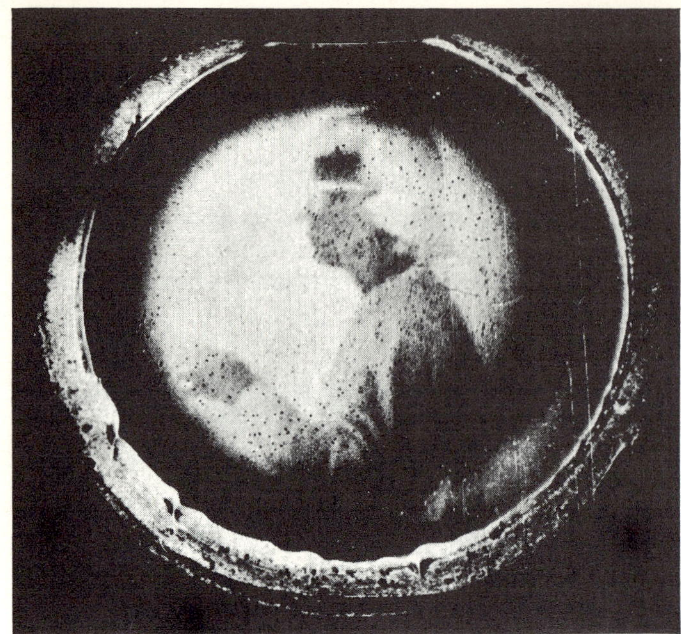

FIG. 201. *This image was formed by focusing the heat radiated from the woman on to a thin oil film. This caused oil to evaporate to a varying degree, according to the amount of heat that fell on the different parts of the film. Remember that heat is radiated in the form of electromagnetic waves, similar to light waves.*

will enable us to investigate and give answers to these two questions.

Blacken the bulb of a laboratory thermometer with indian ink and hang it up beside an identical thermometer which has a bright unblackened bulb. Beam heat from an electric fire on to the two bulbs equally and observe the rise in temperature indicated by them. The blackened bulb absorbs heat better than the bright bulb and shows a higher temperature.

Suspend two identical copper cans beside each other, but polish one can and blacken the other with paint or lamp black. Pour equal amounts of hot water at the same temperature into the cans quickly and record the temperature of the water in each can every minute by means of two thermometers. It will be observed that the black can cools more rapidly than the polished can, showing that it radiates heat more quickly.

Leslie Cube

A more accurate experiment to investigate the comparative radiation properties of different surfaces may be carried out using a hollow copper cube each side of which has a different surface: lamp-blacked by holding it in the smoky flame of a candle; brightly polished; roughened; painted white. Such a cube is known as a Leslie cube, after its inventor who first used it in 1804.

The copper tube (Fig. 202) is filled with water which is maintained at the boiling point, 100° C, by means of an immersion heater. Heat is radiated from the four surfaces at different rates. A device known as a *thermopile* (Fig. 203) is used to detect and measure the rate heat is radiated from the surfaces. This apparatus includes a series of antimony and bismuth bars running from the front to the back of the thermopile and joined to each other end to end alternately. Those junctions at the front of the thermopile are

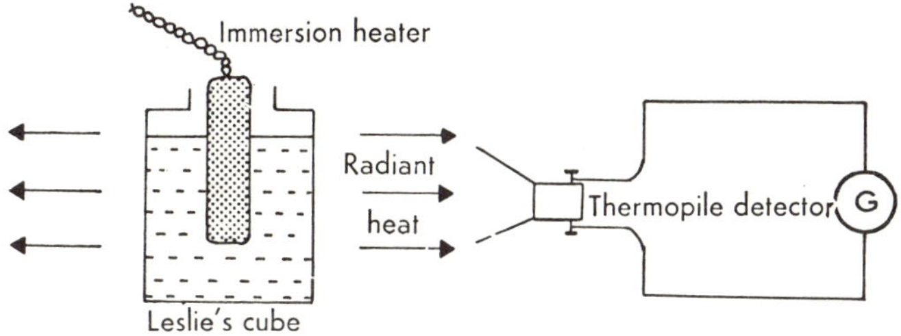

FIG. 202. *Apparatus to measure the rate at which heat is radiated.*

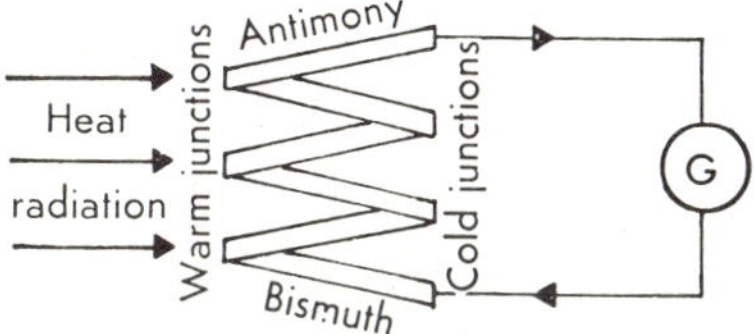

FIG. 203. *Thermopile.*

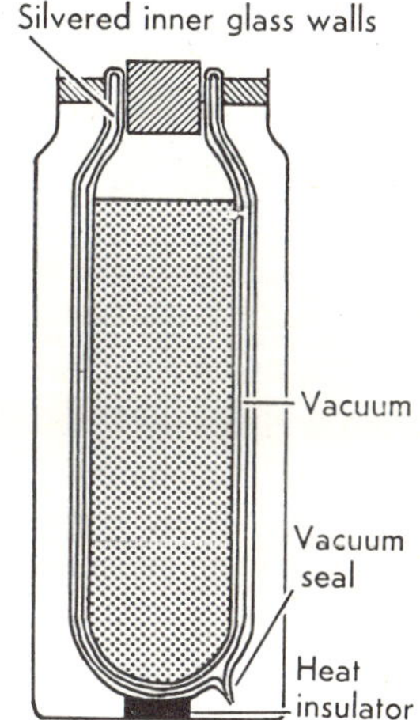

FIG. 204. *Thermos flask.*

heated while those at the back are kept cool. This temperature difference generates an electrical e.m.f. between the ends of the chain of metals, which can drive a current through an external circuit and galvanometer as shown.

When placed at equal distances from the four surfaces of the heated Leslie cube, the thermopile indicates by the deflection of the galvanometer pointer, that the dull-black surface radiates heat best, while the brightly polished surface is the worst radiator of heat.

It may be concluded from experiments such as these that not only are black surfaces better absorbers than brightly polished surfaces, but also they are better heat radiators.

Thermos Flask

In most cases of a hot body cooling, all three types of heat transfer are involved. The common Thermos flask (Fig. 204) attempts to prevent heat losses by eliminating all three methods of heat transfer. Conduction is prevented by exhausting air from the double-walled glass container which is itself a poor conductor. There is then no medium between the two glass walls to enable conduction to take place. Similarly convection losses are removed. The cork prevents free air circulation through the top, and eliminates evaporation which would otherwise cause serious heat loss. The inner surfaces of the double glass walls are silvered to reduce any radiation losses from the inner surface to the outer across the vacuum. As a result very little heat energy can escape from the hot liquid inside the flask and cooling is reduced to a minimum. It is interesting to note that the original vacuum flask, invented by Sir James Dewar, was designed to keep liquid air placed inside it cold, since no heat could enter into the flask.

The different methods of heat transfer that we have considered may be summarized in a table.

Table 9
Methods of Heat Transfer

Conduction	*Convection*	*Radiation*
In solids, liquids, and gases.	In liquids and gases.	Through a vacuum or transparent bodies.
Energy passed from molecule to molecule by vibration. The medium does not move.	Energy carried by the fluid itself moving.	Energy transfer by electromagnetic waves obeying the laws of reflection and refraction like light.
Speed of transfer depends on the temperature gradient and the medium.	Movement of medium is caused by density changes in the fluid resulting from temperature differences.	Speed of transfer is at the speed of light: 3×10^8 metres per second.

QUESTIONS

1. Briefly describe the various processes by which a jar of hot water loses heat. What methods could you use to reduce the rate of heat loss?

2. Explain clearly the differences between the three ways in which heat energy travels. Which are the main methods operating in (*a*) the warming of the sand on the seashore by the sun, (*b*) the boiling of a kettle of water on a hot plate, (*c*) the cooling of a car engine?

3. What do you understand by the thermal conductivity of a material?

A window pane 1 metre square is made of glass 4 mm thick. If the temperature inside the window is 20° C and outside it is 0° C, calculate the heat loss by conduction through the glass. Thermal conductivity of glass is 1 W/m K.

4. Describe carefully why land and sea breezes occur.

CHAPTER 18

SOURCES OF ENERGY

THE progress of man throughout his long evolutionary history has been characterized by the development of more and more complex machines and the continual search for new kinds of power to drive them. In the earliest days all work had to be performed by man himself, but soon simple aids, or machines, like the lever and the wheel were invented, which enabled larger loads to be lifted and moved more easily. The movement and erection of huge slabs of stone like those found at Stonehenge must have been laboriously carried out by the combined effort of many men assisted by the knowledge that it is easier to haul a heavy load up an inclined plane rather than to attempt to lift such a load vertically. Possibly tree trunks were used as rollers to help them.

Mechanical Energy

Animal power was harnessed at an early date in man's evolution, for it was realized that domestic animals like the elephant, camel, ox, and horse could pull heavy loads more readily than man himself. Even today in most parts of the world animal power is used, sometimes quite extensively. Camels or oxen are used to rotate simple wheels driving water pumps, and elephants

FIG. 205. *Animals are still a main source of power in many parts of the world. Here a mule is working an irrigation wheel in Iraq.*

are used to haul timber. Animals like the horse and the camel were man's main means of rapid transport until the latter part of the nineteenth century, and in Europe speed of travel was restricted to that of a horse until a hundred years ago.

The power of water and wind was used much later in man's history and

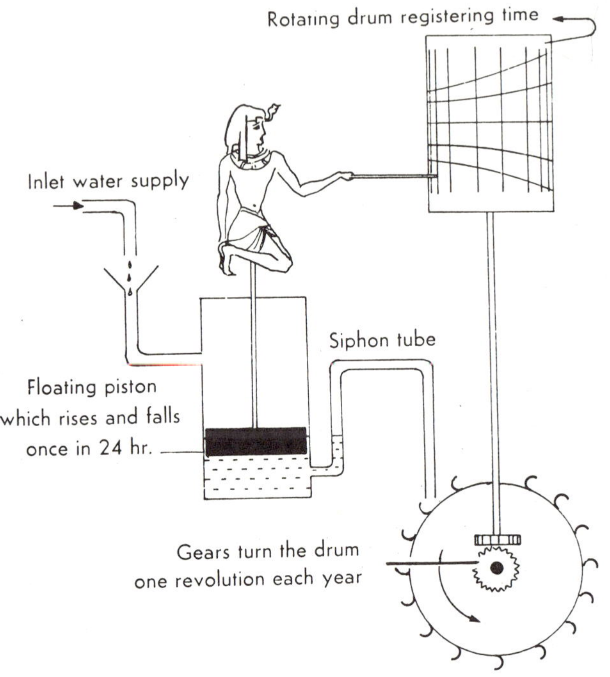

FIG. 206. *The clepsydra, an early form of Egyptian water clock.*

led to the development of wheels and gears. Water power has been used to drive simple machines for many centuries, and the early Egyptian clepsydra (water clocks) depended for their operation on water power (Fig. 206). Medieval village life centred round the water-mill where corn was ground in the production of flour (Fig. 207).

Tracing in more detail the energy changes involved in these water-driven devices, we find that in all cases the mass of water starts at a higher level, possessing potential energy, and falls under gravity acquiring kinetic energy of motion but losing potential energy. This energy is used up in driving the machine. It is thus mechanical energy which is used and transformed from one form to another.

Similarly windmills, which flourished up to the seventeenth century in

FIG. 207. *Whether undershot* (left) *or overshot* (right), *water mills utilize the energy of the water flowing under or over them.*

Britain, and are still used in Holland, harnessed a form of natural energy to do work. The momentum of the wind striking the windmill sails produced motion and rotated machinery. Sails turned very slowly and were heavy, but today more efficiently designed rotating blades are used to harness natural windpower, either to drive water pumps or, sometimes, to drive dynamos for electricity production.

Thermal Energy

As we saw in Chapter 15, man learnt to convert heat energy into useful mechanical work in the eighteenth century. Such an energy transfer—heat to mechanical—had been tried out by Hero of Alexandria in 150 B.C. He devised a simple reaction type steam turbine (Fig. 208), but it had no practical application in those early days.

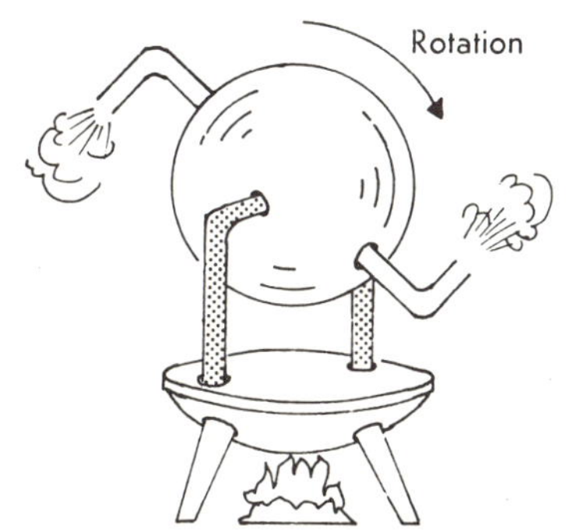

FIG. 208. *Hero's engine.*

The earliest steam engines were very inefficient, converting less than 1 per cent of the input heat energy into useful mechanical work. They relied on atmospheric pressure for their operation, and were called atmospheric engines in consequence. The steam engine invented by Newcomen in 1712 (Fig. 209) was an atmospheric type of engine and was widely used in mines in Britain for pumping out water.

Burning coal was the energy source used to raise steam, which then filled a cylinder closed by a piston. Little mechanical work was done by the expanding steam, as it passed through valve *A*, for it was so arranged that the weight of the beam and pump chain pulled up the piston to the top of the cylinder. At this stage valve *A* was closed and a fine spray of cold water injected through valve *B* into the cylinder. This caused the steam to condense and produced a partial vacuum in the cylinder. Atmospheric pressure then forced down the piston, turned the beam, and worked the pump. The source of energy and power comes from the burning fuel—coal or wood—which is used to change the state of the water to steam.

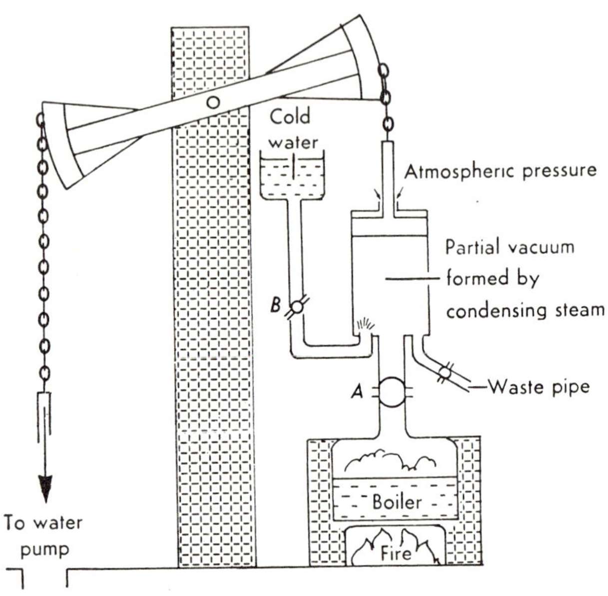

FIG. 209. *Newcomen's atmospheric engine was widely used for pumping water.*

In modern steam engines, which use high pressure steam, coal as fuel is used to produce the heat energy needed to change water into steam. This heat energy (carried by water vapour molecules as molecular kinetic energy) is then transformed into mechanical energy as the piston is pushed backwards and forwards to drive the engine. Some of this mechanical energy will inevitably be wasted by friction and reconverted to heat, but a large fraction of it is employed usefully in doing work.

Chemical Energy

Where does the energy in the wood or coal come from initially? If we examine coal we find that it is made up of the fossilized remains of plants that flourished millions of years ago, when the land was covered by vast fern-like forests. Several countries use peat (the partly compressed remains of vegetation) as a fuel today, and some countries mine a form of coal known as lignite or brown coal, intermediate in form between peat and normal coal. Plants are able by means of the chemical catalyst chlorophyll to trap radiant energy direct from the sun and store it in the form of chemical energy in carbon compounds, like starch, for example, which is found concentrated in the tubers of potatoes. It is this energy which is released when wood, peat or coal is burned. The energy originally came to earth from the sun.

Investigations of the origins of the energy involved in driving a water-wheel lead us to a similar conclusion. The sun's radiant energy evaporates water from seas and lakes, and leads ultimately to the formation of clouds and rain. This energy becomes potential energy as the rain water settles in upland lakes and rivers, and finally changes to kinetic energy as the water flows down under gravity to drive water-wheels or water turbines (Fig. 210).

It can be understood also that the energy we derive from our food comes initially from the sun. Animal foods, such as meat and milk, can be traced back in a food chain to plants which were eaten by the animals. In these plants the energy of the sun was originally stored as chemical energy.

The transfer of the sun's energy through the plants, through the animals, into ourselves, involves a large number of very complex chemical reactions. All chemical reactions either absorb or evolve energy in the form of heat.

The chemical reactions which evolve energy in the form of thermal (heat) energy, are known as *exothermic reactions.* For example, when zinc is dissolved in dilute sulphuric acid, hydrogen gas bubbles off, zinc sulphate is formed in solution, and heat energy is given off:

$$Zn + H_2SO_{4\ aq.} = ZnSO_4 + H_2 + \text{Heat energy.}$$

The many chemical reactions which instead of evolving heat, absorb heat, are known as *endothermic reactions.*

Electrical Energy

The conversion of chemical energy directly into electrical energy explains the action of electrical cells and batteries. As electricity is drawn from an accumulator, or a dry Leclanché cell, to light a torch bulb, chemical energy is used up. In the case of the accumulator the reaction is reversible. In the case of the Leclanché cell, the reaction is not reversible for when the ammonium chloride in the cell is exhausted, no further current of electricity can flow.

Although in these cells and batteries chemical energy is converted into electrical energy, the chemical energy itself involves electrical energy in binding together the atoms which make up the compounds. In general, energy

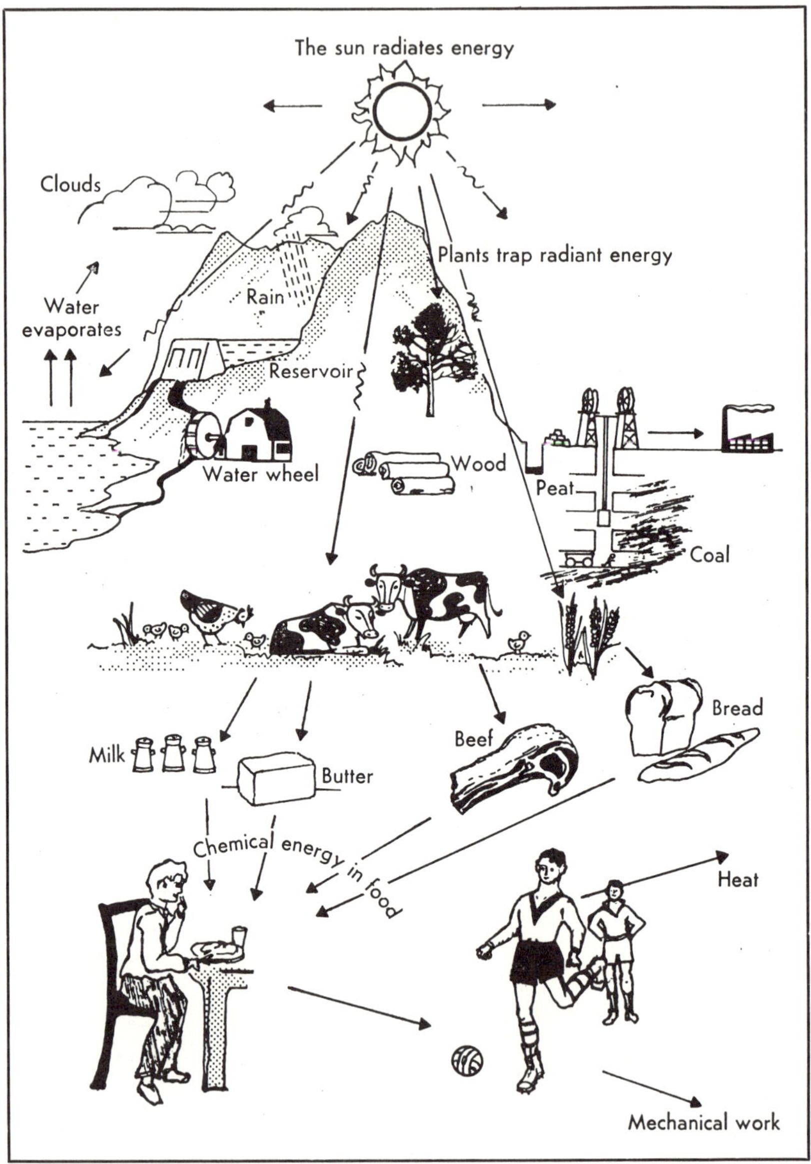

FIG. 210. *Most of the energy available to us radiates originally from the sun as heat and light. Here are a few of the changes this energy undergoes.*

FIG. 211. *Electrical energy is obtained from many sources. On the bleak, wind-swept Orkney Island, a 100-kW wind-powered electric generator has been erected at the top of a 25m tower. Using the energy in winds that can exceed 60 m/s, enough electricity is produced to supply a small town.*

is required to separate these atoms. This can be achieved by passing an electrical current through an electrolyte such as slightly acidulated water; the water is then split up into its component elements, hydrogen and oxygen.

From our everyday knowledge of electric fires and lamps we realize that electrical energy is readily converted into heat or light energy.

It is not so easy to reverse this process and convert heat energy into electrical energy.

Radiant Energy

Since today electrical energy is in such great demand man has struggled hard to harness directly radiant energy from the sun to provide electrical power. The radiation is focused by huge concave mirrors on to tubes containing water to raise steam to work conventional steam turbines driving dynamos, though in normal practice this method is limited to the heating of houses, furnaces and ovens. The rate at which energy is received on earth per unit area is known as the *solar constant*. It is approximately 1400 watts per metre squared. Some houses use this energy to heat water circulating through pipes just below a transparent roof through which the sun's radiant energy can penetrate. The hot water is then used in a domestic central heating system.

Electricity can however be generated more directly by focussing the heat

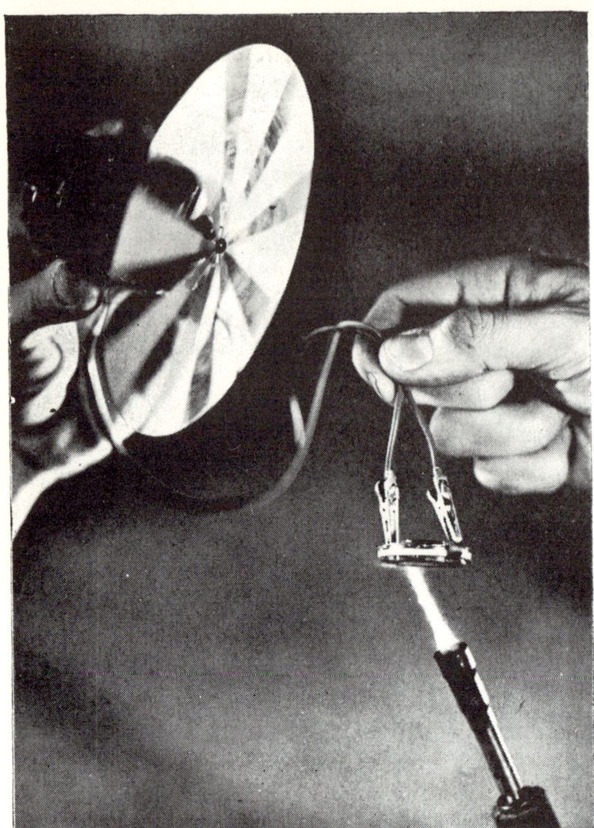

FIG. 212. *A small electric fan is here being driven by a flame applied to a thermo-electric cell.*

rays on to a junction of two different metals, e.g. iron and copper (Fig. 213). Arranged in a suitable circuit such thermo-electric junctions can be used to

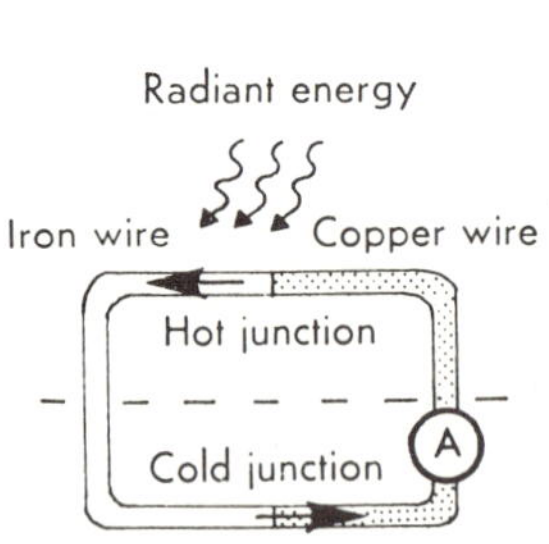

FIG. 213. *Thermo-couple.*

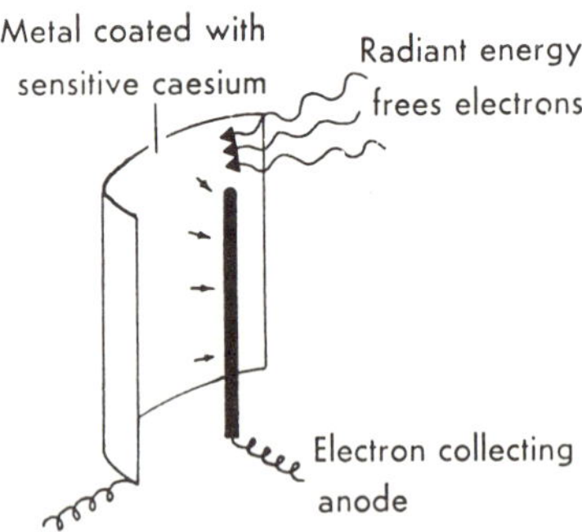

FIG. 214. *Photo-electric cell.*

convert solar energy directly into electrical energy. A more modern direct method of harnessing solar radiation is by the use of a photo-electric cell

(Fig. 214). Such devices are reasonably efficient and are used to provide energy in artificial satellites. The energy of the radiation is able to knock out electrons from a sensitive metal surface, and these freed electrons give rise to an electric current. A photographic light meter works on this principle.

Nuclear Energy

Most electrical power today is derived from either water power or by burning fuels to drive steam turbo-generators. However, an increasingly important fraction is being derived from nuclear power. Even here the electrical power is obtained from standard steam turbo-generators, but the initial heat energy needed is provided by nuclear energy.

Nuclear energy is produced when a chemical atomic nucleus such as that of uranium, splits into smaller atomic parts. This energy liberation is fundamentally different from that involved in normal chemical energy exchanges, for it stems from the nucleus itself rather than the surrounding electrons, and signifies the conversion of matter to energy. That this is possible was first postulated by Albert Einstein at the beginning of the twentieth century. The equation by which Einstein related energy E with mass m, involves also the velocity of light c. It is of paramount importance today.

$$E = mc^2$$

In a nuclear reactor the total mass of the chemical products produced by the splitting (fission) of uranium is slightly less than the original mass of the uranium. This "disappearance" of matter represents the energy liberated as heat and radiation.

Unlike normal chemical fuel burnt to liberate heat energy, a minute mass of atomic fuel will produce a vast quantity of energy. In fact one pound of uranium can liberate about as much heat energy as 1,500 tons of coal.

Recently it has been possible to generate heat energy by a process of fusion of hydrogen nuclei in which the heavier atoms of helium are built up. Heavy hydrogen is used. This is an isotope of hydrogen, the nucleus of which contains an extra neutron. Such a reaction can occur only at very high temperatures such as exist in the sun itself, and indeed similar thermonuclear reactions are believed to be the source of the energy in our sun and the stars. Uncontrolled fusion reactions provide the devastating power of the hydrogen bomb, but man has not yet learnt how to harness this latest nuclear source of power successfully.

Man seems at last to have limitless power and energy within his grasp. It is clear that deposits of natural fuels like coal, oil and natural gas must soon become exhausted. Most countries have insufficient water power to meet modern demands for increasing quantities of electric power, and to date the direct harnessing of solar power has been on a very small scale. It is

FIG. 215. *Calder Hall on the Cumberland coast was the first nuclear power station in the world to produce electricity on a full commercial scale. It comprises four separate reactors, housed in the four large square buildings shown. On either side of the four reactor houses are heat exchangers, eight in all, where the heat from the reactors is used to convert water into steam. This steam is then used to drive turbines which are housed in the two long low buildings connecting the two pairs of reactor houses. Waste steam from the turbines is condensed and returned to the heat exchangers. The cooling water used in condensing this steam is itself circulated through the four giant cooling towers, two at either end of the site.*

quite possible that tidal power (using the energy involved in raising the level of the sea at high tide) will be used in the future. Quite a large tidal plant has been built in north-west France to supply electricity. But the source of man's power in the immediate future will most likely be nuclear power.

QUESTIONS

1. Describe the construction and operation of an early atmospheric steam engine. Why is it so named?
2. By means of a diagram trace back to its original source the energy radiated by an electric fire.
3. Explain why coal is sometimes described as "bottled sunshine."
4. Trace briefly the evolution of power sources employed by man from prehistory to the twentieth century.

FIRST REVISION SUMMARY

FUNDAMENTAL IDEAS

WHAT IS PHYSICS?

Page

PHYSICS is the study of how non-living things act. This study has 9
been progressing throughout history and is progressing now. But the
progress is achieved only when experiment and theory go hand-in-hand. 11

Experiment can produce new facts that will change old theories and
it is for such changes that the mind of a scientist must **always** remain 15
open and ready.

BASIC QUANTITIES

The study of physics involves the measurement of many physical quantities. In order to do this, we must first select the basic quantities and then **choose** and define the units in which to measure them. The first four basic quantities are:

			Page
Time	second (s),	also minute (min) and hour (h)	16
Space	metre (m),	also millimetre (mm) centimetre (cm) kilometre (km)	20
Mass	kilogramme (kg),	also gramme (g) milligramme (mg)	23
Temperature	degree Celsius (°C),	also degree Centigrade (°C), and kelvin (K)	32

We then automatically have units for other quantities:

			Page
Motion	metre per second (m/s),	also kilometre per hour (km/h) centimetre per second (cm/s)	22
Acceleration	metre per second per second (m/s^2)		23
Density	gramme per cubic centimetre (g/cm^3) kilogramme per cubic metre (kg/m^3)		25
Force	newton (N)		28
Energy	joule (J),	also kilowatt hour (kW h)	29
Power	watt (W),	also kilowatt (kW)	30

The range of sizes in each of these quantities that we meet in 34
everyday life is only a **small** range roughly midway between **atomic**
and **galactic** size.

Page

ELEMENTS, ATOMS AND MOLECULES

An element *is a substance which cannot be split into different sub-* 36
stances by normal physical or chemical means.

Evidence for elements: 37

The chemist cannot split some substances. All elements give different ***spectra*** when their vapours give out light.

An ***atom*** *is the smallest unit of an element which is identifiable as the element.* A ***molecule*** is a group of atoms which stay together unless forced apart by physical or chemical processes.

Evidence for atoms and molecules: 38

Law of constant chemical composition 38

Law of multiple proportions 39

Atomic and molecular weights 39

Avogadro's number 40

Brownian movement 40

In Physics we are basically concerned with ***matter*** and ***energy.***
The atomic theory of matter and the concept of energy can explain 41
the following phenomena amongst many others:

The difference between ***solids, liquids,*** and ***gases.*** 42

The pressure exerted by gases. 42

The heat required to change a solid to a liquid or a liquid to 43
a gas.

Other phenomena, all concerned with the interactions between matter and energy, which will be discussed in this course include:

Light, radiant heat, and radio waves. 43

Sound waves and other vibrations of matter. 43

Gravitational forces, magnetic forces, electrical forces, atomic 44
and nuclear forces.

Most of the questions that will be answered by this book are 46
those that might well be asked by any inquisitive person **looking at the world in which he or she lives.**

EFFECTS OF FORCE

MOLECULAR THEORY OF MATTER

Page

In ***solids,*** molecules and atoms are close together and are bound 52
by cohesive forces. They merely vibrate about a fixed mean position. A solid keeps its shape.

In ***liquids,*** molecules are about the same distance apart but they 52
move about and take the shape of the container.

Page

In ***gases,*** molecules are wide apart and the effect of cohesive forces is negligible. They move freely with random motion in straight lines between collisions and exert pressure. A gas can be compressed. 52

Capillarity and surface tensions are explained by adhesive and cohesive forces. 53

Movement of molecules is shown by the ***Brownian motion*** (page 51), by ***diffusion*** experiments (page 57), by ***osmosis*** experiments (page 59), by ***viscosity*** experiments (page 56).

SOLIDS IN MOTION (NEWTON'S LAWS)

Page

Speed is distance per unit time. 61

$$\text{Average speed} = \frac{\text{Total distance}}{\text{Total time}}$$

Velocity is distance per unit time in a stated direction. 61

Vector quantities (such as velocity, acceleration, force) must be added **geometrically.** 61

Scalar quantities (such as mass, volume, energy) may be added **arithmetically.** 62

Acceleration is change in velocity per unit time. **Distance travelled** may be obtained from a ***velocity-time graph.*** 63

Newton's laws of motion: 65

1. *When no resultant force acts on a body, the body remains at rest or continues moving in a straight line at constant speed.*
2. *When a body is acted upon by a force, the rate of change of momentum is directly proportional to the force and takes place in the direction of the force.*

$$\text{Momentum} = \text{Mass} \times \text{Velocity}$$

3. *To every action there is always opposed an equal reaction.*

Unit force is that force which gives to unit mass, unit acceleration.

$$\text{Force} = \text{Mass} \times \text{Acceleration}$$

Unit of force: 75

One ***newton*** is that force which gives to a mass of 1 kg an acceleration of 1 m/s^2.

Work done by a force, is the product of the force and the distance through which the force moves. 75

$$\text{Work} = \text{Force} \times \text{Distance}$$

OCL/PHYS—P*

Page

Unit of work: 75

One ***joule*** is work done when 1 newton acts through 1 m.

Power is work done per unit time:

One ***watt*** is rate of working of 1 joule/second and

1 kW = 1000 W

Parallelogram of forces: *If two forces acting at a point be represented in size and direction by the adjacent sides of a parallelogram, then the diagonal passing through their point of intersection represents their resultant in size and direction.* This theorem can be used to add any vectors. 69

Motion in a circle is an accelerated motion and needs a force acting towards the centre of the circle, called a ***centripetal force.*** 81

DIMENSIONAL CHANGES DUE TO APPLICATION OF FORCES

Hooke's law: *the stretch produced by a force is directly proportional to the force.* This law applies to many solids in the form of springs, wires, cords, etc. Using this law, a force may be measured by means of a spring balance. 86

Potential energy may be stored when a solid is stretched or compressed or placed in a suitable position and then this energy may be changed into useful work. 89

Laws of fluid pressure (Pascal): 91

1. *The pressure is the same at all points which are at the same level in one fluid.*
2. *Fluid pressure on any surface is perpendicular to the surface.*
3. *At any place in a fluid, pressure acts equally in all directions.*
4. *Pressure is transmitted uniformly from one place to another throughout a fluid.*
5. *The difference in pressure between any two places in the same fluid* = hdg.

Pressure of the earth's atmosphere at the surface of the earth is close to 100 kN/m^2 or 10 N/cm^2. The pressure may also be measured by means of a ***barometer*** and is expressed as 76 cm of mercury or 10·4 m of water. 97

Page

Boyle's Law: *The volume of a given mass of gas is inversely proportional to its pressure (temperature constant).* 102

$$P_1V_1 = P_2V_2$$

SOLIDS MOVING ON SOLIDS

Movement of one solid over another produces **frictional forces.** 106

Laws of friction: 108

1. *When the two surfaces are at rest, the frictional force can take any value up to a certain maximum, called the* ***limiting friction.***
2. *When motion occurs, the force of sliding friction* $\boldsymbol{F}$ *is proportional to the normal reaction* $\boldsymbol{R}$*. The ratio of* $\boldsymbol{F/R}$ *is called the* ***coefficient of sliding friction*** μ*.*
3. *The coefficient of sliding friction* μ *is independent of the area of contact and of the relative speed, within fairly wide limits.*

Rolling friction is less than sliding friction; hence the use of wheels for movement. 109

Lubrication allows sliding friction to be replaced by ***liquid friction,*** which is much less than sliding friction. 110

Causes of friction: 111

Intermeshing of surface irregularities causing very high pressures at points of contact.
Adhesion between two kinds of molecules.
Possibly electrostatic forces brought about by rubbing.

SOLIDS IN FLUIDS

Archimedes' Principle: *whenever a solid is totally or partially immersed in a fluid, it experiences an upthrust equal to the weight of fluid it displaces.* 112

Relative density: 113

$$\text{R.D. of solid} = \frac{\text{Weight of solid in air}}{\text{Weight of water displaced by solid}}$$

Law of floating bodies: *a floating body displaces its own weight of fluid.* 114

Use of ***hydrometer*** to measure the R.D. of **liquids:** 115

$$\text{R.D. of liquid} = \frac{\text{Length of hydrometer immersed in water}}{\text{Length of hydrometer immersed in liquid}}$$

When solids move through fluids, frictional forces are produced. 116

Page

Bernoulli's principle: *in the regions of high fluid speed there is a diminution in pressure.* Hence the swerving of a spinning ball, the lift given to the aerofoil of an aircraft and the necessity for **streamlining.** 117

MOTION AT UNIFORM ACCELERATION

Equations of **uniformly accelerated motion:** 122

$$v = u + at$$
$$s = ut + \tfrac{1}{2}at^2$$
$$v^2 = u^2 + 2as$$

Law of conservation of momentum: *when two or more particles collide, or separate, the total momentum in any given straight line remains constant provided no external force acts in that direction.* 125

$$m_1u_1 + m_2u_2 = m_1v_1 + m_2v_2$$

Newton's second law of motion (if mass is constant): 125

Force = Mass × Acceleration

$$P = ma$$

Momentum equation: 125

Impulse of force = Change in momentum

$$Pt = mv - mu$$

Energy by virtue of **position** is known as ***potential energy.*** 129

Energy by virtue of **motion** is known as ***kinetic energy.***

Principle of the conservation of energy: 129

In a closed system the total energy (i.e. the sum of kinetic, gravitational potential, chemical etc energy) remains constant.

MECHANICAL AND HEAT ENERGY

ENERGY, WORK AND MACHINES

Work is said to be done when a force acting on a body moves the body in the direction in which the force acts. 131

Work = Force × Distance moved

Energy is a capacity for doing work and is expended when work is done. *Energy can be stored, changed from one form to another, but it can be neither created nor destroyed.* 132

Potential energy $= mgh$ 134

Kinetic energy $= \frac{1}{2}mv^2$ 134

Page

Power is the rate of doing work: 137

1 ***horse power*** = 746 ***watts***

1 ***watt*** = 1 joule/second

A ***machine*** is a device which enables man to perform a task more conveniently. Examples are **levers, pulleys,** and **winches.** 138

The ***moment*** of a force is its turning effect about a point and is measured by the *product of the force × its perpendicular distance from that point.* 139

Principle of moments: for a body in equilibrium under the action of a number of forces in one plane, *the sum of the clockwise moments about any point equals the sum of the anticlockwise moments about the same point.* 140

The whole of the weight of a body may be considered to act through its ***centre of gravity*** whatever the position of the body. 141

Equilibrium: When a body in **stable equilibrium** is slightly displaced, forces acting on it are such that they return the body to its original position. In the case of **unstable equilibrium** the forces increase any slight displacement. 143

$$\textit{Mechanical advantage} = \frac{\text{Load}}{\text{Effort}}$$ 145

$$\textit{Velocity ratio} = \frac{\text{Distance moved by effort}}{\text{Distance moved by load in the same time}}$$ 146

$$\textit{Efficiency} = \frac{\text{Work output}}{\text{Work input}} = \frac{\text{Mechanical advantage}}{\text{Velocity ratio}}$$ 146

Levers: 147

First order = Load—Fulcrum—Effort.
Second order = Fulcrum—Load—Effort.
Third order = Fulcrum—Effort—Load.

Machines:

Wheel and axle. 147
Winch. 148
Screw Jack. 150
Common sheaved pulley. 150
Archimedes pulley. 150
Weston differential pulley. 151

Page

MEASURING THE QUANTITY OF HEAT

Heat is a form of energy and is produced when frictional resistance is overcome in a working machine. 154

The ***temperature*** of a body indicates its relative hotness or coldness. Heat generally flows from a region of high temperature to one of lower temperature. 154

Equal masses of different substances need different quantities of heat to raise their temperatures through equal amounts. *The heat needed to raise the temperature of* 1 kg *of a substance through* 1 K *is known as its* ***specific heat capacity.*** That of water is 4·18 kJ/kg K. 158

Quantity of heat = Mass × Temperature change × Specific heat capacity

Specific heat capacities of materials are often determined by using the ***method of mixtures.*** In this method a known mass of a **solid** is heated to a measured temperature and then transferred to a weighed calorimeter containing a known mass of water at a recorded temperature. The final temperature of the mixture enables the specific heat capacity of the solid to be calculated. The heat given out by the solid cooling is assumed equal to the heat taken in by the water and calorimeter. 158

$$s = \frac{(m_c s_c + m_w s_w)(\theta_2 - \theta_1)}{m(\theta - \theta_2)}$$

Its units are J/kg K. Sometimes values are given in terms of kJ/kg K.

The specific heat capacity of a liquid can also be determined by the method of mixtures. 160

The ***specific calorific value*** of a substance is *the quantity of heat energy liberated when unit mass of the substance* (*or unit volume in the case of a gas*) *is completely burned.* The units are kJ/kg. 160

MEASUREMENT OF TEMPERATURE

A ***thermometer*** measures the **temperature** of a body and depends for its operation on the expansion of a liquid such as mercury. 162

A ***pyrometer*** is a device for measuring **high temperatures.** 162

Fixed points: 163

Melting point of ice = 0° C or 273 K
Boiling point of water = 100° C or 373 K

Page

Solids expand on heating and the *fractional increase in length per* 166
degree rise in temperature is known as the ***linear expansivity***, α.

Expanded length $l_t = l_o\,(1 + \alpha\, t)$

Linear expansion and contraction in solids may be harmful as in 168
railway lines, bridges, and watch balance wheels, where precautions are necessary to counter it. Expansion and contraction of metals is used in riveting and in thermostats. A ***thermostat*** is a device for 169
maintaining a steady temperature and depends on differential expansion.

In general, liquids expand more on heating than solids, and gases expand even more than liquids.

The cubical expansivity *γ, is the fractional increase in volume of* 170
a substance per degree rise in temperature.

Increased volume $V_t = V_o\,(1 + \gamma t)$

Anomalous expansion of water: though most liquids contract on 170
cooling, water on cooling contracts to 4° C when it achieves a maximum density. Further cooling to 0° C causes **expansion.**

Charles' law: *the volume of a fixed mass of gas at constant pressure* 173
increases by $\frac{1}{273}$ of its volume at 0° C for each kelvin rise in temperature.

$$V_t = V_o\left(1 + \frac{t}{273}\right)$$

This behaviour of gases leads to the postulation of an **absolute** 174
zero of temperature at —273° C or 0 K.

***Absolute temperature*=273° C + *t*° C** 175

Boyle's law: *the volume of a fixed mass of gas at constant temperature* 175
is inversely proportional to its pressure.

$$PV = \text{Constant}$$

Combined gas laws: 175

$$\frac{P_0 V_0}{T_0} = \frac{P_1 V_1}{T_1} = \text{Constant}$$

(temperatures T are measures of **absolute temperature**)

Conversion of volumes to **standard temperature and pressure.** 176

$$V_0 = \frac{P_1 V_1 \times 273}{T_1 \times 760}$$

CHANGE OF STATE

Heat energy supplied to a body does not invariably cause a rise in
temperature. Heat may cause a body to change from the solid to 177

Page

liquid or liquid to gaseous states. **Additional energy** is required by the molecules to bring about this change of state.

The quantity of heat energy needed to change 1 kg *of a substance from the solid to the liquid state at its melting point is known as the* ***specific latent heat of fusion*** *of the substance.* For ice the specific latent heat is 334 kJ/kg. 178

The quantity of heat energy needed to change 1 kg *of a substance from the liquid to the gaseous state at a fixed temperature is known as the* ***specific latent heat of vaporization*** *of the substance.* For water the specific latent heat of vaporization is 2253 kJ/kg. 178

Impurities in liquids lower their freezing points and raise their 179
boiling points. An **increase in pressure** may raise or lower freezing
points and (normally) raises boiling points. Liquids boil at lower 180
temperatures than normal under **reduced pressure.**

When a liquid **evaporates** heat energy is required by the molecules in changing from liquid to vapour and, if this heat is not supplied by an external source, **cooling** of the liquid will result. 184

HEAT ENERGY AND MECHANICAL ENERGY

First law of thermodynamics *expresses the fact that energy is conserved; it can be neither created nor destroyed, but can be changed from one form to another.* 187

Second law of thermodynamics *expresses the fact that heat energy* 190
normally flows from a higher temperature to a lower temperature. In order
to convert heat energy to mechanical work a machine must be used, 191
in which heat energy at a high temperature is changed partly into mechanical work, but a large fraction of the heat energy is rejected at a lower temperature from the exhaust or to the sink.

The **steam engine** and **internal combustion engines** are heat engines employing the principle of the second law of thermodynamics. 192

A **refrigerator** is another type of heat engine, in which heat energy is moved from a lower temperature to a higher temperature (against the natural direction of heat flow), by expending electrical or mechanical energy. 194

Page

KINETIC THEORY

Under normal temperature conditions the molecules and atoms of which a liquid or gas is composed are in a state of **continuous and random motion.** This was first shown by Robert Brown and can readily be seen by the diffusion of coal gas through an unglazed porous pot.

A rise in temperature increases the average speed of the molecules 197
and hence their **kinetic energy.** The kinetic theory of gases predicts their behaviour (if perfect) according to the gas laws of Boyle and Charles.

Molecules exhibit **mutual attraction** and in the case of liquids this 198
attraction gives rise to the phenomenon of **surface tension.** The surfaces of liquids behave like stretched films and tend to reduce their surface areas. Since the liquid molecules are close together such surface tension forces may be considerable.

Evaporation of liquids on supplying heat energy can be understood 199
in the light of the kinetic theory. **Expansion** and **change of state** from liquid to vapour increases the **potential energy** of the molecules. This energy can be supplied by heating when the temperature can be maintained constant, or can be drawn from the liquid itself, when cooling and loss of molecular **kinetic energy** results.

Vapour molecules exert a pressure. The **saturated vapour pressure** 200
of a liquid can be shown to increase with rise of temperature until it finally equals the pressure on its surface, when free boiling begins.

HEAT TRANSFER

Heat energy can be transferred from a high temperature to a lower temperature by ***conduction, convection,*** and ***radiation.***

In general, solids are better conductors of heat than liquids or 201
gases, and metals are the best conductors due to their electron structure. Some solids (non-metals) are very poor heat conductors ***(insulators).*** 202

The **rate of heat flow** across a conductor under constant conditions is 203
given by k × **temperature gradient** × **area of surface,** where k is the ***thermal conductivity*** of the material of the conductor.

The ***Davy safety lamp*** utilizes the good conductivity of copper 203
in avoiding explosions.

Fluids transfer heat more effectively by convection than by con- 204
duction. A rise in temperature causes expansion and a reduction of

Page

density which results in the setting up of a convectional heat flow. Both liquids and gases can be shown to transfer heat by convection.

The domestic hot water system and land and sea breezes can be explained by convection. 206

Heat can be transferred independently of a medium. Hot or luminous bodies give off radiation which, when absorbed by another body, produces a heating effect. 207

Dull surfaces radiate heat better than polished ones and also are better absorbers of radiant heat. This can be shown by the ***Leslie cube*** experiment or by other simple calorimeter and thermometer experiments. 208

SOURCES OF ENERGY

Throughout history man has sought to find means of **harnessing power,** e.g. beasts of burden, water wheels, windmills, steam engines and internal combustion engines. 211

Newcomen's engines at the beginning of the 18th century were very inefficient. More modern **slide-valve engines, turbines** and **internal combustion engines** are more efficient. 214

The ultimate source of all energy is the sun. Green plants can trap the sun's radiant energy and use it in the synthesis of complex chemicals using chlorophyll as a catalyst. It is really such energy that is released when fuels such as coal are burned. Our own food energy can be traced back through a food chain to the sun. Water power can be similarly traced back to the sun as the original source. 215

Attempts are being made to harness radiant energy direct from the sun by mirrors and solar furnaces. Radiant energy can be converted directly into electrical energy by means of ***thermo-electric junctions*** or ***photo-electric cells.*** 217

The sun derives its own energy from nuclear fusion reactions. It is now practicable to use **nuclear fission energy** to produce supplies of electricity. Radioactive atoms can undergo ***fission*** (splitting) into smaller atomic particles with liberation of energy partly in the form of heat and a corresponding loss of mass. 219

Albert Einstein proposed that the relationship between **matter and energy** was given by the equation $E = mc^2$ 219

The fission of uranium 235 provides a chain reaction in which enough **neutrons** are liberated from each uranium atom to promote a further fission. 219

It seems possible that in the future man may be able to use a nuclear ***fusion*** reaction (similar to that of the sun itself) to provide power. 219

LIGHT AND VISION

CHAPTER 19

PHYSICS AS A QUANTITATIVE SCIENCE

THERE follow two descriptions of the same event: the boiling of a kettle of water. They are both scientific descriptions, giving, as far as possible, objective accounts of the event, i.e. accounts about which different observers should agree. If you read them you will notice a fundamental difference between them.

Qualitative Description

A kettle of cold water is placed on a hot stove. At first the kettle feels cool when you touch it, but after a time it feels warm. Later it begins to feel hot, and when you lift the lid and look inside you can see little bubbles in the water. Ordinary water contains small amounts of dissolved gases, mainly oxygen and carbon dioxide. These gases are less soluble in hot water, and so come out of solution to form bubbles of gas when the water is heated.

Later still, clouds of water droplets issue from the spout of the kettle, and inside, the water is bubbling vigorously. The water is said to be boiling. The energy from the stove is used to convert water, at its boiling point, into a gas (steam) at the same temperature. The clouds that you see consist of water formed from the steam as it condenses in the cool air of the room, giving out heat as it does so.

Quantitative Description

(*a*) An electric kettle, which is rated at 2000 watts, contains 1 kg of water at a temperature of 20° C. The kettle is lagged to reduce loss of heat and switched on; the water begins to boil after 177 seconds. The temperature of the boiling water is measured, and found to be 100° C.

(*b*) The procedure in (*a*) is repeated using 2 kg of water, when it is found that the time for the temperature to rise from 20° C to 100° C has increased to 345 seconds.

Amount of electrical energy converted into heat in (*a*) $= 2000 \times 177$

$= 354\,000$ J (joule) ....(1)

Amount of electrical energy converted into heat in (b) $= 2000 \times 345$

$= 690\,000$ J ....(2)

Raising the temperature of the additional kilogramme of water by 80 degrees (80 K) must require an amount of energy equal to the difference between 2 and 1 = 336 000 J..(3)

The value of the *specific heat capacity* of water obtained from these readings

$$= \frac{336\,000}{80} = 4\,200 \text{ J/kg K}$$

The amount of heat needed to raise the temperature of the kettle by 80 K is the difference between (1) and (3) = 18 000 J

Using these readings, the value of the *heat capacity* of the kettle

$$= \frac{18\,000}{80} = 225 \text{ J/K}$$

The most noticeable difference is that the second account contains measurements, and calculations based on those measurements, while the first account is purely descriptive. Account 1 is said to be *qualitative*, while Account 2 is said to be *quantitative*. A qualitative account can be a useful way of describing an event, but usually the results obtained from the calculations in a quantitative account, lead to more useful conclusions. Physics is a quantitative science, and in order to describe things quantitatively, you must make measurements.

In the quantitative description the quantities measured were:

1. The mass of water in the kettle.
2. The temperature of the water (in using a thermometer you are really measuring the length of the mercury thread).
3. The time taken for the water to heat up.

Assuming a power rating for the kettle implies measurement of this quantity, but it will be seen (page 75) that power may be interpreted in terms of mass, length, and time.

Most experiments that you do can be broken down into measurements of these three things. This means that the basic equipment of the physicist is a balance, a ruler, and a stop watch.

You will readily appreciate the need to make your measurements accurately. The ruler is the least accurate of the three basic instruments mentioned: when you measure with a ruler, there are two ways in which you can be in error.

Fig. 216 shows two ways of reading the same ruler which is being used to measure the distance from the block to the line. Two different answers are obtained because the ruler is being looked at from two different directions.

The type of error so obtained is said to be due to *parallax*; you will meet this term again (page 250).

To avoid errors due to parallax, you must be sure to look at the ruler with your eye vertically above the point on the ruler where the measurement is being taken.

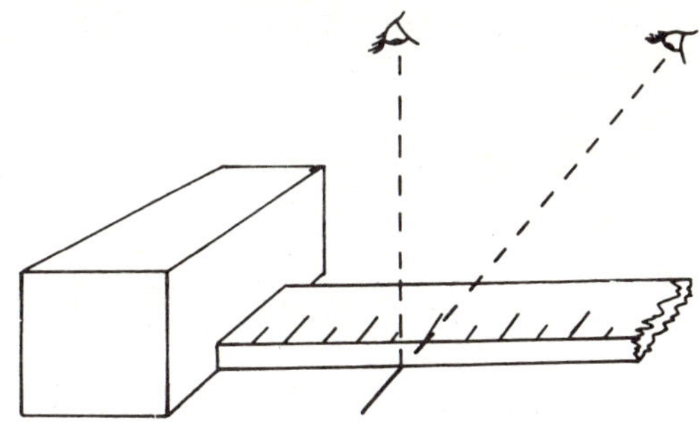

FIG. 216. *Error due to parallax.*

Other errors occurring during this measurement could be caused by failing to place the end of the ruler against the block, or by using a ruler that was not itself accurate.

Random or Personal Errors

Suppose that two people are careful to use the ruler properly; they now ought to obtain the same reading, but in practice they will not necessarily agree. Fig. 217 shows what they see. (This is an ordinary ruler in which numbers refer to cm, although our observers are making their measurements in mm.) They both agree that the measurement lies between 81 mm and 82 mm, but one says that the accurate measurement is 81·3 mm, while the other judges it to be 81·4 mm. This type of error arises because the observers have to exercise judgement and their judgements differ.

FIG. 217. *Error of judgement.*

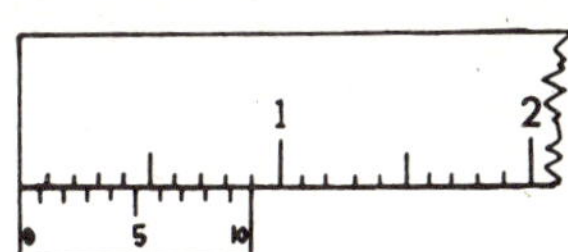

FIG. 218. *Vernier scale.*

The remedy for this type of error is to use a measuring device which will measure to the accuracy required. In the above example, if the two observers had been able to measure to 0·1 of a millimetre, they would not have had to estimate the second decimal place. Two devices capable of such measurements are the *vernier scale* and the *micrometer*.

Vernier Scale

A vernier scale is a small scale which you can use to measure a fraction of a division when reading a ruler. Fig. 218 shows a ruler fitted with such a scale. Notice that although the vernier scale has ten divisions, it is only 9 mm long; each division on the vernier scale is 0·9 mm.

To see how the vernier scale works, you can imagine a vertical ruler (Fig. 219*a*), with the vernier scale represented by 10 wooden blocks, each one

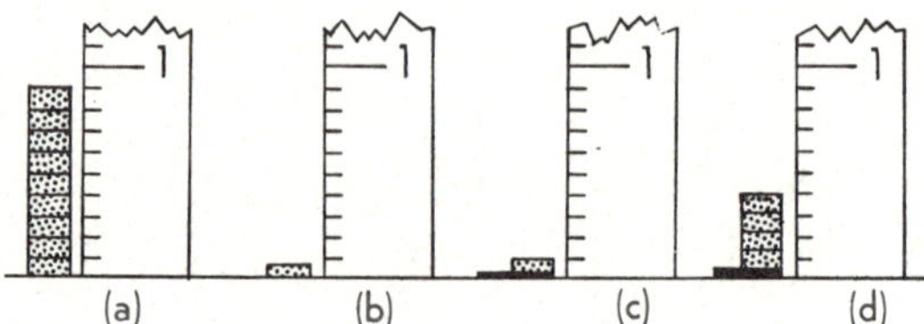

FIG. 219. *How the vernier scale works.*

accurately 0·9 mm high. In Fig. 219*b*, you see that a single block is short of the mark on the ruler. If a plate of thickness 0·1 mm is placed under the block, as in Fig. 219*c*, the block is now coincident with the mark. In this way 0·1 mm can be measured. In Fig. 219*d* a plate of unknown thickness is placed by the rule. When you put blocks on top of it, you find that the top of the fourth block coincides with a mark on the scale, as shown. The plate is making four blocks, each of height 0·9 mm, have a total height of 4 mm; it is clear that the plate must therefore be 0·4 mm thick.

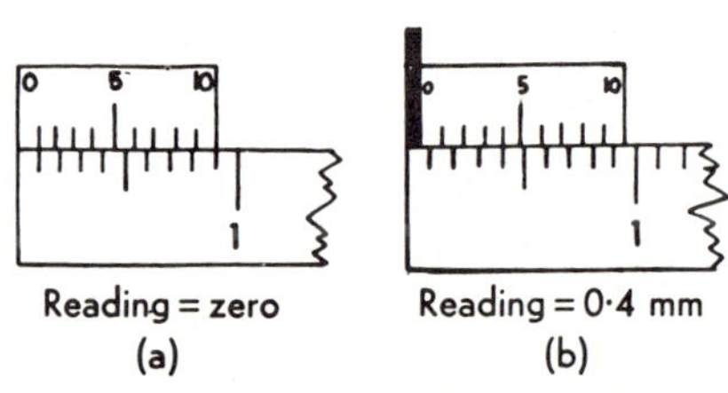

FIG. 220. *Vernier scale readings.*

Fig. 220 shows how this measurement is carried out with a vernier scale. When the vernier is at the end of the ruler, the 0 and 10 marks on the vernier correspond to marks on the ruler. When the vernier is displaced by the

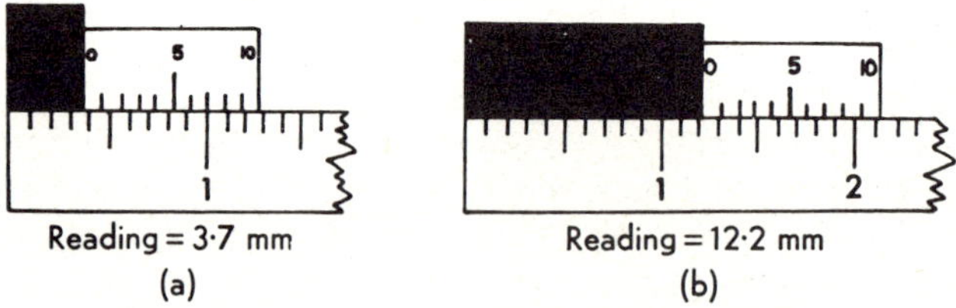

FIG. 221. *Vernier scale readings greater than 1·0 mm*

thickness of the plate (0·4 mm), the fourth mark only on the vernier coincides with a mark on the main scale.

If the vernier is displaced by 1 mm or more, then this amount is read first, and the vernier reading is used to obtain the final figure as shown in Fig. 221.

Vernier Calipers

A pair of calipers is an instrument with jaws, which is used for measuring the thickness of small objects. It is particularly useful for measuring the diameter of round objects.

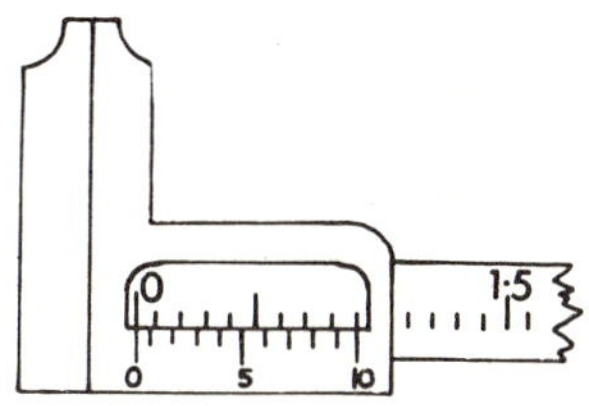

FIG. 222. *Vernier calipers.*

Fig. 222 shows a pair of vernier calipers, in which a scale and vernier scale are used to make the measurement. The jaws are closed, and the vernier scale indicates zero on the main scale. In Fig. 223 the jaws have been opened to admit a cylinder, and you see that the vernier has been moved 11·7 mm along the main scale. If the jaws are closed firmly and squarely on the cylinder, then this reading is the diameter of the cylinder.

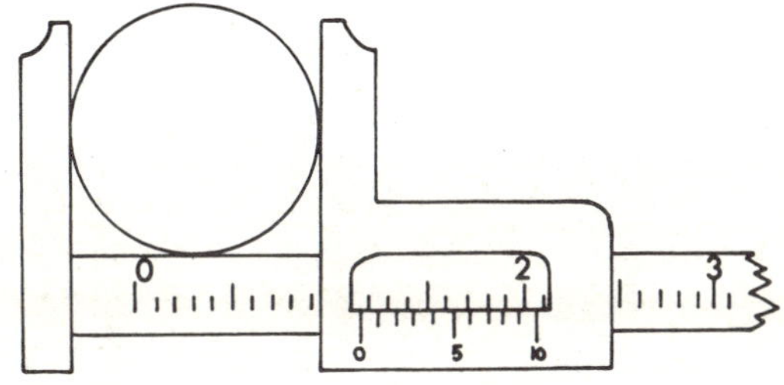

FIG. 223. *Vernier calipers measuring the diameter of a cylinder.*

Zero Error

Suppose that when you close the jaws of the vernier calipers, the scales do not read zero (Fig. 224). The vernier scale reads 0·2 mm, and this reading is called the zero error, because the scale should read 0·0 mm. When these calipers are closed on an object you know that the scales are reading 0·2 mm more than the jaw opening, so you subtract the zero error to obtain the correct reading.

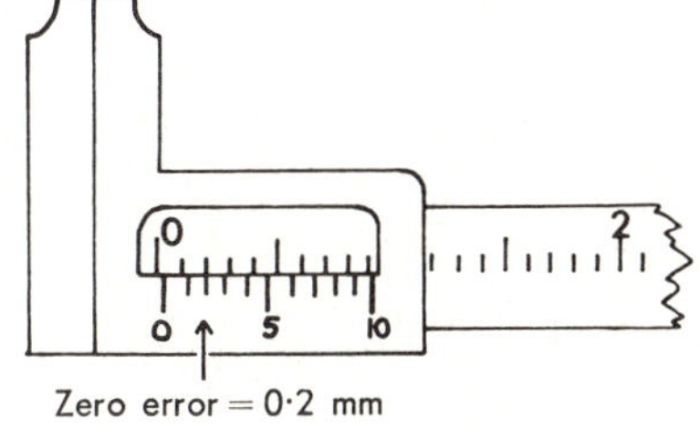

FIG. 224. *Zero error.*

This is an error of the first type (page 234), i.e. it is avoidable; if you fail to allow for the zero error, you are not using the apparatus properly.

Micrometer Screw Gauge

You have probably handled a G-clamp (Fig. 225*a*). You can alter the gap in the clamp by turning the handle. When you turn the handle one complete turn, the screw moves forward a distance P (Fig. 225*b*). The distance P is

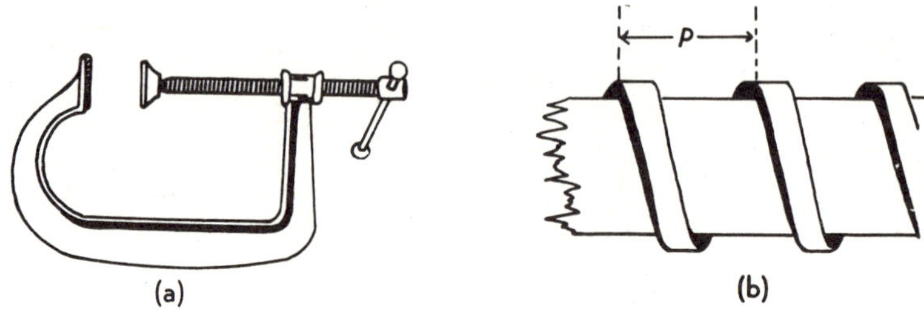

FIG. 225. *G-clamp and the pitch of its thread.*

called the *pitch of the thread.* Assuming the thread to be uniform, if you turned the handle half a complete turn, then the screw would move forward a distance $\frac{1}{2}P$. If you turned the screw $\frac{1}{100}$ of a complete turn, the forward movement would be $\frac{1}{100}P$. You could estimate the gap of the clamp by counting turns, provided you knew the pitch of the screw.

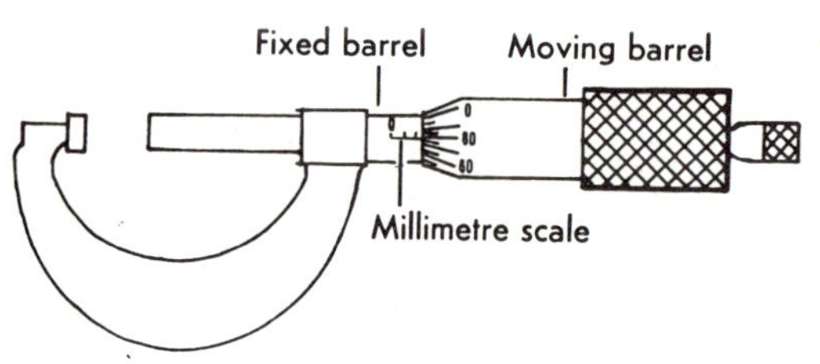

FIG. 226. *Micrometer screw gauge.*

The micrometer screw gauge makes use of a very accurate screw thread. If you wish to measure millimetres, then the screw in the micrometer has a pitch of exactly 1 millimetre. Fig. 226 shows a micrometer.

There are two cylindrical barrels, one inside the other. The inner barrel

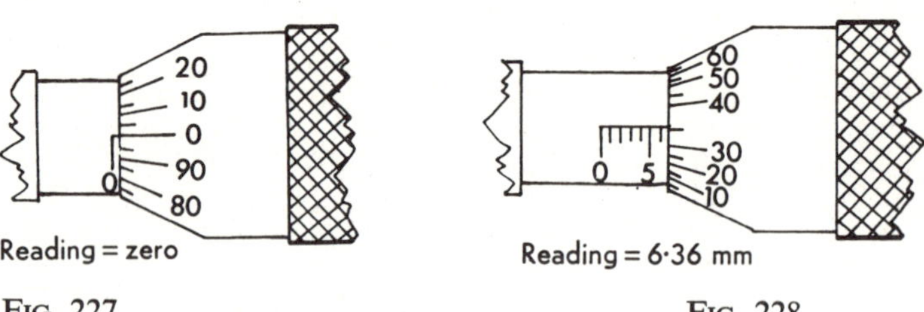

FIG. 227.

FIG. 228.

is fixed, and carries a millimetre scale; the outer barrel is fixed to the screw thread, turns and moves in and out with it, and carries a circular scale around its front edge. When the micrometer is closed, the edge of the outer barrel coincides with the zero of the millimetre scale. The circular scale round the movable scale also reads zero; this scale is read at the line of the millimetre scale (Fig. 227).

As the micrometer is opened, the reading changes. The number of complete turns it is opened corresponds to the number of whole millimetres in the gap, and is read from the millimetre scale. Any fraction of a turn corresponds to a fraction of a millimetre in the gap, and is read on the scale on the outer barrel. This circular scale is divided into 100 parts, so each division represents a movement of 0·01 mm. Fig. 228 shows a typical reading.

If your micrometer does not read zero when fully closed, then you must correct for zero error, as with the vernier calipers.

Magnifying Glass

Vernier calipers and micrometer screw gauges can be made with great precision; their scales consist of fine lines engraved on metal, and are therefore difficult to read. You can read such scales more easily, and with greater accuracy, if you use a magnifying glass (page 292), to make the scale appear larger.

QUESTIONS

1. A rifle bullet, whose velocity is about 300 m/s, breaks two contacts which, respectively, start and stop an electronic timer. If the contacts are placed 6 m apart (with negligible error), what is the maximum allowable error in timing the bullet over this distance, in order to measure its velocity to within one per cent?

2. A pendulum takes 60 seconds to execute 37 complete oscillations. In the time taken for a 0·001 m^3 flask of water to be emptied through a funnel, the pendulum executes 25 swings. What is the average flow of water, in m^3/s, through the funnel?

3. Make a model vernier to fit on a ruler calibrated in centimetres (i.e. your vernier scale will be 9 cm long). To what accuracy will this vernier measure?

4. Make a model vernier, as in Question 3, but take 11 cm instead of 9 cm to divide into 10 equal parts. Confirm that this arrangement will work as a vernier. How should you number the vernier scale?

5. An inch is 25·4 mm. How many kilometres are there in 1 mile?

6. What is meant by the pitch of a thread of a G-clamp? If the pitch of the thread is 4 mm and the handle of the clamp makes $6\frac{1}{4}$ turns, by what distance does the gap of the clamp alter?

CHAPTER 20

PROPAGATION OF LIGHT

WHEN you sit in a darkened room, you can see objects in the room only faintly because there is little light in the room. The darker the room, the less you are able to see. You can easily imagine a completely dark room, in which you are unable to see anything because there is no light at all.

Now switch on, in your darkened room, a single electric light bulb. There is now light in the room, and you can at once see, not only the bulb which is the source of the light, but also other objects in the room which have been made visible by the light from the bulb.

Sources of light are described as *luminous objects*. Electric light bulbs, candles and electric torches are all, when lit, luminous objects. The sun, from which we receive our "daylight," is also a luminous body since it is a source of light. The dials of watches and aircraft instruments, on which the numbers glow because they are faint sources of light, are given the name "luminous dials." The other objects in your room, such as chairs and books, which were not seen until there was a source of light present, are non-luminous objects.

Light is energy in the form of electromagnetic waves, i.e. radiations of the same nature as wireless waves or X-rays. This kind of energy is given out by luminous bodies. Your eye (page 285) is sensitive to this kind of energy, so that when the light from a luminous body falls directly on the eye, you see the luminous body. To see a non-luminous body, such as a book, light from a luminous body must be falling on the book. Some of this light is scattered by the book; it is reflected from the book in all directions so that it seems, indeed, to be coming from the book. When this light enters your eye, you see the book. Light from the source is reaching your eye indirectly by way of the non-luminous body.

How Light Travels

You will have seen a shaft of sunlight coming, through a keyhole or a chink in a curtain, into a darkened room. You may also have noticed that the shaft of light is made visible by tiny dust particles in the air. Remember that light has to enter your eye in order for you to see, so that, if the shaft of light is missing your eye, you will not see it unless it strikes something, in this instance dust particles. The light from a cinema projector on its way to

FIG. 229. *Shafts of sunlight entering windows high up in St. Paul's Cathedral are made visible by the tiny dust particles that lie in their path. Notice how the light appears to be travelling in a straight line.*

the screen is made visible in the same way by smoke particles in the air.

Both the shaft of sunlight and the light from the projector appear to trace out straight lines in the dust or smoke.

You can investigate the way light travels by making fairly large pinholes in three pieces of cardboard. Place the three cards horizontally, at different heights, one above the other. You should then be able to arrange them so that you can see an electric torch pointing upwards through the bottom hole by looking down through the two upper holes. Light from the torch is passing up through all three holes to enter your eye (Fig. 230*a*). You can test whether the holes are in a straight line by hanging a plumb bob, or a small weight from a thin thread passing through all three holes (Fig. 230*b*).

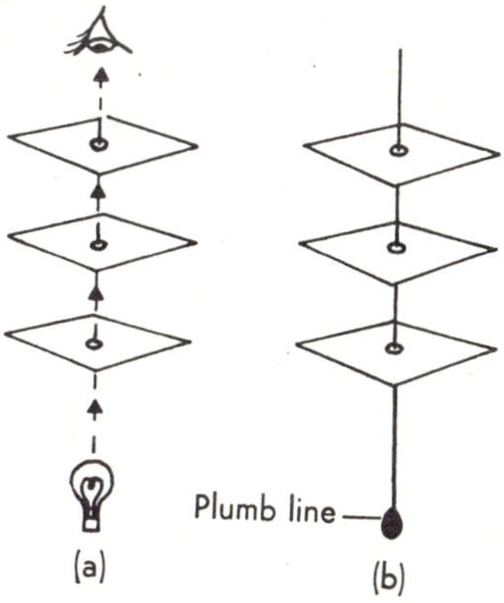

Plumb line

(a) (b)

FIG. 230. *Rectilinear propagation.*

The energy in light waves appears to travel in straight lines and this is sometimes referred to as the *rectilinear propagation of light*. This is only true, as we shall see later, provided there are no materials present which affect the passage of the light by reflection or refraction. It is, furthermore, only approximately true at the best of times for light travels in the form of waves.

FIG. 231. *Under certain conditions two sources of light can be made to interfere with one another and produce darkness as is the case with the dark lines crossing the piece of glass shown here. This phenomenon can only be explained convincingly by the wave theory of light.*

When you watch the behaviour of ripples on water as they join up after passing round an obstacle, it may puzzle you that light, being a wave, travels in straight lines. This is explained by the short wavelength of the light, which limits the extent to which light waves can bend round corners. The bending is very slight but you can tell that it happens if you look with one eye at a light background about two metres away and place your thumb and forefinger close together just in front of this eye, so that you look between them. Close your thumb and finger gradually, and as they just touch, you should see a dark, whiskery "bridge" linking them (Fig. 232).

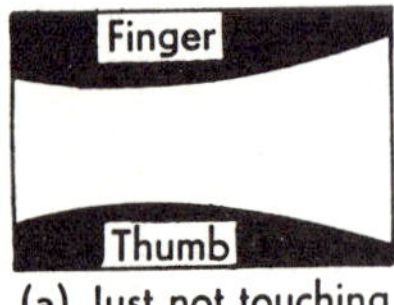

(a) Just not touching

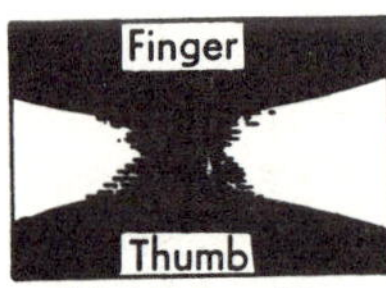

(b) Just touching

FIG. 232.

Where thumb and finger are not touching there is a narrow gap which the slight bending of the light causes to appear larger than it really is, while when the thumb and finger touch, there is no longer any gap at all, hence the abrupt formation of the "bridge".

RAYS OF LIGHT

In all ordinary circumstances you will find that you can regard light as travelling in straight lines, as in your experiment with the pinholes. If these pinholes were made very fine they would define a straight line. In diagrams, therefore, you can represent the direction that light energy is travelling by a straight line. We call this straight line a *ray of light*.

If you go outside on a misty evening with an adjustable flashlamp, you can see the light coming from the lamp as it passes through the mist. You can alter the flashlamp until the light has the appearance of a thick uniform rod. Notice that it is straight, but it is too thick to be represented by a single ray, so

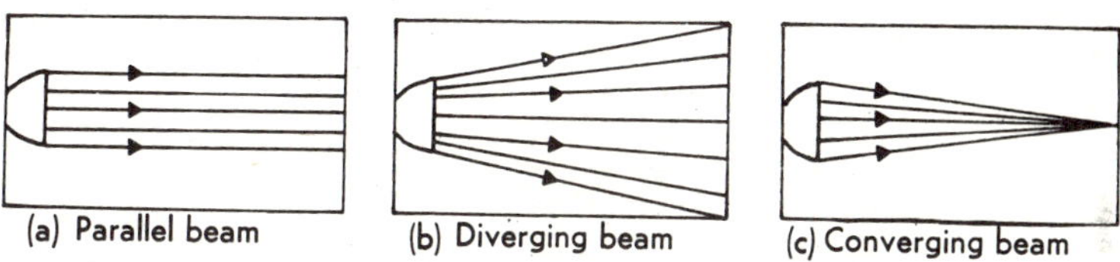

FIG. 233. *Beams of light from a flashlamp.*

it is called a *beam of light*, and is represented by a bundle of rays (Fig. 233*a*). In this particular beam the rays are parallel and it is called a *parallel beam*. If you alter the flashlamp, you can make the beam spread out in a cone. The beam seems to radiate out from a point; this is called a *diverging beam* (Fig. 233*b*). If the beam is made to get narrower as the rays converge towards a point, then you have a *converging beam* (Fig. 233*c*)

Shadows

A candle gives a good shadow of an object provided the object is not too close to it. A pea bulb or small flashlamp bulb produces a better shadow. Referring to Fig. 234 you can easily see why these sources of light give good

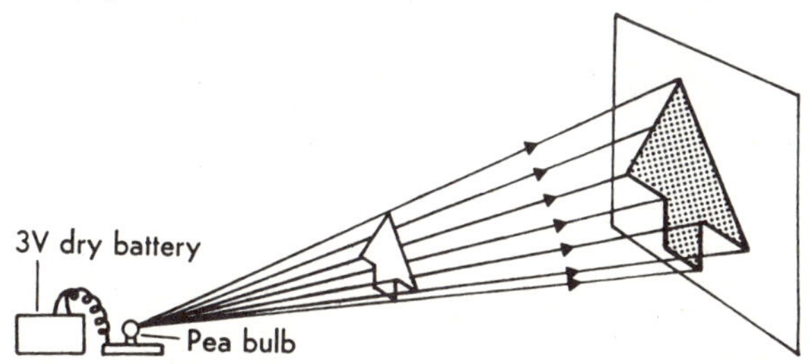

FIG. 234. *Formation of a shadow with a point source of light.*

shadows. They are both small sources so that the light from them comes effectively from a point, the rays radiating out in straight lines in all directions. Any obstacle placed in the way so as to block out some of the rays will leave a space behind it which the light cannot reach. You can see from Fig. 234 that the shadow corresponds to the shape of the obstacle.

You can first examine the shadow formed when the source of light is very small by propping up a ruler 0·5 m from a plain, light-coloured wall. If you throw the shadow of the ruler on to the wall by placing a pea bulb about 0·5 m in front of the ruler, you will see that the shadow *XY* is sharp (Fig. 235*a*). Notice that the shadow is about twice as wide as the ruler, which is to be expected when the wall is twice as far from the bulb as is the ruler.

If you now replace the pea bulb by an unshaded electric lamp, in which there is a "white" light bulb you will find that the appearance of the shadow changes. It is now like 235*b*, having in the centre a completely dark portion *XY*, probably narrower than the ruler itself, and on either side of it an indistinct zone of partial shadow which gives a gradual change from shadow to light, *XW* and *YZ*. These zones are called the *umbra* (*XY*) and *penumbra* (*XW* and *YZ*). If you move the ruler to and fro between lamp and wall you will see the shadow change; the nearer the ruler is to the wall, the more umbra there is, while nearer to the lamp the penumbra predominates, and a point can be reached at which the umbra disappears altogether, provided the lamp is wider than the ruler.

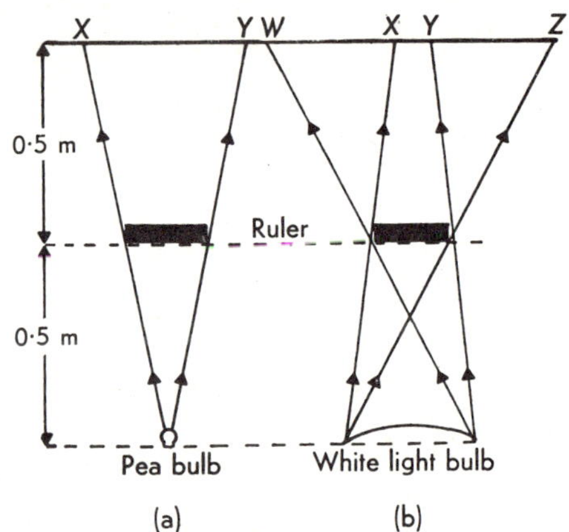

FIG. 235. *Formation of umbra and penumbra.*

The bulb cannot be regarded as a point source of light; it is an *extended source*.

It can be seen from Fig. 235*b* that between points *X* and *Y* on the wall, light cannot reach the wall from any part of the bulb. *XY* is therefore completely dark and is the umbra. To the left of *X*, light from some of the bulb can reach the wall, and the amount of bulb giving light to the wall increases until at *W* light from the whole of the bulb can reach the wall. The regions of *XW* and *YZ* make up the penumbra.

Eclipses of the sun occur when the moon passes between the sun and the earth. The moon's shadow falls on the earth and the eclipse is seen from within the region where this shadow falls. An eclipse can be partial or total.

The sun is an extended source, therefore the shadow of the moon will consist of umbra and penumbra. The diagrams of Fig. 236 (in which the

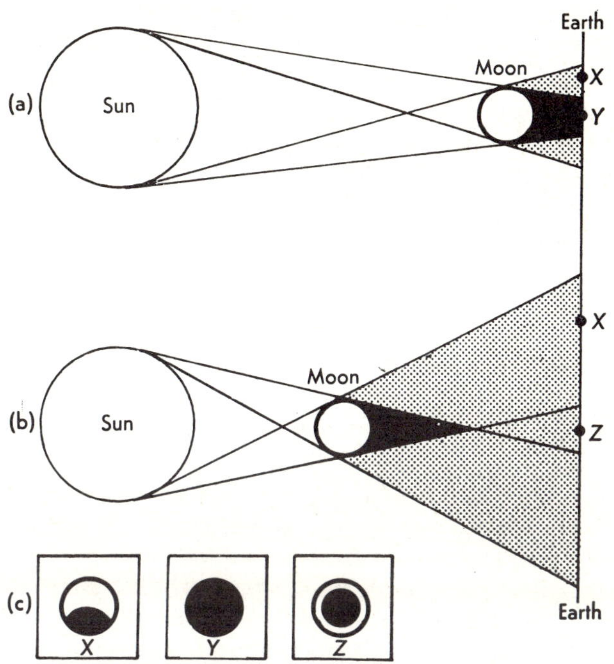

FIG. 236. *Full and partial eclipses of the sun.*

sun, moon and earth are not drawn to scale), show how eclipses occur.

In Fig. 236*a*, *Y* is a point at which the eclipse is total, while at *X* in the penumbra, the eclipse is partial. If the moon is farther from the earth, as sometimes happens since the orbit of the moon is not a perfect circle, then there may be no total eclipse at all, because the umbra does not fall on the earth (Fig. 236*b*). In this case at the point *X* the normal partial eclipse is still visible, while at the point *Z* the entire edge of the sun can be seen but not the centre; this latter is an annular eclipse. Fig. 236*c* shows the types of eclipse visible from the points *X*, *Y* and *Z*.

Pinhole Camera

You can make a suitable pinhole at the centre of a fairly large piece of cardboard (0·2-0·3 m square) with a coarse needle or fine nail. Set this cardboard vertically about 0·25 m from a candle, and have a screen covered with white paper about 0·25 m on the other side of the pinhole from the candle (Fig. 237). If the room is otherwise dark, you can see on the screen an image, X' Y', of the candle flame, and, more faintly, one of the candle. Notice that this image is upside down. If you alter the distance from pinhole to screen you will find that the image X'' Y'' remains sharp; there is no

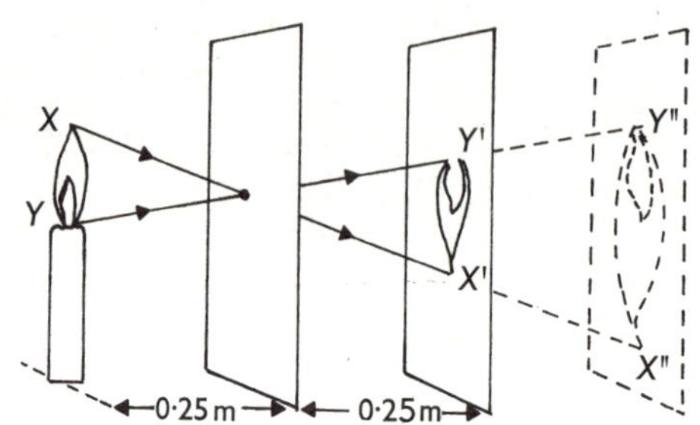

FIG. 237. *Formation of an image by a pinhole camera.*

question of focusing this image. It also changes in size; the size of image depends on the relative distances of object to pinhole and pinhole to screen.

Fig. 237 also shows how the image is formed. Rays of light which leave the candle travel in straight lines. Of the rays from the tip X of the flame, only a few can pass through the pinhole and these will all strike the screen at X' which will be the tip of the image. Similarly, rays from Y on the candle strike the screen at Y' giving X' Y' as the inverted image of the flame because all points between X and Y on the flame will give corresponding image points in this way.

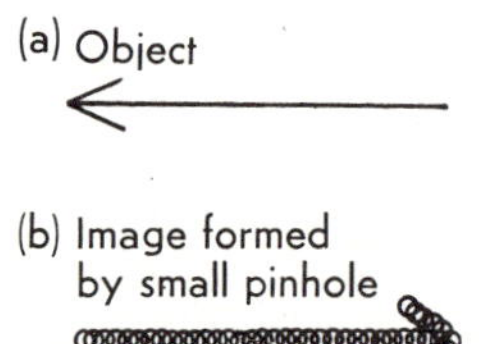

FIG. 238.

If you now move the screen to the position marked with the dotted line you can see that the image is formed in exactly the same way, but is now much larger.

If you leave the pinhole and screen in position and enlarge the pinhole you should notice two changes. Although the image is the same size, it is now brighter because more light gets through the larger hole. It is also blurred because the light from each point on the candle now forms a small circular patch on the screen. These patches make up an image of the candle, rather as a series of dots can make up a newspaper photograph, but because they overlap, the image is indistinct. Fig. 238 shows the effect of the size of the pinhole when the object is arrow-shaped.

From this you will see that the shape of the pinhole does not matter;

FIG. 239. *The principle of the pinhole camera was first mentioned more than 2,000 years ago by the Greek scientist, Euclid, who noted that the image formed by the sun was round even when the pinhole was square. In the form of the Camera Obscura (dark room) shown in the picture, it was used for observing the sun, moon and stars. Copernicus, Brahe and Kepler used it. In the 16th century lenses were introduced; in the early 19th century means were found of recording the image; the modern camera was then developed.*

Fig. 238*c* could just as well be composed of small triangles or small squares as the small circles shown. You can check this by cutting your larger pinhole into a triangle and then a square, but it is better if you keep it less than about 3 mm across.

The pinhole camera is a lightproof box, pierced by a pinhole at one end. Light through the pinhole forms, on the back of the box, an image of the objects at which the camera is pointing. If photographic film is placed at the back of the camera, then it records any image which is formed on it. The pinhole must, therefore, be kept covered when the camera is loaded with film, and only uncovered briefly to take a photograph. In making your pinhole camera, remember the following points.

1. The size of image depends on length of camera and distance of object from camera. All objects are in focus.

2. Sharpness of focus depends on the fineness of the pinhole, but a small hole also means a faint image which requires a long exposure of the film.

3. The pinhole need not be any particular shape.

N.B.: If you make the pinhole *very* small, then you will get a blurred image and not the very sharp image that you might expect. As we have already noted, the light is not travelling in absolutely straight lines so the effect with a very small hole is like that which you saw with your thumb and finger (page 242).

You can make a model pinhole camera by replacing the back of a small box by a piece of ground glass or tracing paper, and making a pinhole at the other end. By pointing the box at bright objects such as windows, you will be able to see, even in daylight, the image which would be formed on the film.

On a sunny summer's day, when you are walking beneath trees, the sunlight filtering through the leaves makes small round patches on the ground. On examination, you will see that these patches are all circular, whereas the gaps between the leaves are more likely to be angular. The small spaces between the leaves are acting as pinholes, so that their shape does not matter, and the circles you see on the ground are all images of the sun.

Sometimes in a dark corridor you will see, on the wall opposite a closed door, an image of a window in the room beyond the door; the keyhole is forming a "pinhole" image.

QUESTIONS

1. If you look closely at the shadow formed in sunlight by a tall vertical post, you will see that the base of the post forms a sharp shadow, while that of the top of the post is less distinct. Draw diagrams to explain why this is so.

2. A rod 30 mm in diameter is mounted horizontally 500 mm beneath and parallel to a fluorescent light tube 40 mm in diameter. A shadow is formed on a table 1 m beneath the rod. What are the relative areas of the umbra and the penumbra?

3. A source of light consists of a narrow vertical slit, illuminated from behind. A square of cardboard, held vertically in front of the slit, casts a shadow on a screen. If the slit is half as long as the edge of the square, and the cardboard is midway between slit and screen, describe the appearance of the shadow.

4. Draw a diagram showing the positions of the sun, moon, and earth during an eclipse of the sun. At what places on the earth's surface would the eclipse be: (*a*) total; (*b*) partial?

5. Describe a pinhole camera. What determines: (*a*) the size of the image; (*b*) the sharpness of the image?

6. Why are small patches of sunlight on the ground beneath trees circular in shape? And in what ways would these patches of light be affected during a partial eclipse of the sun?

CHAPTER 21

MIRRORS AND REFLECTION

IF YOU look in a flat mirror hanging on the wall of a room, you will see an *image* of the room. This image has depth; it is three-dimensional, and it is almost as if you were looking through a window the size of the mirror into a room like the one in which you are. This image seems to lie behind the mirror.

If you approach the mirror, your image appears to approach the surface of the mirror from the other side. When a mirror forms an image of an object, the position of the image depends upon the position of the object.

If you raise your right hand as if to shake hands with your image you observe that your image seems to raise its left hand. Right and left are inter-

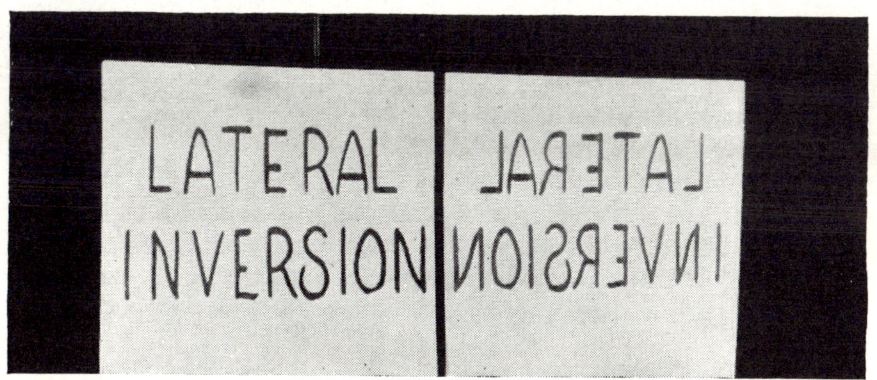

FIG. 240. *The writing* (*left*) *forms a laterally inverted image in the mirror* (*right*). *Have you ever noticed lateral inversion on a piece of used blotting paper?*

changed in the mirror image which is said to be "*laterally inverted.*" Lateral inversion accounts for the strange appearance of mirror writing, seen when ordinary writing is held up to a mirror (Fig. 240).

PLANE MIRRORS

These observations have been made using a flat mirror; mirrors of this type are given the name *plane mirrors*. Let us now find the position of the image in a plane mirror.

You can use plasticine to stand a small plane mirror vertically on a piece of white paper on a deal drawing board. Draw a straight line to mark the position of the mirror and set the back of the mirror accurately on the line. For an object you can stick a pin into the board about 100 mm in front of the mirror. On looking from one side of the object pin *O*, you should see an image *I* of it in the mirror.

When you are looking at the image from a position well to one side of the object pin, carefully stick two pins, *A* and *B*, into the board so that they appear to be exactly in line with the image *I* (Fig. 241). You can repeat this by looking from the other side of the object pin and inserting two more pins, *C* and *D*, again in line with the image. The image *I* lies on the line through pins *A* and *B*, and also on the line through pins *C* and *D*. If you remove the mirror and draw these lines, their intersection will give the position of the image *I*. Join *O* and *I* and measure the angle that this line makes with the mirror; measure also the distances of *O* and *I* from the mirror. You should find that the image formed in a plane mirror lies on the perpendicular from the object to the mirror, and is as far behind the mirror as the object is in front of the mirror.

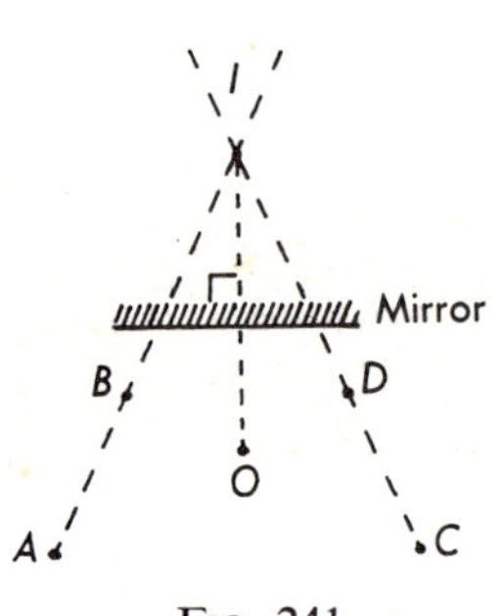

FIG. 241.

This explains the three-dimensional appearance of images in a plane mirror, and at the same time accounts for lateral inversion.

Suppose that you are looking at the letter "L" in a plane mirror. You are holding up a sheet of paper with an L printed on it in front of a mirror so that you are seeing the image of the paper in the mirror. The image of each part of the letter L lies on the perpendicular to the mirror so that the image is laterally inverted. Confirm this for yourself by drawing a diagram.

Method of No Parallax

There is another way in which you could find the position of the image in the plane mirror in the previous experiment. To understand it you can try to solve the following problem. Get someone to stand an object approximately underneath a hanging lamp. A music stand on a coffee table works very well. If you stand near one wall of the room and cover one eye, how can you decide, without moving your feet, whether the music stand is nearer or farther away than the centre of the lampshade?

You see at first the two objects in line, as in Fig. 242*a*. If you now lean to your right, so as to look from a slightly different position, you see either Fig. 242*b*, Fig. 242*c* or Fig. 242*d*. The small plans with each of these figures show why the relative positions of lamp and stand change. If you lean to the right, then whichever of the two is farther away seems to move to the right also.

This movement is known as *parallax*. In Fig. 242*d* there is no relative movement of lamp and stand; the stand is here directly under the lamp. This is *the position of no parallax*.

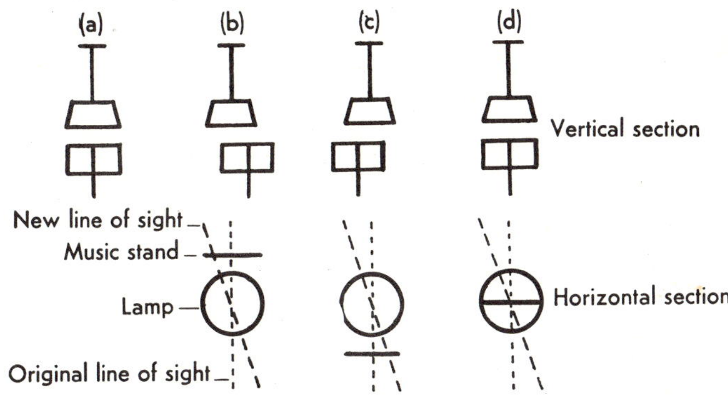

FIG. 242. *Finding the position of no parallax.*

You can use this technique to find the position of the image *I*, which a plane mirror forms of a pin *O*.

Set up the mirror and object pin *O* on your drawing board as before. If you insert at *S* (Fig. 243), a large pin, you should be able to see this pin over the top of the mirror. Since you can see the image *I* in the mirror at the same time, you can adjust the position of *S* until there is no parallax between *S* and *I*. *S* now gives the position of the image *I*.

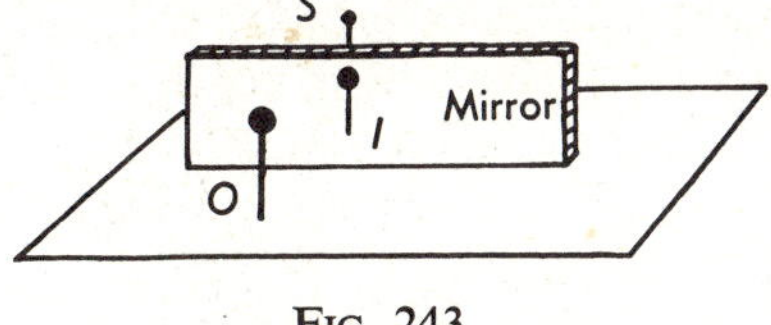

FIG. 243.

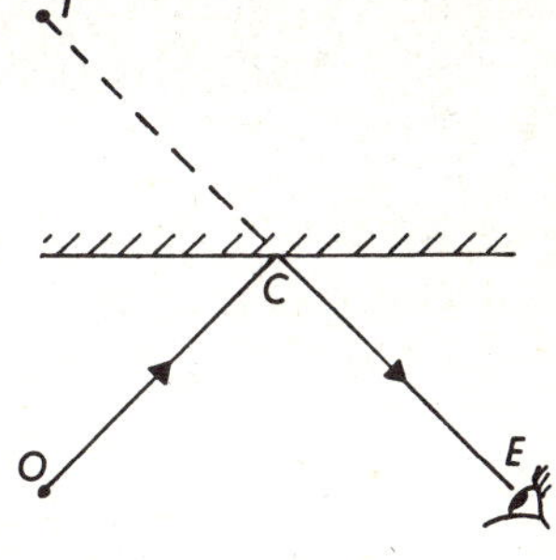

FIG. 244.

Laws of Reflection

Fig. 244 shows a plane mirror forming an image *I* of an object pin *O*. When you place your eye at *E*, you see the image at *I* because rays of light are entering your eye from the direction of *I*. Consider a ray of light entering your eye along *CE*. This ray must be one of the rays leaving the object *O*, because when it enters your eye, you see the image of *O*. Therefore the path of the whole ray must be *OCE*. The change of direction of the ray at the mirror is a *reflection*.

Now try an experiment to show how a ray of light is reflected at a plane mirror. Mount your mirror vertically on a piece of white paper on a drawing board (Fig. 245), and insert two pins *A* and *B* to mark a ray of light at about 45° to the mirror surface; when you look along the line *AB* so that *A* and *B* appear in line, you are looking along the ray of light through *A* and *B*. If you look from the right you should see the images *A′* and *B′* of *A* and *B*. Insert pins *C* and *D* so that they are in line with *A′* and *B′*.

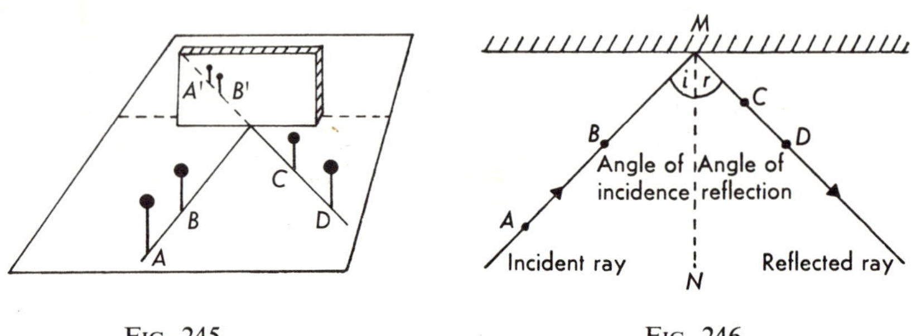

FIG. 245.

FIG. 246.

CD gives the path of the ray *AB* after it has been reflected from the mirror (Fig. 246). *AB* is called the *incident ray*; *CD* is called the *reflected ray*. The line *MN*, which is drawn perpendicular to the mirror at the point of incidence, is called *the normal*. The angle *i* between incident ray and normal is the *angle of incidence* and, similarly, angle *r* is the *angle of reflection.*

After you have removed the mirror, you will be able to draw both rays. Draw the normal to the mirror through the point at which they meet so that you can measure angle *i* and angle *r*.

You can repeat your measurements for rays making angles of incidence of about 15°, 30°, 60°, and you will notice that *i* and *r* are always approximately equal. Small differences are due to experimental error.

Replace the mirror on one of your sheets of paper after the pencil lines have been drawn in. The lines will be reflected like those in Fig. 245, provided the mirror is perpendicular to the board (use a set square). The image *A′ B′* of the line *AB* appears to be a continuation of the line *CD*. This means that the incident ray and the reflected ray can both lie along the paper, provided that the normal to the mirror does too. We may therefore summarize the laws of reflection:

1. *Incident ray, reflected ray, and normal to the mirror at the point of incidence all lie in the same plane.*
2. *The angle of reflection is always equal to the angle of incidence.*

These laws do not compel the ray of light to be reflected in this way;

FIG. 247. *Five images are formed when a thimble is placed between two mirrors inclined at 60° as shown. Investigate for yourself the number of images that are formed when three mirrors are used, placed at angles of 120° to each other.*

they simply describe the way in which we have found light to be reflected.

Types of Image

If you think about the sort of image formed by a pinhole (page 246), you will realize that it is different from the sort of image formed by a plane mirror. The pinhole image is formed on a screen by rays of light passing through the pinhole; the rays of light are in fact present at the image, so this is called a *real image.*

When you look at the image in a plane mirror, you see the image by means of rays of light which have been reflected by the mirror, while the image is behind the mirror (Fig. 244). The rays only seem to come from the image; it is called a *virtual image.*

Uses of Mirrors

You should be familiar with the everyday uses of mirrors; in physics, however, mirrors have special uses because they can increase the accuracy of measurement.

Elimination of parallax. Fig. 248 shows that a reading of a meter will be inaccurate if you do not look squarely at the dial. With the eye at A the reading will be correct. With the eye at *B*, however, there will be an error in the reading due to parallax (compare this error with the error in reading a ruler, page 235).

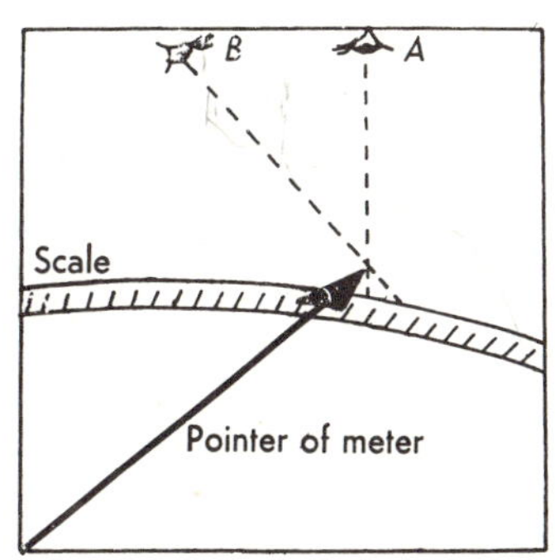

FIG. 248. *Parallax error.*

Since many measurements consist of reading a meter, plane mirrors are often fitted on the dial. When you use this type of meter you will see an image of the pointer in the mirror. If you move your head so that this image seems to lie directly behind the pointer, then you must be looking at right angles

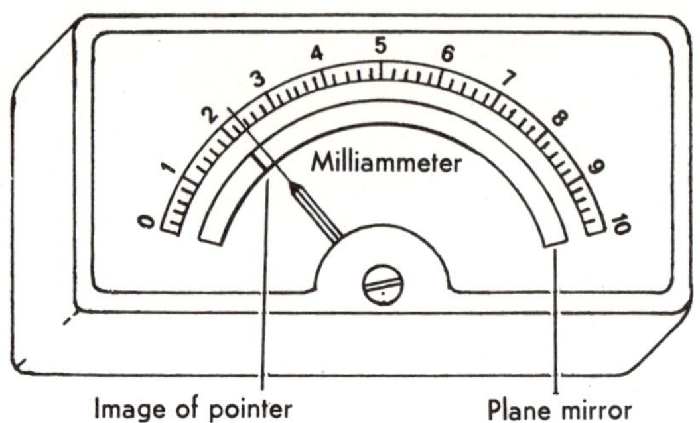

FIG. 249. *Use of plane mirrors to eliminate error due to parallax.*

to the mirror. Why? The mirror is parallel to the dial, so you are in the correct position for taking a reading. Fig. 249 shows a milliammeter with a strip of plane mirror inserted in the dial for this purpose.

Optical lever. You have already found (page 139) that a lever can be used to magnify small movements; the longer the lever the greater can be the magnification.

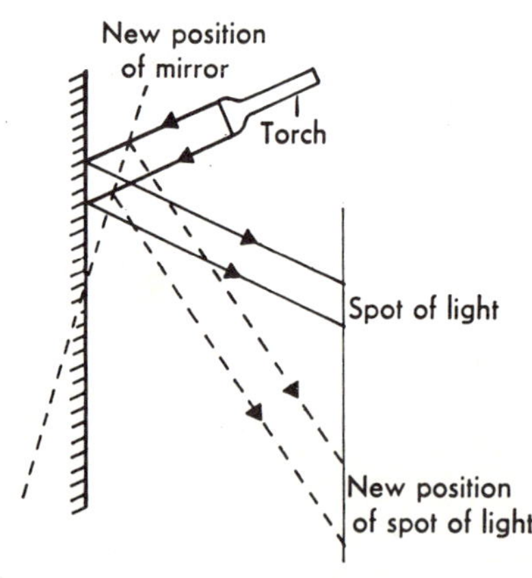

FIG. 250. *Optical lever.*

If you reflect a beam of light from a plane mirror on to a screen (Fig. 250) then you will find that a very small movement of the mirror causes the spot of light on the screen to move appreciably.

Certain instruments, used for making small accurate measurements, often use the principle of the optical lever; for example a sensitive galvanometer which is used for measuring minute electrical currents. In this case the small deflection of the coil is often measured by reflecting a beam of light from a mirror attached to the coil.

Mirrors are things which reflect light well; the reflecting surface is smooth and the mirror can form images. You can see images formed by reflection

FIG. 251. *Images can be formed by the reflection of light at the surface of still water. Notice in this case that the water is not perfectly still and smooth, so the image is slightly distorted.*

at the surface of still water or flat glass. Here the surface is reflecting light like a mirror. This is called *regular reflection.*

You do, on the other hand, see a non-luminous object such as a book because light falling on the book is scattered in all directions so that some of it enters your eye. This scattering is called *diffuse reflection*, and it happens because the book has a dull rough surface. If you look at a piece of paper with a low power microscope, you will find that it has a rough fibrous appearance. This means that rays of light striking the surface will all make different angles of incidence because of the uneven nature of the surface. Fig. 252 shows a parallel beam of light undergoing diffuse reflection from the surface of a piece of paper.

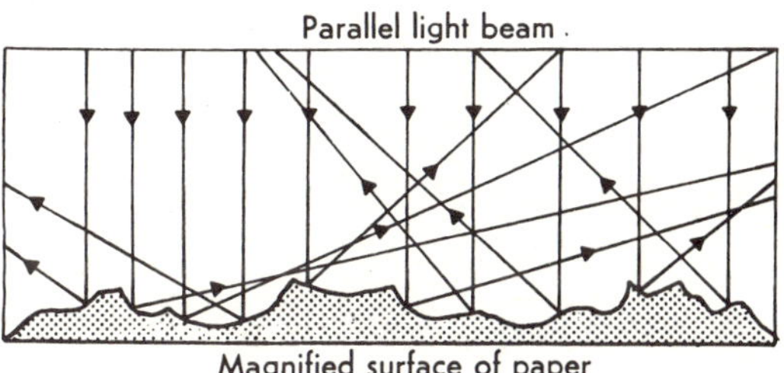

FIG. 252. *Diffuse reflection from a rough surface.*

Slate has a surface which is slightly rough, so that light falling on it normally is scattered, giving diffuse reflection. If the light falls on a flat piece of slate at a large angle of incidence, however, then most of the light is reflected regularly. Many glossy surfaces behave in this way, e.g. shiny paper, in which you can see the reflection of a lamp if you look along the surface of the paper.

The reflecting efficiency of different surfaces also varies. Some surfaces reflect most of the light falling on them, perhaps only in a diffuse way. Such surfaces are bright, e.g. white paper. Other surfaces absorb most of the incident light and appear dark, e.g. black paper, which reflects light poorly. Other substances (see page 303) absorb light selectively.

Ghost Images

Sometimes, when you look at the reflection of a candle or light bulb in a fairly thick mirror in an otherwise dark room, you see faint images on either side of the main image (Fig. 253). The images are easier to see if you are looking at the mirror obliquely so that the angle of reflection is large. These extra images are called *ghost images* and are regarded as a defect of this type of mirror since they interfere with any observations you make in the mirror.

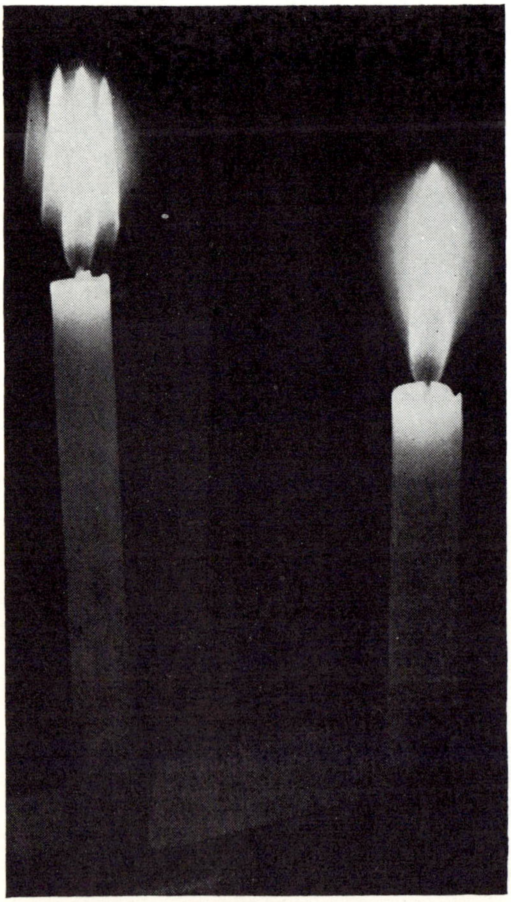

An ordinary mirror consists of a flat sheet of glass, on whose back surface there is a thin layer of silver. This layer of silver is the reflecting surface and it is protected by a layer of paint. When light is to be reflected from the silvered back surface of the mirror, it has first to pass through the glass.

Notice (Fig. 254) that the light changes direction as it passes into the glass. This is explained in a later section (page 263). Some of the light is reflected at the front surface of the glass, giving a faint

FIG. 253. *The candle flame* (*right*) *has formed two ghost images in the mirror to the left of the main image* (*one is rather faint*), *and a third to the right of the main image.*

image above the main image. A little light is reflected back on to the silvered surface as the light leaves the glass, giving an extra image below the main image. This could happen several times to give a number of ghost images on this side of the main image; often only the first ghost image is bright enough to be seen. A thicker mirror gives a wider separation of the images, as you will see if you draw out Fig. 254 with a thicker piece of glass.

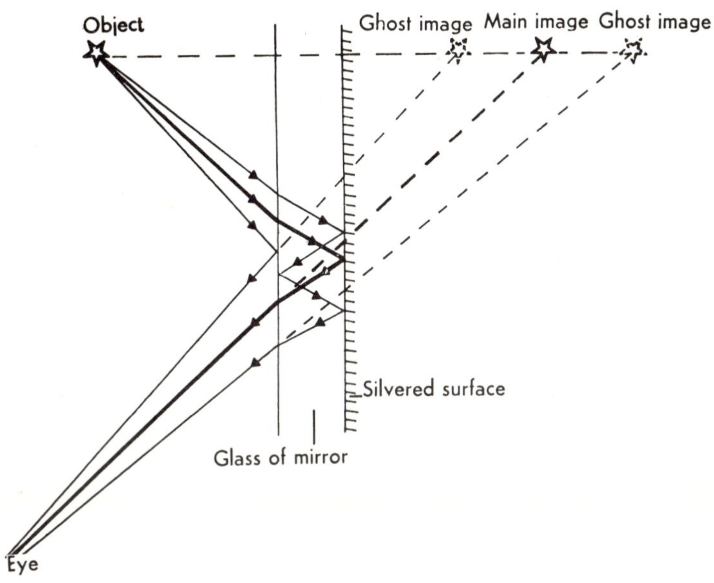

FIG. 254. *Formation of ghost images in a back-silvered mirror.*

When ghost images must be avoided, it is necessary to use either a prism (page 271) or a "front silvered" mirror. A front silvered mirror has its reflecting layer on the front surface of the glass. The layer is of aluminium, which needs no protection against tarnishing; it is deposited on glass so that it is smooth and shiny. Mirrors of this type are expensive and easily damaged.

Curved Mirrors

It is possible to make mirrors of various shapes by silvering appropriately shaped pieces of glass. Such mirrors do not give the same sort of images as the plane mirrors we have been considering. You may be familiar with distorting mirrors.

The most common types of curved mirrors are known as spherical mirrors. They may curve outwards or inwards, for if you take a hollow glass sphere and cut off the ends at *AB* and *CD* (Fig. 255), then at *AB* you could silver the inside surface, and at *CD* the outside surface as shown (we are considering back silvered mirrors so both of these would reflect light coming from the left). Looking at the reflecting surfaces, mirror *AB* curves outwards and is called a *convex mirror*, mirror *CD* curves inwards and is called a *concave mirror*.

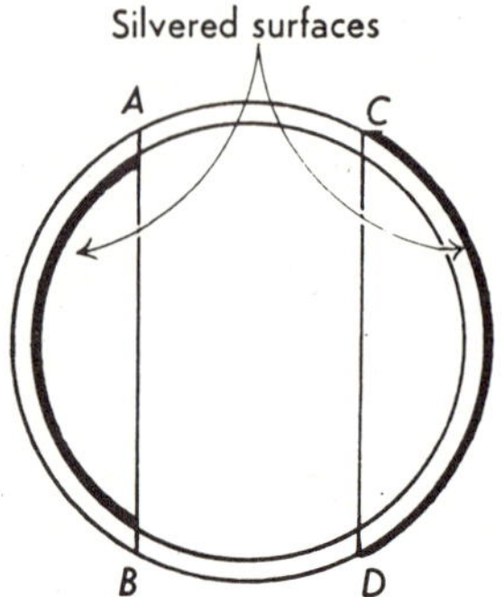

FIG. 255. *Formation of back-silvered mirrors: convex (left); concave (right).*

The laws of reflection apply to a curved mirror, provided that you draw the normal to the surface at the point of incidence of the ray.

Convex Mirrors

A convex mirror is shown in Fig. 256*b*. The light entering your eye comes from a wider range of angles than does the light from a plane mirror the same size (Fig. 256*a*). For this reason, convex mirrors are used as rear view mirrors

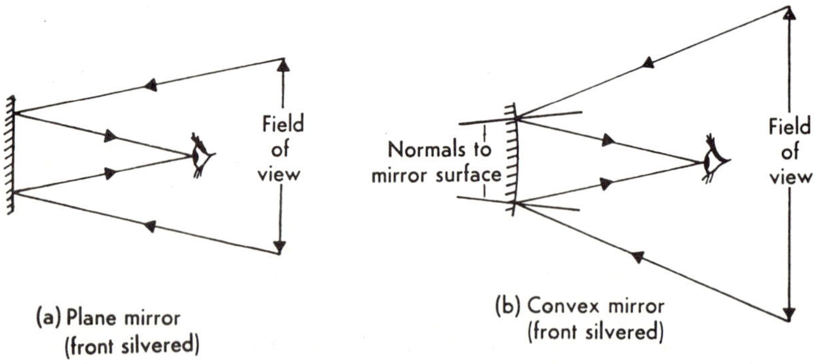

FIG. 256. *Comparison of field of view for plane and convex mirrors.*

in cars, giving a large field of view. The images in these mirrors are smaller than the corresponding plane mirror images because the field of view is larger.

Concave Mirrors

Some special shaving mirrors are concave; if you have seen one you will know that they can give a magnified image. Fig. 257 shows how this happens.

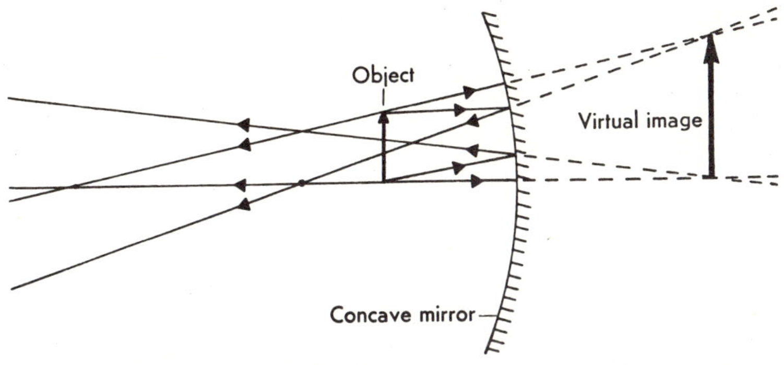

FIG. 257. *Formation of magnified virtual image in a concave mirror.*

If the image is magnified it means that the field of view can only be small. Note that the image is virtual.

Another use of concave mirrors is to produce beams of light. The rays

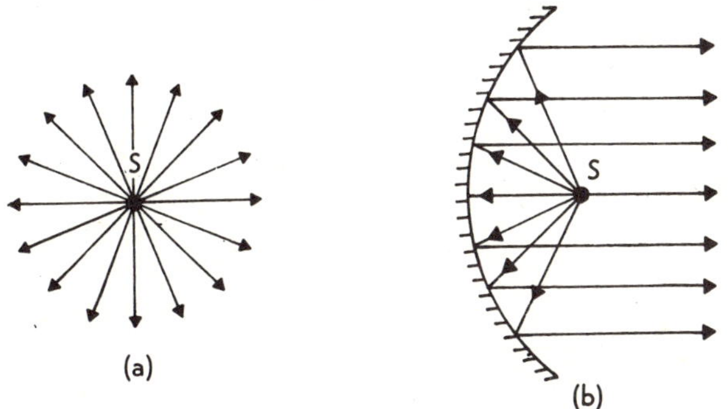

FIG. 258. *Use of concave mirror to form a parallel beam of light.*

from a point source *S* radiate out in all directions (Fig. 258*a*). If a concave mirror is placed behind the source then a position can be found where the rays reflected from the mirror form an approximately parallel beam (Fig. 258*b*).

FIG. 259. *Popularly used as ornamental mirrors to give a far more extensive view of the room in which they are hung than would a plane mirror, convex mirrors find a more practical use as rear view mirrors for cars. In this application the wide field of view can contribute much towards the safe handling of the car.*

If an accurately parallel beam is needed, as for instance in a searchlight, then the reflector must be parabolic rather than spherical (Fig. 260). Note that the parabolic reflector can come right around the source using much more of the light; a spherical reflector could clearly not do this, as you can prove with a sketch. Parabolic reflectors are used in motor car headlamps.

Images in Curved Mirrors

To examine the images formed by convex and concave mirrors look first at your reflection in the back of a well polished soup spoon (an ordinary spoon will do but it is not so spherical).

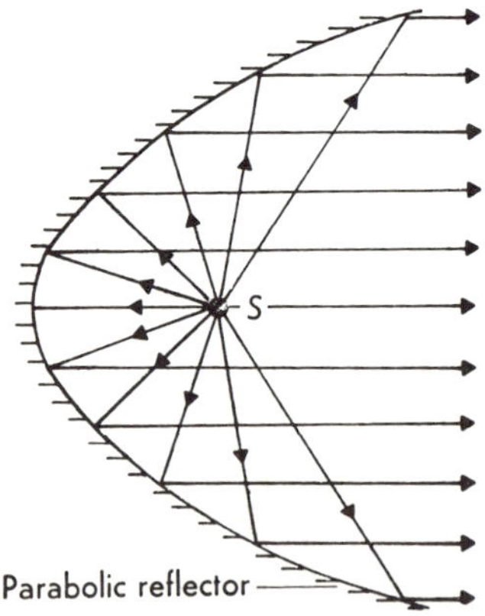

As you move the spoon backwards and forwards you will notice that your image is always erect, virtual, and about the same distance behind the spoon. The back of the spoon is a convex mirror.

S

Parabolic reflector

FIG. 260.

Use the front of the spoon as a concave mirror, holding it at arm's length. Your image is inverted and diminished. Hold the forefinger of your other hand a few inches away from the spoon and you will be able to find a position for it, such that there is no parallax between it and the image of your face. The image of your face is formed in front of the mirror and is therefore a real image. Why? Move your forefinger nearer to the spoon and you will see an image of your finger. This is an erect, magnified image, which is behind the mirror and therefore virtual; this is

FIG. 261. *The searchlight reflector into which the boy is looking is but one useful application of the concave mirror. Smaller versions are often used as shaving mirrors while still smaller concave mirrors are sometimes used by dentists to give a magnified image of the tooth they are examining.*

the kind of image formed by a shaving mirror. When you read Chapter 22 on lenses, you will see that this type of behaviour is similar to that of a convex lens.

QUESTIONS

1. State the laws of reflection of light. If an object is placed before a plane mirror, where does the mirror form an image of the object, and why is this image called a virtual image?

2. Trees by the side of a lake can be seen by reflection in the water only if the water is still: why is this? They appear by reflected light to be upside down: explain this.

3. A man, whose eyes are 1·8 m from the ground, stands 1·8 m from a mirror, whose bottom edge is 1 m from the ground. How much of his legs are cut off in his reflection?

4. Some barbers' shops have mirrors on opposite walls, so that the mirrors are parallel. In them you see yourself many times, successive images being farther and farther away. Draw a diagram to explain this. Do all the images in one mirror appear to be the same (allowing for distance)?

5. When a ray of light from an object is reflected at a plane mirror so as to enter your eye, it is taking the shortest path between the object and your eye. Suggest a rough demonstration to verify this, using a length of string.

6. A ray of light is incident normally on a plane mirror; how is it reflected? Draw a careful diagram to show how the same ray is reflected when the mirror is turned through an angle. Through what angle does the reflected ray turn through?

CHAPTER 22

REFRACTION AND LENSES

If an object is immersed in water and you look at it from above the surface, its appearance changes. Take a straight stick and put it partly in water at an angle of 30° to the surface. The stick will appear bent (Fig. 262).

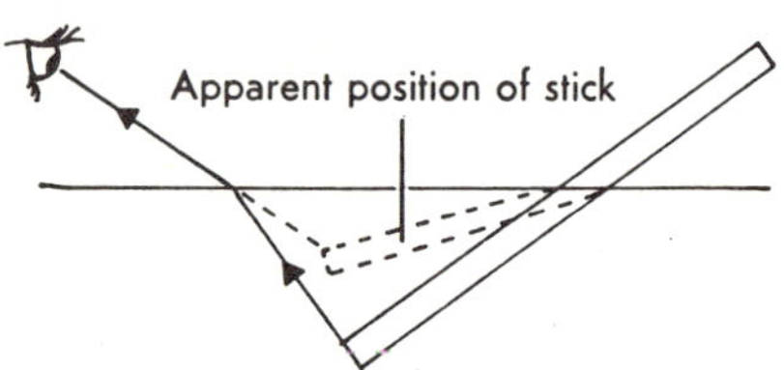

Fig. 262. *Refraction of light.*

Although rays of light can be regarded as travelling in straight lines in any one material, if they pass into a different transparent material, then the rays of light change direction at the boundary between the two materials. This bending of a ray is called *refraction.*

To explain your observations with the stick, you must draw a diagram showing the ray of light with which you see a point on the stick (Fig. 262). This ray is refracted at the surface of the

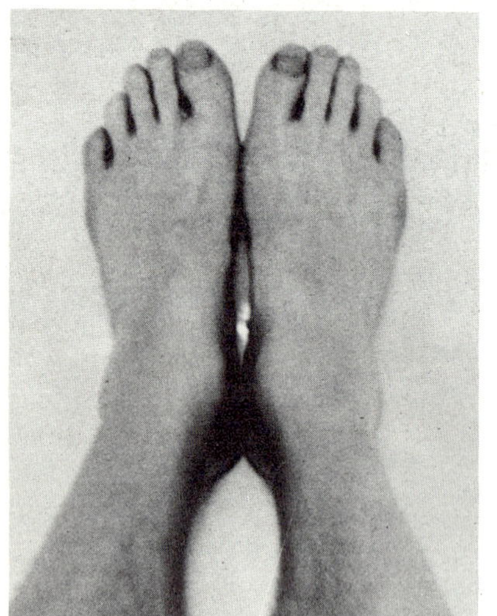

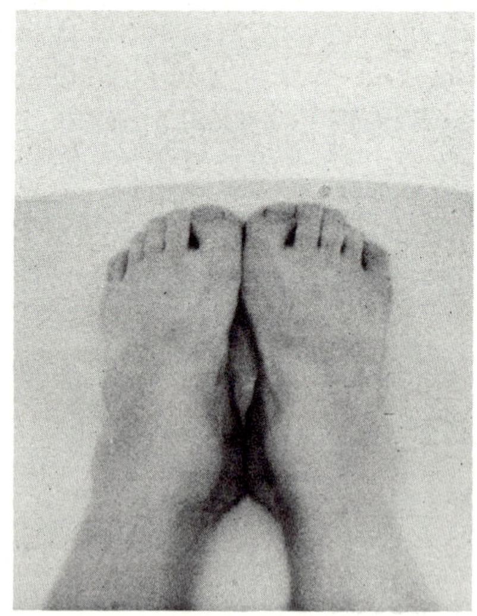

Fig. 263. *Refraction causes the compressed appearance of feet in a bath of water (right). Appearance of feet in empty bath (left).*

water, so that you get a false impression of the direction in which this point on the stick lies.

A transparent material is known as a *medium* for the transmission of light, and Fig. 264 shows the refraction which takes place when a ray of light passes into a medium which is denser than the medium which the ray is leaving. The ray is bent towards the normal (the angle of incidence $\theta_1 >$ the angle of refraction θ_2). When a ray of light passes into a less dense medium, then it will be refracted away from the normal ($\theta_1 < \theta_2$). In both instances, only a ray perpendicular to the surface, i.e. a ray passing along the normal to the surface, will not be deviated.

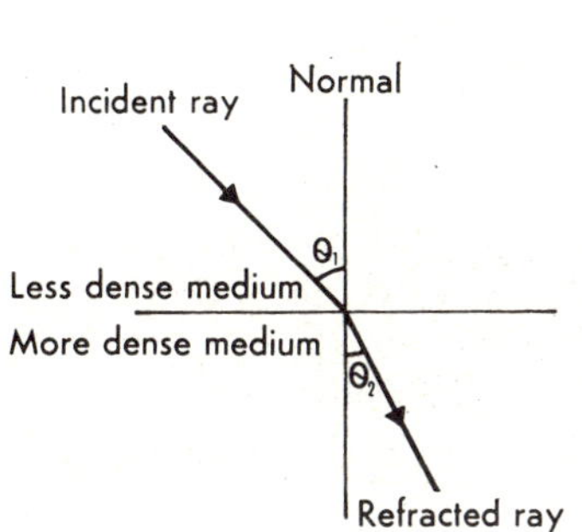

FIG. 264. *Refraction of light.*

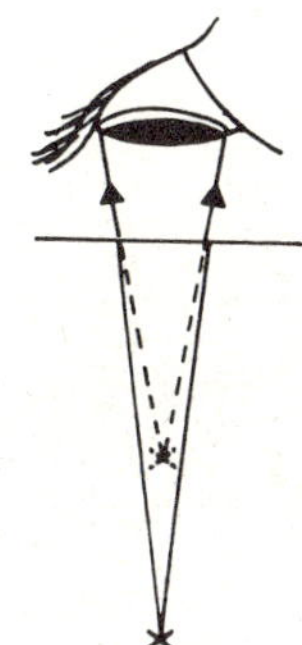

FIG. 265. *Apparent depth.*

The observations you have made all show that distances in a more dense medium do not appear to be as large as they really are. The depth of the end of the stick beneath the surface (Fig. 262) appears less than it really is, and if you check this with a ruler, then the ruler will also appear to be shortened. That you see this apparent depth, rather than the real depth, even when looking vertically downwards, is understood by realizing that, since the pupil of the eye has width, you never see the normal ray alone, but always rays oblique to the surface. Refraction will therefore take place and the cross at the bottom of some water (Fig. 265) will appear higher up than it really is.

THE LAWS OF REFRACTION

1. *The incident ray and the refracted ray lie in the same plane as the normal and on opposite sides of it.*
2. *The ratio of* $\dfrac{\sin \theta_1}{\sin \theta_2}$ *has a constant value for light passing between two given substances.*

θ_1 is the angle between ray and normal in one medium, and θ_2 the angle between ray and normal in the other medium; the light may be travelling in either direction.

FIG. 266. *It can clearly be seen from this photograph how a beam of light passing from air into water is refracted. Notice that the beam of light passing from the water at the side of the glass, is bent away from the normal.*

These laws are really rules describing how light behaves when it is refracted. The second one is called *Snell's Law* after its discoverer. The value of the constant will be different when different materials are used.

Investigating the Laws of Refraction

You will already have observed that the stick appears bent in a plane perpendicular to the surface of the water. This demonstrates the accuracy of Law 1, as does the fact that the following experiment is performed on a flat sheet of paper.

Fasten a plain sheet of paper to a drawing board, and put a glass block or a plastic container filled with water, flat at the centre. Mark the position of the block in pencil. Put two vertical pins at *A* and *B* (Fig. 267), so that the line *AB* makes an angle of about 30° with the normal to the block. Then if you look from such a position that the pins appear to be in line, you must be looking along the ray of light which passes through both pins. Look at the pins *A* and *B* through the block as shown, and insert two pins *C* and *D*, so that all four pins appear to be in line. *ABCD* must therefore be the path of one ray of light, and although it appears to be a straight line looking through the block, viewed from above it is clear that the ray is refracted at both surfaces.

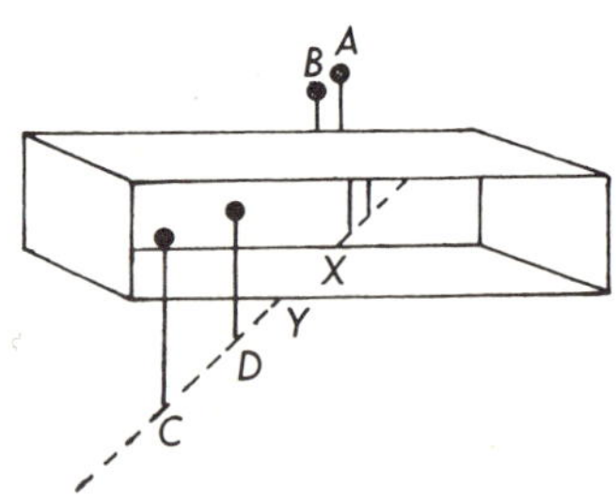

FIG. 267. *Water-filled block.*

Remove the block and draw straight lines through *AB* and through *CD*. If line *AB* meets the block at *X*, and line *CD* meets the block at *Y* (Fig. 267),

XY must be the path of the ray inside the block and may be drawn in with pencil. Replace the block and, with the pins out of the way, look at the pencil line ABX through the block. It should appear to be a continuation of the line CDY.

Measure the four angles shown in Fig. 268, and record their values (Table 10). θ_A refers to the angle measured in air, and θ_W to the angle measured in the water. There are two refractions, so you can measure the angles for each. (Keep the angles correctly in pairs, i.e. θ_A and θ_W in one row of the table must refer to the same refraction.)

Look up the sine of each angle, and divide sin θ_A by sin θ_W, recording all your results in the table. Repeat these measurements, making the angle θ_A have values of about 40° and then 50°. Complete the table.

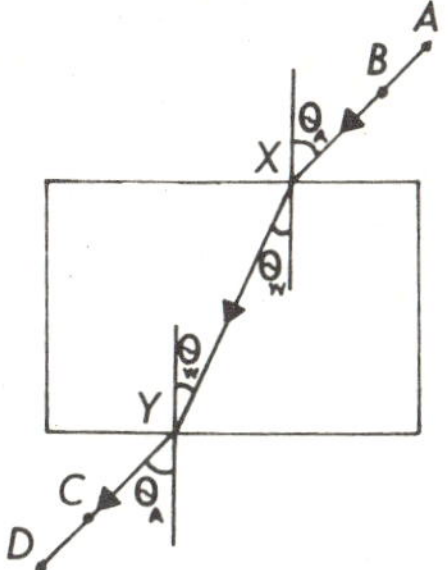

FIG. 268.

The measurements recorded in Table 10 were taken using a water-filled block made from transparent plastic. If you use a solid glass block, then you will measure, in the glass, angles θ_G each one smaller than the corresponding value of θ_W. The ratio $\frac{\sin \theta_A}{\sin \theta_G}$ will therefore be higher than the value of $\frac{\sin \theta_A}{\sin \theta_W}$.

The values of $\frac{\sin \theta_A}{\sin \theta_W}$ are not all the same. There is a slight variation due to experimental errors, but it is clear that they are all estimates of the same quantity with an average value of about 1·33.

Table 10

Investigation of Snell's Law

θ_A	θ_W	sin θ_A	sin θ_W	$\frac{\sin \theta_A}{\sin \theta_W}$
31° 30′	23° 15′	0·522	0·395	1·32
31°	22° 45′	0·515	0·387	1·33
38° 15′	27° 30′	0·619	0·462	1·34
37° 45′	27° 15′	0·612	0·458	1·34
48° 45′	34° 15′	0·752	0·563	1·33
48° 15′	34°	0·746	0·559	1·33

Refractive Index

If you obtain different values for the ratio of $\frac{\sin \theta_1}{\sin \theta_2}$ when you use different pairs of substances, then this constant must depend on the optical properties of the two media concerned in the refraction. If you wish to describe the way a single substance behaves, then you can do so by using, as the other medium, a vacuum. The constant for the refraction is then the ratio of $\sin \theta_{\text{vacuum}}/\sin \theta_{\text{substance}}$ and this constant is called the *refractive index* of the substance. The symbol for the refractive index is *n*.

You would say, for instance, that if a ray of light were passing between a vacuum and glass, that:

$$n_{\text{glass}} = \frac{\sin \theta_{\text{vacuum}}}{\sin \theta_{\text{glass}}}$$

In practice, air behaves optically very much like a vacuum, the ratio of $\sin \theta_{\text{air}}/\sin \theta_{\text{glass}}$ has very nearly the same value as n_{glass}, and this is a useful approximation.

It follows that when a ray of light is passing between a material and air, *in either direction*, the refractive index of the material is the ratio:

$$\frac{\text{Sine of angle between normal and ray in air}}{\text{Sine of angle between normal and ray in material}}$$

The normal referred to here should, of course, be drawn at the point at which the ray crosses the boundary of the material.

In advanced work it is found that the refractive index of a substance is the ratio:

$$\frac{\text{Speed of light in a vacuum}}{\text{Speed of light in that substance}}$$

Rectangular Glass Block

You may have noticed that in the previous experiment the two values of θ_A are approximately the same for each ray. This is because a parallel-sided block is used, making the two angles θ_W (Fig. 268) equal, since they are alternate angles. The two refractions ought therefore to be identical. The ray emerges from the block along a path parallel to the path along which it enters the block, but has been displaced sideways. This sideways displacement increases if the angle θ_A is made larger, as may be verified experimentally or by drawing an accurate diagram, calculating the correct values for θ_W. This calculation is done below for a parallel sided glass block.

Suppose that the glass of the block has a refractive index of 1·5 and that the incident ray makes an angle of 22° with the normal.

$$\text{Now we know that } n = \frac{\sin \theta_A}{\sin \theta_G}$$

$$\therefore \frac{\sin 22^\circ}{\sin \theta_G} = 1{\cdot}5$$

$$\therefore \sin \theta_G = \frac{\sin 22^\circ}{1{\cdot}5} = \frac{0{\cdot}3746}{1{\cdot}5} = 0{\cdot}2497$$

Now use your sine table in reverse to find the angle whose sine is 0·2497; it is an angle of 14° 28′.

$$\therefore\ \theta_G = 14° 28'$$
(this is approximately 14·5°)

If you draw a rectangle *ABCD* to represent the glass block (Fig. 269), you can mark in your ray of light incident at *O*, making the angles to the normal 22° and 14·5° as shown. Continue the refracted ray to the other side of the block and draw in the normal at *N*. The angle here will be about 14·5° if you have been accurate, so that it will emerge at an angle of 22° to the normal. If you extend the line of the emergent ray backwards along *NP*, and measure the distance *PQ* at right angles to the two lines, then *PQ* is the sideways displacement.

Try this calculation for the same block when the incident ray makes an angle of 45° with the normal.

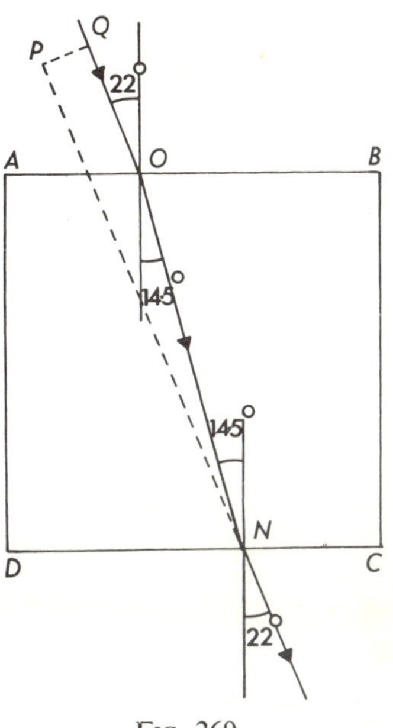

FIG. 269.

Triangular Prisms

Referring back to Fig. 268, if the sides of the block are not parallel, then the two angles marked θ_W will not be equal and the refractions at the two surfaces will not be identical. The extreme example of a block with non-parallel sides is a triangular prism (Fig. 270*a*). Fig. 270*b* shows the passage of a ray light through such a prism made of glass. At *P* the ray is refracted towards the normal, i.e. deviated downwards from its original direction. Because of the inclination of the second face, the ray strikes this face above the normal at *Q*. Passing from glass into air, the ray is bent away from the normal, so that it undergoes further deflection downwards from its original direction. The effect of the prism is an alteration in the

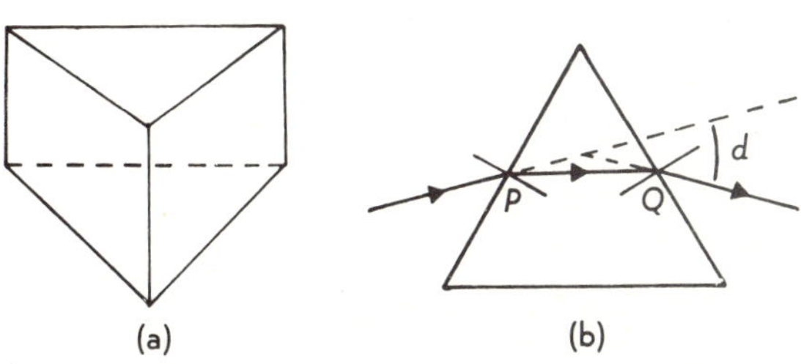

FIG. 270. *Triangular glass prisms.*

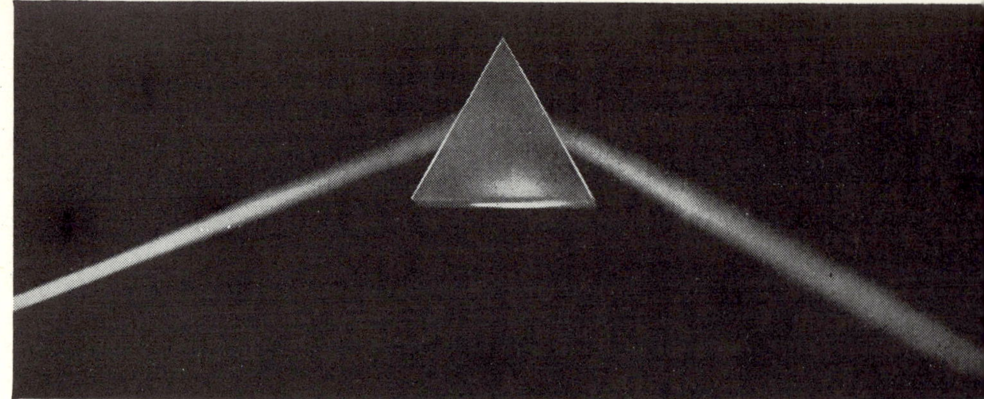

FIG. 271. *The photograph shows how a beam of light from a source at the left is deviated by a triangular glass prism.*

direction of the light ray by an angle *d*, an alteration shown in Fig. 271.

This deviation of the light may be observed using an ornamental prism from a chandelier. To look at an object through the prism you will have to look in a direction more towards the apex of the prism than the object really is. This difference in direction represents the deviation produced by the prism. You will notice too, that objects viewed through the prism become fringed with colours. This aspect of prisms is explained later (page 296).

Mirages and the Critical Angle

The photograph (Fig. 273) shows the type of mirage that deceives the desert traveller who would see a lake amongst the sand dunes.

You will see this type of mirage in strong sunlight when the surface of the ground becomes hot; the layer of air next to the ground becomes hot, expands, and therefore becomes less dense than the layer of air above it. Light from

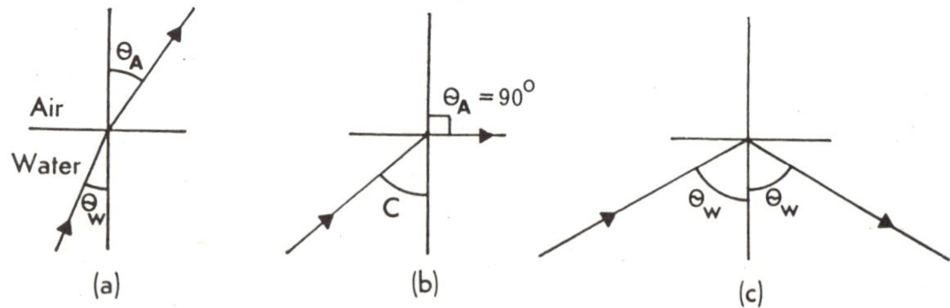

FIG. 272. *Refraction; critical angle; total internal reflection.*

the sky is passing through a more dense medium to strike a less dense medium when you see the mirage.

When light is passing into a less dense medium, it is refracted away from

FIG. 273. *Notice the mirage on right horizon. Such mirages are often seen in this country on a hot day, particularly on long straight roads.*

the medium (Fig. 272*a*). As the angle θ_W increases the angle θ_A also increases, remaining greater than θ_W in accordance with Snell's law. It is possible therefore to increase the angle θ_W to such a value c that the angle θ_A is 90° (Fig. 272*b*). This is clearly the largest angle at which the refracted ray can emerge from the more dense medium; c is called the *critical angle.*

If the angle θ_W is made larger than c (Fig. 272*c*), then the ray of light is completely reflected, obeying the laws of reflection, because it cannot emerge from the water. This is known as *total internal reflection* (note that partial reflection takes place whenever a ray is refracted, for some of it is reflected at the surface).

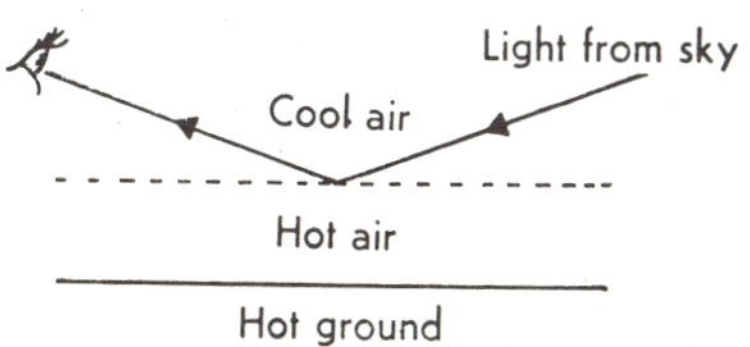

FIG. 274. *Mirage formation.*

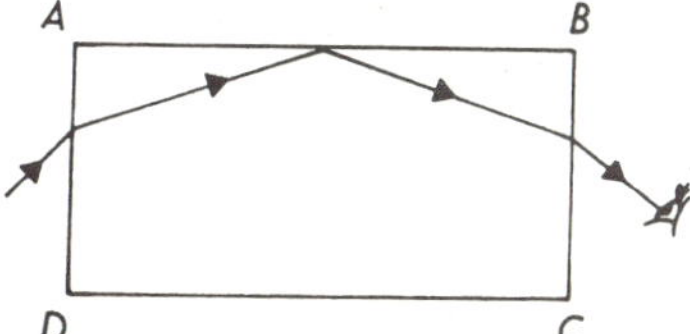

FIG. 275. *Total internal reflection.*

The mirage (Fig. 273) is caused by total internal reflection. Light from the sky is reflected at the less dense layers of air and you see, looking at the ground, an image of the sky (Fig. 274). Realism is lent to the illusion by the variations of the air layers which cause the image to shimmer like water.

Total internal reflection may be observed in the glass block used earlier. Referring to Fig. 275, if you look at face AB through the face BC, it appears mirror-like. This is because you are looking at the face AB at an angle greater than the critical angle, so that the only light you receive from face AB is that

FIG. 276. *A beam of light entering the glass of water from the left, strikes the surface of the water at an angle greater than the critical angle and so undergoes total internal reflection. This can only happen when the light is trying to pass from a more dense medium into a less dense medium, in this case from water into air. This is the principle behind the formation of a mirage, where light attempts to pass from a layer of cool air into a layer of hot air.*

which has been totally reflected at that face (see photograph. Fig. 276).

Conditions for total internal reflection to take place:

1. Light must be incident on the boundary of a less dense medium.
2. The light must be incident at an angle greater than the critical angle.

Relation Between Critical Angle and Refractive Index

The critical angle is related to the refractive index of the material. Consider a ray of light passing from water to air at the critical angle, i.e. the ray emerges from the water at an angle of 90° to the normal (Fig. 272*b*).

$$\text{Then } n = \frac{\text{sin of angle between ray in air and normal}}{\text{sin of angle between ray in water and normal}}$$

$$= \frac{\sin 90^\circ}{\sin c} = \frac{1}{\sin c} \qquad (\sin 90^\circ = 1)$$

therefore for water and air, since the refractive index of water is $\frac{4}{3}$,

$$\frac{1}{\sin c} = \frac{4}{3} \therefore \sin c = \frac{3}{4} \therefore c = 48^\circ\ 36'$$

Reflecting Prisms

If you calculate the value of the critical angle for glass, taking $\frac{3}{2}$ as an average value for the refractive index, you will find that $c = 41^\circ\ 48'$. This is less than 45° and makes it simple to construct reflecting prisms. These are made in the shape of triangles having two angles of 45° each and one right angle.

FIG. 277. *Binoculars and field glasses use a system of prisms to invert laterally and vertically the image formed by the lens system. A ray of light entering the objective (bottom left) passes into the top prism where it undergoes total internal reflection at two surfaces, leaving the prism in a direction opposite to that in which it entered. The inverted ray then passes into the bottom prism where a similar inversion takes place and the ray is reflected into the eye-piece (top right).*

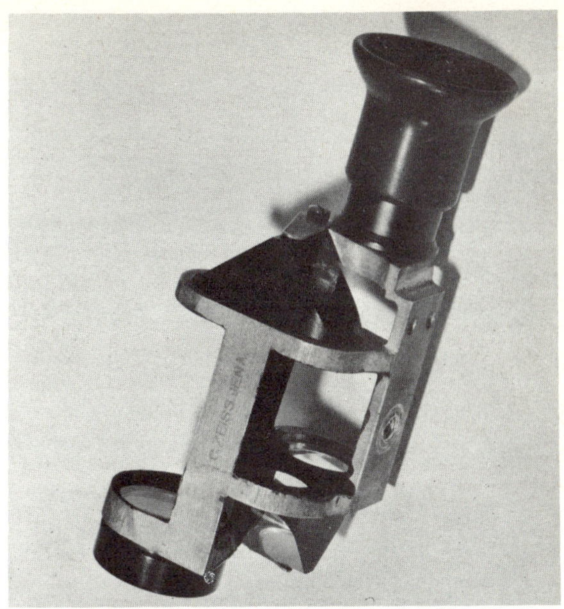

In Fig. 278 a ray of light entering normally through one of the short sides strikes the hypotenuse at 45° ($> c$) and is totally reflected to leave the third side normally. Two prisms used in this way would make a periscope such as is used in tanks and submarines.

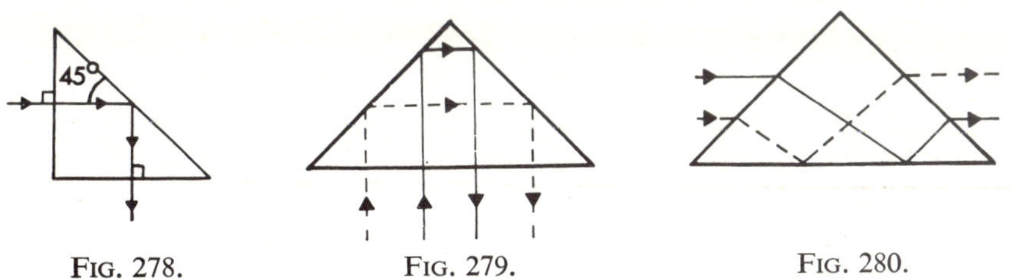

FIG. 278. FIG. 279. FIG. 280.

If the ray of light enters the hypotenuse normally (Fig. 279), then it is totally reflected from one short side on to the other where it is again reflected totally, to emerge from the hypotenuse at right angles. By drawing a second ray we see that a prism used in this way will invert an image. Two such prisms at right angles to one another can invert an image both laterally and vertically and are therefore used to correct the image in prismatic binoculars: their arrangement is shown in Fig. 277. You will see that the prisms also make the instrument short in length.

Fig. 280 shows another way of using the prism, to produce inversion without changing the direction of the light. This is the one use of the reflecting prism in which refraction occurs.

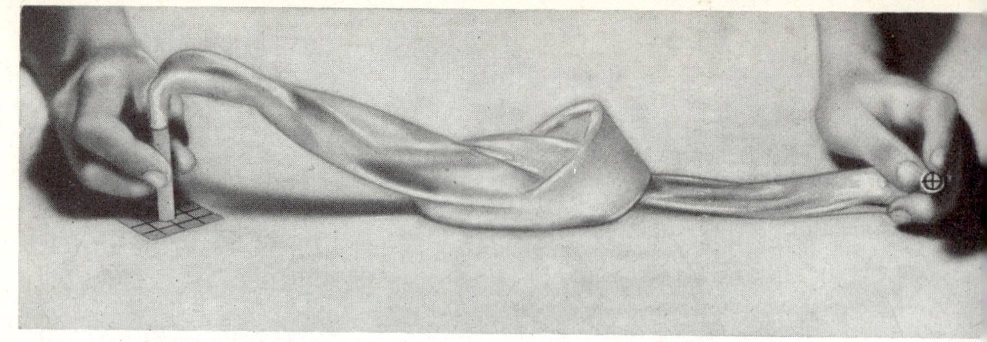

FIG. 281. *Not all light guides are in the form of tubes, the one shown being made from a bundle of glass fibres along which images are transmitted, even if the bundle is tied in a knot as here. Notice the image of the cross at the right end of the fibrescope, transmitted from the left end.*

Light Guides

A ray of light can be trapped inside a jet of water in an ornamental fountain because it undergoes total internal reflection and cannot escape as long as the surface remains smooth. Small ripples allow a little light to escape from the surface and as a result the jet appears luminous. If a transparent plastic rod is used in place of a jet of water, the surface is quite smooth, no light escapes and the rod may be used to guide a beam of light along a desired path.

The latest advances in this field involve the use of bundles of glass fibres for guiding light. As Fig. 281 shows, images can be transmitted in this way; each fibre carries light from one point on the object (left) and when these are put together, the image is seen (right).

Using instruments based on this principle, doctors can look inside the body; pilots can see inside jet engines in flight; engineers can see inside machinery without taking it apart.

LENSES

If you examine a magnifying glass of fairly low power, you will find that it curves outwards, like part of a sphere, on both sides. Such a glass can be used to magnify, for instance, the print of this page. Now stand in a fairly dark room on the far side from the window, hold up the magnifying glass, and let the light from the window pass through the glass on to a sheet of paper held 100-200 mm behind the glass. Adjust the distance between paper and glass until a clear coloured picture of the window is seen on the paper. Note that it is upside down. This picture is called an *image*.

Action of a Lens

A prism causes a ray of light to deviate from its original direction; the larger the angle of the prism, the more the light deviates, while a parallel-sided block of glass causes no change at all in the *direction* of the ray. Fig. 282 shows how you could arrange prisms of differing angles and a block so as to cause five parallel, but separated, rays to converge to a point *P*. You could combine the prisms in a single piece of glass which would look like Fig. 283*a*, having ten faces.

To deal with a beam of light, which contains an infinitely large number of rays, then the piece of glass must likewise have an infinitely large number of faces; it becomes curved (Fig. 283*b*). In three dimensions the surface of the lens is spherical like the magnifying glass you have examined.

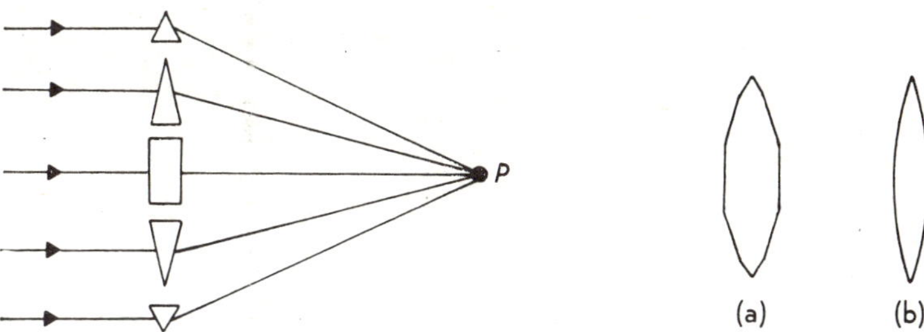

FIG. 282. *Formation of convex lens.*

FIG. 283. *Convex lens.*

Formation of an Image

Rays of light from a distant point reach us in the form of a nearly parallel beam (Fig. 284).

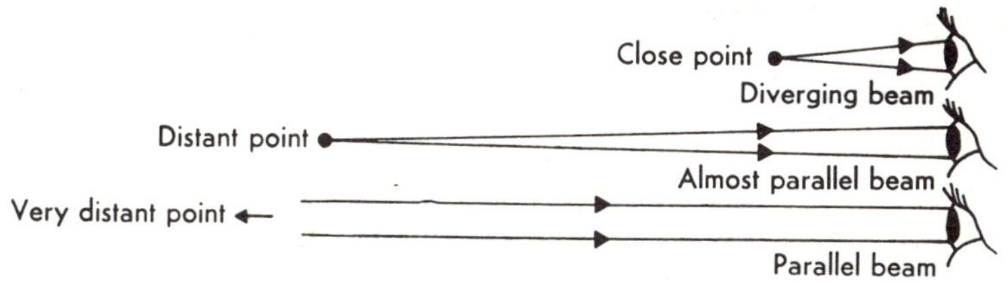

FIG. 284. *The form taken by beams of light rays from a distant point.*

When you look at a distant object, rays of light from each point on the object reach your eye in the form of a nearly parallel beam. Since the points are in different places, however, each of these beams is coming from a different direction.

The ray of light from one of these parallel beams which passes through the centre of the lens (where it behaves like a parallel block) is undeviated. The lens, however, causes each of the other rays, which make up the parallel beam, to converge to a point. Fig. 285 shows how the image of a window is formed. You can see from this drawing why the image is completely inverted.

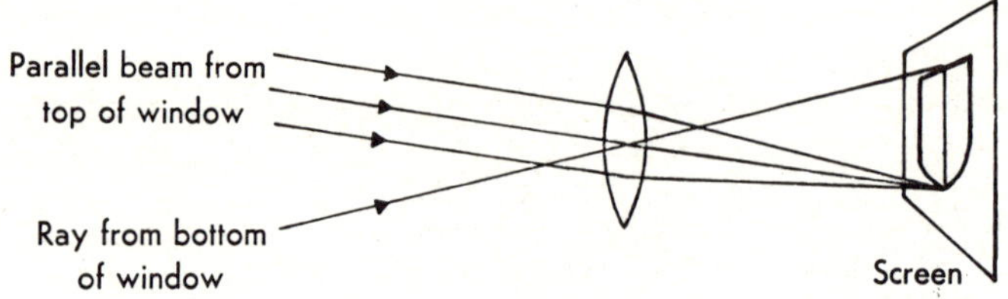

FIG. 285. *Formation of the image of a distant window with a convex lens.*

It is useful to call the very midpoint of the lens O (Fig. 286) the *optical centre*, and the line drawn through O perpendicular to the lens is known as the *principal axis* of the lens. Rays of light parallel to the principal axis will, after passing through the lens, converge to a point on the principal axis F, known as the *principal focus*. The distance OF is the *focal length* f of the lens. A parallel beam of light which is not parallel to the principal axis will be made by the lens to converge to a point, not at F, but in the plane through F perpendicular to the principal axis. This is the *focal plane*. The image of the distant window formed in the last experiment would be approximately in the focal plane of the lens.

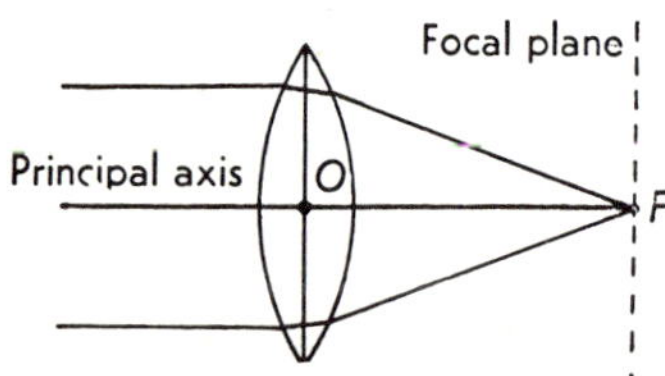

FIG. 286. *Convex lens.*

Near Objects

Repeat the experiment forming an image of a distant window and measure the distance between the sheet of paper (the image) and the lens. This is approximately equal to the focal length of the lens. Why?

Now use as an object, not the window, but a flashlamp the glass of which is covered with tissue paper (some ink marks on the tissue paper will assist in focusing the image). Place the lamp about 300 mm from the lens. If the paper is moved away from the lens, a position can be found giving a sharp image of the flashlamp. Measure the image distance again. A lens can form an image of a close object, and this image will *not* be in the focal plane.

The formation of such an image can be understood by drawing an accurate diagram known as a *graphical construction*.

Having drawn the principal axis XY, denote the position of the lens by a straight line AB (Fig. 287). Mark, to scale, the position F of the principal focus. There will be another principal focus F' the same distance on the other side of the lens since the lens acts in either direction.

Using the same scale, mark in the position of the object, representing it by an arrow CD. Putting one end of the object on the principal axis simplifies the construction, for we know that the corresponding end of the image will lie on the axis also. The point C of the object gives out rays of light in all

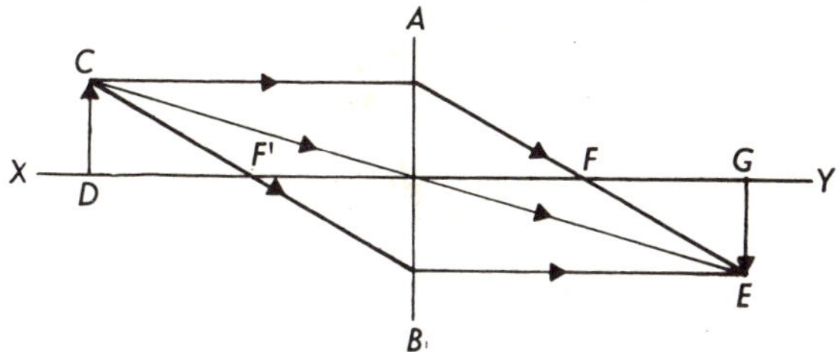

FIG. 287. *Formation of the image of a near object with a convex lens.*

directions, and to form an image, those rays which pass through the lens must be made to converge to a point. Three of these rays are easy to trace through the lens. They are:

1. The ray which passes through the centre of the lens and is undeviated.
2. The ray of light parallel to the principal axis, which is deviated so as to pass through F, the principal focus.
3. The ray of light through F', which emerges from the lens parallel to the principal axis. This follows from the definition of principal focus, bearing in mind that light rays are reversible.

If your diagram is accurately drawn, these three rays will meet at a point such as E.

This is therefore the image of the point C, since all rays from C through the lens converge at E, and the complete image of CD is EG. It is inverted and a different size from the object.

In carrying out your construction, it is sufficient to use any two of the three rays, although the third is useful as a check of your accuracy. You need not draw the height of the object to scale; it is an advantage, in fact, if you exaggerate the height slightly.

Now place an object 240 *mm from a convex lens of focal length* 100 *mm and by an accurate drawing find the position, nature and magnification of the image formed by the lens.*

A suitable scale is 1 : 4, i.e. 10 mm on the diagram will represent a length of 40 mm. Draw the axis and lens (Fig. 288). Mark in the principal foci (F, F') 25 mm ($\frac{1}{4} \times 100$) either side of the lens. Draw a vertical line AB, 60 mm ($\frac{1}{4} \times 240$) in front of the lens to represent the object. Make this line 20 mm high. Draw in rays BO through the centre of the lens, and BC parallel to the axis to meet the lens at C. Ray BC is refracted to pass through F, meeting BO produced in E. DE is therefore the image.

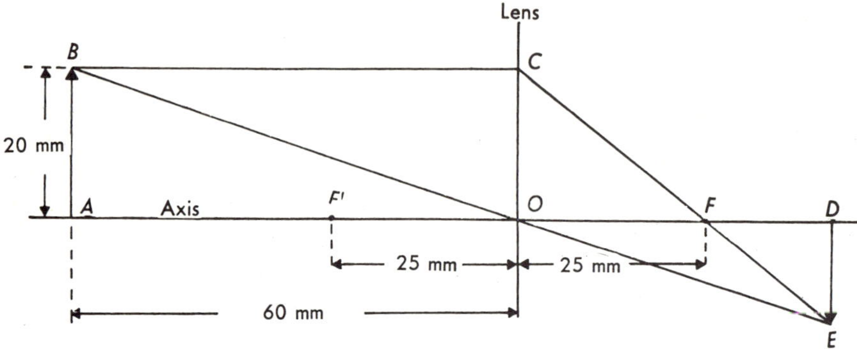

FIG. 288. *Finding the nature of the image formed by a convex lens.*

The distance OD is 43 mm. Therefore, remembering our scale, the image is formed $4 \times 43 = 172$ mm from the lens. From the diagram we see that the image is *real* and *inverted.*

Measure DE. It is 14·3 mm. The object had a height of 20 mm therefore the size of the image is $\frac{14{\cdot}3}{20}$ times the size of the object.

$$\text{Magnification} = 0{\cdot}715 \text{ times}$$

Lens Formula

The position of the image formed by a given lens depends on the position of the object.

If the distance of object and image from the lens are u and v respectively, and f is the focal length of the lens, calculations may be performed using the formula:

$$\frac{1}{f} = \frac{1}{u} + \frac{1}{v}$$

The quantity $\frac{1}{f}$ is called the reciprocal of f, e.g. $\frac{1}{2}$ or 0·5 is the reciprocal of 2.

A strong lens has a short focal length so that $\frac{1}{f}$ will have a large value.

This quantity is used to compare the strength of lenses and if f is in metres, it is known as the *power of a lens* in *dioptres.*

Investigating the Lens Formula

Find the rough focal length of your magnifying glass by forming the image of a window on a piece of white cardboard. Now use the covered flashlamp (page 274) as a luminous object, and form its image on a piece of white cardboard with the magnifying glass.

Measure u, which is the distance from flashlamp to lens, and then v, the distance from lens to image, and record these distances in a table (Table 11). Repeat this several times for different values of u, but the flashlamp should not be closer to the lens than about $1\frac{1}{4}$ times the focal length; you will see why when you do the experiment.

Complete the table by working out the reciprocals of u and v, i.e. $\frac{1}{u}$ and $\frac{1}{v}$, and add them together for the last column.

Table 11

Investigation of the Lens Formula

u in metres	v in metres	$\frac{1}{u}$	$\frac{1}{v}$	$\frac{1}{u}+\frac{1}{v}$
0·425	0·13	2·35	7·69	10·04
0·31	0·145	3·23	6·90	10·13
0·22	0·18	4·55	5·55	10·10
0·16	0·255	6·25	3·92	10·17
0·145	0·30	6·90	3·33	10·23

From Table 11 we find that the average value of $\frac{1}{u}+\frac{1}{v} = 10{\cdot}1$

According to the lens formula $\frac{1}{u}+\frac{1}{v}=\frac{1}{f}$

$$\therefore \frac{1}{f} = 10{\cdot}1$$

But the rough focal length measured for the lens was 0·095 m.

$$\therefore \frac{1}{f} = 10{\cdot}5$$

The readings in Table 11 are for a particular lens with a particular focal length. You will of course get different values by using a different lens.

A similar experiment may be carried out using a pin as the object.

Set up a pin in a cork in front of the lens (Fig. 289), at a distance about double the focal length ($u = 2f$). From the formula you will see that the image distance v should also be about double the focal length, so you must look for the image at I. It will help to locate the image if you place a second pin P in line with the lens and the first pin, at about the right distance from the lens. Place your eye in the same line as shown, and look at the pin P. The inverted image I, should now be visible.

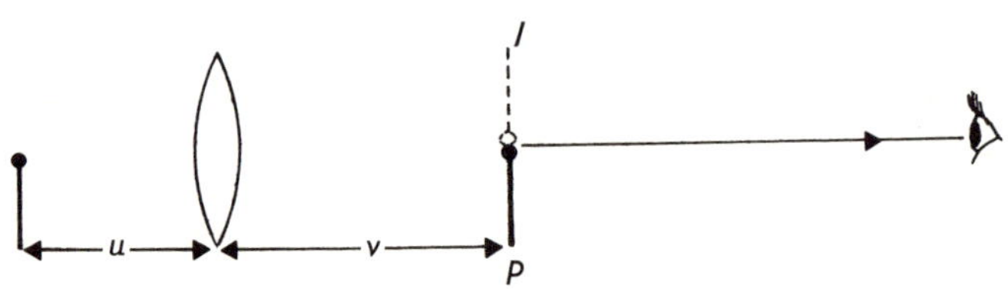

FIG. 289. *Method of investigating the lens formula experimentally.*

Move P until it is in a position of no parallax (page 250) with the image I, when its distance from the lens will give the image distance v. You can take a series of readings using different values of u, recording them as before.

Magnification

You will notice in carrying out the previous experiment that the size of the image varies as u and v change. Fig. 290 shows a lens with object AB and image CD, and the ray through the centre of the lens is marked. This ray forms two triangles, AOB and COD, which may be shown to be similar triangles. It follows that:

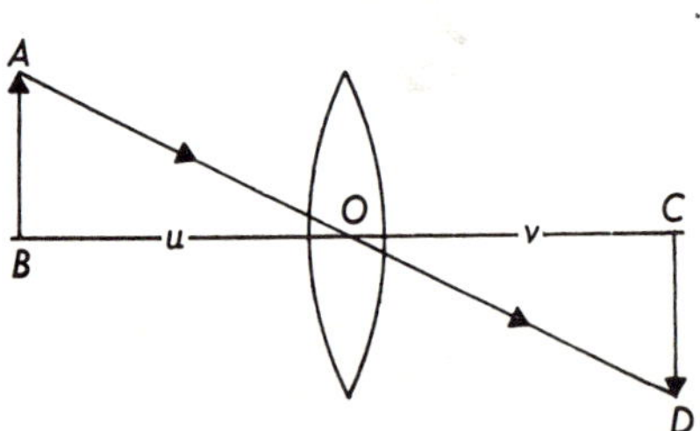

FIG. 290. *Magnification=v/u.*

$$\frac{CD}{AB} = \frac{CO}{BO}$$

$$\text{Magnification } m = \frac{\text{Size of image}}{\text{Size of object}}$$

$$\therefore m = \frac{CD}{AB} = \frac{CO}{BO}$$

$$\therefore m = \frac{v}{u}$$

Projection Lantern

Images can be formed on a screen by a projection lantern; this is done when colour transparencies are shown on a screen by a "projector." The object is a transparent slide XY (Fig. 291), which is brightly illuminated by the lamp L. The light from the lamp is concentrated at XY by the concave mirror M and the condensing lens C. The objective lens O of the lantern now forms a real image on the screen S. The distance from the lantern to the screen is much greater than the distance between the slide and the

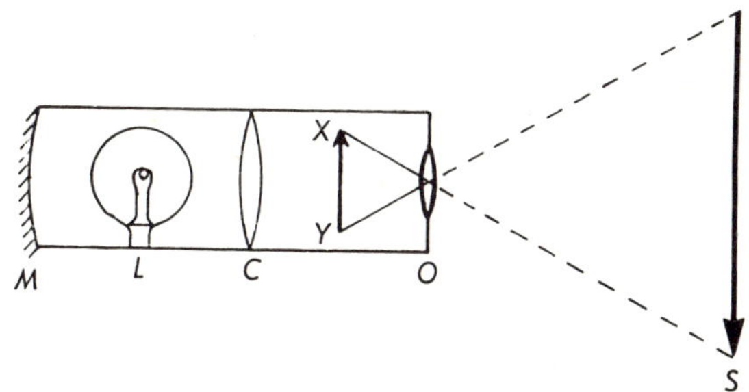

FIG. 291. *Formation of an image by a projection lantern.*

objective, so that the image can be made much larger than the slide. The image is also inverted so that the slide must be inserted reversed and upside down in order to get an image the right way up.

The objective O is usually not a single lens, but a compound one made up from two lenses, giving greater efficiency. It is movable, to make it possible to focus the image on a screen at any distance.

Camera

In a camera, a lens L is used to form an image on a sensitive film F, contained in a light-proof box (Fig. 292). The exposure is made by opening a shutter S and the film may be processed to give a permanent record of the image.

A box camera is used for taking pictures of distant objects. Their images are formed in the focal plane of the lens, and the film is placed behind the lens at a distance equal to the focal length f. These images are obviously diminished.

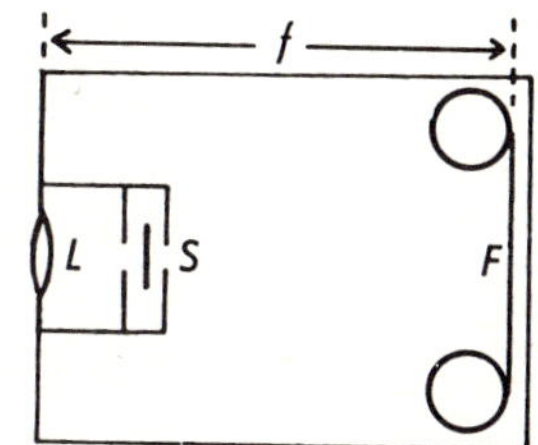

FIG. 292. *Camera.*

If "close-ups" are to be taken, then the position of the lens must be changed, moving it away from the film, in order still to form the image on the film. With a box camera, the focal length of the lens is sometimes changed by fitting an auxiliary "close-up lens," but this method of adjustment is effective for only one fixed object distance. Here are three refinements that are made to the simple camera.

1. The shutter has a variable speed to give exposures of different lengths.
2. A variable stop (*aperture*) is fitted to adjust the effective diameter of the

lens, and regulate the amount of light admitted at an exposure.

3. The lens is movable, to enable an object at any distance to be brought into focus.

You can compare a camera which uses a lens with the pinhole camera described on page 246. In the lens camera, the distance from lens to film is critical, but this type of camera has the advantage that a sharp image can be obtained using a large aperture. This makes it possible to use short exposures. A good lens gives a better image than even a very fine pinhole. Why should this be?

Virtual Images

As the object is brought close to the focal point of the lens, i.e. when u is not much bigger than f, then v becomes large, and the image is a long way from the lens.

Fig. 293 shows the construction in which the object is placed actually at F_2. Light emerges from the lens in parallel beams. Now we have regarded light coming from a very distant object as arriving at the lens in parallel beams, so that the light in Fig. 293 must be going away to form an image at a large distance from the lens. We say that this image is formed at *infinity*.

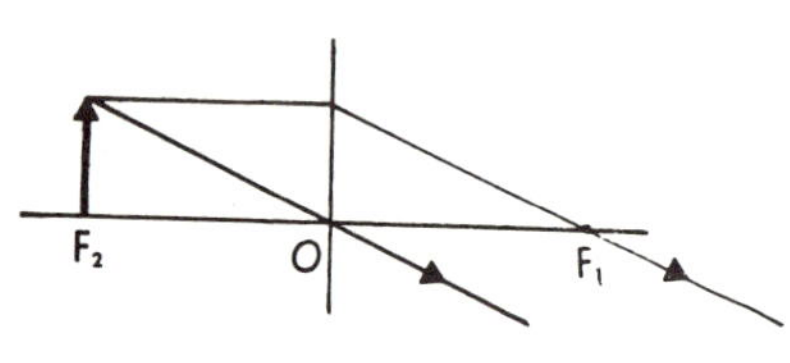

FIG. 293. *Formation of image at infinity.*

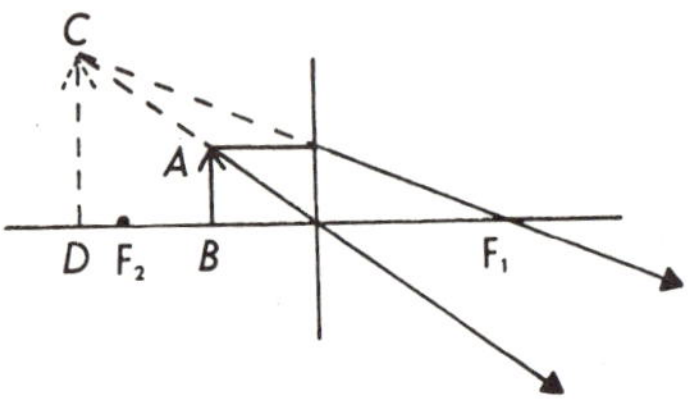

FIG. 294. *Formation of virtual image.*

If the object is now placed inside the focal length (Fig. 294), then the light leaving the lens diverges, and cannot therefore form a real image. The light from A is made to appear to come from C, and thus a virtual image is formed.

For this type of lens (a convex lens), if the object distance is greater than the focal length, the image is real, if less, then the image is virtual.

The magnifying glass is a simple convex lens used as in Fig. 294 with the object AB between principal focus and lens. The virtual image CD is magnified, and has the further advantage that it is the right way up, or erect, which is important, for instance, when reading printed matter.

When we had a real image, we found that we could use the equation

CALCULATIONS WITH A VIRTUAL IMAGE

$$\frac{1}{f}=\frac{1}{u}+\frac{1}{v}$$

Suppose that an object is placed 30 mm from a lens of focal length 60 mm, so as to form a virtual image. Then substituting in the equation

$$\frac{1}{60}=\frac{1}{v}+\frac{1}{30}$$

$$\frac{1}{v}=\frac{1}{60}-\frac{1}{30}=\frac{1-2}{60}=-\frac{1}{60}$$

$$\therefore v=-60 \text{ mm}$$

The minus sign occurs because the image is virtual and we can use the equation if we adopt the convention that: *distances measured to a real object or image are positive, while those measured to a virtual object or image are negative.*

A negative answer tells us that we have calculated the distance to something virtual.

An object is placed 80 mm from a lens of focal length 50 mm. Where will the lens form an image of the object and what is the nature of the image?

You are given that $u=80$ and $f=50$

Substituting in the lens formula $\frac{1}{50}=\frac{1}{80}+\frac{1}{v}$

$$\therefore \frac{1}{v}=\frac{1}{50}-\frac{1}{80}=\frac{8-5}{400}=\frac{3}{400}$$

$$\therefore v=\frac{400}{3}=133 \text{ mm}$$

The image is real and magnified, and therefore inverted.

A lens, which is 100 mm away from an object 10 mm high, forms an image at a distance of 200 mm on the same side as the object. What is the focal length of the lens? How big is the image? You can see by drawing a diagram that this must be a virtual image:

$$\therefore v=-200 \text{ and } u=100$$

Substituting in the lens formula $\frac{1}{f}=\frac{1}{100}-\frac{1}{200}=\frac{2-1}{200}=\frac{1}{200}$

$$\therefore f=200 \text{ mm}$$

$$m=\frac{v}{u}=\frac{-200}{100}=-2$$

The object is 10 mm high, therefore the image is 20 mm high and virtual, as indicated by the minus sign.

Concave Lens

Fig. 282 showed how you could arrange prisms so as to make five parallel rays of light converge to a point. If you were to arrange the same prisms so as to make the five parallel rays diverge from a single point *P*, then you would use the arrangement of Fig. 295. Fig. 296 shows the lens derived from this arrangement of prisms. It is thinner at the centre than at the edges and is

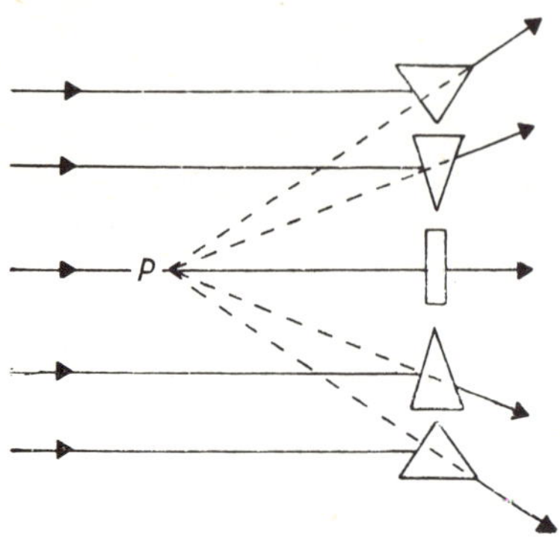

FIG. 295. *Formation of concave lens.*

FIG. 296. *Concave lens.*

called a *concave lens*. Its surfaces are spherical, although they curve inwards. Referring to Fig. 295, you can regard *P* as the principal focus of the lens. For a concave lens, the principal focus is the point from which rays parallel to the principal axis appear to diverge after passing through the lens. Concave lenses have a *virtual focus*. Why?

Because of their behaviour, you can refer to concave lenses as *diverging lenses;* in the same way, convex lenses are often called *converging lenses*.

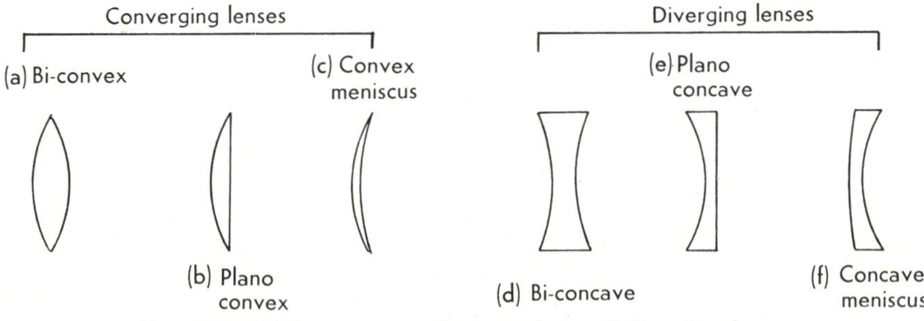

FIG. 297. *Different types of converging and diverging lenses.*

Other Shapes of Lens

The two types of lens considered so far have both been symmetrical and are

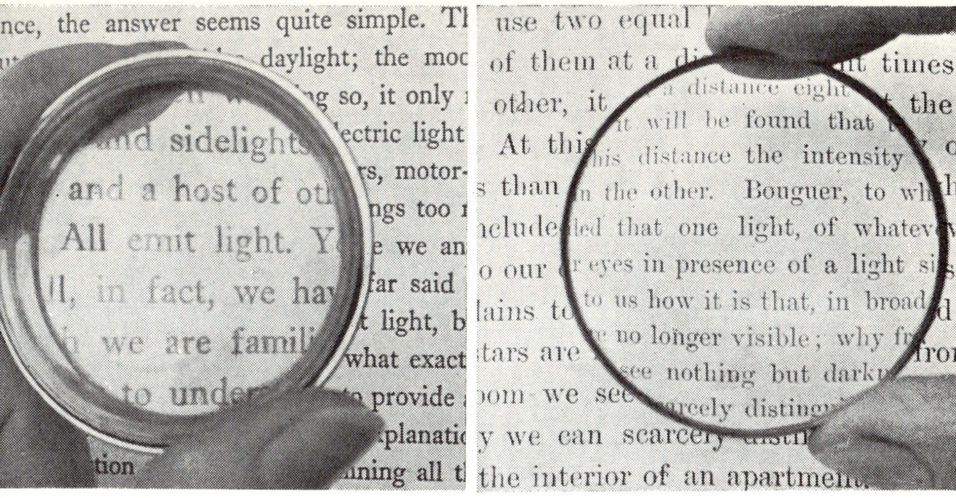

FIG. 298. *Compare the images formed by the convex lens (left) and the concave lens (right), with the original printing.*

shown in Fig. 297*a* and Fig. 297*d*. Other shapes of lens are found to be more suitable for certain uses. You will see that the converging lenses are all thicker at the centre, while the diverging lenses are all thinner at the centre. Later, on page 291, you will find that the meniscus type of lens (Fig. 297*c* and Fig. 297*f*) is often used in spectacles, so that the eye, as it moves, will always be looking normally through the lens.

See if you can draw the prisms of Figs. 282 and 295 (which correspond to lenses *a* and *d* of Fig. 297 respectively), so that they correspond to lenses *b*, *c*, *e* and *f* of Fig. 297.

QUESTIONS

1. Some native fishermen shoot fish in streams with bows and arrows. When doing so, they do not aim directly at the fish, for they would then miss. Draw diagrams to explain why. Where should you aim in order to hit the fish?

2. State the laws of refraction. Define refractive index. Fig. 299 shows a prism, with a refracting angle of 50°. A ray of light strikes one face, making an angle of 45° with the normal as shown: calculate the angle of refraction θ_1, if the refractive index of the material of the prism is 1·6. By means of an accurate drawing, find the angle θ_2, at which the ray strikes the second face of the prism. Then calculate the angle θ_3, at which the ray emerges from the prism. Through what angle θ_4 has the ray been deviated?

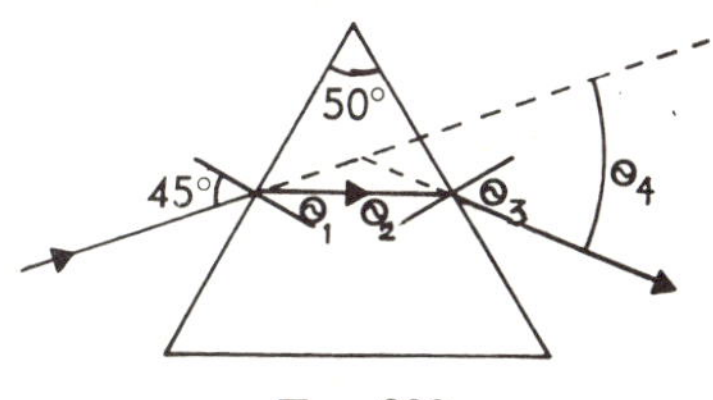

FIG. 299.

3. Explain what is meant by: *critical angle*; *total internal reflection*. Fig. 300 shows a semi-circular glass block of a type used in experiments on total internal reflection. A ray of light strikes the inside of the flat face at an angle of 35° to the normal. (*a*) What is the angle θ, between ray and normal in air, if the refractive index of the glass is 1·52? (*b*) What must be the angle in the glass before total internal reflection occurs? (*c*) What would the angle be for polystyrene, whose refractive index is 1·6?

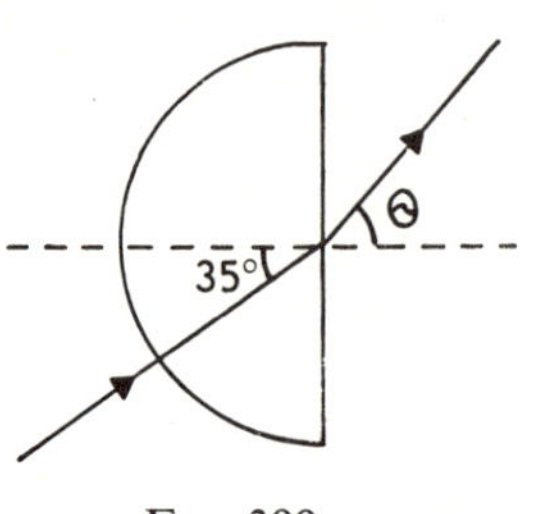

FIG. 300.

4. Draw a diagram of the rays of light which enter the eye of a fish swimming in water with a smooth surface; find the apparent change in position of objects outside the water. How does the fish see parts of the pond bed twice?

Small air bubbles in water appear clear at the centre and silvery round the outside when you examine them. Why?

5. Solve by graphical construction the following problems:

A convex lens of focal length 160 mm is placed 400 mm behind an object 50 mm tall. Where must a screen be placed to receive the image of the object, and what is the smallest screen that could be used?

The object in the previous problem is moved 280 mm nearer the lens. Where is the image formed now, and what is its nature?

An object is 150 mm from a convex lens which forms a real image on a screen 300 mm from the lens. What is the focal length of the lens?

6. Check your answers to Question 5 by calculations.

7. Explain by referring to a convex lens the terms: *principal axis; principal focus; focal length.*

Calculate the position of the image and determine its size and nature when an object 20 mm high is placed: (*a*) 180 mm from a convex lens of focal length 120 mm; (*b*) 240 mm from a convex lens of focal length 480 mm.

8. A projection lantern is set up with its objective lens 5 metres from a screen, and a slide 0·1 m square gives an image on the screen 1 metre square. How far is the slide behind the objective and what is the focal length of the objective?

9. A camera, which has the film 0·08 m from the lens when set for distant objects, is to be used to photograph an object 1·5 m away. How much must the lens be moved, and in which direction?

10. How is an image formed by: (*a*) a pinhole; (*b*) a lens? Why is the image produced by a camera with a lens "better" than that produced by a pinhole camera?

CHAPTER 23

THE EYE AND OPTICAL INSTRUMENTS

IN AN earlier chapter (page 234) we discussed why in physics you need to make measurements: we showed that you must be accurate in your measurements and that you can use instruments to help you to achieve this.

Most measurements that you take will depend, in part, upon your sense of sight; measuring a length with a ruler or reading a meter for example. The eye is therefore the most important of optical instruments, being the organ of sight and responsible for transmitting to your brain the visual impressions that enable you to "see" objects.

THE EYE

When discussing the camera (page 279) we considered the requirements to be met in order to record an image on film. Similar requirements must be met in your eye in order to transmit images to your brain.

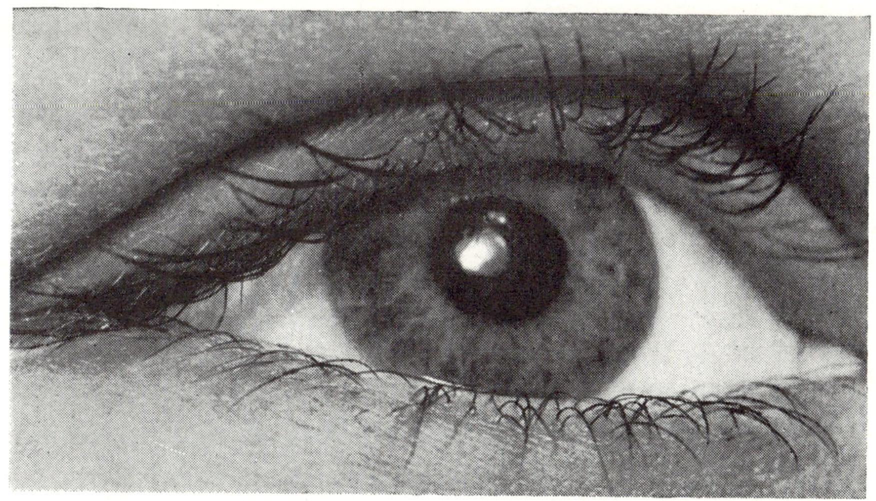

FIG. 301. *The eye is almost spherical in shape. It is mounted in a socket in the skull called an "orbit", and can be moved about in this socket by muscles connecting the outside of the eyeball to the back of its socket. The eyelid can cover and protect the eye.*

There follows a list of the principal components of the eye. It will help you to study these, in conjunction with Fig. 302, before reading about how the eye works.

Aqueous humour. A watery transparent fluid which fills the front compartment of the eye between the *cornea* and the *crystalline lens.*

Blind spot. The area where the *optic nerve* joins the retina. This spot contains no *rods* and *cones* and is in consequence insensitive to light.

Choroid. The tissue lining the *sclerotic*; it is black in colour so that it may absorb stray light, and so prevent internal reflection and the consequent blurring of the image falling on the *retina.*

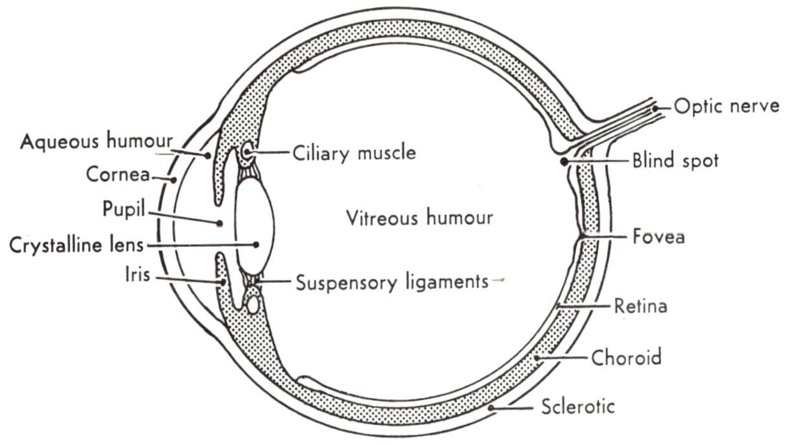

FIG. 302. *Horizontal section through left eye.*

Ciliary muscle. A ring-shaped muscle which runs around the thickened portion of the *choroid.* It is used to alter the focal length of the *crystalline lens.*

Cones. Cone-shaped nerve endings situated on the *retina.* They are sensitive to colour and detail, but only if the light is strong.

Cornea. The tough transparent portion of the *sclerotic* through which light enters the eye; the cornea is more curved than the rest of the sclerotic giving a bulging appearance to the front of the eye.

Crystalline lens. A transparent convex lens made up from jelly-like living cells. The curve of its front is considerably less than the curve of its back, a feature which tends to decrease distortion of the image.

Fovea. A small depression on the *retina*, containing only *cones* and lying on the axis of the eye. The area around the fovea contains a large number of *cones* and is known as the *yellow spot.* An object being viewed is automatically focused on to this spot.

Iris. A coloured diaphragm lying behind the *cornea* and partly covering the *crystalline lens.* The iris is part of the *choroid* and has a circular hole in the middle called the *pupil.* Muscles control the size of this hole according to the amount of light falling on the eye. It is to the colour of the iris that we refer when talking of the colour of a person's eyes.

Optic nerve. Nerve fibres from the *rods* and *cones* pass across the inside surface of the *retina* and combine to form the optic nerve. This leaves the back of the eyeball sheathed in *sclerotic membrane* and passes to the brain.

Pupil. The hole at the centre of the iris which appears black because the interior of the eye is black.

Retina. A membrane containing light-sensitive cells shaped like *rods* and *cones.* It forms the inside lining of the eye.

Rods. Rod-shaped nerve endings situated on the retina. They are sensitive to dim light but are unable to detect colour or detail.

Sclerotic. The tough outer covering of the eye; white and opaque except at the front of the eye where it becomes the transparent *cornea.* Part of the opaque scelerotic, "the white of the eye," is visible.

Suspensory ligaments. These ligaments attach the *crystalline lens* to the thickened part of the *choroid* which contains the *ciliary muscles.*

Vitreous humour. A transparent jelly filling the larger part of the eye between the *crystalline lens* and the *retina.* The vitreous humour together with the *crystalline lens* and the *aqueous humour*, form a compound lens system.

How the Eye Forms an Image

Fig. 303 shows parallel rays from a distant point entering your eye. The cornea is curved so that it acts as a lens, refracting the light to make it converge; the light is further refracted by the crystalline lens which is denser than the humours surrounding it so that the image of the point is formed on your retina. The rods and cones in this part of your retina are stimulated, electrical impulses are produced, and the optic nerve transmits these to the brain, so that you are aware of the nature of the image formed on your retina.

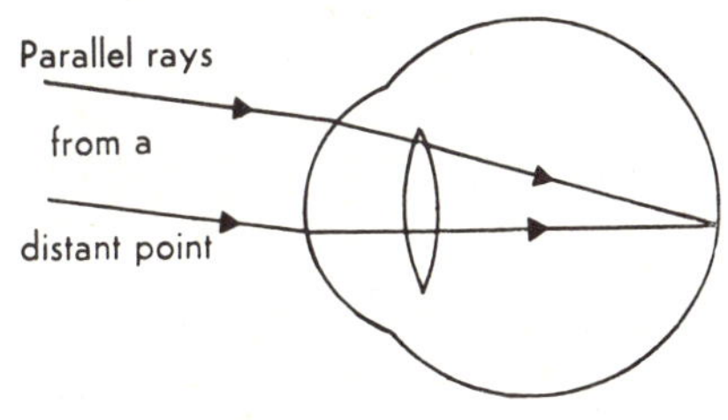

FIG. 303.

Each point on a distant object will give a parallel beam at your eye from a different direction. Each of these beams is made to converge to a different point on the retina, building up an image of the whole object (see the convex lens page 274). Notice that, as in the camera, the image formed in your eye is inverted (Fig. 304). You brain is accustomed to interpreting this sort of image on your retina, so that the tree in Fig. 304 will seem to you to be the right way up.

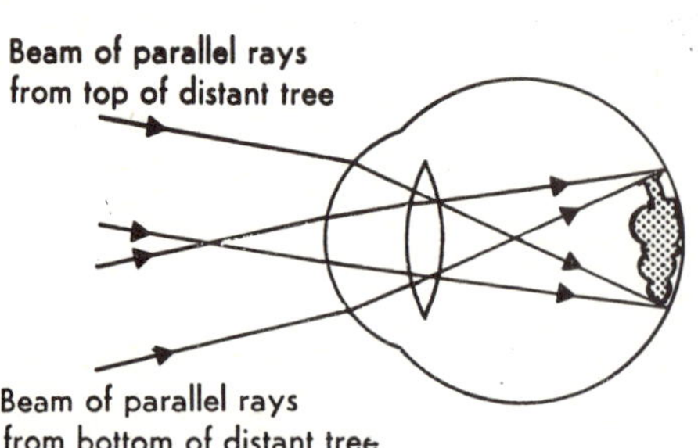

FIG. 304.

When you wish to look at a particular object, you turn your eye so that the image of this object falls on the fovea. This is the most sensitive part of your retina, so that you now see this object most clearly. The iris adjusts the size of your pupil according to the intensity of the light entering the eye. In dim light your pupil widens to admit more light; in brighter light your pupil is made to contract in order to reduce the amount of light entering the eye.

If you stand close to a mirror in a fairly dark room so that you can look at the reflection of the pupil of one of your eyes, you will see that it is large. Shine a flash lamp obliquely into this eye, and you will notice that the pupil immediately grows smaller.

Seeing Near Objects

The image (Fig. 304) is formed by sets of parallel rays from points on a distant object. Points on a near object give out rays of light which diverge as they enter your eye and these will not form an image on the retina unless your eye adapts itself.

The crystalline lens is elastic but is stretched and slightly flattened by the ligaments which fasten it to the eyeball (Fig. 305*a*), When the ciliary muscle contracts, the ligaments are relaxed allowing the lens to shrink and grow thicker. The lens now has a shorter focal length, and can make the diverging rays focus on the retina (Fig. 305b). Note that the cornea is more curved so that more refraction takes place here as well.

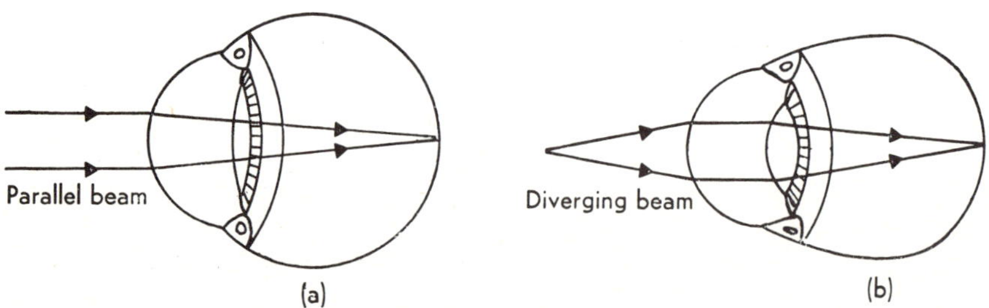

FIG. 305. *Accommodation of the eye to see a near object.*

Accommodation

This adjustment of the eye for near vision is called *accommodation.* If you look at this page with one eye and bring the book closer to your eye, you will reach a point when you can no longer see the print clearly however hard you try. The lens in your eye is accommodated to its fullest. The closest point to your eye at which you can see objects clearly is called your *near point.* Similarly, the farthest point at which you can see objects clearly is called your *far point.* For distant vision the lens in your eye is relaxed, i.e. unaccommodated so that the far point in a normal eye is at infinity.

With a disease known as a cataract, the crystalline lens becomes opaque and must be removed from the eye by surgery. The eye can still form an image although it is not focused on the retina and appears blurred. Spectacles can restore distinct vision but since accommodation is no longer possible, two pairs are required, one for near and one for distant vision.

Blind Spot

Since the portion of the retina at which the optic nerve enters the eye contains neither rods nor cones, images falling on this part of the retina are not detected. The existence of this spot can be easily demonstrated: close

FIG. 306.

the left eye and look at the cross in Fig. 306. Although you are not looking directly at the black circle, you will be able to see it. Move the book slowly towards your eyes. At a certain distance you will find that the circle seems to disappear. Its image will have fallen on your blind spot.

Since the optic nerve leaves each eye on the side of the eye nearest to the nose, the images in both eyes never fall simultaneously on the blind spots. This positioning of the blind spot accounts for the effect observed in Fig. 306. The right eye, being focused on the cross, receives the image of it on the yellow spot. Since the image in the eye is reversed, the image of the black circle is focused on the side of the right eye nearest to the nose; the image eventually becomes focused on the blind spot (it appears to vanish). If however you were to try this demonstration by focusing your left eye on the cross, you will find that there is no position in which the circle appears to vanish; in this case the image of the circle is being formed on the side of the eye farthest away from the nose where there is no blind spot.

Binocular Vision

Hold an object such as a large dice (Fig. 307) as near as you can comfortably see it. Notice that it appears three dimensional. If you close your eyes alternately so that you are looking at the dice first with your left eye and then with your right, you will realize that your two eyes do not see quite the same thing. Fig. 307 illustrates this. You will also notice that when you look at the dice with one eye only, it appears flat and like a good drawing. The three dimensional effect that you see with both eyes results from the ability of your brain to assimilate the two different images received simultaneously from your eyes.

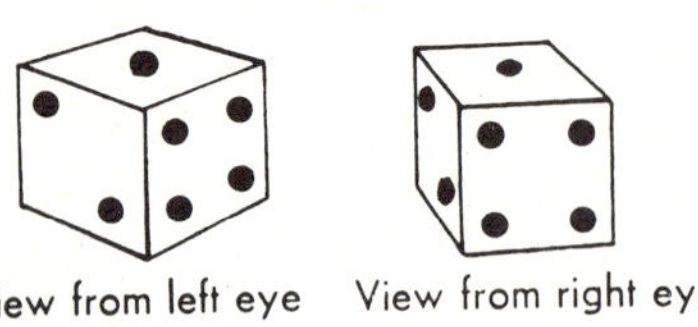

FIG. 307. *Binocular vision.*

Defects of Vision

An optical defect of the eye occurs when the eye fails, in some way, to focus an image properly on the retina. Optical defects can usually be corrected by wearing spectacles.

Long sight (*hypermetropia*). A long-sighted eyeball is small, or its lens is weak, so that when the lens is relaxed and unaccommodated, distant objects are brought to a focus at a point behind the retina (Fig. 308*a*). To see a distant object the lens must be accommodated, and this means that the lens cannot be accommodated sufficiently to see really close objects. Fig. 308*b* shows how a converging lens can be used to correct this defect, bringing the distant object to a focus on the retina of the unaccommodated eye.

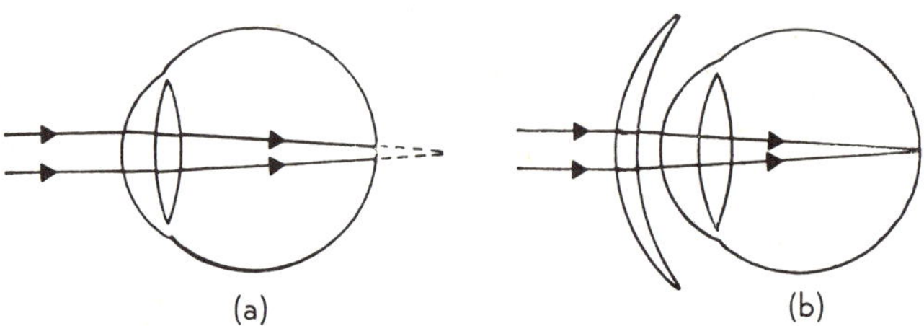

FIG. 308. *Correction of hypermetropia with a convex meniscus lens.*

Short sight (*myopia*). A short-sighted eyeball is too large, or its lens too strong, so that while near objects can be seen perfectly well, when the lens is relaxed completely, i.e. as thin as it can be, distant objects are brought to a focus in front of the retina (Fig. 310*a*). The far point is not at infinity. A diverging lens used as a spectacle, makes the parallel rays from a distant

Fig. 309. *The correction of vision defects with the aid of lenses is not a modern practice. Here is the earliest picture known that shows the use of spectacles. It was painted in Italy in 1352.*

object diverge as if they were coming from a nearer point which the eye is able to see clearly (Fig. 310*b*).

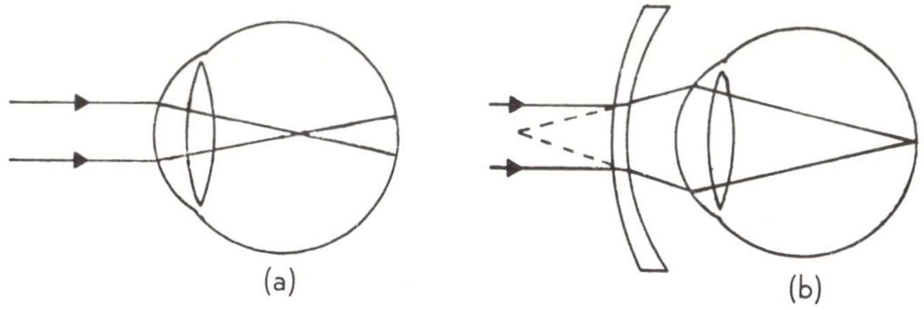

Fig. 310. *Correction of myopia with a concave meniscus lens.*

OPTICAL INSTRUMENTS

We have already mentioned the use of a convex lens as a magnifying glass to give an enlarged virtual image (page 280). To understand how this helps the eye, you must realize that the size that an object seems to be, depends on the size of its image formed on the retina of your eye.

The size of your retinal image increases as an object is brought closer to your eye; objects appear larger the closer they are to your eye. There is a limit to this sort of enlargement, however, because you cannot make the distance between eye and object less than your near point distance (page 289), otherwise you will not see the object distinctly.

Magnifying Glass or Simple Microscope

Fig. 311 shows, for comparison, the size of the image on your retina when you are looking at an object with your naked eye, and with the aid of a magnifying glass. The magnifying glass can be used to form the virtual image at your near point, *A*; your eye now looks at this image so that the image on your retina is larger.

Examine a strong magnifying glass. Place it immediately in front of your eye and look at a piece of ruled paper so as to get a clear image. If you take away the magnifying glass without moving either your head or the paper, then you will probably find that the paper is closer than your near point and you cannot see it clearly; you will be aware of the lines, although they are blurred, and will see that they seem the same distance apart as they did when you were looking through the magnifying glass.

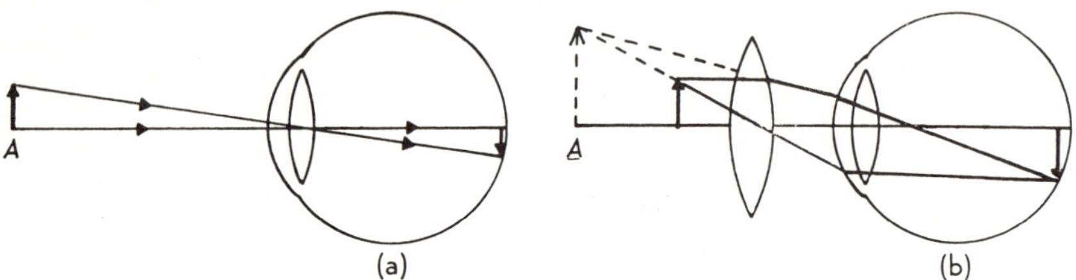

FIG. 311. *Action of a magnifying glass or simple microscope.*

Compound Microscope

This is really a simple magnifying glass, called the *eyepiece*, which is used in the manner described above, but which is preceded by another stage of magnification.

This additional stage is provided by the *objective lens*, a convex lens of very short focal length. Referring to Fig. 312, you will see that the objective can be positioned to give a real image $A'B'$ of the object AB. If you were to put a screen at $A'B'$, the image would be formed on it. If the light is allowed to pass through this image, as in the diagram, it falls on the eyepiece. You are using the eyepiece as a magnifying glass to examine the image $A'B'$, so that the final image $A''B''$ can be many times the size of the object.

The final image is inverted; this is no disadvantage to anyone trained to use a microscope. The eyepieces and objectives used in microscopes are seldom single lenses, but are made up of two or more lenses.

Try now to assemble your own compound microscope. A good object for you to examine is the filament of a small electric torch. When switched on this will provide you with a good bright object.

Select as your objective lens a hand magnifier labelled 12x. Place it about 30 mm above the torch and hold a piece of greaseproof paper about 150 mm above that. With a little adjustment you will be able to form a clear image of

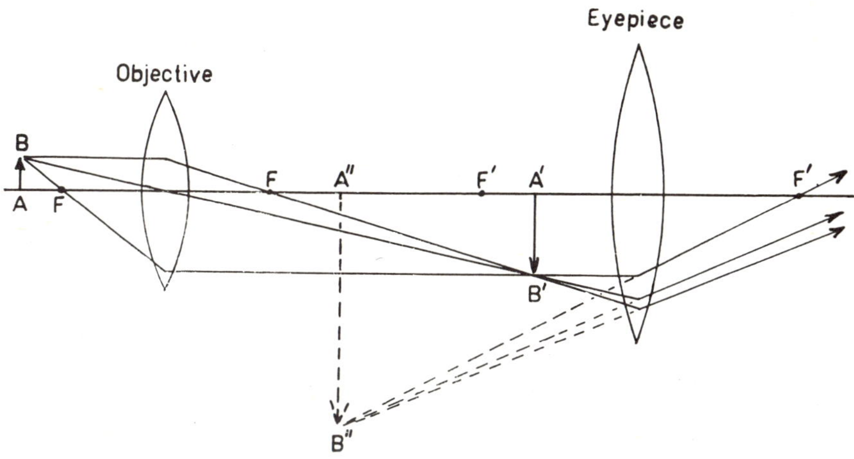

FIG. 312. *Action of a compound microscope.*

the filament on the paper. This will be your intermediate image which you must now look at, using an eyepiece.

For an eyepiece select a small magnifying glass with a diameter of about 40 mm and a focal length of 100 mm. Position this lens above the greaseproof paper until, by looking through the lens, you can see a clear image of that on the greaseproof paper. You can now dispense with the paper. This should leave you with a final image in which the coils of the filament can be easily distinguished. You will notice the following defects:

1. Distortion of the image, called spherical aberration.
2. Slight coloration of the image, called chromatic aberration.
3. The field of view is small.

In the manufacture of compound microscopes these defects are eliminated by making both objective and eyepiece from several lenses.

Astronomical Telescope

The astronomical telescope consists, like a compound microscope, of a convex eyepiece and a convex objective.

Telescopes are used for looking at distant objects. The objective of a telescope consists of a convex lens, which forms in its focal plane an image

$A'B'$ of a distant object AB (Fig. 313). Clearly, the longer the focal length of the objective, the larger will be this intermediate image.

As in the compound microscope, you can look at this intermediate image through an eyepiece which acts as a magnifying glass, giving an enlarged final image $A''B''$. Telescopes are normally adjusted to form this image at infinity, unlike Fig. 313. The image is inverted; this is irrelevant in astronomy, but allowed for in the construction of binoculars (page 271).

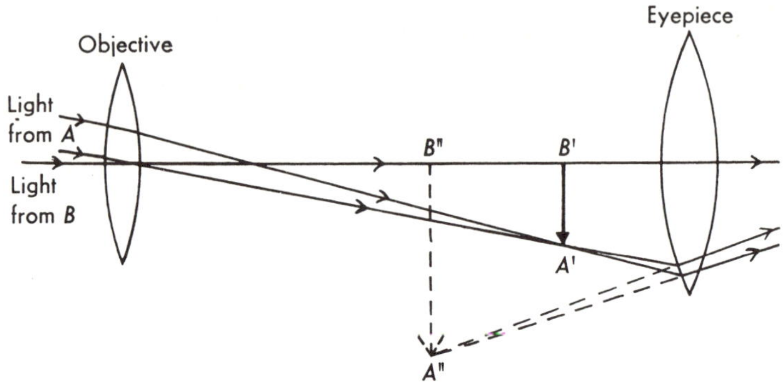

FIG. 313. *Action of an astronomical telescope.*

In making an astronomical telescope you will find it easier to mount your lenses in the ends of a tube. Use as your objective a convex lens of long focal length (about 500 mm). Find the focal length of your objective accurately (page 277) and mount it in the end of a cardboard tube which has been cut to a length exactly equal to the focal length of the lens. When you point this tube out of the window and look at the open end, you should see the inverted intermediate image there.

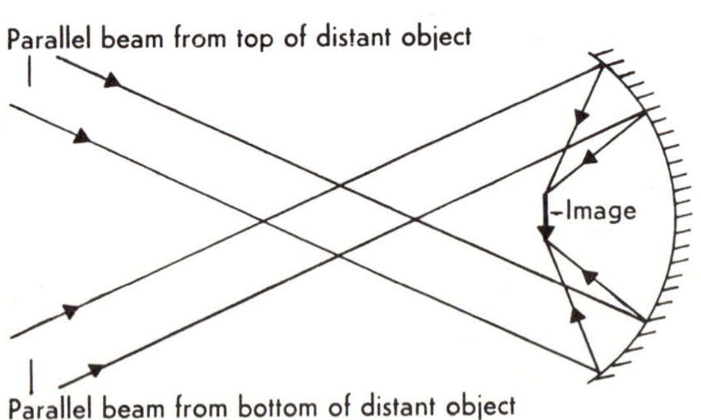

FIG. 314. *Action of a reflecting telescope.*

Select as your eyepiece a simple magnifying glass with diameter of 40 mm and a focal length of about 100 mm. Find the focal length of this lens accurately and mount it in a second cardboard tube which should be wide enough to slip easily over the first and should have a length equal to about twice the focal length of the eyepiece. This means that you will be able to see the intermediate image most comfortably when the first tube is fitted about half way into the second. You will be able to slide the outer tube backwards or forwards in order to focus the instrument.

Reflecting Telescope

In larger astronomical telescopes a concave mirror is used, instead of a convex lens, to form the intermediate image. Fig. 258 showed how a concave mirror could be used to form a parallel beam of light from a point source. Such a mirror should therefore bring a parallel beam of light to a focus. Parallel beams in different directions will be focused at different points, building up an image as a lens does (Fig. 314). This intermediate image is then magnified by an eyepiece as in the ordinary telescope. A plane mirror is used to reflect the light forming the intermediate image so that the eyepiece may be positioned more accessibly.

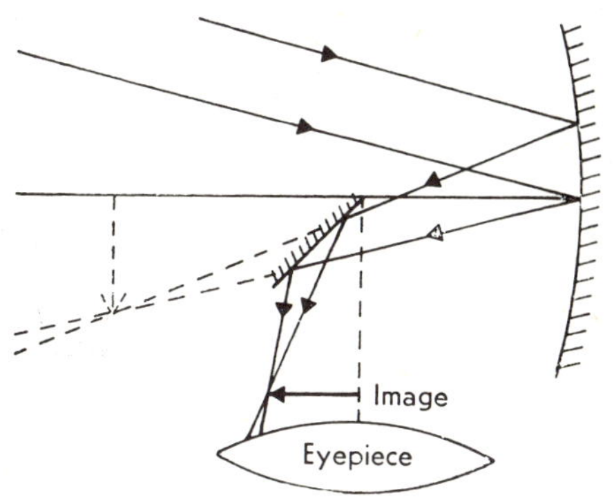

FIG. 315. *Reflecting telescope.*

QUESTIONS

1. Describe the structure of the eye. What are the chief ways in which vision may be defective?

A form of defect, known as astigmatism, occurs when the cornea is not spherical, but curves more sharply in one direction than another. How will this affect the focal length of the eye for light in the plane in which the curvature is: (*a*) greatest; (*b*) least? What would the eye see if it were looking at a wagon wheel with large radial spokes?

2. How does a magnifying glass work? What sort of image is formed in normal use?

3. A magnifying glass held close to the eye is used to form an image at the near point (250 mm from the eye) of an object 10 mm high. How far from the lens must the object be if the lens has a focal length of: (*a*) 50 mm; (*b*) 100 mm? How high is the image in each case?

4. Describe a compound microscope. Why can much greater magnifications be obtained with a compound microscope than with a single lens?

5. An astronomical telescope is focused on a ship on the horizon. What adjustment to the eyepiece must be made in order to see clearly a building 50 metres away? What sort of adjustment would be necessary if the telescope were a reflecting one?

CHAPTER 24

COLOUR

WE HAVE already considered the deviation of a ray of light as it passes through a glass prism (page 268). It was noted that objects viewed through a prism appear, not only displaced, but also fringed with colours. Let us now consider further the coloured effects produced by a triangular glass prism.

Refraction Through a Triangular Prism

In order to investigate this refraction we require a light source in the shape of a narrow vertical line. This source is most conveniently provided by a clear 40-watt light bulb arranged on its side so that from the point of view of the prism being used, the filament of the bulb is a narrow vertical line of light.

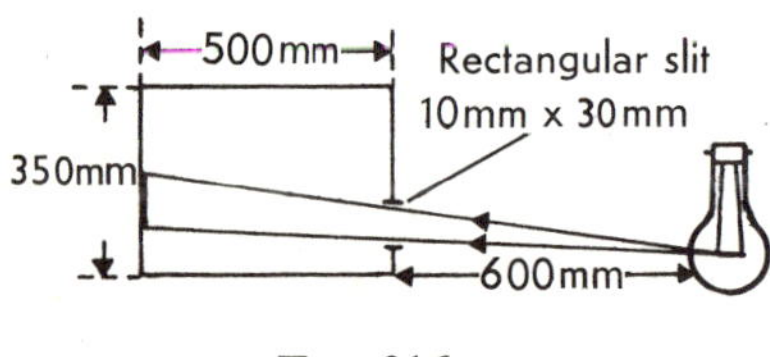

FIG. 316.

The light source is set up facing the side of a cardboard box in which has been cut a rectangular slit 10 mm wide and 30 mm high (Fig. 316). A small patch of light will be seen on the back of the box. Note that this light, received directly from the source, is white.

If a triangular prism is now placed inside the box close to the rectangular slit (Fig. 317), the beam of light, now falling on the prism, will be deviated by it so as to strike the long side of the box.

To show the deviated light more clearly, stick a piece of white paper where the beam strikes the side of the box. Note that the patch of deviated light

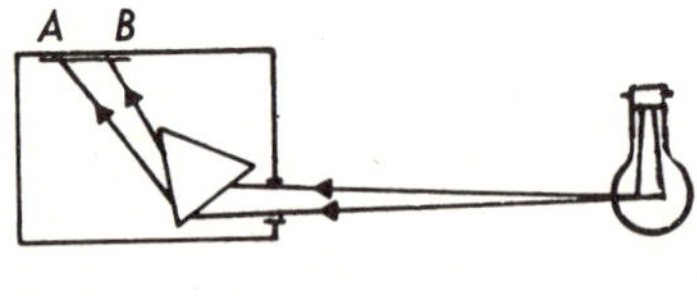

FIG. 317.

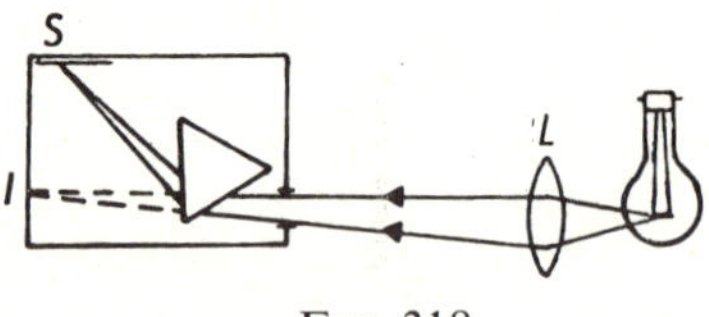

FIG. 318.

now has coloured edges; the edge A being coloured red and orange, while the edge B is coloured blue and violet. The central portion of the patch of light is still white. If you now turn to the coloured plate at the front of the book, you will see that the patch of light obtained with the prism is similar to the *impure continuous spectrum* (Frontispiece, Fig. 1).

Now place a convex lens of medium size and low power between the light source and the box. Position it so that an image of the filament is formed at *S*, where the patch of light previously fell, while light passing over the top of the prism forms an approximately focused image of the filament at *I* (Fig. 318). The image formed at *I* by undeviated light is a normal white image while that at *S* is wider and is formed from a sequence of colours running across it similar to the *pure continuous spectrum* (Frontispiece, Fig. 2).

Formation of Pure and Impure Spectra

The filament of the bulb in your experiment was like a narrow vertical slit.

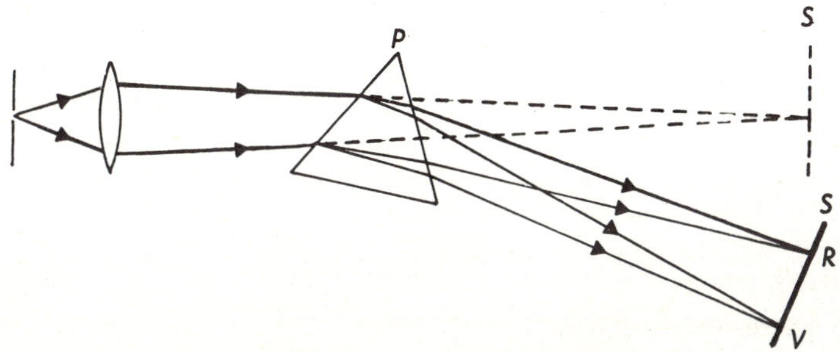

FIG. 319. *Use of prism and lens to form a pure spectrum.*

Fig. 319 (dotted lines) shows how the lens would form an image of this slit on the screen *S*. The prism *P*, however, deviates the light by refraction at the two faces, so that the screen must be moved to receive the new image. The actual position of the image on the screen depends on the amount of deviation caused by the prism, which in turn, depends upon the refractive index of the material of the prism. Since the different colours of light form images at different points on the screen, we can conclude that:

1. *White light is really a mixture of coloured light.*
2. *The refractive index of a material is slightly different for light of different colours.*

The point *R* in Fig. 319 is the position of the image formed by red light. The prism has a higher refractive index for violet light which is therefore deviated more, so as to form an image at *V*. Other colours form images between *R* and *V* and there is a continuous range of colours with gradual transitions from shade to shade. For convenience, it is usual to name six colours in the spectrum: red; orange; yellow; green; blue; violet; in that order, but you will realize that each of these names will cover a variety of shades. Try to pick out these colours in the coloured image of the filament in your experiment when using the lens (Frontispiece, Fig. 2).

An impure spectrum was obtained when you did not focus the filament with a lens. Fig. 320 shows what happens: red light from the source falls this time on the screen to give a red patch of light and not a straight line image. Violet light also forms a patch of light, but because it is refracted more, the two patches do not coincide.

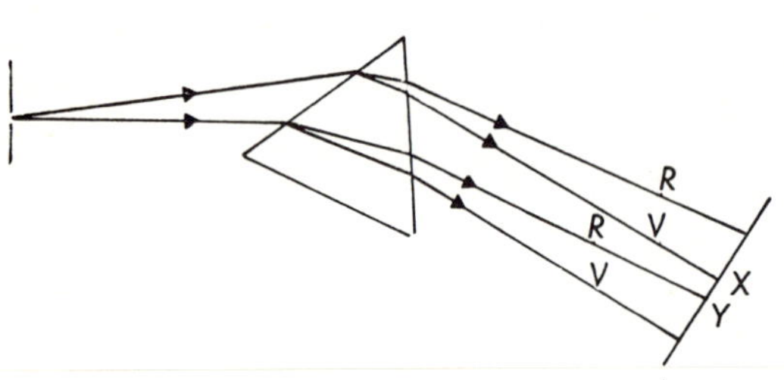

FIG. 320. *Formation of impure spectrum.*

Other colours give similar patches of light falling between these two extreme positions. Above X therefore, you see the colours yellow, orange and red, but they are not very intense. Below Y, the colours green, blue and violet are visible. The region XY will receive all the colours of the spectrum and appears white. We may conclude that just as white light may be split into its constituent colours to form a spectrum, if these colours are recombined, white light is produced.

Newton's Experiments on Spectra Formation

The above explanation of the formation of a spectrum was given by Sir Isaac Newton, based on some experiments which he carried out at Cambridge in 1666, using a large prism made of glass.

For his source, Newton used a small hole in the shutter of his window, which gave him a parallel beam of sunlight, so that the spectrum he obtained was not completely pure. He performed two experiments on his spectrum.

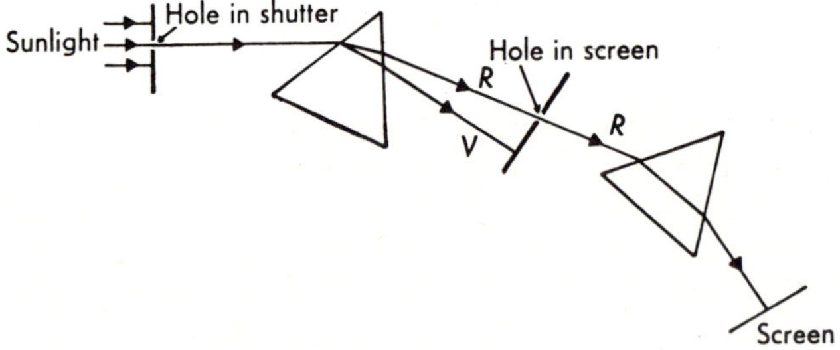

FIG. 321. *Newton's first experiment on spectrum formation.*

1. By making a hole in the screen, Newton was able to pass a single colour from his spectrum through a second prism (Fig. 321). No new colours were

formed on the second screen, showing that the first spectrum contained all possible colours.

2. Newton used a reversed prism, of similar shape to the first one (Fig.

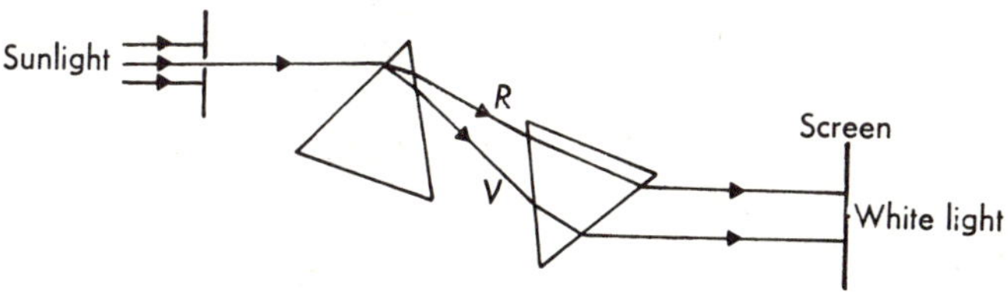

FIG. 322. *Newton's second experiment on spectrum formation.*

322), to bring all the colours of the spectrum together again on a second screen. He found that white light was formed.

Complete Continuous Spectrum

We have already described the pure spectrum (Frontispiece, Fig. 2) as continuous because there are no gaps in it; but the word has greater significance. We have also described light as electromagnetic waves. There are many different types of electromagnetic wave, which differ only in frequency. Those waves whose frequencies lie within the range that we can see make up the "*visible spectrum.*" The frequency of a light wave determines its colour. The violet end of the visible spectrum is the highest frequency that we can see; the red end is the lowest.

If you hold your hand near a lighted light bulb, you will feel the radiant heat which is given off as well as light. If you searched for radiant heat in a pure spectrum you would find it concentrated outside the red end of the visible spectrum.

Radiant heat is transmitted by *infra-red rays*, which are electromagnetic waves of a lower frequency than light waves. At lower frequencies still, electromagnetic waves are used in radar and radio.

If you can find a piece of fluorescent material, try holding it so that your pure spectrum (page 297), falls on it. When you move the material to a dark corner of the box, you will see that the part which glows most strongly was that which was held outside the violet end of your visible spectrum. There must be invisible radiations here too, and these are called *ultra-violet rays;* they affect photographic film, and are electromagnetic waves of a higher frequency than light waves. At higher frequencies still there are *X-rays* and, higher still, the *gamma-rays* of nuclear physics. Not all the radiations of the complete continuous spectrum are given out by the lamp filament, but the word "continuous," usefully describes the nature of the radiation that is

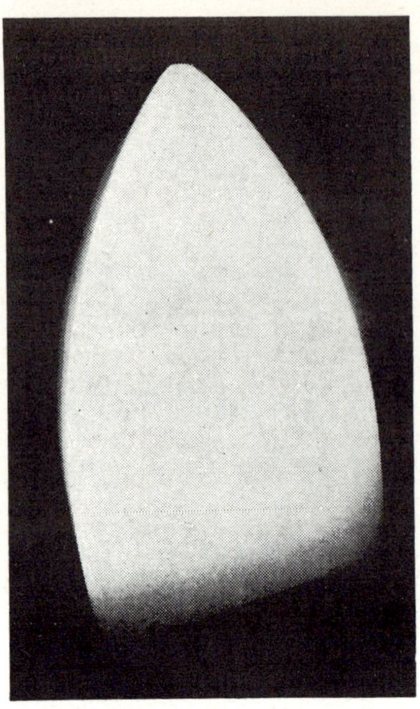

FIG. 323. *The iron (left) was photographed with ordinary film, and then with a film sensitive to infra-red radiation (right). Was the iron hot or cold when the two photographs were taken?*

given out. This type of spectrum is characteristic of hot sources, such as the filament.

Other Types of Spectra

Sprinkle a little table salt (sodium chloride) into the flame from a gas fire, gas ring or spirit burner. You will notice that the flame becomes deep yellow.

Next, take a copper wire and hold it in the flame until it just becomes red. Dip the wire into a solution of ammonium chloride and return the wire to the flame. You will notice this time that the flame becomes an intense blue colour, turning to green.

The flame colours seen in these two experiments differ from the white light given out by a hot solid. In the first case, the sodium chloride vaporizes and the yellow light is given out by excited sodium atoms. In a similar way, the green light given out in the second experiment is due to the presence of vaporized cuprous chloride, which is formed in the flame from ammonium chloride and copper.

Sodium vapour lamps are sometimes used as street lamps; you may be familiar with their striking yellow colour. If such a lamp is used as the source light in an arrangement for forming pure spectra, then instead of a continuous spectrum, all that is seen is a single yellow line. No colours other than yellow are given out by the sodium vapour (Frontispiece, Fig. 3). Mercury vapour can be used as a filling for lamps in the same way. The light is here a very bright greenish blue, and the spectrum formed from it contains a number of colours of varying intensity (Frontispiece, Fig. 4).

Light from single atoms and monatomic ions gives these "line spectra" (slit images in single colours with gaps in between). Molecules tend to give more complicated spectra, but all these types of spectrum are distinctive of the substances producing them. They are, in consequence, of much use in chemical analysis. At a simple level, they are used in flame tests to identify certain elements.

Sodium, as we have seen, imparts a deep yellow coloration to the flame; calcium, a brick red colour; strontium, deep red; potassium, lilac; barium, green.

In advanced analytical work, the entire visible spectrum of the sample under investigation is examined. Sometimes, even the lines formed in the infra-red and ultra-violet areas of the spectrum are taken into account.

Filters and Coloured Light

To continue our investigation into the nature of coloured light, it is now necessary to acquire "colour filters" of the type used for projecting coloured spotlights in the theatre. Filters for the three *primary colours* (red, green and blue) are required.

If theatrical filters are not available, red, green and blue sheets of cellophane paper will make a satisfactory substitute, provided that the colours are pure and deep.

If you look through a green filter, everything appears either green or black.

Now introduce the filter into the experiment for producing a pure spectrum with a glass prism (page 297). Place the filter between the light source and the convex lens (Fig. 319). It will be noticed that the continuous spectrum disappears except for the green area. If the green of the filter is impure, however, other colours might remain in the spectrum; a certain amount of red light might be seen, for example.

This simple experiment shows us that the green filter owes its green colour to the fact that it absorbs from white light all colours except green. It does not "dye" all the white light green as you might think, otherwise the whole spectrum would be seen coloured green. The filter merely absorbs all light except the green as a beam of white light passes through it. Repeat this experiment with the red and blue filters.

With the aid of colour filters we can now produce light of a given colour. Let us now try the effect of mixing these different colours.

The main piece of apparatus used for experiments in colour mixing can easily be made from a cardboard box (Fig. 324). M_1 and M_2 are small mirrors about 100 mm by 80 mm. The inside of the box, opposite the slits, is covered with white paper. The flaps forming the top of the box are left on, since they can be partly closed in order to make it darker inside the box.

Stick the three colour filters over the slits in the box and place a "white" (not "pearl") electric light bulb about 150 mm in front of the slits. The bulb should be level with the slits.

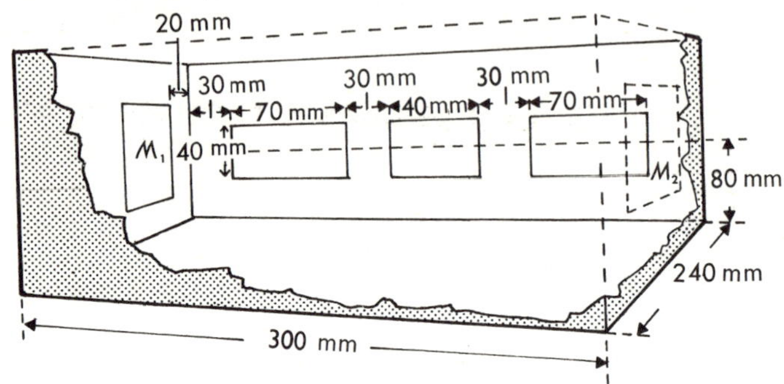

FIG. 324. *Apparatus for investigating the mixing of coloured light.*

Coloured light from all three filters should now be falling on the paper inside the box; this can be checked by covering over two filters at a time.

When all three colours mix on the white paper (Fig. 325), white light is seen; it may be tinged with blue, for instance, if the balance of the three colours is not quite right. Hold your finger down at X to cast a shadow in the "white" light. It is a curious shadow because there are three separate sources of light. Notice the colours of the shadows. Put your hand over the blue filter so that only red and green light reaches the screen. The screen appears yellow. In the same way, green light and blue light give peacock blue, while blue and red give magenta (Frontispiece, Fig 5). You should now be able to see how the coloured shadows were formed.

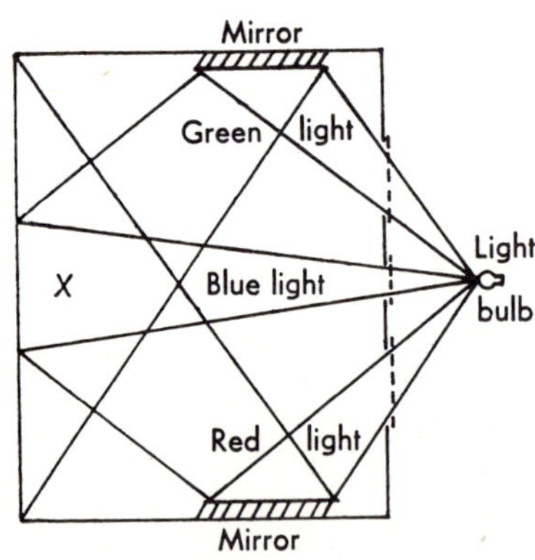

FIG. 325. *Mixing of coloured light.*

The mixing of coloured light (Frontispiece, Fig. 5), suggests the colour triangle (Fig. 326), which is a useful way of predicting the colour which results when two or more colours are mixed. The triangle is equilateral, with red, green and blue, called the three primary colours, at the corners. Mixing red and green light gives the secondary colour yellow, which is therefore placed halfway between red and green along one side. Magenta and peacock blue are similarly placed on the other sides. Mixing all three primary colours gives an appearance of white, so white occupies the centre of the triangle.

If you were to mix green light with red light which was three times as strong as the green light, then this would give the point *O*, midway between yellow and red; the colour you would see would be a shade of orange.

Points within the triangle represent the addition of white light to the colour at the edge. The point *P* lies between red and white, and is therefore pale red, i.e. pink.

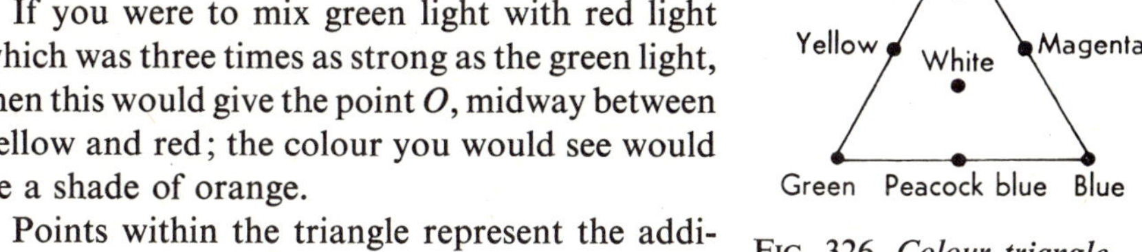

FIG. 326. *Colour triangle.*

Notice also that white light can be produced in several ways: for example, by mixing yellow, magenta and peacock blue, or by mixing 1 part green with 2 parts magenta (1 part red and 1 part blue).

Paints and Pigments

You have, no doubt, mixed coloured paints together, and may be troubled that what we have been saying about coloured light seems to contradict your experience. Why does paint appear coloured? If you look at a white object in white light, it seems white because it reflects all the light falling on it. A red object appears red in white light because, of all the colours falling on it, the pigment in the paint reflects red light only and absorbs the other colours. Viewed in red light, a white object appears red and a red object appears red. Check this with your red filter by placing a white object and a red object in white light and looking at them through the filter. Repeat with the blue filter. The white object appears blue in blue light, whereas the red object, which absorbs blue light, appears black.

In mixing coloured light we found that a mixture of red and green light entering the eye gave the sensation of seeing yellow. Mixing red pigment, which ideally absorbs all colours but red, with green pigment, which should absorb all colours but green, one should obtain black; owing to imperfect pigments, brown is obtained instead. Blue pigments reflect some green and violet light as well as blue, while yellow pigments reflect red and green light. When yellow and blue pigments are mixed, therefore, the only colour not absorbed by one or other of the pigments is green, so a green paint is obtained (Frontispiece, Fig. 6).

Look at a coloured picture through your red, blue and green filters (have the picture illuminated by strong sunlight). It will be clear what colours the pigments in the coloured ink are reflecting.

You can liken the action of pigments to that of filters. Pigments reflect certain colours, absorbing the rest, while filters transmit certain colours, absorbing all others. This is a property of the molecules of the chemicals contained; you have already seen that molecules are capable of emitting light of definite colours when heated in a flame (page 300). It is well known that chlorophyll is a green pigment contained in plants; it is the agent responsible for "photosynthesis," the conversion of carbon dioxide and water to carbohydrates. A green pigment is one which absorbs light of other colours, and chlorophyll, therefore, absorbs light from both the blue and the red ends of the spectrum. This absorbed light provides energy for the process of photosynthesis.

Colour Vision

In the eye, the nerves in the retina end in sensitive cells called rods and cones. As we have already learnt, the cones are responsible for our sense of colour. While colour vision is not yet completely understood, the theory due to Young, that there are three types of cone corresponding to the three primary colours, is plausible. Thus colours which we see would be the result of the relative stimulation of cones sensitive to red, blue and green light, say. It is possible that colour vision is like this; colour blindness would be due to defects in one or more sets of cones, so that certain colours are not perceived correctly. It seems probable, in fact, that colour vision is more complex, and the full explanation would require more than three types of cone.

QUESTIONS

1. Draw a diagram of an arrangement of apparatus with which you could form a pure continuous spectrum. How would the spectrum be altered if you placed a green filter in front of the light source? What would be the appearance of a strip of blue paper if held right across your continuous spectrum?

2. If you looked at a stick of white chalk through a prism which was held with its refracting edge parallel to the chalk, what would be the apparent position of the chalk? Describe the appearance of the chalk.

3. If you had a red flower with green leaves, what would be its appearance when viewed through: a red filter; a blue filter; a yellow filter?

4. What is the colour of the flame produced by a mixture of strontium and barium salts? If you formed a pure spectrum from the light from such a flame, what would you see?

VIBRATIONS AND WAVES

CHAPTER 25

PROGRESSIVE WAVES

ALL of us have watched rain drops falling into a puddle and observed the waves that spread out in circles from each drop. Most people when asked to mention other types of waves would think of sound waves, but the physicist would add to the list light waves, radio waves and many more. What have all these in common that we should call them waves? First, in each something is varying rhythmically, e.g. in water waves the height of the surface, and in sound waves the pressure of the air. Secondly, with this rhythmic variation is associated a transfer of energy from the source to other bodies.

We can illustrate the fact that waves transmit energy by two simple examples. If a model yacht has become becalmed in the middle of a pond, you may attempt to move it by throwing a stone at it. If this hits the yacht, it will gain energy directly from the stone. If, however, the stone falls into the water alongside, waves will be set up in the surface which in turn will disturb the yacht. In this case the energy of the stone has been transformed into energy of water waves, part of which has been transferred to the yacht.

In a similar way you may shout to a friend to attract his attention. The energy of your vocal chords is changed into energy of sound waves which spread out in all directions. Some of this energy reaches your friend and affects the nerves in his ear.

It is with sound waves that we are primarily concerned in this section. However, as these are produced by vibrations of the air which are invisible,

FIG. 327. *Waves generated by a stone falling into water.*

we shall begin our study of wave motion by examining water waves. This will lead on to a study of sound waves and their production by vibrating strings, columns of air, and musical instruments. We shall then consider the characteristics of musical notes, and the recording and reproduction of them. Finally we shall discuss the application of the ideas of wave motion to other branches of physics.

TRANSVERSE WAVES

You can study the behaviour of water waves at home by pouring a little water into a shallow dish or sink. When you move your finger up and down in the surface with a steady rhythm, waves will be generated which spread out in circles. If you use a horizontal ruler instead of your finger to generate the waves, you will find that they move forward in straight lines. The appearance of the water surface is as shown in Fig. 328, while in Fig. 329, the elevation, or view through one side of the dish, is represented.

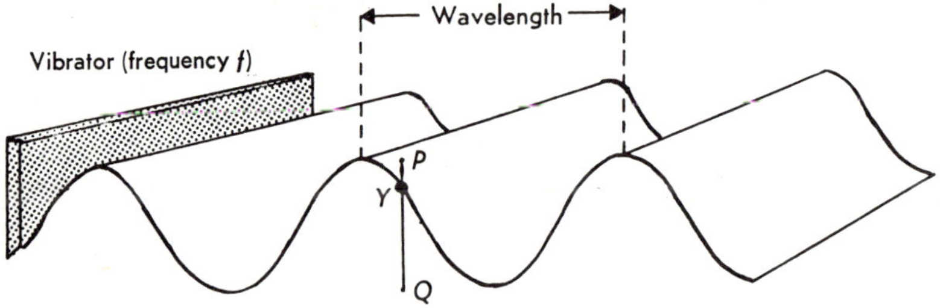

FIG. 328. *Generation of transverse waves in water.*

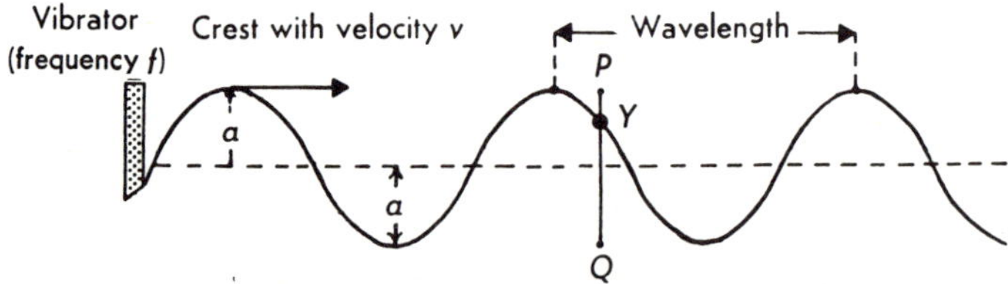

FIG. 329. *Generation of transverse waves in water (sectional view).*

If you fix your eyes on one crest, you will see that it moves forward with a steady velocity, while remaining separated from adjacent crests by a fixed distance. It appears as though the water moves forward as a whole, but you can show that this is not so by scattering a few small pieces of cork on the surface. You will see that they move up and down but do not move forward.

The size and motion of a wave is described in terms of its frequency, wavelength, period, velocity, and amplitude. The following definitions of these terms should be studied in conjunction with Fig. 328 and Fig. 329.

Frequency (f) is the number of vibrations of the generator per second. In the case of Fig. 329 the generator is a ruler, one vibration of which is defined as one complete up-and-down movement. Frequency may also be defined as the number of complete waves generated per second, or as the number of crests passing any line (such as PQ) per second. Frequency is measured in hertz, abbreviated to Hz, one Hz being a frequency of one complete vibration per second.

Wavelength, represented by the Greek letter λ (lambda), is the distance between adjacent crests.

Period or *periodic time* (T) is the time taken for one complete vibration of the generator. It should be noted that frequency $f = \frac{1}{T}$, or $fT = 1$.

Velocity of a wave (v) is the velocity with which any crest moves.

Amplitude (a) is the height of the crest above the surface of the water when at rest; in other words it is the maximum displacement from the undisturbed position of any point on the surface.

Now consider any point on the wave such as Y in Fig. 329. Since the ruler is generating one vibration in T, it follows that Y travels up and down along the line PQ, making one complete vibration in T.

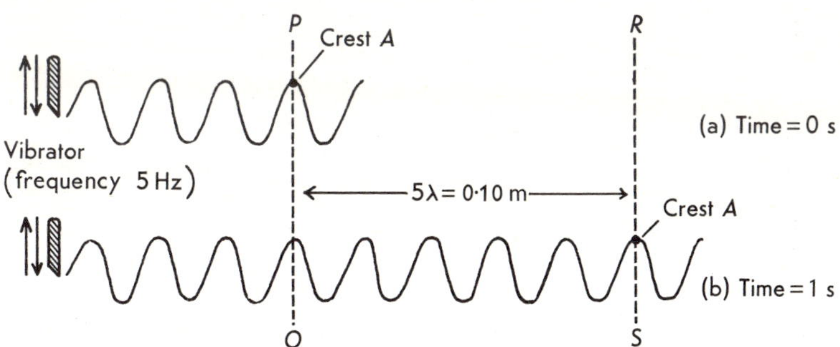

FIG. 330. *Velocity, wavelength, and frequency of a transverse wave.*

To find the relationship between the velocity, wavelength and frequency, study Fig. 330 in which the frequency is taken as 5 Hz and the wavelength as 0·02 m. Fig. 330*a* shows the shape of the water surface at a particular instant and Fig. 330*b* the same surface one second later. In this time five waves have been generated so that the crest A, which was on the line PQ at the start, has moved to RS, a distance of 5 wavelengths or 0·10 m. The velocity of the crest A is therefore 0·10 m/s. In general:

$$\text{Velocity of the wave} = \text{Frequency} \times \text{Wavelength}$$
$$v = f\lambda$$

This formula is true for all waves and is most important. You can check it for yourself in the case of radio waves by referring to the table of frequencies and wavelengths given in the *Radio Times:*

Programme	*Wavelength*	*Frequency*
Radio 1	247 m	1214 kHz
Radio 2	1500 m	200 kHz
Radio 3	464 m	647 kHz
Radio 4	330 m	908 kHz

Check for yourself that if you multiply the frequency by the corresponding wavelength, you obtain the same value in each case for the velocity, namely 300 000 000 m/s.

We defined the velocity of the wave as the distance moved by a crest in one second. You may be able to see this more clearly in the case of a wave moving along a rope. Tie one end of a long rope to a post and hold the other end in your hand. Jerk this end upwards and you will see a hump travelling down the rope which may bounce back at the post and return to you again. It is quite clear that the rope itself has not moved to the post and back, but that particles of the rope have vibrated up and down in such a way that the crest—the position of maximum displacement—has travelled down the rope and returned.

The Siren

Waves in a string or on water can be seen and are therefore fairly easy to understand, but sound waves in air are invisible and much more difficult to explain. Let us think first of how a musical note is produced by a siren. The type used in a physics laboratory, called Caignard de la Tour's siren, is shown in Fig. 331. Air from the bellows is blown into the cylinder *A* from which it may escape through holes *B* in the top. This is only possible when these holes are opposite some corresponding holes in the disc *C* which can rotate about a vertical axis. So if we spin the disc, puffs of air will come through every time the holes are opposite to each other. Actually the holes are set at an angle, so that the air passing through them drives the disc round. At slow speeds the siren produces a puffing noise, but as the pressure of the air increases, the disc will turn faster and the individual puffs blend to give a musical note whose pitch rises with the speed of rotation

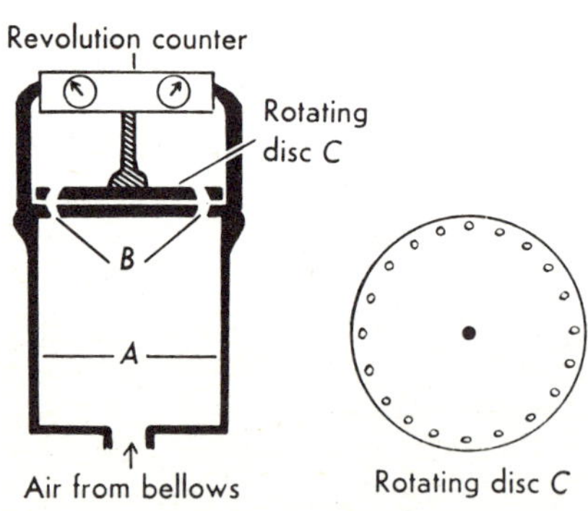

FIG. 331. *Caignard de la Tour's siren.*

of the disc. The frequency of the note is the number of compressions emitted every second, which is equal to the number of holes in the disc multiplied by the number of revolutions per second. The disc is provided with a revolution counter so that with the aid of a stop watch, the time for a given number of revolutions may be measured. Hence the speed of rotation of the disc and the frequency of the note may be calculated.

You may produce an effect similar to that of a siren from your bicycle. Turn it upside-down and drive the back wheel round with the pedals. Push one corner of a piece of card or a folded newspaper into the wheel near the rim. The spokes will make the card vibrate backwards and forwards and so generate a series of compressions. These will form a musical note whose pitch will rise as the speed of rotation is increased.

LONGITUDINAL WAVES

We have seen that a musical note is generated when the air is compressed a fixed number of times per second. How do these compressions travel to the ear? Consider a player blowing a trumpet. Does he generate a compression and produce a wind which blows it to the ear? Certainly this would act in the right direction since the sound is loudest along the axis of the instrument. However, he would have to blow very hard indeed, and the energy given to the air would soon be dissipated.

If we think once again of a wave moving along a rope, we remember it is not the whole rope which moves carrying a rigid crest with it, but the particles of the rope that vibrate so that the rope bends and the crest moves along it. In a similar way in a sound wave the particles of air vibrate, but in the direction of motion of the sound, so that the compression moves without the aid of a wind.

This type of wave in which the vibrations take place along the direction of motion of the disturbance is called a *longitudinal wave*, while other types in which the vibrations are perpendicular to the direction of motion, are called *transverse waves.* Longitudinal waves can be set up in solids as well as in air; for instance, the high-pitched squeak which sometimes occurs in cleaning windows, is due to longitudinal vibrations of the glass.

It should be noted that the compression is not the whole of the longitudinal wave, any more than the crest is the whole of the transverse wave. Just as in the transverse wave each particle vibrates so that a crest is followed by a trough, so in a longitudinal sound wave the air vibrates backwards and forwards so that there is a continuous variation in pressure from a maximum to a minimum. Where the pressure is above normal, there is said to be *compression* and where it is below, there is *rarefaction.* These two terms are often used when what is meant is the *point of maximum compression* or the *point of maximum rarefaction.*

Generation of Longitudinal Waves

Fig. 332 represents the generation of a longitudinal wave in air by a piece of springy steel which is mounted vertically with its lower end clamped. The upper end is made to oscillate horizontally, thus setting the air in vibration and producing waves which move in a horizontal direction. The period of the vibrator, that is the time for one complete oscillation, is taken to be 0·012 s.

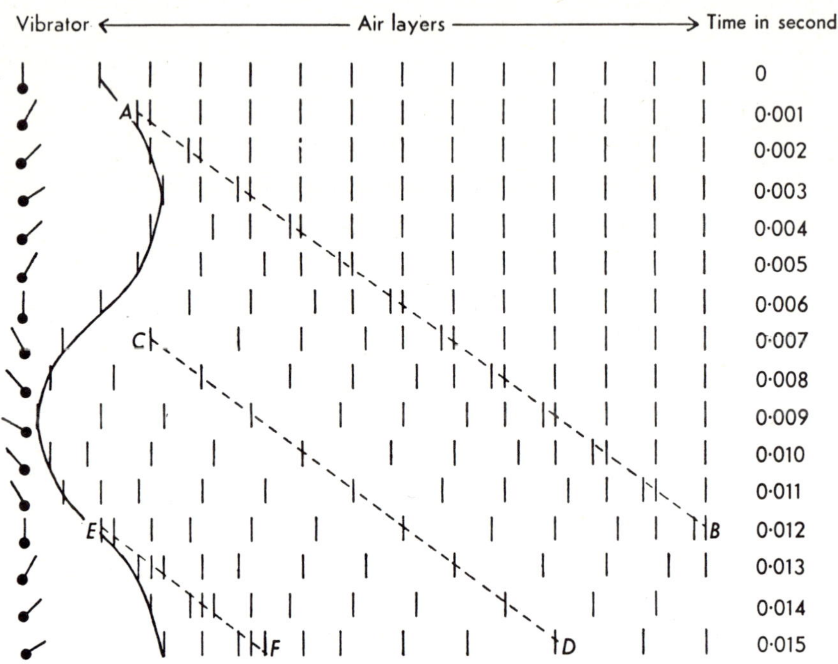

FIG. 332. *Generation of longitudinal waves in air.*

Each line in Fig. 332 shows the state of the air at intervals of 0·001 s. The vertical strokes in each line represent layers of air, which in the first line (at zero seconds) are shown in their undisturbed or equilibrium positions. The following lines show the layers of air in their instantaneous positions at the time indicated.

By studying Fig. 332 it becomes apparent that each layer of air oscillates about its equilibrium position with a period of 0·012 s. The line joining the consecutive positions of the first layer (*AE*) shows this point clearly, but it will help you to trace the movement of some other layers for yourself.

You will notice in each line of Fig. 332 that there is a point where the layers of air come close together. These points which are joined by the lines

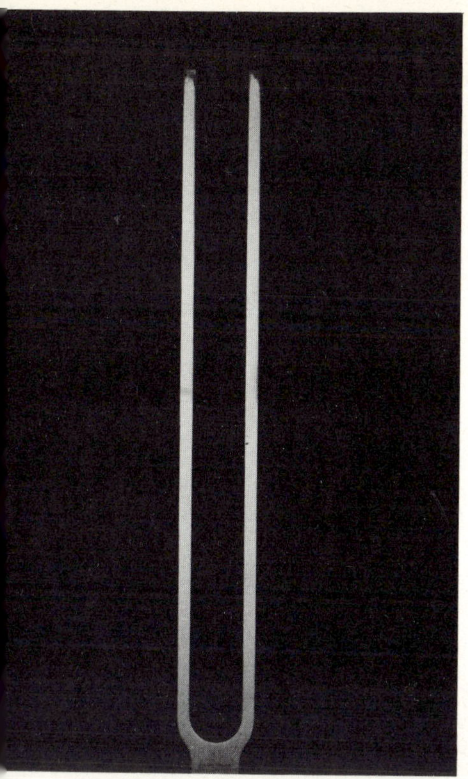

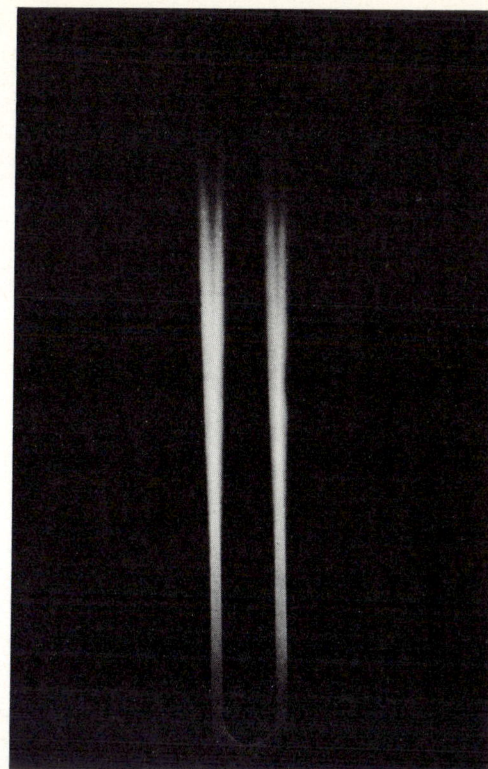

FIG. 333. *A common form of vibrator for the experimental generation of waves is the tuning fork shown stationary (left) and vibrating with large amplitude (right). Tuning forks are used as frequency standards.*

AB and *EF*, are the *points of maximum compression.* You will also notice that the layer of air undergoing maximum compression is in every case in its equilibrium or undisplaced position (the second layer at 0·001 s, the third layer at 0·002 s, etc.).

From 0·007 s onwards, *points of maximum rarefaction*, where the layers are far apart, also become apparent along the line *CD*. Once again you will notice that the layer of air undergoing maximum rarefaction is in its equilibrium position (the second layer at 0·007 s, the third layer at 0·008 s, the fourth layer at 0·009 s, etc.).

We have already learnt in the case of the transverse wave that the distance between successive crests is the wavelength. Similarly, the wavelength of a longitudinal wave is the distance between successive points of maximum compression. Referring once again to Fig. 332 you will see in the line for 0·012 s that two points of maximum compression occur. The distance *BE* between these two points is the length of the wave generated in the air by the vibrating steel.

Comparison Between Transverse and Longitudinal Waves

To make clear the similarity between longitudinal and transverse waves we

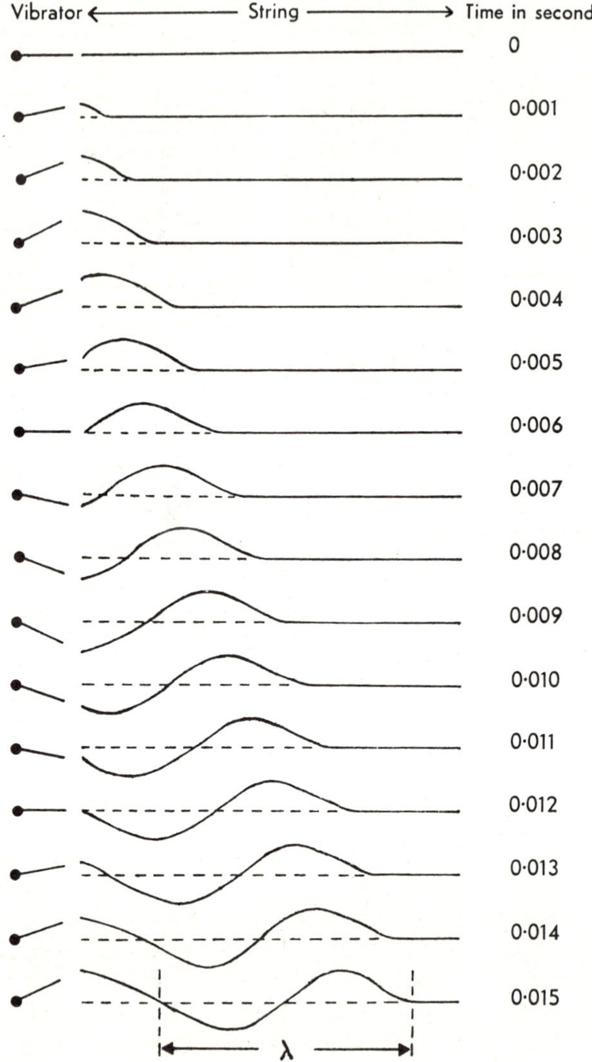

FIG. 334. *Generation of transverse waves in a string.*

will once again use the piece of springy steel. This time, however, we will mount it horizontally with one end clamped and the other end attached to a string (Fig. 334); this end is made to vibrate vertically with the same period as before, so that it generates waves in the string, the crests of which move off to the right.

The first line in Fig. 334 shows the vibrator and spring in their undisturbed positions. Subsequent lines show the shape of the string at intervals of 0·001 s, so that the whole diagram represents the generation of the transverse wave. This diagram should be compared carefully with Fig. 332. A summary of the comparison between transverse and longitudinal waves follows:

	Transverse wave	*Longitudinal wave*
Period	Each particle of the string vibrates at right angles to the direction of motion of the wave: e.g. in Fig. 334 vertical vibrations of period 0·012 s produce a wave moving horizontally.	Each layer of air vibrates in the same direction as the motion of the wave: e.g. in Fig. 332 horizontal vibrations of period 0·012 s produce a wave moving horizontally.
Crests and compressions	A crest appears at any point every 0·012 s and moves to the right. Crests occur at points where particles of the string are at maximum displacement from the equilibrium position.	A compression appears at any point every 0·012 s and moves to the right. Compressions occur at points where layers of air are in their equilibrium position.
Troughs and rarefactions	A trough appears at any point every 0·012 s and moves to the right. Troughs occur at points where particles of the string are at maximum displacement from the equilibrium position.	A rarefaction appears at any point every 0·012 s and moves to the right. Rarefactions occur at points where layers of air are in their equilibrium position.
Frequency	The rate at which crests pass any point is the frequency f.	The rate at which compressions pass any point is the frequency f.
Wavelength	The distance between successive crests is the wavelength λ.	The distance between successive compressions is the wavelength λ.
Velocity	The velocity of the wave is the velocity of a crest. $v = f\lambda$	The velocity of the wave is the velocity of a compression. $v = f\lambda$

Representation of a Longitudinal Wave

As the diagrams of Fig. 332 are difficult to draw, the longitudinal wave is represented by a curve like Fig. 335*a* (this is the representation of the wave of Fig. 332 at time 0·012 s). The diagram implies that a layer of air whose undisturbed position is *P*, is displaced to *Q*, where $PQ = PQ'$. The displace-

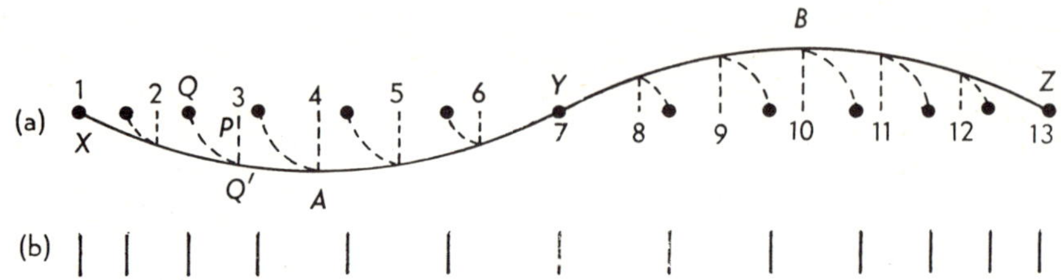

FIG. 335. *Representation of a longitudinal wave in transverse wave form.*

ments along *XY* are drawn backwards because the curve is below the axis, while those of *YZ* are drawn forwards. For clarity, the positions of the layers of air originally at points 1 to 13, are shown in Fig. 335*b*. Note the agreement with the positions shown in Fig. 332 at 0·012 s.

Comparing Fig. 335*a* with Fig. 335*b*, we can see the connection between pressure and displacement:

Points	*Pressure*	*Displacement*	*Gradient of Curve*
X and *Z*	Maximum compression	Zero	Maximum negative
Y	Maximum rarefaction	Zero	Maximum positive
A and *B*	Normal pressure	Max. displacement	Zero

TRANSMISSION OF SOUND WAVES

From what we have already learnt it has become clear that air or some other medium is necessary for the passage of a sound wave. We may demonstrate this fact with the aid of the apparatus shown in Fig. 336.

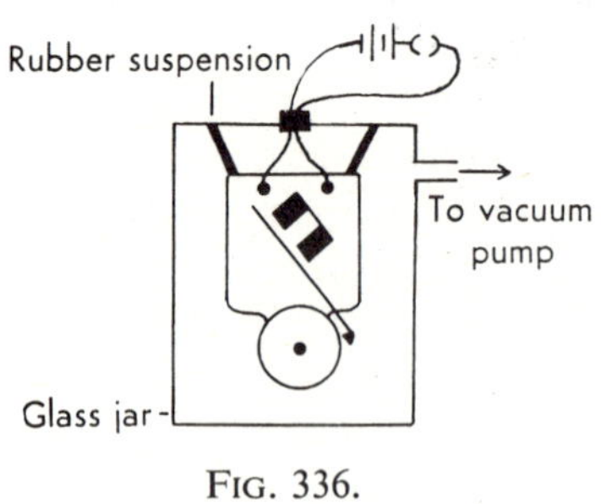

FIG. 336.

An electric bell is suspended by strips of rubber in a closed glass jar (rubber is used because this will not transmit the vibrations of the bell). While the jar contains air, the bell can be clearly heard, but if the air is removed by a vacuum pump, the sound becomes inaudible (although the bell can be seen to be working) because the vibrations cannot be transmitted through a vacuum. For the same reason, if there was an enormous explosion on the moon we might be able to see it but we certainly could not hear it.

Velocity of Sound

Sound travels in air with a velocity of about 340 m/s, which is much less than the velocity of light (3×10^8 m/s). Therefore, if you are watching a cricket match from a distance, you will see the batsman strike the ball before you hear the sound of the collision. In the same way, we see lightning before we hear the thunder which is the noise of the explosion caused by it. The speed of light is so great that we may assume that the interval between the thunder and lightning represents the time taken by the sound to reach us, at a speed which is approximately equivalent to 1 km every 3 seconds. So if the lightning appears 7·5 seconds before the thunder is heard, we can estimate that the storm is 2·5 km (or 1·5 miles) away.

We can use this principle to measure the velocity of sound. An observer A watches another man B, at a distance x away, fire a gun. A measures with a stop watch the time between the flash being visible and the sound audible. If this time is t and the velocity of sound is v, then:

$$v = \frac{x}{t}$$ The velocity will be in m/s.

The value of t will however be small, and will depend on the speed of reaction of the observer. Such personal error can be eliminated by using electrical devices to record the arrival of the light and sound.

The value for the velocity of sound will be affected by three other factors.

1. *The strength and direction of the wind.* The effect of a steady wind in the direction AB or BA may be eliminated by repeating the experiment in the opposite direction, with A firing the gun and B observing. Two values of the time, t and t', will be obtained and the average of these two values is then used in the above equation. Such a method, however, does not allow for gusts or cross-winds.

2. *The humidity of the air.* If the air is damp, it is less dense than dry air at the same temperature and pressure. This decrease in density tends to increase the velocity slightly. Allowance could be made for this change by calculation, if the humidity of the air along AB were known and remained constant during the experiment.

3. *The temperature of the air.* The velocity of sound increases as the temperature rises. In fact it can be shown to be proportional to the square root of the absolute temperature T so that if v_o, v_t are the velocities at 273 K and T respectively—

FIG. 337. *The first accurate determination of the velocity of sound in water was carried out in 1826. Two boats were moored about nine miles apart on Lake Geneva. At the boat (left) the bell in the water was struck and some gunpowder simultaneously ignited. Since it was night the observer in the boat (right) could clearly see the flash and was able to detect the arrival of the sound by means of a large ear trumpet sealed by a membrane and submerged in the lake. The time interval between the arrival of the flash and that of the sound was measured. The velocity of sound in water could then be calculated and was found to be more than four times its velocity in air (about 1600 m/s).*

$$\frac{v_t}{v_o} = \sqrt{\frac{T}{273}}$$

It is because of this variation of velocity with temperature that the speed of supersonic aircraft is given in terms of the velocity of sound. For example, *Mach* 2 means a speed of twice that of sound, which at the earth's surface might be 2×340 m/s, but at the lower temperatures existing at a height of 10 000 m, is only 2×270 m/s.

Reflection of Sound

Sound waves are reflected by any hard surface, so that if you clap your hands at some distance from a rock face or a high building you may hear the echo of the sound you made. The interval between the sound and its echo is the time that the wave has taken to travel to the reflecting surface and back. If the velocity of sound is known, and this time measured, the distance of the reflecting surface from the source of the sound may be calculated.

This principle is used in echo-sounding, by which the depth of water beneath a ship may be determined. An underwater transmitter on one side of the ship emits a pulse of energy (similar to a shout) which is reflected by the sea-bed and detected by a receiver on the other side of the ship. The time interval between the pulse and its echo is recorded electrically and the depth calculated. Weak echoes may be obtained from shoals of fish, so that drifters use this apparatus to help them in their work.

The same principle is used in locating aircraft by radar, but here the energy is emitted in radio waves of very short wavelength. It is said also that blind people are helped in finding their way by the echoes of their footsteps, which they hear reflected from obstructions.

In a concert hall, echoes may ruin the reception of sound. A member of the audience *A* (Fig. 338) will receive sound from the source *S* both by the

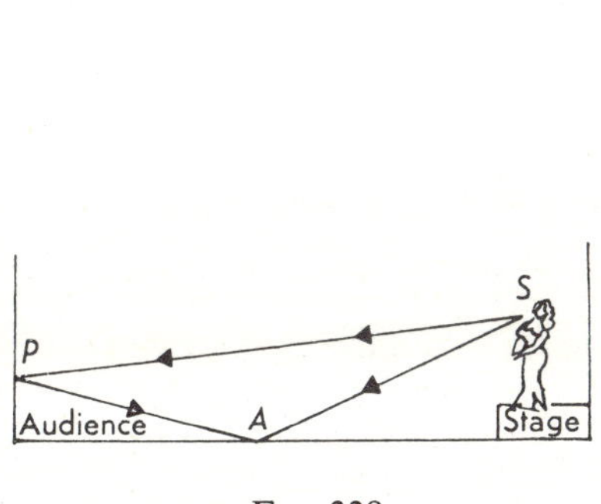

FIG. 338.

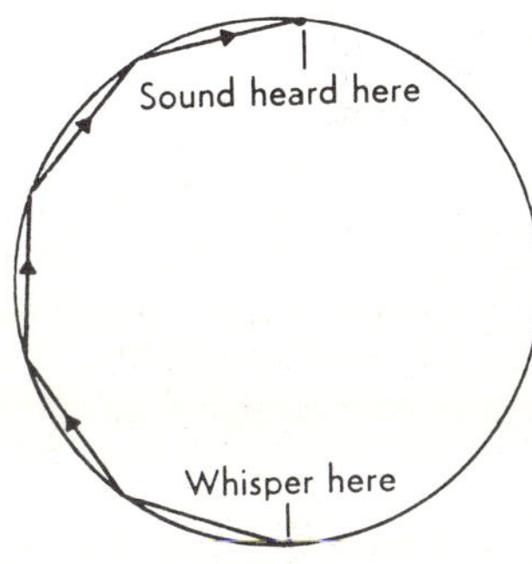

FIG. 339.

direct path *SA* and by reflection from the back wall along the path *SPA*. Because *SPA* is longer than *SA*, the reflected sound will arrive later than that by the direct path and the two together will produce a blurred effect. For this reason it is important to prevent echoes being formed and modern concert halls are therefore designed so that there are no large flat reflecting surfaces. As a further precaution, the walls and ceiling are covered with absorbent plaster or other materials which reflect little sound. It is noticeable that an empty hall is more difficult to speak in than a crowded one, because the audience absorbs sound waves much better than the floor or empty seats.

A remarkable demonstration of the reflection of sound is given in the Whispering Gallery of St. Paul's Cathedral in London. This gallery is just inside the dome so that its wall forms an unbroken circle of hard surface. A whisper close to it is reflected many times and can be clearly heard on the opposite side 36 m away (Fig. 339).

FIG. 340. *To hear sound from the opposite side of the Whispering Gallery in St. Paul's Cathedral it is necessary to listen close to the wall. Why?*

QUESTIONS

1. Explain, by means of an example, what is meant by the statement that waves transmit energy.

2. Explain the formation of echoes. Give one example of their use and describe one instance in which they are a nuisance.

3. Describe how the velocity of sound has been measured in open air. What factors are likely to affect the result obtained?

4. A ship, sailing up a river, uses its echo-sounder to find the depth of the channel. The instrument records the return of the signal 0·02 s after its transmission. How deep is the channel if the velocity of sound in water is 1600 m/s?

5. A signal sent out by a radar station is reflected by an aircraft 90 km away. How long will it be before the signal returns to the station if the velocity of the signal is 3×10^8 m/s?

6. Explain carefully the difference between transverse and longitudinal waves, giving one example of each.

7. Explain what is meant by (*a*) wavelength, (*b*) frequency and (*c*) velocity of a wave. Deduce the connection between them.

8. If the wavelength of Radio 2 with a frequency of 200 kHz is 1500 metres, what is the wavelength of the V.H.F. station broadcasting on 90 MHz? (1 kHz = 1000 Hz; 1 MHz = 1 000 000 Hz.)

9. Describe a siren and show how you would use it to measure the frequency of a tuning fork. If the disc of the siren has 12 holes and rotates at 1200 rev/min, what is the frequency of the note produced?

CHAPTER 26

STATIONARY VIBRATIONS IN STRINGS

WHEN a violinist plays a note, he moves his bow across a string tightly stretched over a bridge on his instrument. The string vibrates with its characteristic frequency and transmits these vibrations to the body of the violin. This in turn sets the air in contact with it into vibration and generates sound waves which you can hear.

As the violin bow moves across the string, it alternately sticks and slips, thus pulling the string to one side and then releasing it. This is the same sort of motion as that of your hand, when you jerked one end of the rope up and down. You remember that this motion produced an incident wave which travelled down the rope, was reflected at the post, and returned to you. If you had moved your hand up and down with constant frequency, you would have filled the whole length of the rope with incident waves, which would have become mixed up with the reflected waves, thus producing a confused disturbance. The technical way of saying this is that the incident and reflected waves *interfere*. In certain circumstances, however, the result of the interference is a fixed pattern called *a stationary vibration*. This is the type of vibration produced by the violinist in his string.

Experimental Formation of Stationary Waves

By using an electric bell instead of your hand to generate the waves you can study the nature of stationary waves more carefully. Remove the gong from the bell and tie a thin piece of string to the clapper *A* (Fig. 341). Pass the string over a nail *B* and hang from it a small weight to keep it taut. When you switch on the bell you will see the confused pattern formed by interference between the incident waves generated by the clapper and those reflected at the nail. Making sure that at all times the weight hangs freely, adjust the length *AB* until you find a position in which the string vibrates steadily in one loop (Fig. 341). In this position the string is said to be vibrating with its *fundamental frequency*, which is the same as that of the bell.

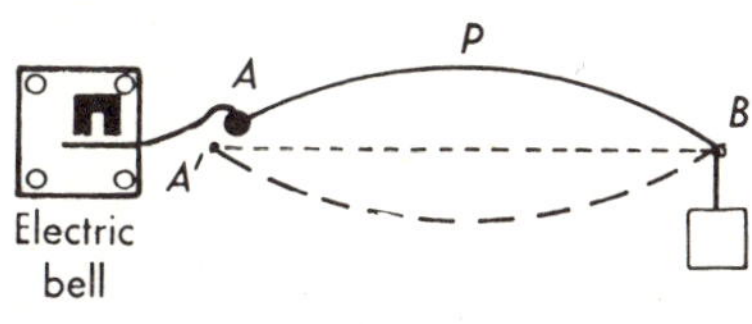

FIG. 341. *Stationary wave-formation.*

The motion of the violin string is similar to this, but we can learn more about it from experiments with a spiral spring curtain rail of the type sold by most household stores.

Stretch about six feet of this spring between two hooks A and B, pluck it and note the vibration (Fig. 342). By plucking the wire you have generated a wave which has travelled down the wire and been reflected at the end B. Reflection has also occurred at the end A, and all the waves formed have added together to give the vibration shown.

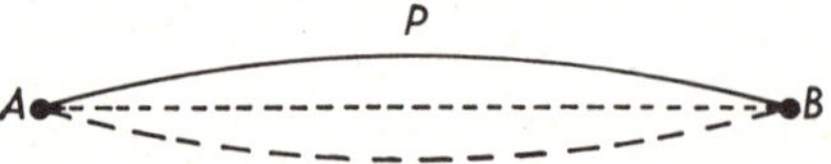

FIG. 342. *Stationary wave-formation.*

In both these vibrations, you can see that each point, whether on the spring or on the string, moves up and down with the same frequency, while the amplitude of the vibration varies from nothing at the ends A and B to a maximum in the middle P. The points which are at rest (A and B) are called *nodes*, while the point of maximum amplitude (P) is called an *antinode*. (In the bell experiment, the loop is not quite complete and the antinode is just beyond the clapper at A'). This type of vibration is called a stationary vibration because the crests—the points of maximum amplitude—do not move along the string. This particular vibration with one loop is known as the *fundamental or first harmonic* and occurs when the string is half a wave-length long.

Frequency of a stationary vibration

How does the frequency of a vibrating wire depend on the properties of the wire? We have already stated that the length of a loop l is half a wave-length

$$\text{or } \lambda = 2l$$

We also know that velocity $v = \lambda f$

Then it can be shown that the velocity of a wave moving along a wire whose mass per unit length is m when the tension in the wire is T, is given by:

$$v = \sqrt{\frac{T}{m}}$$

$$\text{so that } \quad f = \frac{v}{\lambda} = \frac{1}{2l}\sqrt{\frac{T}{m}}$$

We do not often need this formula at this stage, but we do need to note that for a given wire under a fixed tension:

Frequency is inversely proportional to the length

$$f \times l = \textit{constant}$$

Sonometer

The relationship between the length l of a string and the frequency f of its fundamental vibration is usually proved in the laboratory with a sonometer. This, as shown in Fig. 343, consists of a sounding box, on which is stretched a steel wire supported by two fixed bridges, A and B. A movable bridge C is provided so that the portion AC of the wire may be set into vibration by plucking it. The length of AC is adjusted until the note produced by it is the same as that of a tuning fork of known frequency. This is repeated with other forks and it is found that:

$$\textit{Frequency} \times AC = \textit{constant}$$

As it is not always easy to get the wire exactly in tune by ear the final adjustment may be made by the following method.

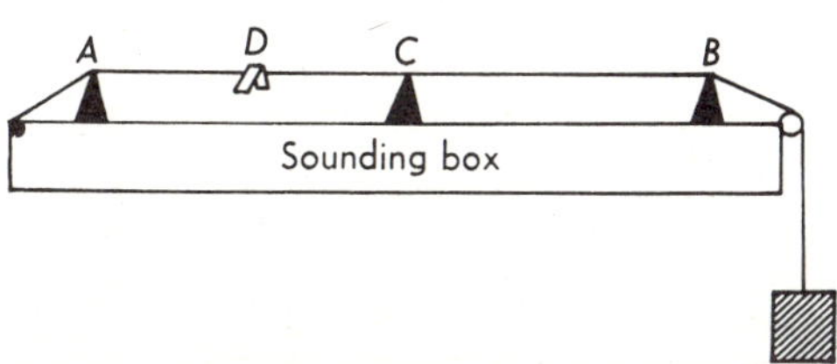

FIG. 343. *Sonometer.*

Sound the fork and place it on the bridge at A. This will produce what are known as forced vibrations of the wire. These can be detected by means of a small paper rider, D, balanced on the wire as shown. If the wire is nearly in tune, the rider will move on the wire. When, however, the frequency is exactly right, resonant vibrations will be produced in the wire, and the motion of the rider will become much more violent.

Our experiments have shown that if the length of the wire AC increases its frequency decreases. We can also use the sonometer to discover the effect of changing the tension, or of replacing the wire with one of a different material or diameter. We shall find that the frequency is increased if the tension is increased or if the wire is changed to one whose mass per unit length is less.

Let us recapitulate what we have learnt about the vibrations of a wire.

1. The stationary vibration is formed by interference between the incident and reflected wave.
2. A loop is half a wavelength long.
3. The frequency is inversely proportional to the length (for a given wire under a given tension).
4. The frequency of a given length may be increased by increasing the tension or by using a lighter wire.

All these facts are used in the construction of stringed instruments. The violinist shortens his string by pressing it on to the fret board with his fingers

FIG. 344. *The 46 strings of a modern harp differ in length, tension, and thickness. Tension is adjusted by means of the pegs along the top of the instrument.*

when he wishes to play a higher note. The stringed instruments of lower pitch—the cello and double bass—use heavier and longer strings than the violin. In the piano, as in the harp, the lower notes use longer wires and these are loaded to make them heavier.

Harmonics

So far we have only considered the fundamental vibration of the string which occurs when only one loop is formed. If in the experiment with the

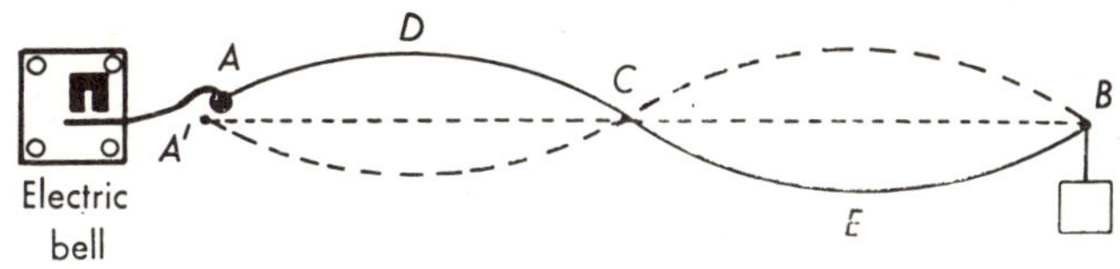

FIG. 345. *Formation of the second harmonic in a stretched string.*

bell you double the length of the string, you will obtain the vibration shown in Fig. 345. The frequency, which is that of the bell, remains constant, so the length of the loop is unchanged, but there are now two loops. A node has appeared at the midpoint *C*, while *D* and *E* are antinodes. Note that the displacements at these two points are in opposite directions (one up and one down). By increasing the length of the string you could obtain vibrations with three or more loops all of the same length.

To generate a double loop in your spring, pluck it at *X* while gently steadying it at its mid-point *C* with your finger (Fig. 346*b*). Since the length of the loop is halved, the frequency will be doubled. Similarly, by steadying the spring at a point *D*, one third of the way along it (Fig. 346*c*) you can obtain

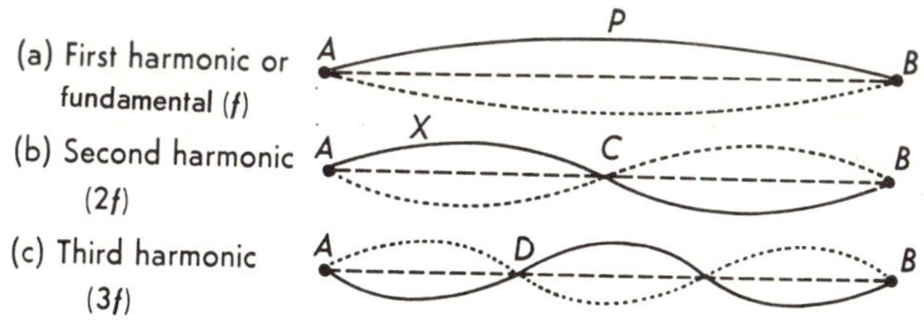

FIG. 346. *Formation of first three harmonics in a stretched spring.*

a vibration with three loops, and you can go on in the same way to form four, five or six loops. In each case as one more loop is added, the length of each loop decreases and the frequency can be seen to increase.

The notes given out by a wire vibrating in this way are known as *partial tones* because the wire is vibrating in parts. In simple vibrations, such as those described, the frequency of the partial tones is in the ratio of 1 : 2 : 3 etc. and these tones are called *harmonic tones*, or *harmonics*. They are sometimes known as overtones, but as there is some confusion in the use of this term, it will be avoided in this book. The first harmonic, produced when the wire vibrates as one loop, is clearly very important and is called the *fundamental tone*.

It is possible to get more than one of these vibrations to take place at the same time. Hang a weight of about half a pound at the mid-point of your spring, and pluck it in the usual way. You will see the weight move up and down with the fundamental frequency, while, at the same time, loops are formed on either side in which the wire is vibrating with twice this frequency.

Harmonics in Musical Instruments

This type of vibration, with more than one frequency present, happens in all musical instruments, but before discussing this, it is necessary to know a little about musical notation.

You are all no doubt familiar with the musical scale "doh, ray, me, fah, soh, lah, te, doh." This is called the *diatonic scale* and is the basis of European music. A diatonic scale can start from any musical note. For convenience musical notes are labelled with the first seven letters of the alphabet. With C as the starting note the scale is C, D, E, F, G, A, B, C. Such a group of eight notes is known as an octave.

An ordinary piano keyboard comprises six such octaves, which we can illustrate by considering the seven notes of C together with their frequencies, as used in physics. In an orchestra the frequencies would be slightly higher.

Note	C_1	C	c	c′	c″	c‴	c⁗
Frequency (Hz)	32	64	128	256	512	1024	2048

The note c′ with a frequency of 256 Hz is known as Middle C.

When you strike the key of the note c′ on the piano, you set the corresponding wire vibrating with the frequency of the fundamental (the first harmonic) of 256 Hz, as well as with the frequencies of other harmonics such as 512, 768 and 1024 Hz. You can demonstrate this point by using the fact that 512 Hz is not only the frequency of the second harmonic of c′, but also the frequency of the first harmonic (fundamental) of c″. If the wire of c′ vibrates with a frequency of 512 Hz (as well as with its fundamental frequency of 256 Hz) then the frame of the piano will also vibrate with a frequency of 512 Hz and will make the wire of the note c″, whose fundamental it is, vibrate vigorously. This in fact can be shown to be correct by carrying out the following procedure:

1. Depress the key of the note c″ so that the damper is removed from the wires, allowing them to vibrate when set in motion.

2. Strike the key of the note c′ so as to set up vibrations in the piano of 256 Hz, 512 Hz, etc.

3. Release the key of the note c′, thus returning the damper to the strings to stop them vibrating.

4. The note of frequency 512 Hz will be heard sounding loudly. This sound must come from the wires of the note c″ (fundamental frequency 512 Hz), which has been made to vibrate by the frequency of the second harmonic (512 Hz) of the note c′.

In the same way it is possible to detect the presence of higher harmonics in the vibration of the wire of the note c′.

The presence of partial tones or harmonics gives to a note its quality, or *timbre*, and it is largely by the harmonics that we are able to distinguish between notes of the same frequency played on different instruments. For instance, a tuning fork gives a pure tone, a harp sounds only the lower harmonics while a violin produces a wide range of high harmonics. The famous makes of violin, such as the Stradivarius, produce more of these high harmonics than ordinary instruments and consequently give the rich tone for which they are so highly valued.

QUESTIONS

1. Describe, with a diagram, how the particles of a violin string vibrate when it is sounding (*a*) its fundamental, (*b*) its second harmonic.

2. Describe the sonometer and show how, given a set of tuning forks, you could use it to prove that the frequency of a vibrating string is (*a*) inversely proportional to its length, (*b*) proportional to the square root of its tension.

3. The following readings were obtained for the length l of a sonometer wire tuned to a fork of frequency f:

f in Hz	256	324	384	426	480	512
l in m	0·60	0·48	0·40	0·36	0·32	0·30

Show a graph or otherwise that l is proportional to $\frac{1}{f}$

4. The following readings were obtained for the tension T of a sonometer wire when 1 m of it was tuned to a fork of frequency f:

f in Hz	256	324	384	426	480	512
T in N	31	51	72	88	110	125

Show by a graph or otherwise that f is proportional to $\sqrt{T}$.

5. Explain what is meant by the harmonics of the vibrations of a string. What effect have these on the sound produced by the string?

CHAPTER 27

STATIONARY VIBRATIONS IN TUBES

YOU must at some time have tried to produce a musical note by blowing across the end of your fountain-pen top. You can do the same thing with a bicycle pump, and with this you are able to show how the pitch of the note depends on the length of the tube. With the handle right out, blowing across the other end will produce a note near to c′. If you push the handle about halfway in, the note will rise by about one octave to c″ and if you again halve the length of the air space, the note rises a further octave to c‴.

Experiments with Open and Closed Tubes

This type of tube is called a closed tube because one end is blocked (by the pump washer). You can also produce a note from an open tube, such as the outer case of a ball-point pen. If you can drill a hole in the middle of such a tube you will be able to produce two notes, one with the hole covered and a higher note (its octave) with the hole open.

Finally you can compare the open and closed tubes, if you change the former into the latter by placing your hand over the end. You will find that the closed tube gives the lower note—approximately one octave lower than an open tube of the same length.

While it is possible to get some ideas of the vibrations in tubes with this crude apparatus, a recorder will show the effects much better. When you blow into the mouthpiece (Fig. 347), air hits the sharp edge *A* in such a way

FIG. 347. *Mouthpiece of a recorder.*

as to set up stationary vibrations in the column of air in the instrument. Because this is a much more efficient way of setting up the vibrations, a much clearer note is obtained.

The mouthpiece, which is a short open tube, should be detached from the instrument. If you blow into this you will hear a note whose pitch will fall about one octave when you close the other end of the tube *B* with your hand, thus showing the relationship between the fundamental frequencies of open and closed tubes of the same length. If you now blow harder, the second harmonic is heard while the fundamental (or first harmonic) is hardly audible. You will find that the note rises by an octave if the tube is open and by $1\frac{1}{2}$ octaves if the tube is closed.

If you reassemble the instrument you will find that you can make almost any note by opening suitable holes. With all the holes covered, you will sound the lowest note for this particular instrument. To raise this note by one octave, you will find that you have to open a hole about halfway down the tube.

So far we have learnt that:

1. The fundamental note of the closed tube is one octave below that of an open tube of the same length.
2. The note produced by an open tube rises by one octave when a hole is opened in the middle.
3. The second harmonic of an open tube is one octave higher than the fundamental, and, of the closed tube, $1\frac{1}{2}$ octaves higher.

Now let us try to work out a theory to explain our experiments, remembering as we learnt before, that a rise of pitch of one octave is equivalent to doubling the frequency of a vibration.

Formation of Stationary Vibrations

In Chapter 26 we showed that transverse stationary vibrations were formed as a result of the interference between the incident wave and the wave reflected at the fixed end of a string. Longitudinal stationary vibrations in pipes are set up in a similar way, but reflection may take place at either an open or a closed end. It is clear that the latter must be a node (a point of no motion) in the stationary vibration, since the air is not free to move at this point.

It would therefore be reasonable to assume that the open end would be an antinode, especially as the air is completely free to move at this point. We may also argue, in terms of pressure, that the open end is a point where the pressure is atmospheric, i.e. normal.

But on page 314 we showed that the presssure is normal at points of maximum displacement. So the open end of the tube is an antinode, that is, a point of normal pressure and maximum displacement.

When we drew diagrams of the stationary vibrations of a string, we saw that there was one loop between two nodes and that this distance was half a wavelength. We use similar diagrams to illustrate the vibrations of air in a tube, but we must be very careful how we interpret what these mean. You will recall that on page 314 we learnt how to represent a longitudinal wave by drawing displacement diagrams in which the longitudinal displacement was drawn perpendicular to the wave. We will now use this method to illustrate the behaviour of vibrating columns of air.

In Fig. 348, the x-axis represents distance along the tube and the y-axis displacement along the tube.

In Fig. 348*a* the simplest case is represented—a closed tube with a node at the closed end and an antinode at the open end. This indicates that the air which is normally at point *P*, is vibrating backwards and forwards between

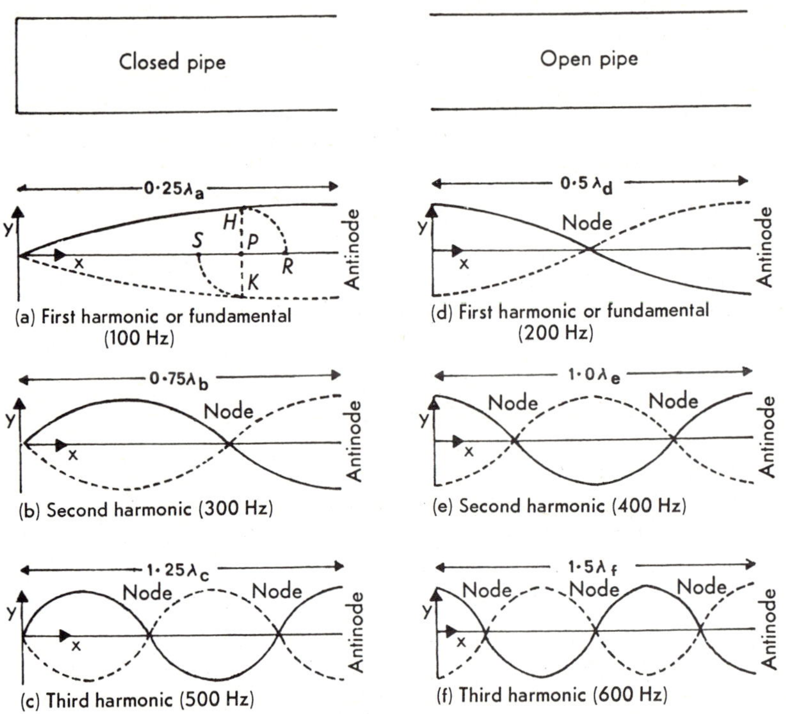

FIG. 348. *Formation of harmonics in closed and open tubes.*

R and *S*, where $PR = PH$ and $PS = PK$. Note at the closed end a point of no motion—a node—where the pressure changes from maximum to minimum; at the open end a point of maximum motion—an antinode—where the pressure is constant.

Fig. 348*d* represents the vibration in an open tube. This will have an antinode at each end and for the fundamental vibration will be half a wavelength long. The closed tube then has a wavelength twice as long as the same tube when it is open, i.e. the frequency of the closed tube is half that of the open tube or its note is one octave lower. This we have already discovered by experiment. It is also clear that the shorter the tube, the shorter the wavelength and therefore the higher the frequency as we found in our experiments.

Harmonics

So far we have been dealing only with the simplest type of vibration of the air in the tube, the one which we call the fundamental or first harmonic. Other vibrations are possible provided the open end always remains an antinode and the closed end a node. The possible vibrations with their frequencies (taking the fundamental frequency of the open tube as 100 Hz) up to the third harmonic are shown in Fig. 348.

In each case as we move from one harmonic to the next, we add an extra node and an extra loop to the vibration. The fundamental vibration of the closed tube is represented by half a loop, so that the second harmonic will be 1·5 loops long, or will have a frequency three times that of the fundamental (Fig. 348*b*). In the case of the open tube, the fundamental is represented by two half loops, so that the second harmonic shows two loops and has a frequency twice that of the fundamental (Fig. 348*e*). This confirms what happened when you produced the second harmonic of the recorder mouthpiece by blowing hard. For a closed tube the note rose by 1·5 octaves and in an open tube by one octave. The third harmonics are represented in Fig. 348*c* and Fig. 348*f*.

Lastly, what happens if we open a hole? This will become an antinode like an open end. Hence the open tube with a hole in the middle has an antinode at each end and in the middle and is one wavelength long (Fig. 348*e*). So introducing the hole halves the wavelength and therefore doubles the frequency of the note, i.e. produces the octave. This is what we found in our experiment.

Resonance Tubes

We can use a resonance tube and tuning fork to confirm in the laboratory our theories of the vibration of air in tubes. The tube is usually about 0·05 m in diameter and 1 m long and is arranged so that its length may be varied. This is done either by raising or lowering the tube in a vessel of water, or by closing the bottom of the tube with a bung and adding water inside the tube. A tuning fork is set in vibration and moved across the open end of the tube. If the fork has one of the natural frequencies of the tube, then the air will be set into resonant vibration and a louder sound will be heard. The amount of water in the tube is adjusted until the greatest increase of sound is heard as the fork is moved over the tube. To save time in carrying out the experiment, we can estimate beforehand the approximate lengths for resonance, assuming a reasonable value for the velocity of sound.

This experiment corresponds to the one you carried out with the electric bell (page 323). In both experiments you have a fixed frequency, and you adjust the length of the vibrating body to be in resonance with that frequency. Just as with the bell, you could change the length of the string so as to produce one, two or more loops, so you can adjust the length of the resonance tube to give 0·5, 1·5 or 2·5 loops (Fig. 349). In each case, as shown in the figure, the closed end is a node and the open end an antinode.

For the note c′, the wavelength is about 1·28 m, so that the initial position

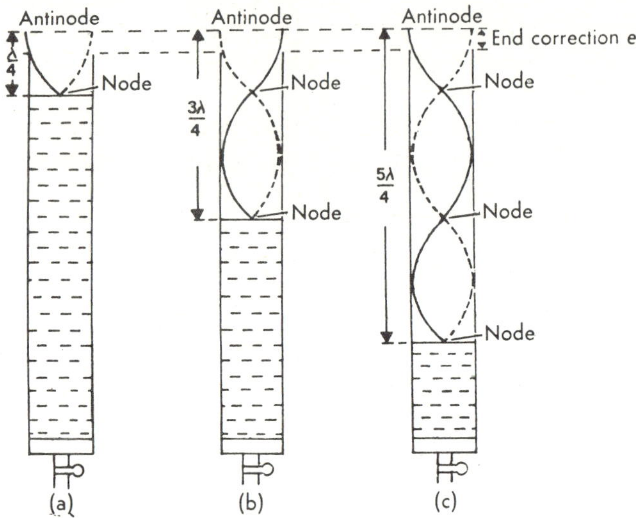

FIG. 349. *Use of resonance tube to investigate vibrations of air columns.*

of resonance would be a column length of 0·32 m or 0·25λ (Fig. 349*a*), while that for the second position would be 0·96 m or 0·75λ (Fig. 349*b*). On the other hand the octave of c′ (the note c″) has a wavelength of about 0·64 m, so that three positions of resonance would be detectable with a tube 1 m long: 0·16 m or 0·25λ (Fig. 349*a*); 0·48 m or 0·75λ (Fig. 349*b*); 0·80 m or 1·25λ (Fig. 349*c*). In practice the resonance of the third position will be rather difficult to hear.

Suppose the tube is of the same length as the closed tube shown in Fig. 348. This tube would then give resonance with forks of frequency 100, 300 and 500 Hz. In other words, if the air is set in vibration in the tube, all these frequencies will be heard, their ratio being 1:3:5. If the tube were open and had a fundamental frequency of 100 Hz, its second and third harmonics would be 200 and 300 Hz, or a ratio of 1:2:3. Thus the harmonics of a closed tube are of higher frequency than those of an open tube, so that the former has a shriller tone. This is shown in the organ where two similar sets of pipes—open and closed diapason—give different tones because one set is open and the other closed.

Velocity of Sound

From our measurements with the resonance tube we can calculate the velocity of sound. Suppose, for example, we find that the shortest length of a closed tube which will resonate with a 512 Hz tuning fork is 0·16 m. From Fig. 349, we know that this length is equal to 0·25λ, so that the wavelength generated must be 0·64 m:

$$\begin{aligned}\text{Velocity of sound} &= \text{Frequency} \times \text{Wavelength}\\ &= 512 \times 0{\cdot}64 \text{ m/s}\\ &= 328 \text{ m/s}\end{aligned}$$

This is not quite the same as the velocity of sound in the open air, for the diameter of the tube affects the result, while the velocity itself depends on the temperature and humidity of the air.

End Correction

We have said that in calculating the velocity of sound from measurements taken with a resonance tube, the diameter of the resonance tube must be taken into consideration. This is because the antinode is formed slightly above the level of the open end of the tube, at a distance of approximately 0·6 times the radius of the tube. The distance $0{\cdot}6r$ is known as the end correction.

There is a method for calculating the velocity of sound with a resonance tube in which the end correction e is eliminated. First find the shortest length of tube which will resonate with the tuning fork (say l_1). Increase this length about three times and then accurately adjust it so that resonance with the tuning fork is again obtained. This will be the second position of resonance (distance l_2, say). With reference to Fig. 349, and remembering the end correction we may say:

Distance of first position of resonance $= l_1$

$$\therefore \frac{\lambda}{4} = (l_1 + e)$$

Distance of second position of resonance $= l_2$

$$\frac{3\lambda}{4} = (l_2 + e)$$

Subtracting,

$$\frac{\lambda}{2} = (l_2 - l_1)$$

$$\therefore \text{Velocity of sound} = \text{Frequency} \times 2(l_2 - l_1)$$

Wind Instruments

In wind instruments, the air in a pipe is set in vibration with its natural frequency by blowing either at a sharp edge or through a reed which is made to vibrate. The first method is used in the flue organ pipe (Fig. 350). The jet of air from the slit S is made to strike the sharp edge of the lip L and so sets the air in the pipe in vibration. The top of the pipe may be open or closed. A similar effect is obtained in the flute and piccolo.

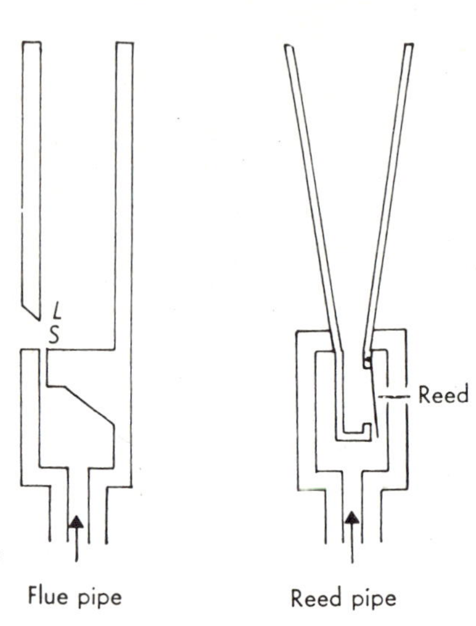

FIG. 350. *Organ pipes.*

In the reed organ pipe (Fig. 350) the air makes a thin strip of metal called a *reed* vibrate so that it opens and closes the pipe and sets the air in it vibrating. In the clarinet and oboe the reed is made of cane, while in the cornet and bugle the player uses his lips as the reed.

In order to change the frequency of the note, it is necessary to change the length of the pipe producing it. This is done in the trombone by using a movable U-tube; in the cornet by adding tubes through a series of pistons; and in the clarinet and flute by opening holes and so changing the effective length of the tube.

In all these instruments, the material of the pipe itself is also made to vibrate and this affects the tone, i.e. the number and strength of the harmonics. Thus a flute gives a fairly pure note with few harmonics, while a metal instrument gives far more. The shape of the instrument also has an effect on the production of harmonics. With all these factors producing an effect it is extremely difficult to calculate what will happen in a particular instrument, so that the construction of musical instruments remains an art rather than a science.

The human voice may be regarded as a wind instrument. Air from the lungs is forced through the vocal chords into the resonant cavity of the mouth. The movement of the chords controls the nature of the sound—whether it starts explosively or not, how long is lasts, and whether it stops suddenly or gently decays. The pitch of the note and the number of harmonics present depend on the size and shape of the resonant cavity of the mouth. In the simplest case when the tongue lies along the base of the mouth the result is similar to organ pipes of frequency approximately 500 Hz for men, 700 Hz for women and 800 Hz for children. By varying the size and shape of the opening of the mouth and the position of the tongue we are able to produce the more complicated sounds necessary for speech.

FIG. 351. *The organ at the Royal Festival Hall, London, contains 8,000 different pipes. Both flue pipes and reed pipes are used in a very wide variety of shapes and sizes. Only a small proportion of these pipes are visible in the photograph.*

QUESTIONS

1. Explain with reference to experiments with a tuning fork and a tube of variable length what is meant by resonance.

2. Explain what is meant by a closed tube and an open tube. A closed tube and an open tube of the same length are made to sound their fundamental notes. Draw diagrams indicating the nature of the vibrations of the air in the two tubes and compare the frequencies of the two notes.

3. The shortest length of a closed tube which is in resonance with a fork of 380 Hz is 0·22 m. What is the approximate velocity of sound in the tube? Why is the answer only approximate? At what length would you expect to find the next position of resonance?

4. With a fork of frequency 520 Hz the first two positions of resonance of a closed tube occur with lengths of 0·15 m and 0·47 m. Explain what is meant by this statement and calculate the velocity of sound in the tube and the value of the end correction.

5. A closed tube has an end correction of 0·015 m and it can be lengthened or shortened. What are its lengths when it resonates in its first two positions with a fork of frequency 660 Hz, if the velocity of sound in the tube is 330 m/s?

6. What would be the answer to Question 5 if the tube were open?

CHAPTER 28

WAVES AND THE EAR

IN THE last two chapters we have seen how musical notes containing one or more frequencies may be generated by the oscillation of air in pipes or the vibration of a string. In this chapter we shall discuss the transmission of these vibrations through the air to the ear and impression they convey to the hearer.

Displacement-time Curves

When a violin string is set in vibration with its fundamental frequency, the body of the instrument will also vibrate with this frequency and thus set the large mass of air in contact with it moving in the manner shown on page 310 (Fig. 332). Here the actual positions of particles of air are shown at intervals of one-twelfth of the period.

We may interpret this diagram in graphical form by turning it through a right-angle so that the time scale lies horizontally along the x-axis from 0 to 0·015 s, while the corresponding displacements of the layers of air are shown vertically along the y-axis. The *displacement-time curve* for the first layer of air is then approximately indicated by the line drawn to connect the consecutive positions of the first layer. If the layers had been represented by points instead of lines, an accurate *sine curve* would have been obtained as shown in Fig. 335 and repeated in Fig. 352*a*.

As we have seen in Chapter 26 it is possible for a violin string to vibrate with other frequencies besides its fundamental. Suppose that we assume that the second harmonic (of twice the frequency of the fundamental) gives to the particles of air a vibration of half the amplitude of the fundamental. The variation with time of the displacement of the particle of air for this motion is shown in Fig. 352*b*. If the third harmonic has one quarter of the amplitude of the fundamental, then this would produce the displacement shown in Fig. 352*c*. To find the total displacement of the particle of air we add all these together, e.g. at one instant (0·0025 s) the displacements are represented by AB, CD and EF so that the total displacement at this instant is: $GH = AB + CD - EF$. The resultant curve made up in this way is shown in Fig. 352*d*.

This type of curve can be obtained from a musical instrument playing a fixed note in front of a microphone connected to a cathode ray oscilloscope.

The vibrations of the air are converted by the microphone into a varying electric potential which is fed to the oscilloscope. This contains a television tube on which a straight horizontal line is continuously traced by a fine jet

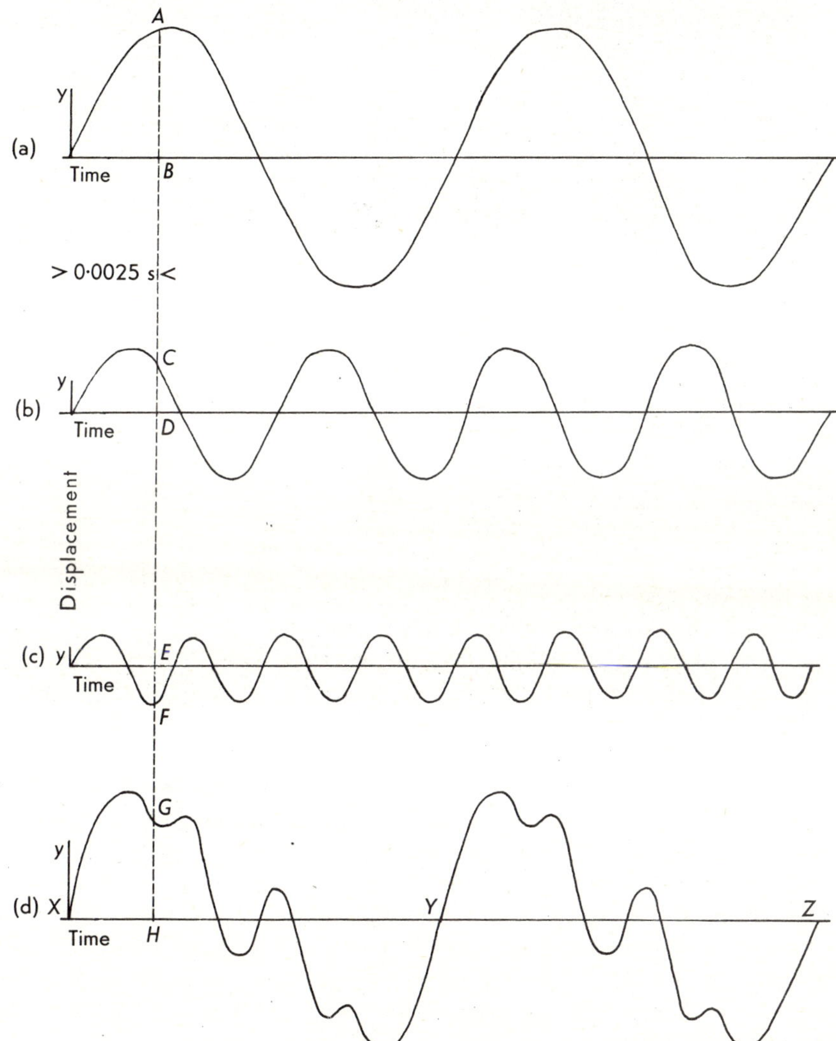

FIG. 352. *Displacement-time curves for a vibrating violin string.*

of electrons. The impulses from the microphone are applied so as to make this jet move vertically, tracing out the characteristic curve of the note

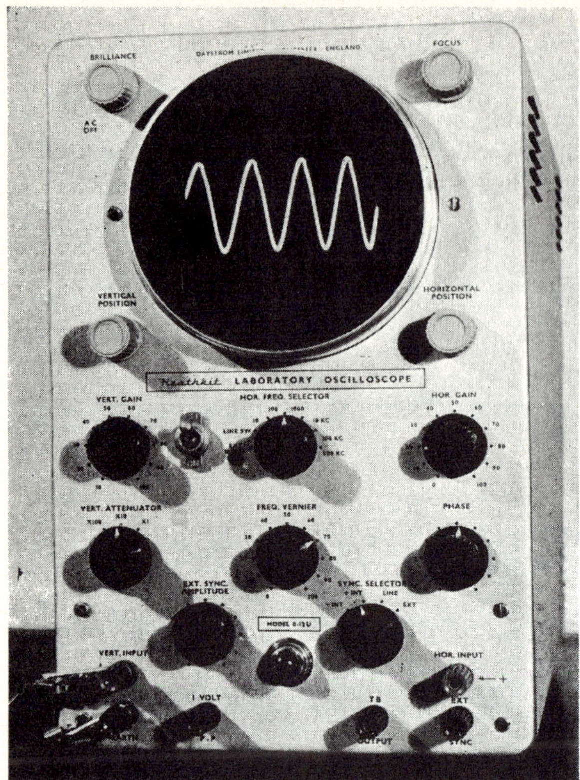

Fig. 353. *The type of cathode ray oscilloscope normally found in a laboratory is here seen tracing the displacement-time curve for a sound of frequency 400 Hz (a note slightly higher in frequency than G above Middle C). The curve is an accurate sine curve from which we may deduce that the note concerned is a pure fundamental with no harmonics present. This was the case, for the note was generated by a tuning fork of standard frequency.*

(Fig. 353). The oscilloscope traces obtained by sounding the same note on a flute, clarinet, oboe and tuning fork, are shown in Fig. 354. It is possible to analyse such traces and determine the frequencies and amplitudes of the harmonics that are present. In the case of Fig. 354 we can see that the flute gives a fairly pure note while the oboe produces a large number of harmonics.

Properties of a Musical Note

When air vibrating in this manner impinges on our ear we interpret the result as a musical note. In particular, we say that it has a certain pitch, a certain quality or timbre and a certain loudness.

Pitch. The pitch of the note is the sensation produced by its frequency. Although the graph of Fig. 352*d* is not a sine curve it does repeat itself—the section XY is the same as YZ. Thus it has a fixed frequency which, by comparison with Fig. 352*a*, can be seen to be the same as the fundamental frequency. It is clear then that the pitch of the fundamental is unaltered by the addition of the harmonics (except in cases where the higher harmonics have a very much greater amplitude than the fundamental).

Timbre. This is the word used to describe the quality of the note, which is due to the presence of harmonics. As we have seen, these change the shape of the displacement-time curve.

Loudness. This is the sensation produced on us by the rate at which energy is supplied by the wave. This rate is called the *intensity* and the greater it is, the louder we say the sound is provided the frequency is an audible one.

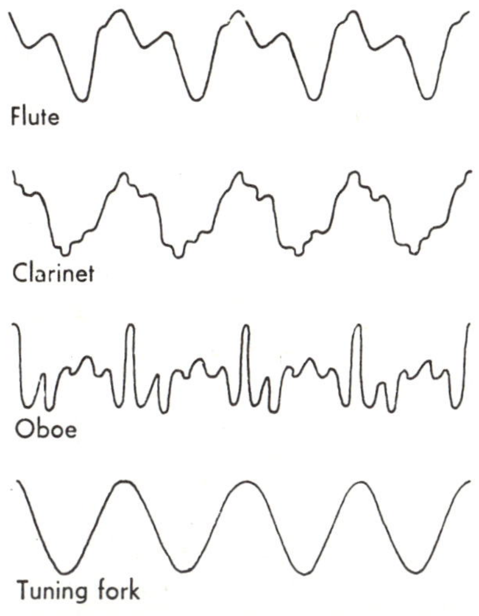

FIG. 354. *Displacement-time curves.*

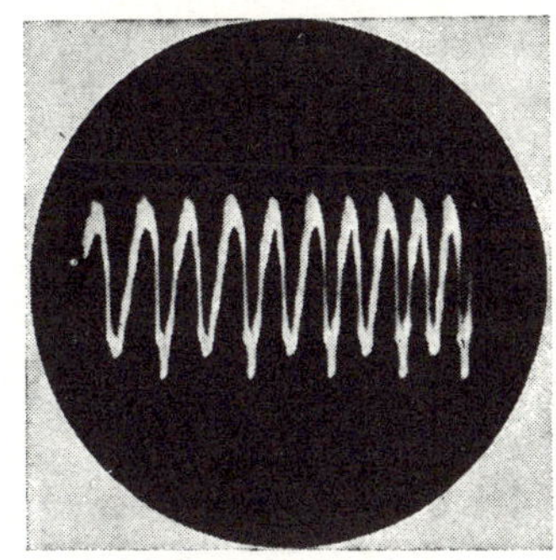

FIG. 355. *The curve shown was produced by the human voice singing Middle C into the microphone attached to the oscilloscope (Fig. 353). As is to be expected, the curve is not an accurate sine curve since the human voice cannot produce a pure sound, harmonics always being present.*

Loudness and Intensity

The intensity produced by a wave at a point is measured by the energy in the wave which crosses an area of 1 m^2 in one second. The ear is so sensitive that it can detect a sound whose intensity is one million millionth (10^{-12}) of a watt/m^2. This is approximately the same rate of flow of energy as would be produced by the heat energy from a 1-kW electric radiator 8000 km away in a vacuum.

This does not mean that the ear can detect a change as small as this in a loud note. The minimum change that can be detected is found to depend on the level of the sound concerned, that is the ear is able to detect a difference in loudness of two notes if the ratio of their intensities is above a certain value. To make this clear, consider how sensitive your eye is when detecting the number of matches in a pile. You would probably be able to detect immediately a difference of one between four and five, but not the difference between forty and forty-one. You might, however, be able to distinguish between forty and fifty or between four hundred and five hundred. If these figures are correct we can say that the eye can only detect a difference between two piles if the ratio of the numbers in them is greater than 5 : 4. In a similar way the ear can only detect a difference in loudness between two notes if the ratio of their intensities is bigger than about 11 : 10.

Because of this dependence on ratio, we fix a scale of intensity levels based on it as shown by a comparison of three sounds *A*, *B* and *C* below:

Relative Intensity	*Scale of Intensity Levels*
Intensity of $B =$ $10 \times$ Intensity of A	Intensity level of *B* is 1 bel (or 10 decibels) above *A*
Intensity of $C =$ $10 \times$ intensity of B	Intensity level of *C* is 1 bel (or 10 decibels) above *B*
Hence intensity of $C =$ $100 \times$ intensity of A	Hence intensity level of *C* is 2 bels (or 20 decibels) above *A*

We fix the lowest level of this intensity scale by making it that of a sound which can just be heard, i.e. an intensity level of 0 decibels is assigned to a sound which generates a power of 10^{-12} W/m^2. The intensity level in decibels then gives a measure of the relative loudness of sounds.

The values of some everyday sounds in decibels are given below:

Intensity (10^{-12} *W/m*^{2})	*Source*	*Intensity level* (*decibels*)
1	Just audible	0
10	Very quiet country	10
100	A whisper at 1 m	20
1000	A quiet city street	30
1 000 000	Normal conversation	60
10 000 000	Busy traffic	70
1 000 000 000	Pneumatic drill at 6 m	90

The figures given above are approximately true for sounds whose frequencies lie between 200 and 4000 Hz. Outside this range the ear is much less sensitive so that the zero of the scale would have to be altered. A person with good hearing can detect sounds whose frequencies lie between about 20 000 and 30 Hz though the upper limit falls as the age of the person increases. Some animals, in particular dogs, can hear higher frequencies, so that a supersonic whistle which is inaudible to human beings may be used to call a dog.

It is interesting to note that any sound more than 12 decibels above another in intensity will completely mask it. Thus a pneumatic drill at a distance of 6 m completely obliterates the sound of traffic, but normal conversation is possible in a busy street.

The Ear

We have seen that the displacement-time curve for a sound wave may have a complex shape. We should therefore expect that the ear which receives the vibrations of the air and transmits signals to the brain in terms of loudness, pitch, and tone must be a very complicated structure. Its method of working is not completely understood but a broad outline of its construction is given here.

As shown in Fig. 356, the ear has three main parts—the outer, middle and inner ear. The outer ear is a tube with a flexible end known as the *drumskin*,

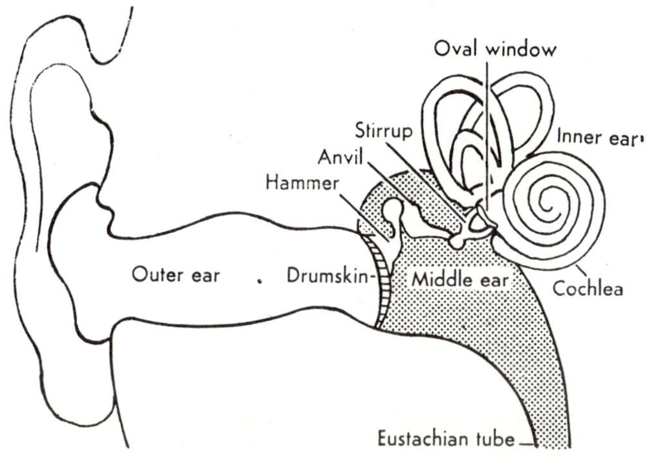

FIG. 356. *Principal components of the human ear.*

which is made to vibrate in time with the oscillations of the air produced by a sound. The vibrations of the drum are transmitted from the outer to the inner ear by a system of levers built up from three bones. The three bones are known as the *hammer*, *anvil* and *stirrup*. They transmit the vibration as far as the *oval window*, which is the outer boundary of the inner ear. The lever system is situated in the middle ear which is connected to the mouth by way of the *Eustachian tube*. The purpose of this tube is to maintain an approximately equal pressure on either side of the drumskin. When, for example, an explosion is about to take place, you should open your mouth, thus equalizing the pressure on either side of the drumskin and preventing it from being damaged.

The part of the inner ear concerned with hearing is called the *cochlea* (meaning a snail), because it is coiled up like a snail's shell. It is almost completely divided into two parts by a partition known as the *basilar membrane* which contains nerve endings through which information is sent to the brain. The whole of the cochlea, on both sides of the partition, is filled with a substance known as *perilymph fluid.*

To make the action of the cochlea clear Fig. 357 shows it in diagrammatic form, straightened out. Some idea of the size of the cochlea is given by basilar membranes that have been removed by surgery, and uncoiled: they are found to be about 30 mm long, 0·16 mm wide at one end and 0·52 mm wide at the other. The membrane has been found to be made up from a very large number of fibres stretched across its width, under tension. In position in the cochlea, the membrane takes the form of a spiral of two and a half turns.

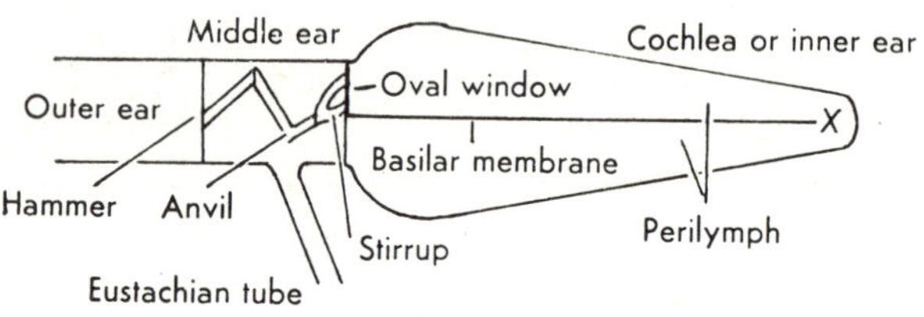

FIG. 357. *Straightened representation of the cochlea.*

When a pressure change is transmitted through the middle ear to the oval window, it travels through the perilymph fluid to the end of the cochlea, passes through the gap X in the basilar membrane and continues back along the cochlea, on the other side of the membrane. In doing so it makes the membrane bend and the nerve endings in it are in consequence stimulated. In some way that we do not understand, these nerves are able to convey to the brain information about the pitch, loudness and timbre of the sound.

QUESTIONS

1. A musical note can vary in pitch, intensity, and quality. What physical changes in the vibrations of the air produce these differences?

2. Sketch the displacement-time graph that you might obtain for the vibrations of the air due to a note sounded by (*a*) a tuning fork (*b*) the human voice (*c*) an oboe. Discuss and explain the differences.

3. How does the loudness of a note depend on its intensity? Explain how a decibel scale of intensity levels is used to measure the loudness of sounds.

4. Give a broad outline of the structure of the ear. Why is it a wise precaution to open your mouth if you know that a loud bang or explosion is about to take place?

CHAPTER 29

TRANSMISSION AND RECORDING OF SOUND

THE transmission, reception, recording and reproduction of sound is a commonplace activity of modern life. We all use the telephone and the radio, or go to the cinema, and most of us are familiar with gramophones and tape recorders. In all these instruments sound is converted into electrical or magnetic impulses, transmitted to some other place, and either reproduced immediately as sound or stored for reproduction later. There is a vast amount of technical detail about the way in which each of these processes is carried out, but in this chapter we shall deal only with the outline of the principles involved.

Telephone

If you have an old pair of headphones—perhaps left over from a crystal set—you can demonstrate for yourself the principles of the telephone. Separate the two earphones, place them in different rooms and connect them in series with a 4- or 6-volt battery. If you now speak into one earphone, your voice can be heard from the other.

Obviously only electrical currents can pass down the wire, so that the first earphone has converted sound into electrical impulses, and the second has changed these back to sound. Unscrew the top of the earphone and look inside. You will find that there is a small but powerful electromagnet below

FIG. 358. *The top of an earphone has been unscrewed (left), the diaphragm slid out (right), exposing the two ends of a small electromagnet (centre).*

the springy disc of metal which is called the diaphragm. If you slide this out gently, you will find that it sticks to the poles of the electromagnet, showing that it is made of magnetic material.

If you speak into the earphone *A* (Fig. 359) vibrations of the air make

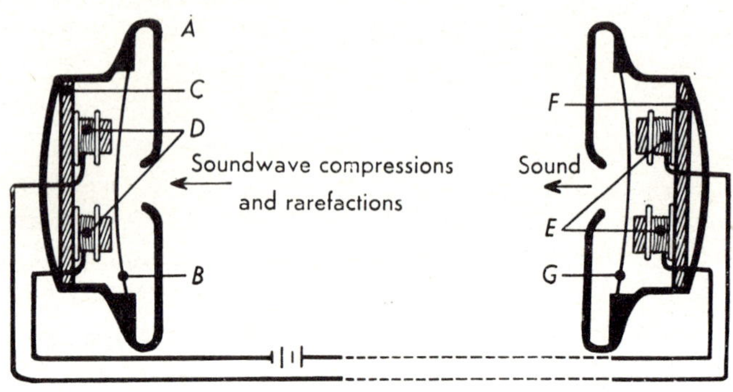

FIG. 359.

diaphragm *B* oscillate and so change the magnetic field in the electromagnet *C*. According to the principles of electromagnetic induction, this induces a change of current in the coils *D* and therefore in coils *E* which are connected to them. This changes the strength of the electromagnet *F* and forces the diaphragm *G* to vibrate with the same frequency as the diaphragm *B* of the first earphone.

Loudspeaker

This type of receiver is used in all telephone circuits, but if higher power is required a loudspeaker (Fig. 360) is employed instead. This consists of a stiff paper cone, suspended at its edges from a baffle board, and made to vibrate by the force on a current in a small cylindrical coil mounted in a radial magnetic field. When the current in the coil increases the cone is pulled back, and when it decreases the cone is allowed to move forward. Hence, if the current varies so that it has a maximum value 256 times in each second, the cone will vibrate with this frequency and sound c′ (middle C). However, if the frequency is very high, the cone may be too heavy to move quickly enough. For this reason, in modern high fidelity equipment the higher frequencies are fed to a smaller loudspeaker which can reproduce them, while the lower frequencies are reproduced by the main speaker.

It is important that only vibrations from the front of the cone should be heard. The back will also produce vibrations of the air of the same frequency, but out of step with those from the front. When waves from the back and the front of the loudspeaker reach the ear simultaneously, they may tend to cancel each other out. In order to prevent this the loudspeaker is mounted on a baffle board so that vibrations generated by the back of the cone cannot reach the ear.

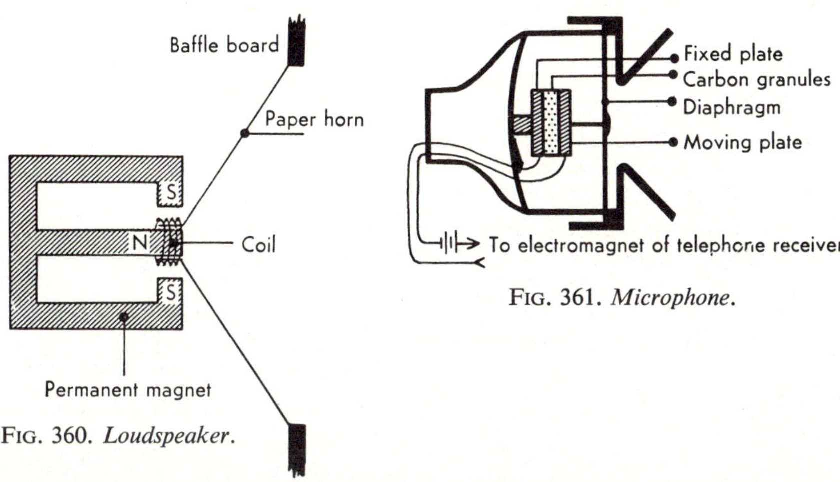

FIG. 360. *Loudspeaker.*

FIG. 361. *Microphone.*

Microphone

When you used your headphones as a telephone, you made one of the earphones act as a microphone. This is not very efficient and is replaced in most telephonic circuits by a carbon microphone (Fig. 361), which is robust and reliable. In this instrument the sound waves falling on the diaphragm make it move backwards and forwards, so changing the pressure on the carbon granules enclosed between a plate attached to the diaphragm and a fixed plate. Since carbon conducts electricity, current will always flow between the two plates though the resistance is high. But when the granules are compressed, the resistance drops considerably. The current will therefore increase and so will the strength of the electromagnet in the telephone to which the microphone is connected. In this way the diaphragm of the telephone is made to vibrate with the frequency of the sound falling on the microphone.

Speech transmitted by this microphone can be easily understood, though the voice is distorted because the higher frequencies are suppressed. For reproduction of music, a moving coil microphone or crystal microphone is used. The former is like a small loudspeaker (Fig. 360) working in reverse. Here the sound waves fall on a small paper cone and cause the coil to move in the magnetic field, thus inducing currents in it. The crystal microphone depends on the fact that when certain crystals are subjected to pressure an electromotive force, which may be as much as 1 volt, is produced between opposite faces of the crystal. In the microphone, the sound waves cause a thin crystal to bend, thus producing a sufficient pressure to provide the electromotive force required.

Gramophone

So far we have dealt with the reception and immediate reproduction of sound. To conclude this chapter we shall describe briefly the methods by which sound is recorded and reproduced from a gramophone, a film and a tape recorder.

In the earliest method of making gramophone records the actual vibrations of the air operated a cutting tool which made a wavy spiral groove in a wax

FIG. 362. *Here, magnified, is the groove of a gramophone record. The wavy nature of the groove causes the stylus to vibrate as it travels along the groove. This vibration generates a varying electric current which is fed through an amplifier to a loudspeaker. For stereophonic records two separate signals are required from the same groove. To achieve this the groove is made to waver horizontally and vertically, although in practice this system is turned through an angle of 45° so that the groove wavers in two diagonal directions at right angles to each other. The vibrations of the stylus in such a groove have two distinct components which are separated and fed to two amplifier/loudspeaker systems.*

disc. From this disc copies were made, which, when put on the gramophone, drove the needle of the pick-up sideways and so, through a system of levers, conveyed the mechanical motion to a diaphragm at the end of a horn thus reproducing the original sound waves.

In the 1920's this process was greatly improved when electrical energy was used for both the recording and reproduction of the sound. The sound waves now had only to operate a microphone (of the type already described) and produce a varying current, which could be amplified and applied to the stylus to cut the record. Similarly, on reproduction of the sound, the tone arm used the vibrations of the needle to move a coil in a magnetic field (as in the loudspeaker) and produce a varying current. This, after passing through a suitable amplifier, could be used to operate a loudspeaker. More recently crystal pick-ups, using the same principle as the crystal microphone, have become common. At the same time, reduction of the width of the groove and of the stylus has permitted slower speeds of rotation and longer playing time on the record.

Sound Film

The sounds to be recorded on a film, like those for a gramophone, are first converted into electrical currents by a microphone and amplified. They are

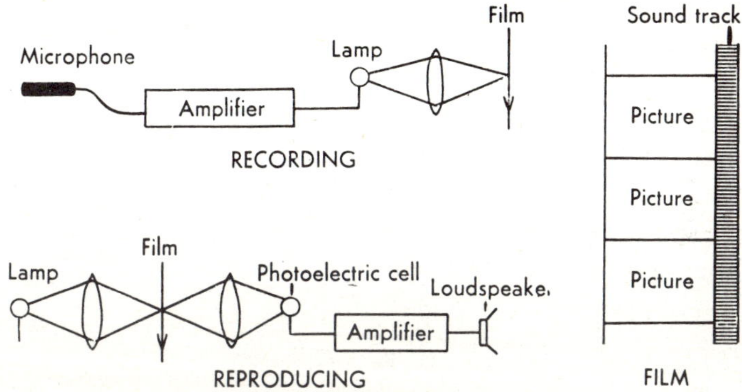

FIG. 363. *Recording and reproduction of a sound film.*

then used to control either the intensity with which a light shines on a negative, or the area covered by the light. When sound is to be reproduced from the processed film, light is passed through the film and allowed to fall on a photo-electric cell which converts variation of light intensity into variations in an electric current. This is amplified, fed to a loudspeaker and reproduced as sound (Fig. 363).

Tape Recorder

The tape used for recording purposes is made from paper or cellophane impregnated with small particles of iron oxide, which can be magnetized. The tape is made to pass very close to the gap in two electromagnets. The first is called the erase head because electric currents in this remove any magnetization in the tape. The second electromagnet is used both for recording and reproducing sound and is called the record and playback head.

To record, the erase head is switched on and the recording head fed with amplified current from the microphone. The tape, which is demagnetized by

FIG. 364. *This early form of tape recorder head shows clearly the record/replay electromagnet (left) and the erase electromagnet (right). When switched on the tape travels from right to left while the two brushes in the foreground move in to hold the tape firmly against the electromagnet.*

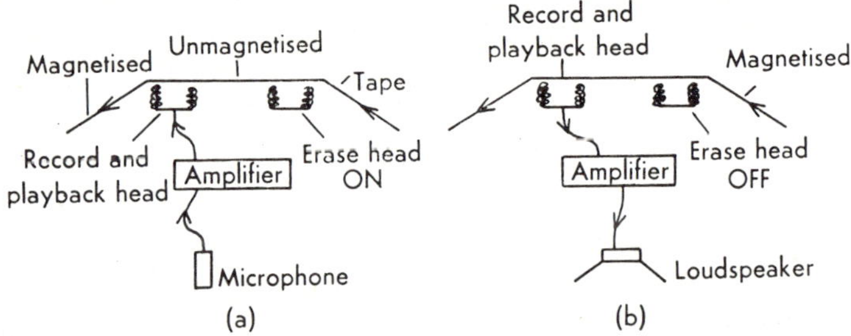

FIG. 365. *Recording and reproduction of a magnetic tape.*

the erase head, is now magnetized by the record head so that the degree of magnetization varies with the frequency of the sound at the microphone (Fig. 365*a*). To reproduce the recording, the instrument is operated with the erase head off and the playback head connected through an amplifier to a loudspeaker (Fig. 365*b*). The variation in magnetization of the tape causes changes in the current in the electromagnet, and these produce sounds of the same frequency at the loudspeaker.

QUESTIONS

1. Describe the action of one type of telephone or loudspeaker.

2. Describe the action of one type of microphone.

3. When someone is talking on the telephone, energy is conveyed from the muscles of his throat to the ear of the person listening. Trace the changes in energy that take place in this process.

4. Compare the action of a gramophone and a tape recorder in reproducing sound.

CHAPTER 30

WAVE PROPERTIES AND THE ELECTROMAGNETIC SPECTRUM

SOUND waves have been the subject of the last three chapters, but you will remember that in the first chapter of this section we investigated the properties of transverse waves. In this chapter we return to the study of these waves in discussing the electromagnetic spectrum, which includes radio waves. light waves, X-rays and gamma rays.

THE ELECTROMAGNETIC SPECTRUM

Sound waves are audible in a certain range of frequencies from approximately 30 Hz to 20 000 Hz. A vibration of less than 30 Hz is felt rather than heard, while one whose frequency is above 20 000 Hz is called supersonic, because human beings cannot hear it.

In a similar way, a certain frequency range of the electromagnetic spectrum is detected by the human eye as light, but just as the pitch of a note varies with its frequency, so the colour of the light varies through the optical spectrum from high frequency violet to low frequency red. Vibrations just outside this range may be detected by a photographic film and are called *ultra-violet* at one end and *infra-red* at the other end of the spectrum. Radiation of lower frequencies than the infra-red may again be detected by human beings, but in this case as *heat*. Longer wavelengths still, between about 0·01 m and 10 000 m, are called *radio waves*, those used for sound broadcasting lying between 3 m for V.H.F. and 1500 m for Radio 2.

At the other end of the electromagnetic spectrum waves of higher frequency than the ultra-violet are known as *X-rays*. These are very penetrating, so that if your hand is placed between an X-ray tube and a photographic plate your flesh will allow the X-rays to pass and a shadow image of your bones will appear on the plate.

Of higher frequency still are *gamma rays*, which are the most dangerous of the radiations from radioactive material such as that produced by a nuclear bomb. The electromagnetic waves of the highest known frequency are called *cosmic rays*, because they appear to originate in outer space.

To summarize what we have already stated: we split up the electromagnetic spectrum into radio waves, infra-red radiation, the optical spectrum (usually divided into red, orange, yellow, green, blue, and violet), ultra-violet radiation, X-rays, gamma rays, and cosmic rays. All these are of the same nature, but are given different names simply because of their different properties. All of them have the same velocity, and for each there is a rhythmic variation of electric and magnetic forces with a frequency which varies from 30 000 Hz for long radio waves to more than 10^{20} Hz for cosmic rays.

Radio Waves

What evidence is there for these statements? Why for instance do we say that a radio transmission is in the form of waves? If we were talking about waves on water, we should expect the surface to rise and fall with the appropriate frequency. We do not expect our radio aerials to vibrate vertically in response to the electromagnetic wave, but we do find that the potential of the aerial oscillates with the frequency of the transmission and by tuning the radio set to this frequency, we are able to receive one programme and reject others. This corresponds to our experiment with the sonometer (page 321). You may remember that when the wire was adjusted to the correct length, it would vibrate vigorously when a tuning fork was struck and placed on the bridge. Similarly, when you have tuned your radio circuit to the right frequency (by turning the knob and so adjusting the capacity) you will receive maximum response from that frequency and hardly any from all the others. In high-frequency transmissions—when the wavelength is 10 m or less —the aerials also have to be tuned. Like the sonometer wire the aerial is made half a wavelength long, so that stationary vibrations of potential are set up in it.

Light Waves

We have been able to give some indication that radio transmissions behave as waves, but it is much easier to demonstrate the wave nature of light. The idea that light might act like waves was first put forward by a Dutchman, Christian Huygens, in 1678. At that time, Isaac Newton held the theory that a light ray was a stream of particles or corpuscles as he called them. There was fierce argument between the supporters of the two theories, but it was not until much later that experiment gave a verdict in favour of the wave theory. More recently this situation has changed and we now believe that in some conditions light behaves rather like a wave, while in others it resembles a stream of particles.

WAVE PROPERTIES

In Chapter 25, it was suggested that you could generate waves in a shallow dish or sink and study their behaviour. If you illuminate the surface of the water with an ordinary electric lamp, the crests and troughs of the wave act like a number of lenses and produce a series of light and dark lines on the bottom of the sink. In the ripple tank, which is commonly found in schools,

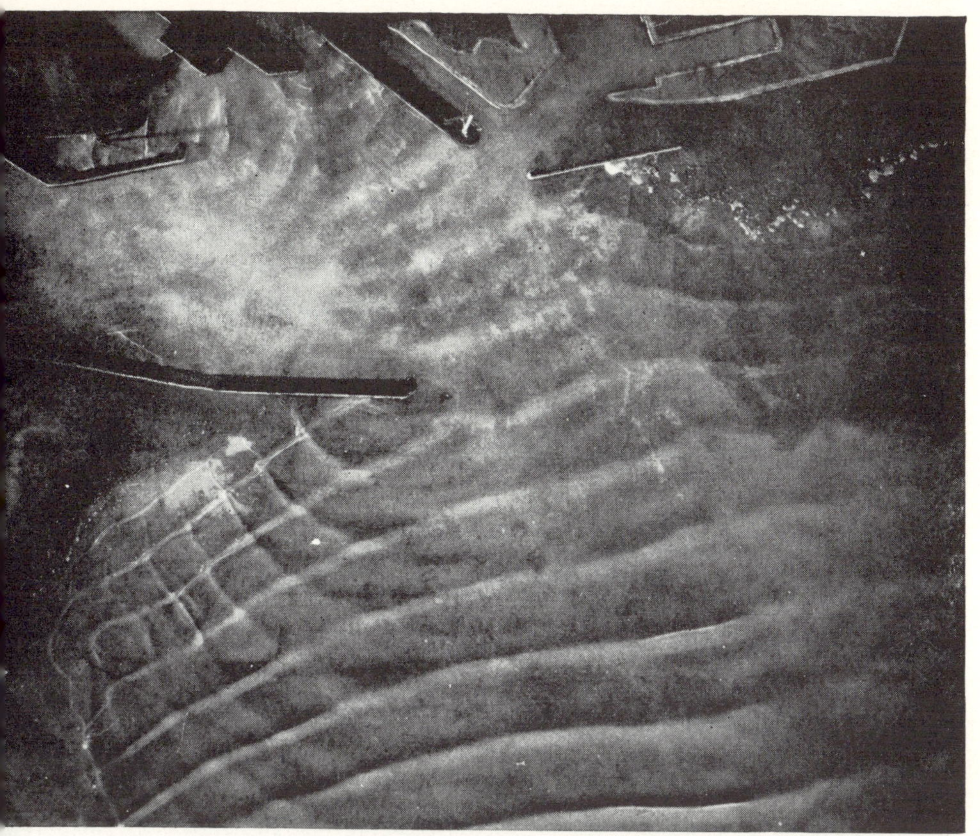

FIG. 366. *In this tank at the Hydraulics Research Station, Berkshire, waves from the bottom right-hand corner of the photograph strike the breakwater (left centre) and are reflected towards the bottom left-hand corner.*

these lines (which correspond more or less to the crests and troughs) are projected on to the ceiling or a wall.

You may find it difficult to follow the pattern of the waves in the sink, but this is partly due to the fact that the waves die out rapidly. The pattern may also become confused because of unwanted reflections from the sides. To prevent these from appearing place pieces of cloth round the edges of the sink.

Lastly, it is often possible to get better results from a short burst rather than a continuous succession of waves.

Reflection and Refraction

Assuming you have obtained waves which you can see, it is not difficult to show reflection at the surface of a piece of metal which stands on the bottom and protrudes above the surface. (If you use wood instead of metal you will have to put a weight on it to hold it down.) Fig. 367*a* shows a straight wave (generated by moving a ruler up and down in the surface), incident at about 45° on a straight reflecting surface and being reflected at the same angle.

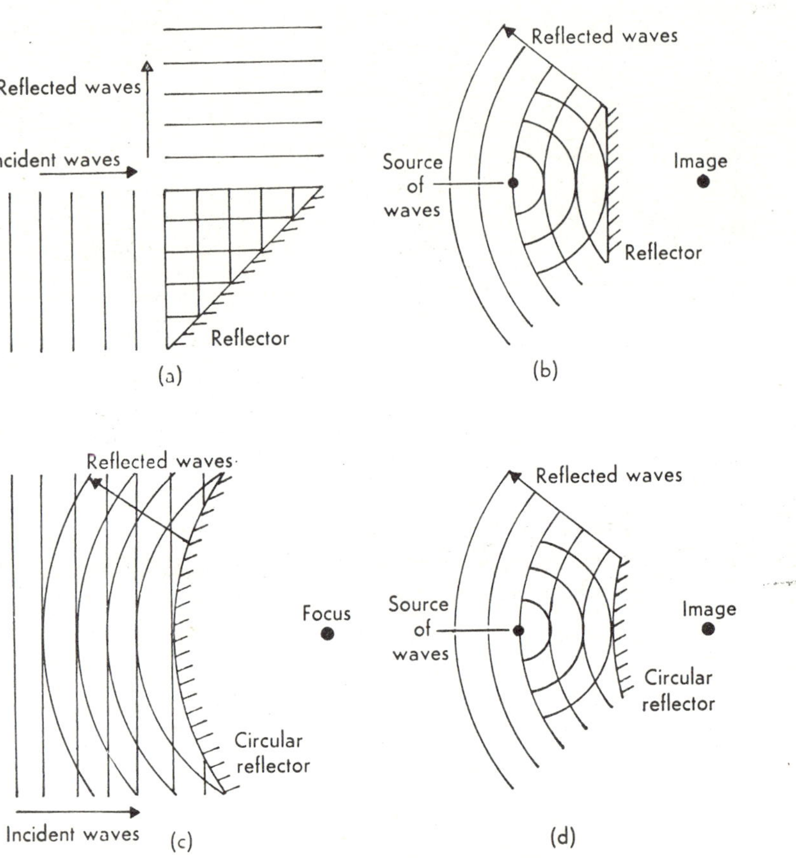

FIG. 367. *Reflection of straight and circular waves.*

In Fig. 367*b*, a circular wave (generated by moving a pencil up and down in the surface) has been reflected by the same obstacle. Notice that the reflected waves are parts of concentric circles whose common centre is the image of the source in the reflector. You can also reflect waves from a curved surface such as a bent piece of brass (Fig. 367*c* and Fig. 367*d*). In Fig. 367*c*, the plane wave, which is another way of representing rays parallel to the axis, is diverged by the convex surface as if it came from the focus. In Fig. 367*d*, circular waves are shown reflected from the convex surface as if they came from the virtual image on the other side.

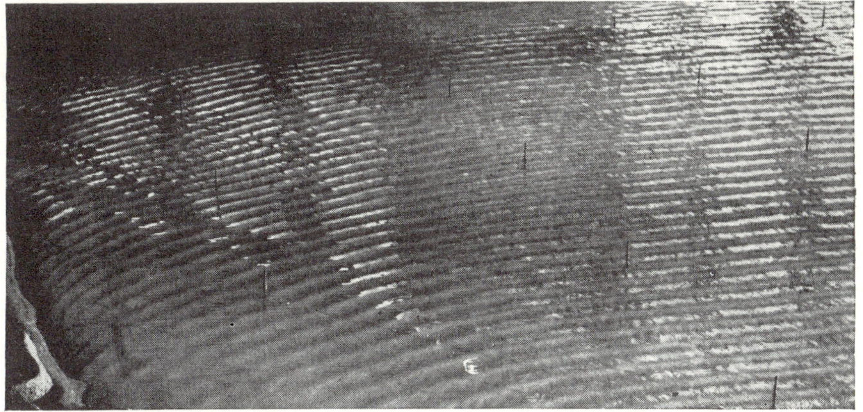

FIG. 368. *Straight waves from the bottom right-hand corner undergo refraction on reaching the shallow area (left). Here, the velocity of the wave is decreased and as a result the wave curves slightly.*

In these experiments we have used our crude ripple tank to demonstrate that the wave theory produces the same results as the particle theory as far as reflection is concerned. Refraction of light, which is due to the fact that the velocity of light depends on the medium in which it is travelling, can also be demonstrated, for the velocity of the ripples depends on the depth of water.

The method is illustrated in Fig. 369 where the effect of a triangular piece of tile, with about 1 mm of water above it, is to change the velocity of the waves so that a circular wave becomes almost straight. The depth of water above the obstacle is rather critical and this is not an easy effect to show convincingly.

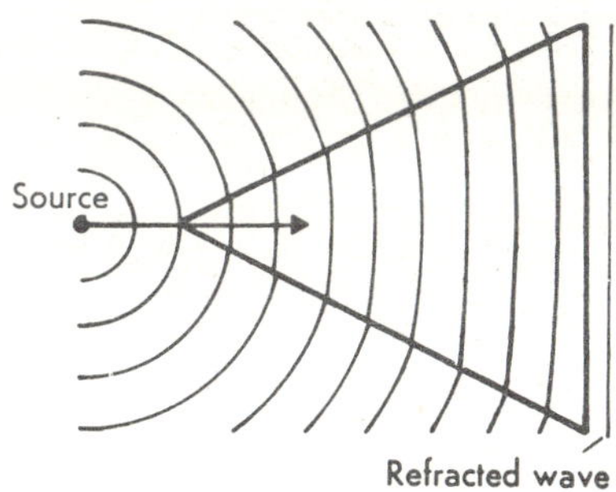

FIG. 369. *Refraction.*

Interference and Diffraction

So far we have seen that a series of waves would be reflected and refracted in just the same way as a stream of particles. There are, however, peculiarities of the behaviour of light which cannot easily be explained except by the wave theory. One of these is the interference of light.

You may remember that in Chapter 26 we explained the formation of stationary waves in a string as the result of the interference of the incident and reflected wave. We can show the same sort of thing in water waves with a tuning fork. Strike the fork and dip it gently into the surface. You will see a number of curved lines between the prongs (Fig. 370). What has happened is that each prong has sent out circular waves which have interfered so that along certain lines there is maximum disturbance of the surface. You can actually hear this interference if you hold the fork vertically and rotate it

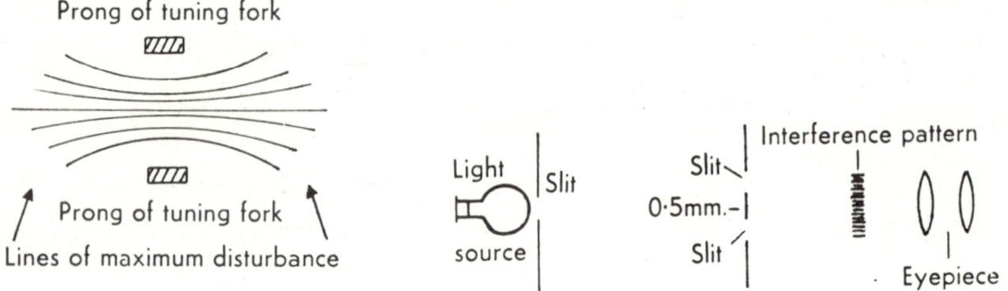

FIG. 370. *Interference.*

FIG. 371. *Interference of light waves.*

slowly by your ear: the sound rises and falls in intensity as the positions of the nodes pass your ear.

We can demonstrate the same effect with light, with the apparatus shown in Fig. 371. The two sources, corresponding to the two prongs of the tuning fork, are two very fine slits about 0·5 mm apart, made by scratching lines on an exposed photographic plate with a needle. These are illuminated by a motor car headlamp bulb with its filament parallel to the slits.

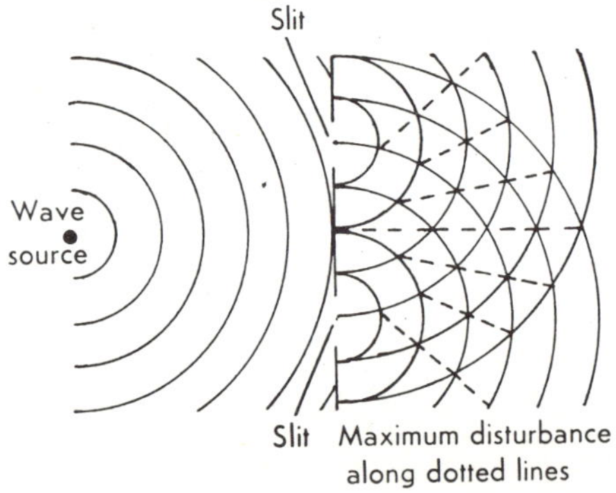

Looking from the other side, through a low-power eyepiece, you can see not two slits but one patch of light crossed by dark lines. If the light is a wave motion this is easily explained. The light from each slit has spread out as a cylindrical wave and these have interfered to form darkness at certain places. This is inexplicable on the particle theory, since those particles from the source passing through one slit, would be totally unaffected by those passing through the other.

FIG. 372. *Interference of water waves.*

You can demonstrate this effect in the ripple tank by generating circular waves and allowing them to pass through two slits close together (Fig. 372). As in the case of the tuning fork you will see lines between the slits indicating the position of the stationary waves. But this experiment also suggests that

FIG. 373. *Straight waves from the top left-hand corner of the photograph undergo diffraction on reaching the end of the breakwater. Notice how circular waves spread out from the tip of the breakwater as if the tip were a new point source.*

a ray of light passing through a slit spreads out in all directions. This is known as *diffraction* and can easily be demonstrated. Generate straight waves and

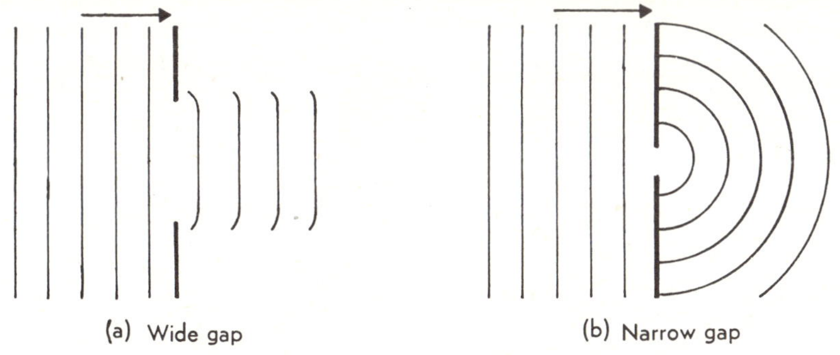

FIG. 374. *Diffraction of waves passing through a narrow gap.*

allow them to pass through a gap as wide as possible. You will find that the waves passing through remain straight, with a little bending at the end (in what we should call the *geometrical shadow*) (Fig. 374*a*). Now make your gap as small as possible. The straight wave has disappeared and become a circular wave spreading out from the small gap (Fig. 374*b*).

This means that waves bend round corners and if light behaves like a wave, then it should do so too. But, as we have seen in the experiment, the effect is only noticeable when the slit is of a size comparable with the wavelength. Since the wavelength of light is between 4 and 7 ten millionths of a metre, it is understandable that in most cases we can neglect the effect.

Huygens suggested that each point on the wave front should be regarded as a new source, from which waves spread out in all directions. These waves then interfere in such a way that in most cases we can say that light travels in straight lines, but with very fine slits more complicated effects are produced. For instance, if you hold up your hand to the light and look through the gap between two fingers when they are close together, you will see a bright slit crossed by one or more dark lines. This is due to interference between light from various parts of the gap, similar to the interference we have been discussing between the light from two separate slits.

In this chapter we have been able to show that there are many cases in which light behaves like a wave. This is true of all parts of the electromagnetic spectrum and so we refer to *electromagnetic waves*. As stated before, however, this is not a complete picture and sometimes an electromagnetic disturbance must be treated as a stream of particles. The arguments for this are beyond the scope of this book.

QUESTIONS

1. Describe briefly what you mean by the electromagnetic spectrum.

2. Why is light refracted when it passes from one medium such as air into another such as glass?

3. Why do we refer to light *waves*? Describe some phenomenon which shows light behaving like waves.

4. Explain what you mean by the interference and the diffraction of light. Illustrate your answer by reference to the behaviour of water waves in a ripple tank.

5. If the velocity of electromagnetic waves is 3×10^8 m/s, calculate the wavelengths of:

(*a*) A radio station of frequency 1088 kHz

(*b*) A radar station of frequency 3000 MHz

(*c*) Yellow light of frequency 5×10^{14} Hz

(N.B. 1 kHz = 1000 Hz; 1 MHz = 1 000 000 Hz.)

SECOND

REVISION

SUMMARY

SECOND
REVISION SUMMARY

LIGHT AND VISION

PHYSICS AS A QUANTITATIVE SCIENCE *Page*

THERE are two ways of describing an event: a ***qualitative*** 233
description contains **observations** of fact but without measurements; a ***quantitative*** description contains **measurements** taken as part of the observation.

To **measure** a quantity you express it in terms of a suitable **unit,** e.g. a length could be measured as so many centimetres.

Basic units are *mass, length, time*; all other units can be expressed 234
in terms of these three, and all measurements are really measurements of these quantities.

Avoidable errors in taking measurements include incorrect use of 234
instruments, e.g. ***parallax.***

Unavoidable errors occur when two observers estimate the same 235
quantity; they are likely to differ. These "personal errors" may be reduced by using more sensitive instruments.

Vernier scale is a small **slide scale** attached to a main scale; if the 236
vernier divisions are each $\frac{9}{10}$ as long as the main scale divisions, you can read to $\frac{1}{10}$ of a main scale division. Read by the **co-incidence of lines** on main scale and vernier. ***Calipers*** can be made with a vernier scale. 237

The micrometer uses an accurate **screw thread;** in a fraction of a 238
turn the screw moves forward by the same fraction of its **pitch.**

Zero error can occur with vernier calipers and micrometers if they 237
do not read zero when closed; you must **adjust your readings** to allow for this.

A magnifying glass can be used to magnify a scale so that you can 239
see it more easily.

THE PROPAGATION OF LIGHT

To see an object, **light from the object** must enter your eye. **Luminous** 240
objects give out light; they are ***sources*** of light. **Non-luminous objects** can only be seen when light from a source falls on them and is **reflected** so as to enter your eye.

	Page
Light travels in straight lines; you cannot see round a corner unless the light is reflected or refracted.	241
Rays are straight lines drawn to represent the direction in which light is travelling.	242
Beams are bundles of rays; beams may be **parallel, converging,** or **diverging.**	243
A point source causes a **sharp shadow,** giving the outline of the object casting the shadow.	244
An extended source causes a shadow that gives an outline that is not sharp. The ***umbra*** is the region of total shadow; the ***penumbra*** is the region of partial shadow in which light is received from **part** of the source.	244
In ***eclipses*** the sun acts as an extended source; the shadow cast by the **moon** consists of ***umbra*** (total eclipse) and ***penumbra*** (partial eclipse).	245
In a ***pinhole camera,*** an image is formed on a screen by light from an object passing through a pinhole. Light from each point on the object, passing through the pinhole, can strike the screen in only one place; a real **inverted image** is formed.	246
Size of the image depends on the ratio of the distances between pinhole and screen, and object and pinhole.	247
Sharpness of the image depends on the size of the pinhole; **the image is always in focus. The shape of the pinhole** does not affect the image.	247

MIRRORS AND REFLECTION

Diffuse reflection occurs when light is **scattered** from a **rough surface** (at small glancing angles some regular reflection takes place if the surface is not too rough).	255
Regular reflection takes place at **smooth surfaces,** e.g. mirrors; rays are reflected in accordance with the laws of reflection.	249
Laws of reflection:	251

1. *The incident ray, reflected ray, and normal to the mirror at the point of incidence, all lie in the same plane.*
2. *The angle of reflection always equals the angle of incidence.*

Images occur when the light which enters your eye seems to come from an "object" which is **not in the same place** as the object from which the light originates; images are ***real*** if the light passes **through** the image (real images can be formed on a screen); images are ***virtual*** if the light only **appears** to come from the image.	253

Page

Locating images: 250

1. Tracing two or more rays from a point on the image using an image pin and two object pins. The point of intersection of the rays is the position of the image.
2. The ***method*** of ***no parallax*** uses a search pin to coincide with the image, so that when the point of view is changed, no displacement occurs between search pin and image.

The image in a plane mirror lies on the perpendicular from object to mirror, as far behind the mirror as the object is in front. The image is **virtual** and **laterally inverted.** 250

A ***concave mirror*** is a portion of a hollow sphere; the reflecting surface is the inside of the sphere so that the mirror is **dish-shaped.** 259

A ***convex mirror*** has the reflecting surface on the outside of the sphere, so that the mirror **bulges outwards.** 258

The image in a concave mirror can either be real, inverted, magnified or diminished, or virtual, erect and magnified. 260

The image in a convex mirror is always virtual, erect and diminished. 260

Uses of mirrors:

Plane mirrors are used for the elimination of parallax in meters, and as optical levers. 253

Convex mirrors are used as driving mirrors. 260

Concave mirrors are used as shaving mirrors, or as reflectors in lamps to produce beams of light (a parabolic mirror is required to produce a good parallel beam). 261

Ghost images are caused by reflections at the **front surface** of the mirror glass, and are avoided by using a **front-silvered mirror.** 256

REFRACTION AND LENSES

A ***medium*** is a **transparent material** in which light can travel. 263

Refraction is the **change of direction** which occurs when a ray of light passes into a different medium. 262

A ray of light is bent **towards the normal** when it passes into an optically **denser medium,** and bent **away from the normal** on passing into a medium which is optically **less dense.** 263

Laws of refraction: 263

1. *The incident and refracted rays lie in the same plane as the normal and on opposite sides of it.*
2. *The ratio of the sine of the angle between ray and normal in one medium, to the sine of the angle between ray and normal in the other medium, has a constant value* ***(Snell's law).***

Page

The ***refractive index n*** of a material is the value of the ratio: 266

$$\frac{\sin \theta_{\text{vacuum}}}{\sin \theta_{\text{material}}}$$

where the angle θ is measured between ray and normal for light passing between the material and a vacuum.

The ***apparent depth,*** when you look into a denser medium, seems 263
less than the ***actual depth.***

Rays of light passing through a **parallel-sided** glass block emerge 266
parallel to their directions on entering the block (two **similar** but **opposite** refractions).

Rays of light passing through a **prism** are **deviated** away from the 267
apex of the prism.

Total internal reflection occurs when light is incident on the 269
boundary of a **less dense medium** at an angle **greater** than the ***critical angle c.***

The ***critical angle*** is the greatest angle of incidence for which 269
light can still pass into the less dense medium (angle in the less dense medium = 90°).

The value for the critical angle c, for light incident on the boundary 270
between the medium and a vacuum is given by:

$$n = \frac{1}{\sin c}$$

Mirages are caused by reflection at a layer of less dense air. 268

Reflecting prisms are glass prisms, with two angles of 45°, which 270
can be used in place of mirrors in periscopes or binoculars (c for a glass-air surface is less than 45°).

Convex lenses are formed by two **spherical** surfaces (one may be 272
flat); they are **thicker** at the **centre** than at the edge ***(bi-convex,***
plano-convex, and ***convex meniscus).*** 282

Rays of light parallel and close to the ***principal axis,*** after passing 274
through the convex lens are made to converge so as to pass through the ***principal focus.***

Light from each point of a **distant object** gives **a parallel** beam at 274
the lens and is made to **converge** to a point in the ***focal plane,*** forming an inverted image there.

Light from each point of a **close** object gives a **diverging** beam at 274
the lens and is made to converge to (or diverge from) a different point so that an image is formed which is **not** in the ***focal plane.***

In **experiments** you can use a **screen** to find the position of a **real** 274
image, or you can use a **search pin** and the ***method of no parallax.*** 278

Page

In a **graphical construction,** two of the following rays can be used: 275

1. The ray through the centre of the lens.
2. The ray parallel to the principal axis.
3. The ray through the principal focus.

Distances can be calculated by using the formula: 276

$$\frac{1}{f} = \frac{1}{u} + \frac{1}{v}$$

where f is the **focal length;** u is the **object distance;** v is the **image distance.** Distances to **real** objects and images are **positive,** while distances to **virtual** objects and images are represented as **negative.**

Magnification is given by: 278

$$\frac{\text{Size of image}}{\text{Size of object}} = \frac{v}{u}$$

A ***magnifying glass*** is a converging lens used to give a virtual erect and magnified image. 278

In a ***projection lantern,*** a converging lens is used to form, on a screen, a real image of a brightly illuminated slide. 278

In a ***camera,*** a converging lens forms a real image on the sensitive film; a **large aperture** makes possible a **small exposure time.** 279

Concave lenses are formed by two **spherical** surfaces; they are **thicker** at the **edge** than at the **centre.** ***(bi-concave, plano-concave,*** and ***concave meniscus)*** 282

Rays of light parallel and close to the ***principal axis*** after passing through the concave lens are made to **diverge** from a **point** (the ***principal focus,*** which is therefore **virtual**). 282

THE EYE

Sclerotic is the tough white outer covering of the eyeball. 287

Cornea is the transparent portion of the sclerotic at the front of the eyeball; it bulges outwards. 286

Crystalline lens is a convex lens made of transparent living cells, and located behind the cornea. 286

Vitreous humour is a clear jelly filling the eyeball behind the lens. 287

Aqueous humour is a clear fluid between the lens and the cornea. 286

Choroid is the tissue lining the sclerotic; it is coloured black to absorb stray light. 286

Ciliary muscle is a ring shaped muscle running round the thickened portion of the choroid. 286

Suspensory ligaments join the lens to the thickened portion of the choroid. 287

Page

Iris is the part of the choroid covering the front of the lens; it 287
contains muscles to control the size of the pupil.

Pupil is a hole at the centre of the iris; it appears black. 287

Retina is a membrane lining the eye and containing light sensitive 287
nerve cells.

Rods are nerve cells that are sensitive to dim light. 287

Cones are nerve cells that detect colour and detail. 286

Optic nerve connects nerve fibres from the rods and cones to the 287
brain.

Blind spot is where the optic nerve joins the retina (**binocular** 289
vision ensures that blind spots do not coincide). 290

Fovea is a depression in the retina at the centre of the yellow spot 286
on the axis of the eye; it contains only cones.

Yellow spot is the most sensitive area of the retina; it is small in 286
area, coloured yellow and lies on the axis of the crystalline lens.

The **curved front of the cornea** and the **crystalline lens,** together act as 287
a **convex lens** to form images on the retina. In a normal **relaxed** eye,
images of **distant** objects are formed on the retina; the **ciliary muscle**
contracts and the lens system is made **stronger** in order to form images
of **close** objects on the retina; this contraction of the ciliary muscle is 288
called ***accommodation.*** 289

The muscles in the iris regulate the size of the **pupil (aperture)** which 288
determines the amount of light admitted into the eye.

The ***near point*** is the **closest point** to your eye at which you can 289
see an object **clearly.**

Long sight (hypermetropia): the lens system is too weak or the 290
eyeball too short, so that even with the eye accommodated, only
distant objects can be seen; it is corrected by **convex spectacles.**

Short sight (myopia): the lens system is too strong or the eyeball 290
too long so that even with the eye relaxed, only **near** objects can be
seen; it is corrected by **concave spectacles.**

OPTICAL INSTRUMENTS

The ***simple microscope (magnifying glass)*** is a convex lens used 292
to form a **virtual image;** you can see an object clearly when it is closer
than your near point; the image on your retina is larger and the
object appears magnified.

In a ***compound microscope*** the objective is a **very strong** (short 292
focal length) **convex lens** which forms a **real magnified image** of the
object; the **eyepiece** is a **magnifying glass** used to look at the intermediate image.

Page

In an ***astronomical telescope*** the objective is a **convex lens** which forms **real images** of distant objects in its focal plane; a long focal length gives larger images. The **eyepiece** is a **magnifying glass** for enlarging the intermediate image. 293

In ***reflecting telescopes,*** a **concave mirror** is used, instead of a convex lens, to give real images of distant objects. A **plane mirror** is used to reflect the light so that these intermediate images are formed outside the telescope barrel, and an eyepiece (magnifying glass) is used to examine them. In telescopes, light is gathered from a large area and concentrated into a small area, so that dim sources become visible. 295

COLOUR

When objects are viewed through a prism they appear not only **displaced,** but fringed with **colour.** 296

White light consists of light of different colours. 297

Refractive index of a material depends on the **colour** of the light used; as a result, a **prism** can separate the colours from white light to form a **spectrum.** 297

A ***pure spectrum*** is obtained when a **lens is used** to form an image of the light source on a screen; the colours in a pure spectrum are **clear and distinct.** 297

An ***impure spectrum*** is formed when a **lens is not used;** colours overlap, are not so intense and the middle area of the spectrum may appear white. 297

Colours recognized in the spectrum are: **red; orange; yellow; green; blue; violet.** Red is deviated least by a prism, while violet is deviated most. 297

Newton's experiments: 298

1. When **one colour** of the spectrum is passed through **another prism,** no new colours are formed.
2. If a **second, reversed prism** is used, the colours of the spectrum can be recombined to form white light.

Light is the visible section of the ***electromagnetic spectrum.*** Light possesses higher frequencies than **radio waves;** different coloured light has different frequencies. 299

Infra-red radiation is **invisible radiant heat;** its frequencies lie between those of radio waves and those of light. 299

Ultra-violet radiation is responsible for **photochemical changes;** its frequencies are higher than those of light. 299

A **hot substance** gives a continuous spectrum, as formed from white 299

Page

light. In a broader sense, **continuous** means the whole range of electromagnetic radiations.

Excited atoms give ***line spectra*** containing certain colours only; the colours present are characteristic of the particular atoms. 300

Excited molecules give more **complex spectra** but they are still characteristic of the particular molecules. 301

The ***primary colours*** are **red, green** and **blue.** Light of these colours added together, gives the **appearance** of white. 301

A pair of ***primary colours*** added together gives a ***secondary colour:*** 303

> **Red** light + **green** light appears **yellow**
> **Blue** light + **green** light appears **peacock blue**
> **Blue** light + **red** light appears **magenta**

Pigments give colour to paints by **reflecting** light of certain colours only and **absorbing** all other colours. 303

Filters **transmit** light of certain colours only, **absorbing** all other colours. 301

If filters or pigments are combined, only the **colour common to both** will be **transmitted** or **reflected,** all other colours will be **absorbed** by the filter or pigment. 302

VIBRATIONS AND WAVES

PROGRESSIVE WAVES

Waves transmit energy; they can be of two kinds: 305

> ***Transverse waves,*** in which each particle vibrates **perpendicular to the motion of the wave** so that a crest moves forward. **Waves on water** and **waves in a string** are examples of this type. 306
>
> ***Longitudinal waves,*** in which each particle vibrates **in the direction of the wave** so that a compression moves forward. **Sound waves in air** are an example of this type. 309

Constants of a wave: 313

> ***Frequency*** = number of vibrations of any particle per second.
>
> ***Wavelength*** = distance between adjacent crests or compressions.
>
> ***Velocity*** = velocity of a crest or compression.
>
> ***Amplitude*** = maximum displacement of any particle.

Page

Transmission of sound:

Sound travels in air but not in a vacuum. 314

Its **velocity** is measured by the interval between the flash and sound of a gun, but this is affected by humidity, temperature and wind. 315

Reflection of sound is shown by ***echoes.*** 316

STATIONARY VIBRATIONS IN STRINGS

Stationary vibrations in strings are set up by **plucking** the string, and formed by **interference** between the **incident wave,** and **waves reflected at the ends.** These vibrations appear as **loops** ($\lambda/2$ long) with ***nodes*** at each end. 319

Fundamental vibration: 320

$$\textbf{One loop length} = \frac{\lambda}{2} \qquad \textbf{Frequency} = \frac{1}{2l}\sqrt{\frac{T}{m}}$$

where l is the length of the string, m is its mass per unit length and T its tension.

Sonometer experiments with a given string confirm: 321

$$f \propto \frac{1}{l}\ (T \text{ constant}) \qquad f^2 \propto T\ (l \text{ constant})$$

The ***first harmonic*** is the **fundamental,** frequency f.

The ***second*** (third, fourth) ***harmonic*** has **two** (three, four) **loops:** 323 the **length** of the string is $\frac{2\lambda}{2}\left(\frac{3\lambda}{2}, \frac{4\lambda}{2}\right)$ and the **frequency** is $2f$ ($3f$, $4f$).

The string of a musical instrument vibrates with the frequency of the fundamental and of the higher harmonics at the same time. It is these higher harmonics that give **quality** or ***timbre*** to the note. 324

STATIONARY VIBRATIONS IN PIPES

Stationary vibrations in pipes are set up by **blowing** across the end of the pipe, and formed by **interference** between the **incident wave and waves reflected at the ends.** These vibrations are represented by **loops** $\left(\frac{\lambda}{2}\text{ long}\right)$ as in transverse waves. 326

A **closed end** is a ***node,*** i.e. no motion of the air but maximum change of pressure. 328

An **open end** is an ***antinode,*** i.e. maximum amplitude of motion but no change of pressure. 328

Fundamental vibration: 328

Open tube: antinode at each end; node in middle; length $\frac{\lambda}{2}$, frequency f.

Page

Closed tube: antinode at one end; node at the other end: length $\frac{\lambda}{4}$; frequency $2f$. 328

The ***first harmonic*** is the **fundamental vibration.**

The ***second*** (third, fourth) ***harmonic*** in an **open tube** has **2** (3, 4) **loops;** the tube is of **length** $\frac{2\lambda}{2}\left(\frac{3\lambda}{2}, \frac{4\lambda}{2}\right)$ and the frequency is **2** (3, 4) times the frequency of the fundamental. 329

The ***second*** (third, fourth) ***harmonic*** in a **closed tube** has $\frac{3}{2}$ ($\frac{5}{2}$, $\frac{7}{2}$) **loops;** the tube is of **length** $\frac{3\lambda}{4}\left(\frac{5\lambda}{4}, \frac{7\lambda}{4}\right)$ and the **frequency** is **3** (5, 7) times the frequency of the fundamental.

Resonance tube experiments show: $f \propto \frac{1}{l}$ **Velocity of sound** $= f\lambda$ 330

End correction is the short distance between the open end and the antinode. 331

WAVES AND THE EAR

Sounds from musical instruments make the air vibrate simultaneously with the frequency of the fundamental and of the higher harmonics, each with its own **amplitude.** This may be shown by a trace on a ***cathode ray oscilloscope.*** 334

Pitch: depending on the **frequency** of the **fundamental.** 336

Loudness: depending on the **amplitude** of the **vibration.** 337

Quality: depending on the **number** and **amplitude** of the **harmonics.** 336

The ***loudness*** of a note is measured by its **intensity level in decibels.** 338

The vibrations of the air are transmitted by the ***ear drum*** through a series of **bones** to the ***inner ear*** where they produce nervous impulses which are analysed by the brain in terms of pitch, loudness, quality. 339

TRANSMISSION AND RECORDING OF SOUND

The ***microphone*** converts the **vibrations of the air** due to a sound wave into a **varying electrical current.** Common types are: 343

Carbon: the **pressure of the air** on a cell of carbon granules changes its **resistance** and hence the current in it.

Moving coil: **the motion of the air** sets a **coil vibrating in a magnetic field** so that currents are induced in it.

Crystal: the **pressure of the air** on a crystal induces an **e.m.f.** **across it.**

Page

Varying currents may be converted into sound by:

Telephone: the current controls the **strength** of an electro-magnet which attracts a **diaphragm.** 341

Loudspeaker: the current, passing through a **coil** (attached to the cone) **in a magnetic field,** causes the coil to move. 342

Gramophone: **vibrations of the needle** as it moves along the groove are converted into **currents** using the principle of the moving coil or crystal microphone. 344

Sound films are **recorded** by allowing the current from a microphone to control the **amount of light** falling on to a portion of the film. To **reproduce** this sound track, light passed through it is allowed to fall on to a **photoelectric cell,** producing a variable current which is amplified and passed to a loudspeaker. 345

Tape recorder: current from a microphone varies the **strength of an electromagnet** which is the **recording** head. The head **magnetizes** the tape as it passes. To reproduce the recorded tape it is passed in front of the same electromagnet which now acts as a **playback** head. This causes currents to be induced in the electromagnet which, after amplification, operate a loudspeaker. 345

WAVE PROPERTIES AND THE ELECTROMAGNETIC SPECTRUM

The ***electromagnetic spectrum*** comprises (in order of **decreasing wavelength** or **increasing frequency**): 347

Radio waves.

Infra-red radiation.

Light (red, orange, yellow, green, blue, and violet).

Ultra-violet radiation.

X-rays.

Gamma rays.

Cosmic rays.

Wave theory of light: in general, light treated as a wave obeys the laws of **geometrical optics.** But when it passes through **narrow slits** it shows: 348

Diffraction: from a single slit light spreads out in all directions. 351

Interference: in certain circumstances **light from two adjacent** slits can give darkness. 352

These effects are confirmed by experiment and illustrated with a **ripple tank.** 348

ELECTRICAL ENERGY

CHAPTER 31

ELECTRICITY AT REST

IT HAS been known for many centuries that some substances, such as amber, when rubbed with others, such as silk or fur, possess the property of attracting light objects such as pith, bran or small pieces of paper. When in this state the substances were said to be *electrified* or charged with electricity. The word is derived from the Greek word "elektron" for amber. It was not until the sixteenth and seventeenth centuries, however, that a systematic study of such phenomena was undertaken.

Electrical Forces

The basic facts about electrification can be learnt from a few simple experiments. Rods of various materials such as dry glass, perspex, polythene, ebonite, sulphur or sealing wax are rubbed with dry silk or fur. The rubbed rod is suspended in a stirrup of paper and hung from a silk or nylon thread. If another rubbed rod is brought near the suspended rod it will be seen that the rods either attract or repel each other (Fig. 375). Two similar rods rubbed with the same material always repel. Ebonite rubbed with fur will attract dry glass which has been rubbed with silk and also Perspex which has been rubbed with fur. If the materials used are very dry, and the atmosphere is not too humid, it will be found that the rubber will always attract the rod which it has electrified. Such experiments show that electrified objects can be divided into two classes. Members of each class repel others of the same class, but attract objects in the other class. Originally these classes were called vitreous (electrified glass) and resinous (electrified resin). Now we call them *positive* and *negative* respectively.

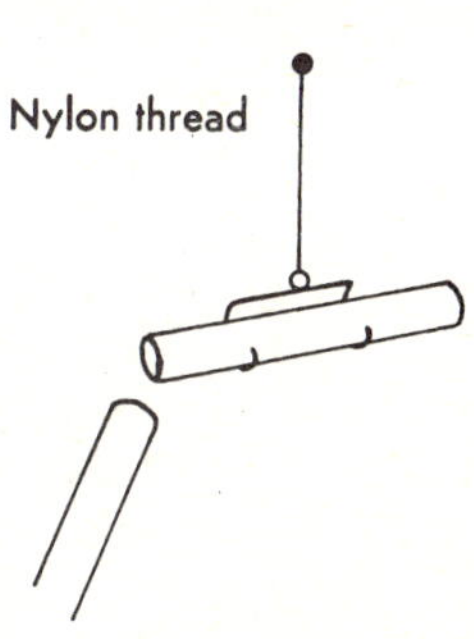

FIG. 375.

Ebonite, sealing wax, sulphur and polythene all become negatively charged when rubbed with fur. Perspex, on the other hand, becomes positive when rubbed with fur, The use of the words positive and negative implies that when two charges are "added" the total charge is the algebraic sum of the two.

That is, equal positive and negative charges will annul each other when added. We shall see later that this is so and we shall explain the meaning of "addition" when applied to charges of electricity.

Rods of metal cannot be electrified by rubbing unless they are fitted with a plastic or ebonite handle. When this is done, and if the metal is not touched with the hand during the rubbing, the metal rod can be charged and is usually found to be negative. These experiments lead to the simple rule that "*Like charges repel, unlike charges attract.*" The fact that the rubber and the rubbed materials always have opposite charges suggests that in the rubbing something is transferred from one to the other, and that the total charge is zero. An experiment to support this view will be described later.

The forces between two charged bodies decrease very rapidly as the distance between them increases. This may be shown quantitatively by coating a ping-pong ball with aluminium paint, hanging it up by a nylon thread and charging it by means of an electrophorus (page 373). Another similarly coated and charged ball is stuck to a perspex rod and brought near the first. It will push the suspended ball to one side, and the nearer it approaches, the more the suspended ball is moved (Fig. 376). Quantitative experiments show that if the distance between the balls is halved, the force is quadrupled. In other words, *the force varies inversely as the square of the distance.* (*Inverse square law.*)

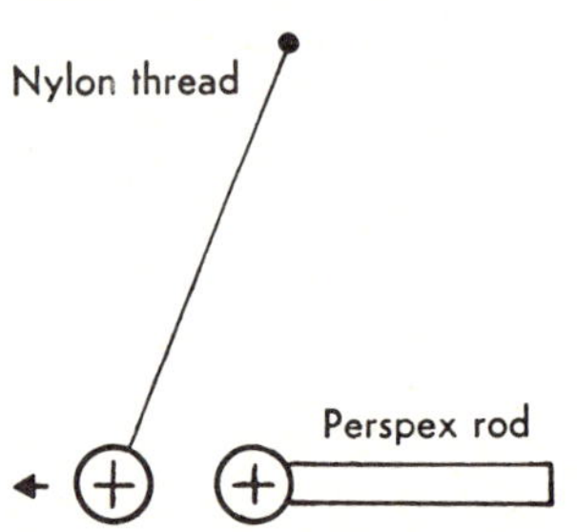

FIG. 376. *Like charges repel.*

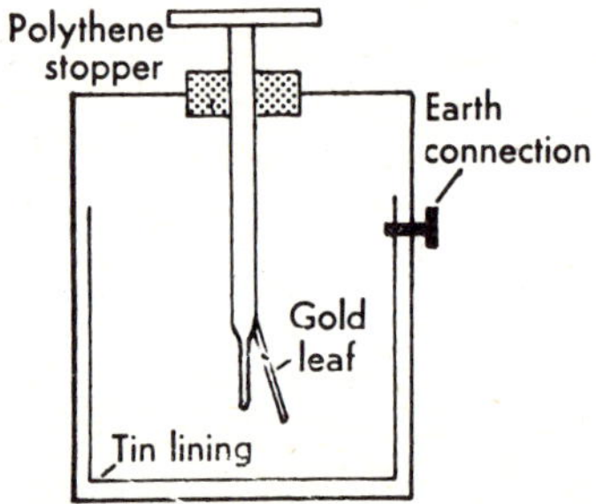

FIG. 377. *Gold-leaf electroscope.*

Gold-leaf Electroscope

Electroscopes are instruments used for detecting electric charges. A very simple form of this instrument is called the gold-leaf electroscope. Essentially it consists of a metal rod with a flat cap or knob at its upper end and a flat metal plate at its lower end. The upper end of a very thin strip of gold leaf or Dutch metal is fixed to the plate (Fig. 377). The rod passes through a stopper made from polythene or paraffin wax, and is mounted in a box with glass windows so that the gold leaf is shielded from draughts. The inside of the box is often lined with tin foil so that an earth connection can be made.

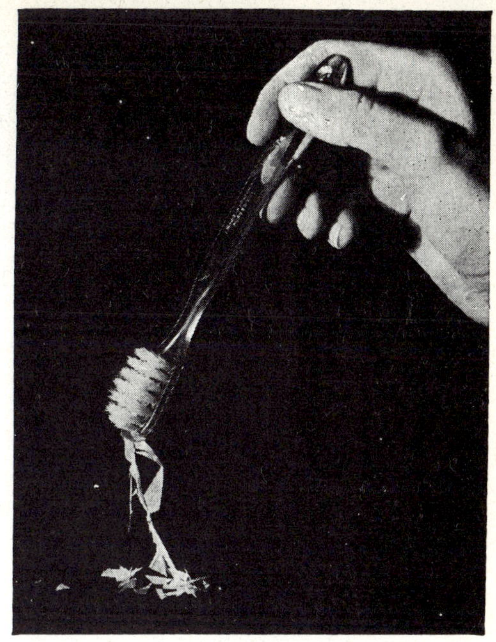

FIG. 378. *A plastic toothbrush can easily be charged by rubbing it with a piece of cloth; it will then attract small pieces of paper. The toothbrush has been negatively charged so that when it is brought near the paper, some of the negative charge in the paper is repelled to earth. The paper becomes positively charged as a result and since unlike charges attract each other, the paper is attracted by the negatively charged toothbrush.*

Conductors and Insulators

We shall now describe some simple experiments to illustrate an important distinction; that between electrical insulators and electrical conductors.

Touch the cap of the electroscope with an electrified object: the gold leaf is repelled from the plate, the electroscope deflects, and the electroscope is said to be charged. Touch the cap of the electroscope with the finger: the leaf falls. Recharge the electroscope and touch the cap with a piece of metal fixed to a polythene or ebonite handle, being careful not to touch the metal: the leaf falls slightly. A ping-pong ball coated with aluminium paint and suspended by a nylon thread can be used for this experiment. Remove the ball out of contact with the cap and touch the cap with the finger to discharge the electroscope. Now touch the cap with the ball. The leaf will deflect, showing that the ball has become electrified. If the experiment is repeated using two balls in contact instead of the single ball, it is found that the fall of the leaf is greater, suggesting that a greater proportion of the charge is lost. Now charge the electroscope and touch the cap with a piece of uncharged polythene or ebonite (the polythene or ebonite should be "wiped" with the hand to discharge it): the deflection remains unaltered. Touch the cap with a piece of wood: the deflection slowly decreases to zero. Next lay a piece of damp cotton over the cap so that it touches the case and touch the cap with an electrified object: the leaves do not deflect.

To account for the results of these and similar experiments we suppose that the electrification is free to move about in some materials (*conductors*) and not in others (*insulators*). When an electrified object is allowed to touch the metal cap, metal being a good conductor, electricity is transferred to the cap and spreads down to the plate and leaf. These being similarly charged repel each other and the leaf deflects. The electricity is prevented from escaping by the polythene stopper, polythene being a good insulator. If a conducting object like the ping-pong ball touches the cap, electricity spreads from the electroscope on to the ball. Thus less electricity is left on the electroscope and the divergence of the leaf decreases.

When the cap is touched with a finger, the charge is shared with the body and the earth, both of which are conductors, and the electricity remaining on the electroscope is negligible. Wood is a bad conductor, or, what is the same thing, a good insulator, and hence the electricity spreads through it only slowly. If a charged ebonite rod is touched at one end with the hand, the electrification disappears from the part touched, but not from the rest of the rod. If, however, a charged metal object on an insulating handle is touched with the finger, the charge disappears from the whole of the metal. These facts are easily verified with the electroscope.

It will be noticed that the substances which can be electrified by friction when held in the hand and rubbed, are all insulators. If they were conductors, the electrification would escape through the hand as fast as it was produced. This is why a metal rod has to be held by an insulating handle if it is to be electrified by friction. Any damp object will conduct electricity even if it is an insulator when dry. This accounts for the necessity of using dry substances.

The fact that when two bodies are charged by rubbing them together the charges produced are opposite in sign is readily explained if we suppose that all substances contain both positively and negatively charged particles. In an uncharged body these charges are equal and uniformly distributed so that uncharged bodies exert no resultant electrical forces on each other. When a body is charged, it has an excess of particles of one kind or another. In rubbing the two bodies together some of the particles are removed from one body to the other. To simplify explanations we shall assume for the present that it is only the negative charges which can be thus transferred from one body to another in solids. For reasons which are beyond the scope of this book we believe that this is what does in fact happen.

ELECTROSTATIC INDUCTION

It will have been noticed in the electroscope experiments that when a charged insulator is brought up to an electroscope, the leaf deflects *before* the charged body touches the cap. Also that the deflection when the charged body is removed after touching the cap, is less than that obtained when it was brought near to the cap. This can be explained as follows. When a charged body (suppose it has a negative charge) is brought near the cap of an uncharged electroscope, some of the negative charges in the metal are repelled to the leaf and plate and thus there is a deflection (Fig. 379*a*). This means that the cap now has an excess of positive charge. If the negatively charged body is removed without touching the cap, the charges on the cap and plate remix and the electroscope is still uncharged. If the charged body

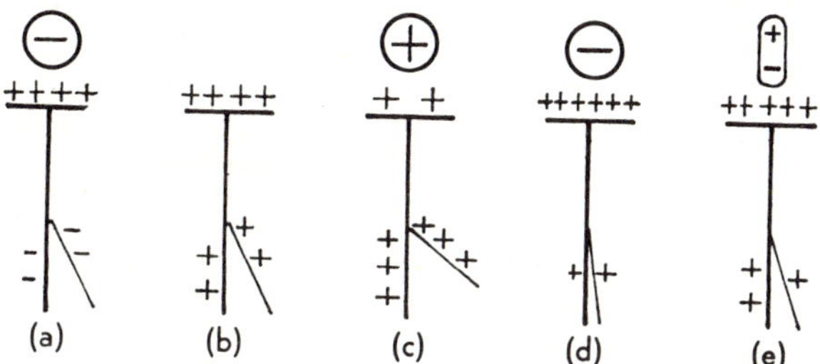

FIG. 379. *Investigation of electrostatic induction with a gold-leaf electroscope.*

is an insulator, when it touches the cap, only the charges from the part actually touching the cap can pass to the cap. These neutralize some of the positive charges on the cap, and when the charged body is removed the electroscope has an excess of negative charges, but as these now spread over the whole electroscope, the charge left on the leaf and plate is in general not as great as before.

If an electroscope is charged, say positively (Fig. 379*b*), and a positively charged object is brought *near* the cap, the deflection of the leaf increases because the positively charged object attracts still more negative charges from the leaf to the cap, leaving the leaf more positive and the cap less positive (Fig. 379*c*). If a negatively charged object is brought near the cap, the object repels some negative charges in the cap down to the leaf, making the cap more positive and the leaf less positive (Fig. 379*d*). Thus the deflection decreases.

If an uncharged conductor such as your hand is brought near the cap of a positively charged electroscope, deflection is seen to decrease slightly. This is because the positive charge on the electroscope pulls negatively charged particles of the conductor towards the end of the conductor nearest to the electroscope. This leaves the more remote end of the conductor positively charged. The negative end of the conductor, being nearest to the electroscope, has more effect on it than does the positively charged end and, in consequence, the deflection of the electroscope falls slightly (Fig. 379*e*).

To detect the sign of an unknown charge we need two electroscopes, one charged negatively and one positively. The unknown charge is brought *near* to each in turn. The electroscope whose deflection increases when the charge is brought near has the same kind of charge as the object has.

This separation of charge on a conductor caused by the presence of a charged body is called *electrostatic induction.* The charges on the conductor are called *induced charges*; the charge on the body is called the *inducing charge.* It must be emphasized that electrostatic induction cannot occur in insulators as the charges are not free to move.

An experiment to show the occurrence of electrostatic induction can be performed with two metallized ping-pong balls *A* and *B*, suspended by nylon threads. These are hung so that they touch one another, and a negatively charged ebonite rod is brought near one of them. The two balls are separated while the ebonite is still near (Fig. 380). On testing with charged electroscopes as above, *A* is found to be negatively charged and *B* positively. If the two balls are now allowed to touch when the ebonite is removed, and if each is now tested on an uncharged electroscope, both balls are found to be uncharged. Their charges, being equal and opposite, have exactly neutralized each other. It should be noted that the charges on *A* and *B* are not necessarily equal to that on the ebonite rod, and in general they are not.

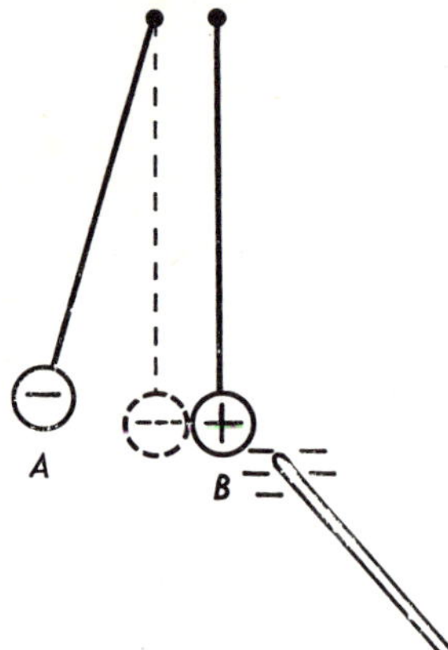

FIG. 380. *Electrostatic induction.*

This experiment, besides illustrating electrostatic induction, also supports the view that *an uncharged conductor contains equal quantities of positive and negative charge.* It also shows that positive and negative charges given to the same conductor will tend to cancel each other.

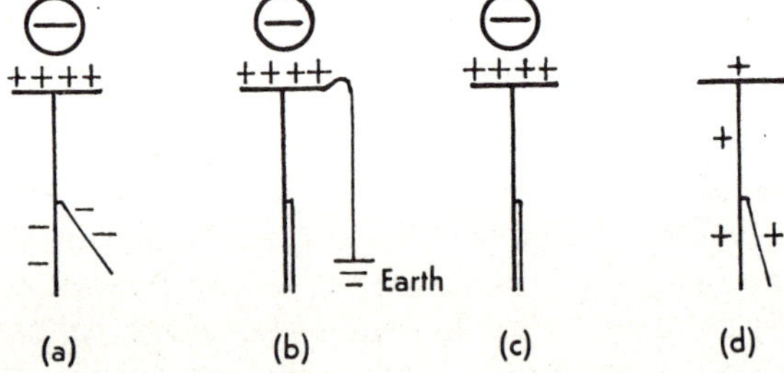

FIG. 381. *Positively charging an electroscope by means of a negative charge.*

Another interesting example of electrostatic induction is the method of charging an electroscope positively using a negatively charged object, or vice versa. Suppose a negatively charged ebonite rod is brought near the cap of an electroscope, the leaves diverge (Fig. 381*a*). If now, while the ebonite rod is still near, the cap of the electroscope is touched with the finger (i.e. *earthed*), the leaf collapses (Fig. 381*b*). This means that the electroscope behaves as the conductor *B* in the previous experiment and the earth as conductor *A*.

The earth connection is now broken (Fig. 381*c*). The negative ebonite is now preventing negative charges in the leaf from rising to neutralize some of the positive charge on the cap.

When the ebonite is removed, negative charges can come from the leaf to the cap, but the electroscope has an excess positive charge and the leaf deflects (Fig. 381*d*).

Electrophorus

The electrophorus is a simple device for obtaining a charged conductor by means of electrostatic induction. A plate of insulator, such as ebonite, is

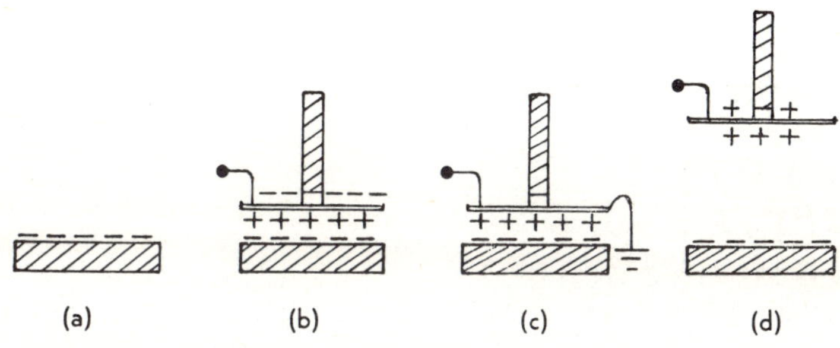

FIG. 382. *Charging a conductor by means of an electrophorus.*

charged by rubbing with fur (Fig. 382*a*). A metal plate, held by an insulating handle, is brought over and very close to the ebonite. The negative charge on the ebonite induces a positive charge on the lower face of the metal and a negative charge on the upper face (Fig. 382*b*). The metal plate is now earthed by touching it with the finger and the negative charges on the upper side of the plate flow to earth, being repelled by the charged ebonite (Fig. 382*c*). The finger is removed leaving the metal plate with a positive charge. The plate is now lifted away and the charges on it are redistributed so that the whole plate has a positive charge (Fig. 382*d*).

Usually there is a metal knob attached to the plate and, if this knob is held near an earthed conductor such as the finger, a spark will jump and the plate becomes discharged. Air is normally an insulator but if two opposite electric charges are strong enough and close enough, the air becomes split up into positive and negatively charged particles called ions. These ions move towards the oppositely charged conductor; both positive and negative charged ions move, and discharge the conductors. This flow of charge is accompanied by a flash and a noise and is called a *spark*. A lightning flash is a large spark jumping between a charged cloud and the opposite charge which the cloud induces on the earth directly beneath it.

Apart from leakage from the charged ebonite to earth, there is no limit to the number of times the plate of the electrophorus may be charged in this way from a single rubbing of the ebonite. The energy represented by the spark obtained when the metal plate is discharged comes from the work done when the metal plate is pulled away from the oppositely charged ebonite.

Faraday's "Ice Pail" Experiments

These simple but significant experiments were first performed by Faraday. A deep metal can is placed on the cap of an electroscope whose case is earthed. The can Faraday used was one which was used for transporting ice in his laboratory and this gives the experiments their name. A charged body is lowered into the can. The metallized ball on a nylon thread can be used. The leaf deflects because the charged ball induces an opposite charge on the inside of the can, and a similar charge on the outside of the can and the electroscope (Fig. 383). It is found that the deflection of the leaf is the same wherever the charged body is moved provided it is not near the mouth of the can. If the ball is allowed to touch the inside of the can near the bottom, the deflection of the leaf does not change (Fig. 384). If the ball is now withdrawn, the leaf remains unaltered, while the ball, if tested on another electroscope, is found to be completely discharged. The discharged ball can be lowered into the can again and allowed to touch the inside of it without affecting the leaf or becoming charged itself. Thus there appears to be no charge left on the inside of the can. If there had been, the ball would have shared it and retained a charge itself. Furthermore, since the leaf's deflection is the same before and after the charged ball was allowed to touch the inside of the can, then the electroscope's charge is the same as before.

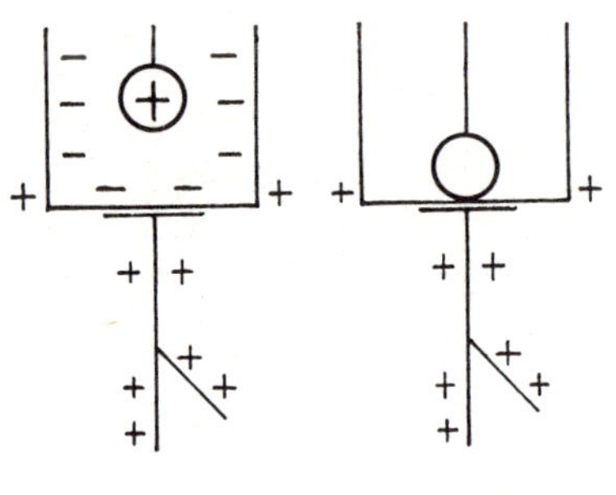

FIG. 383. FIG. 384.

The original charge on the ball and the induced charge on the inside of the can must have neutralized each other exactly and these must therefore have been *equal* as well as opposite. Since the can and electroscope were originally uncharged, the positive charge on the electroscope and the outside of the can must be equal to the charge on the ball.

These results are only observed if the ball is not allowed to touch the can near its opening. Faraday suggested that if the can were completely closed the following would be true:

1. If there is a charged body inside the can, the induced charge on the inside of the can would be equal and opposite to that on the body. (This is the only occasion on which the induced charge and the inducing charge are equal.)
2. If there is no charged body inside the can, there is no charge on the inside of the can however highly the outside of the can is charged.

These results provide a means of transferring the *whole* of the charge on a conductor to a hollow can. The conductor is touched on the inside of the can near the bottom. The whole of the charge which was on the conductor is now transferred to the outside of the can. If the conductor is touched on the outside instead of the inside of the can its charge is *shared* with the can, and some of it remains on the conductor.

Suppose we have two hollow cans, and wish to give them charges in the ratio 3 : 4, we can do it as follows. Take seven equal metallized ping-pong balls on nylon threads and hold them in contact. Charge all the balls (say from an electrophorus). Since they are all similar they will all have equal charges. Then touch the inside of one can with three of them and the inside of the other with the remaining four.

One of Faraday's results has an important practical application in *electrical screening*. Since a hollow metal can can have no charge on the inside, however highly the outside may be charged, objects can be screened from the effect of outside charges by enclosing them in a metal box. This method of screening is frequently used in constructing radio circuits.

The method of the ice pail experiment can also be used to test whether two charges are equal in magnitude. If each charge is lowered in turn into a can on an electroscope, equal deflections show equal charges. This method can also be used to demonstrate that when electrification is produced by rubbing, the charges produced are equal. A can on an electroscope is lined with fur. An ebonite rod is placed with one end on the bottom of the can against the fur and rotated. No deflection of the leaf is observed. This is because the charges produced have equal and opposite inductive effects. If the ebonite rod is removed, however, the leaf deflects owing to the charge on the fur.

Charges can be "added" successively to a can by touching variously charged conductors on the inside. If the can is on an electroscope it can readily be shown that opposite charges "added" to the same can tend to annul each other.

Potential and Capacity

If two hollow cans, one large (*A*) and one small (*B*), are mounted on similar electroscopes and given equal positive charges (ping-pong method page 375), it is found that *B*'s electroscope has a larger deflection than *A*'s, although the charges are equal. If the two cans are now connected together by a wire held by an insulating handle, the deflections of the electroscopes become equal, electricity therefore passing from *B* to *A*. Thus the transfer of electricity from one conductor to another, when they are connected, does not depend solely on their charges. The factor which does determine whether charge flows is called the *electric potential.* If, on connecting a conductor *B* to another conductor *A*, negative charge flows from *A* to *B*, *B* is said to be at a higher potential than *A*, i.e. negative charge flows from low to high potential. In the experiment described, though *A* and *B* have equal charges, *B* has the higher potential. All conductors on which electricity is at rest must have the same potential at all points—otherwise electricity would flow. Different conductors require different amounts of electrical charge to raise their potentials by the same amount (i.e. so that when connected no flow of charge occurs). These quantities of charge are taken as a measure of what are called the *electrical capacitances* of the conductors. In the experiment described *A* has a larger capacitance than *B*.

In practice the earth is taken as the zero of potential. This is because the earth is a conductor and therefore has the same potential everywhere. Also it is very large so that any charges we may take from it or give to it will not affect its potential appreciably. Note that an isolated negatively charged conductor will have a potential below that of the earth since on connecting it to earth its negative charge flows to earth. Thus it has a *negative potential.*

Distribution of Charge on a Charged Conductor

When a conductor is charged, the electrification resides on its outside surface. This has been shown to be so if the conductor is hollow. It is also true if the conductor is solid. In a general way this may seem to be necessitated by the fact that the charges tend to move as far away from each other as possible. The charge, however, does not spread evenly over the surface of the conductor. It tends to collect at the more sharply curved places.

This may be shown as follows. A pear-shaped conductor is mounted on an insulating stand and charged (Fig. 385). A small metal disc *P* on an insulating handle is placed flat on the conductor at *X*. It picks up the charge at that part of the surface. It is then held inside a can on an electroscope and the deflection noted. Then the experiment is repeated at *Y* and *Z*. It is found that the deflection obtained for *Y* is greater than at *X* and greater still at *Z*. Leakage of charge would tend to make the deflections smaller so that the result cannot be explained in this way.

If a pin is attached to a conductor and the conductor charged, it will be found that the conductor loses its charge. This is because the charge concentrates on the point of the pin, ionizes the air in its neighbourhood and thus escapes. If a powerful electrical machine such as a Van der Graaf generator is available, this may be connected to a pin in a dark room. A glow will be seen in the neighbourhood of the pin point. Also a candle flame placed near the pin point is blown aside by the current of ionized air leaving the point.

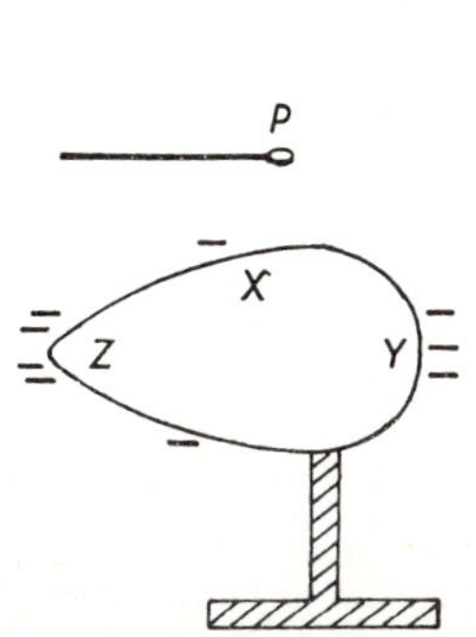

FIG. 385. *Concentration of charge.*

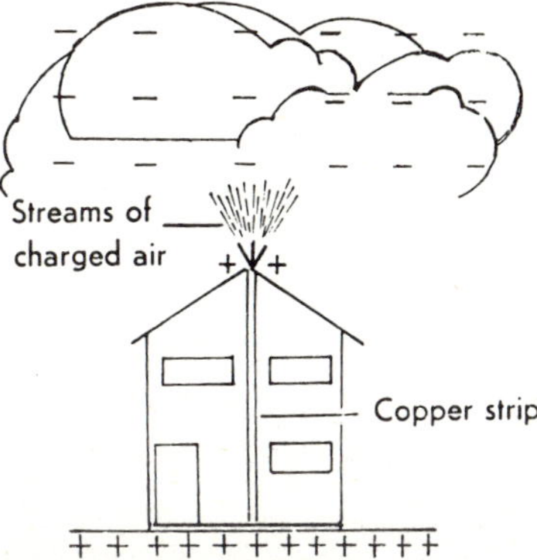

FIG. 386. *Lightning conductor.*

Lightning Conductors

This action of points is used to protect houses from lightning discharges. In a thunderstorm the clouds become highly charged with electricity, though how they become so is a matter of some doubt. Clouds usually, though not always, have a negative charge. The charged cloud induces an opposite charge on the earth underneath it, and if the charges are large enough a spark passes between cloud and earth. The discharge usually takes place between the cloud and the nearest point on earth so that tall buildings, trees and church spires are likely to be struck. The danger of this can be reduced by fixing a metal rod, with points at the top, to the building and connecting this to the earth by copper strip (Fig. 386). The charge on the ground flows on to the points and escapes from them. Thus the points generate a stream of charged air flowing upwards to the cloud with a charge opposite to that of the cloud, thus tending to reduce the cloud's charge. If a flash does occur, the lightning conductor provides a conducting path to earth, and the current, taking the line of least resistance, takes this path.

It is worth mentioning that it is better not to shelter under a tall tree in a thunderstorm. A hedge may not be so effective in keeping off the rain but it is less dangerous. The tops of mountains are particularly dangerous and

FIG. 387. *The spires of the two churches shown have both been struck by lightning. That on the left escaped serious damage since most of the current was carried to earth by the conductor shown. That on the right had no lightning conductor and in consequence the current destroyed most of the spire.*

should be avoided. It is also said that an umbrella with a metal shaft may act as a lightning conductor and encourage lightning to strike.

QUESTIONS

1. Why does a charged body attract an uncharged conductor?

2. A strong positive charge is brought towards the cap of a negatively charged electroscope. The deflection of the electroscope first decreases to zero and then increases again. Explain this.

3. Suppose you have electrified a plastic comb by rubbing it with a silk handkerchief. How would you determine the sign of the charge produced on the comb?

4. In a high voltage laboratory the observer works inside an earthed cage of fine wire mesh. Suggest a reason for this.

5. A simple form of electrophorus can be made using an old gramophone record, a light aluminium disc from the base of a cake tin and a candle. See if you can make it work.

6. If you are provided with an electroscope, a metal can, and two unequal insulated conducting spheres each positively charged, how could you discover which of the spheres has (*a*) the greater charge, (*b*) the higher potential?

CHAPTER 32

MAGNETISM

BATTERIES and dynamos are really devices for separating electric charges. To show this it is convenient to use a gold leaf electroscope whose leaf is partially screened by lining part of the case with tin foil and connecting this to a terminal. If a high tension battery of a few hundred volts has its positive terminal connected to the cap of the electroscope and its negative terminal to the foil, a deflection is obtained. The electroscope can be shown to have a positive charge by removing the battery and bringing a charged glass rod up to the cap, when the deflection of the leaf will increase.

If the battery is connected the other way round, the electroscope becomes negatively charged. When the battery is first connected with its positive terminal to the cap of the electroscope, negative charge flows on to the foil, and from the cap to the positive of the battery. Chemical action inside the battery maintains the charges on the terminals, and the flow of charge ceases when the potential of the cap is the same as that of the positive terminal, and that of the foil is the same as that of the negative terminal of the battery.

Electric Currents

If the battery is left connected to the electroscope and a piece of cotton is laid across the cap and allowed to touch the foil terminal, the electroscope still deflects, though less than before. If the battery is disconnected, the electroscope quickly discharges. This is because the cotton is a conductor, though a poor one, and charge can flow through it from cap to foil. When the battery is connected it maintains the charges on the cap and foil in spite of the flow of charge through the cotton. This flow of charge constitutes an electric current, though in this case the current is exceedingly small. Instruments capable of measuring electric currents when charge flows through them can be made; they are called *galvanometers* or *ammeters* (page 427). The unit of current is called the *ampere* (symbol A). The principles of these instruments will be described later.

The fact that a current is a flow of electric charge can be easily demonstrated: a metal can is placed on an insulating support and connected to one terminal of a sensitive galvanometer. The other terminal of the galvanometer is connected to earth. When a charged ebonite rod is brought near the can, the galvanometer shows a slight momentary deflection. The arrows in Fig. 388

show the direction of flow of the negative charge from the can to earth.

Conventionally we suppose a current to flow out of the positive terminal

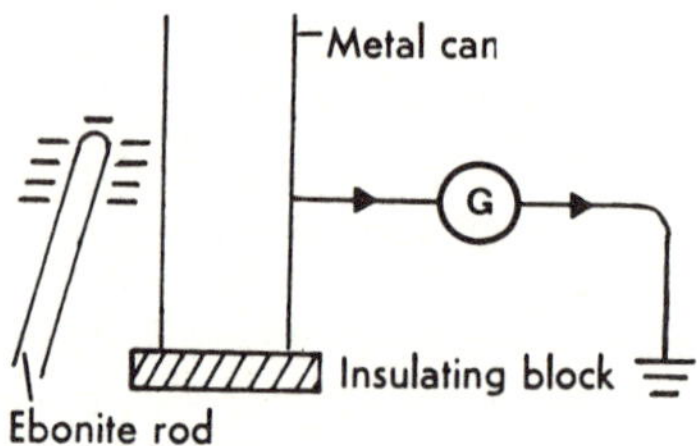

FIG. 388. *Apparatus to show that current is a flow of electric charge.*

of the battery and back into the battery through the negative terminal. In fact the charges which actually flow are negative and these flow in the opposite direction to the current.

MAGNETIC FIELDS OF CURRENTS

A compass needle which is free to rotate about a vertical axis will set in a north-south direction. The end pointing north is called the *north pole* and that pointing south the *south pole*. This is said to be due to the earth's magnetic field. Any part of space in which a compass needle experiences forces tending to make it set in a particular direction is said to be characterized by a *magnetic field*. The direction of the magnetic field is the direction that the compass takes up if it moves freely and is said to be from the south to the north end of the compass needle.

Currents in a Straight Wire

Electric currents produce magnetic fields in their neighbourhood. This was discovered by Oersted in the early nineteenth century. Oersted's experiment can be easily reproduced. The terminals of a cell such as a torch battery are connected by a copper wire through a switch so that when the switch is on a current flows in the wire from the positive carbon terminal (+) to the negative zinc terminal (—). If a compass needle is freely suspended near the wire, it will be deflected when the current is switched on, tending to set at right angles to the wire. The earth's magnetic field, however, tends to prevent the needle from turning so that it is completely perpendicular to the wire. Fig. 389 shows the directions in which the compass needle moves when the wire is running in a north-south direction, (*a*) below the compass needle, and (*b*) above the compass needle.

There is another way in which comparatively strong magnetic fields can be demonstrated. If iron filings are placed in magnetic fields they tend to set with their lengths along the field direction. The reason for this will be made clear later. To demonstrate the magnetic field due to a current, a piece of cardboard is mounted horizontally and a wire is made to pass vertically through the cardboard (Fig. 390). Iron filings are sprinkled on the cardboard and a strong electric current is made to flow along the wire. The cardboard is lightly tapped to enable the filings to move, and it is found that the filings arrange themselves in circles round the wire. The direction of the field can be found by placing a small compass on the cardboard. The relative directions of current and field can be remembered by Maxwell's "*corkscrew rule*": *if a corkscrew is screwed in the direction of the current, the magnetic field direction is in the direction of motion of the thumb*. Alternatively, looking along the wire in the direction of the current, the magnetic field is in clockwise circles round the wire (Fig. 391). The lines in which the filings set, that is lines showing at all points the direction of the magnetic field, are called *magnetic lines of force*.

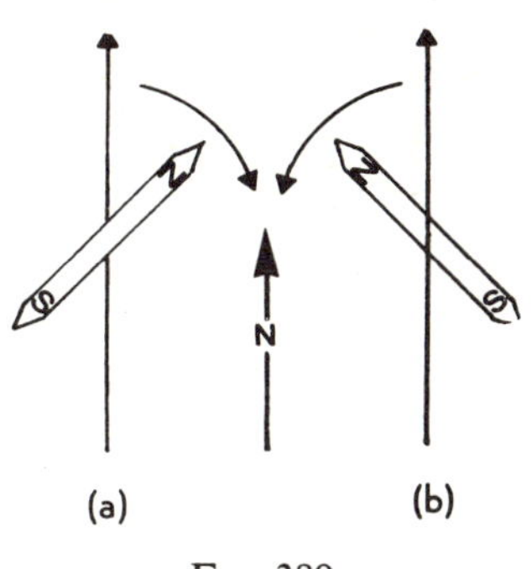

FIG. 389.

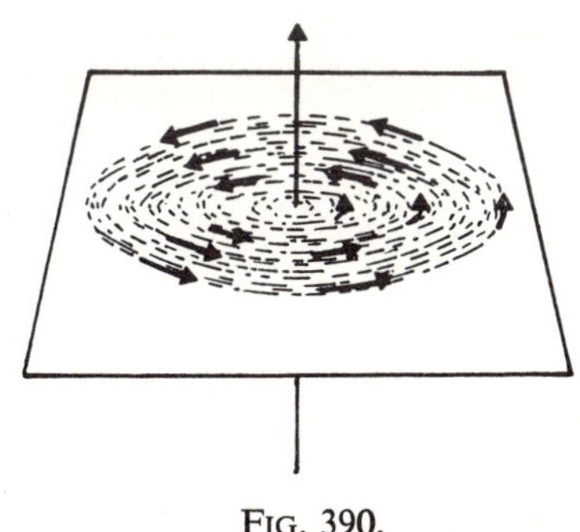
FIG. 390.

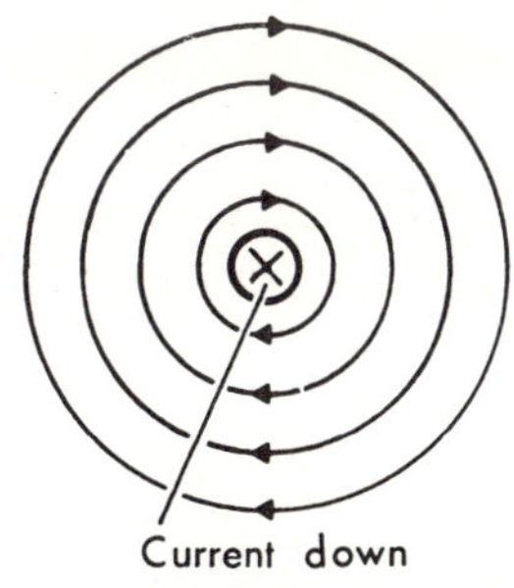

FIG. 391.

Currents in a Circular Coil

The shape of the magnetic field due to the flow of charge along a wire depends on the shape of the wire. We have so far examined the field due to a long straight wire. If the wire is bent into the form of a flat circular coil, the magnetic lines are different (Fig. 392). These can be demonstrated in the same way as those due to a straight current. The coil is arranged so that half is above the card and half below (Fig. 393). Where the wire passes through the card, the field is in rings. These are not exact circles owing to the presence of the current travelling the opposite way at the other point where the wire passes through the card. At the centre of the circle the field is perpendicular to the plane of the coil.

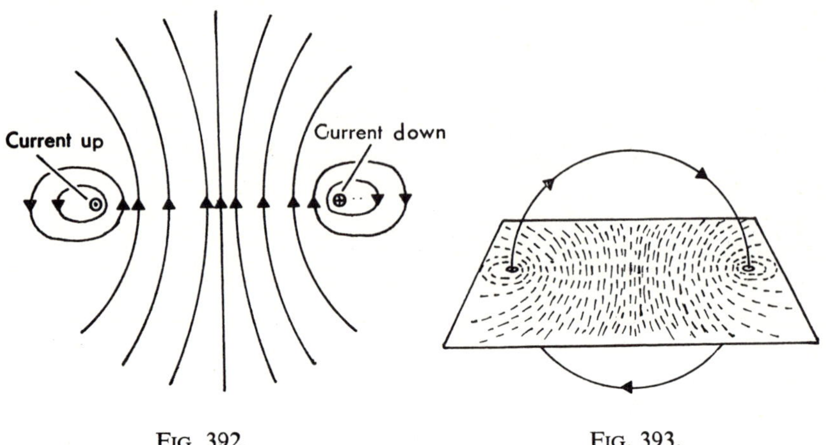

FIG. 392.

FIG. 393.

Another important example of a magnetic field due to a current is that of the *solenoid.* A solenoid is obtained by winding a coil of wire on a cylindrical former (Fig. 394). The relative directions of field and current for such a coil are shown in Fig. 395. If the coil is viewed along its axis, and if the current

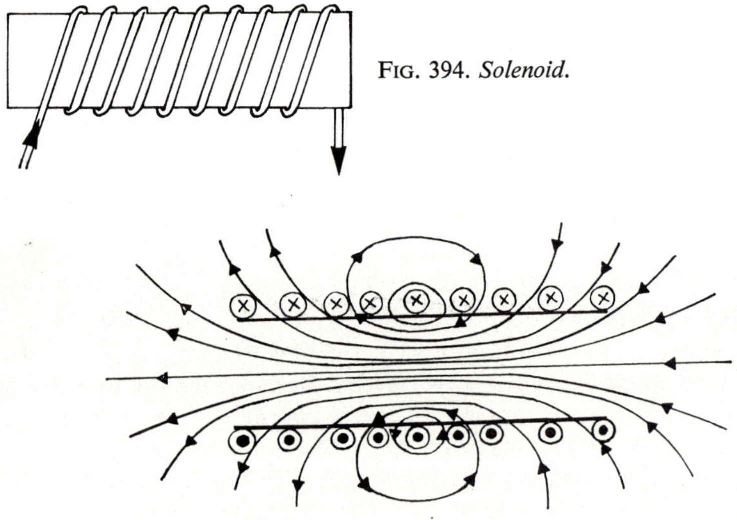

FIG. 394. *Solenoid.*

FIG. 395. *Relative directions of field and current for a solenoid.*

at the near end flows in an anticlockwise direction, the lines of magnetic force emerge from that end. Viewed from the other end, the current will be flowing clockwise and the lines of magnetic force will enter this end.

These experiments demonstrate a very important physical fact. They show that *an electric charge in motion produces a magnetic field.*

Action of Magnets

An experiment which is of great use in helping us to understand the action of magnets can be performed with very simple apparatus. A circular loop of copper wire or, better still, a circular coil of insulated wire, has its ends passing through a large cork. Plates of copper and zinc are fixed to the ends of the coil, and the cork is floated in a dish of dilute sulphuric acid (Fig. 396).

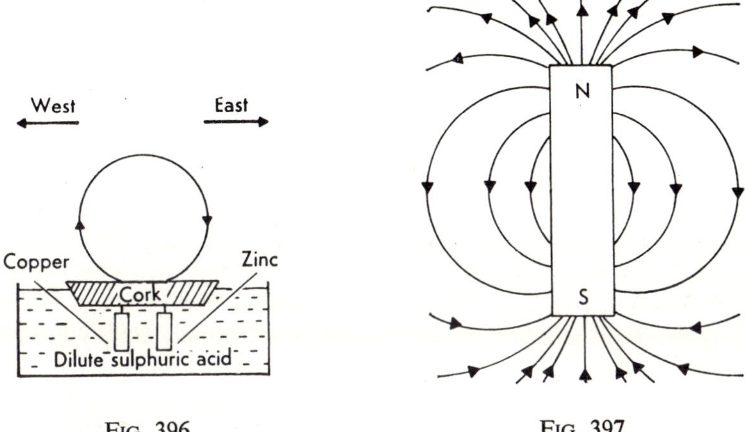

FIG. 396.

FIG. 397.

The zinc and copper plates in the acid act as a weak cell and a current flows through the coil from the copper back to the zinc. The coil always sets itself with its axis north and south and so that the line of force through its centre is parallel to, and in the same direction as the earth's lines of magnetic force. The coil thus acts in much the same way as a compass needle.

If a bar magnet is placed underneath a sheet of paper, and iron filings are sprinkled on top of the paper, when the paper is tapped the filings arrange themselves in a pattern along the directions of the lines of force (Fig. 397). These directions can be verified with a small compass needle placed on the paper.

These lines of force are very similar to those of a solenoid in which a current flows, and this suggests that the magnetic properties of a magnet are due to charges inside the atoms or molecules which are moving in closed paths. We shall call these *molecular magnets* or *molecular currents.* Nowadays we have a great deal of evidence pointing to the fact that inside atoms, negative charges called *electrons* circulate round the positive nuclei of the atoms in much the same way that planets move round the sun. It is possible, too, that the negative charges spin about an axis as the earth does. These rotating and spinning electrons would act as circular currents. If these rotating charges rotate in parallel planes and all in the same sense, i.e. all clockwise or all anticlockwise, their combined field would be very much like that of a solenoid. This can be seen in a general way as follows.

For the sake of simplicity we shall consider a current flowing round a square. The field of such a current would be at right angles to the square at its centre, just as in the case of a circular coil. Suppose we have two such circuits carrying equal currents and of the same size side by side (Fig. 398*a*).

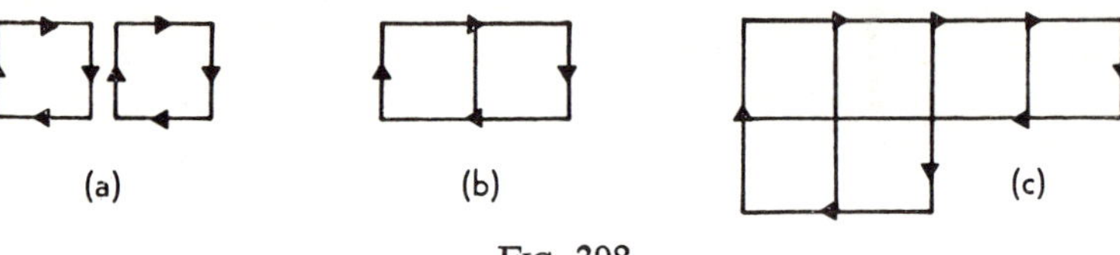

FIG. 398.

It can be seen that where the circuits "touch," the currents are in opposite directions and will cancel, so that the two circuits are equivalent to the same current flowing round the circuit boundary (Fig. 398*b*).

In this way, by placing many such circuits together, the resultant effect is that of a current flowing round the external boundary, the internal currents cancelling (Fig. 398*c*). Remembering that the molecular currents are very small in size, we can imagine the cross section of a magnet to be built up of many millions of such elements. The resultant effect will be that of a current flowing round the boundary of the cross section, and the field of the magnet will be equivalent to a current flowing round the surface with the inside empty. This is equivalent to a solenoid-like coil of the same cross section as the magnet (Fig. 399).

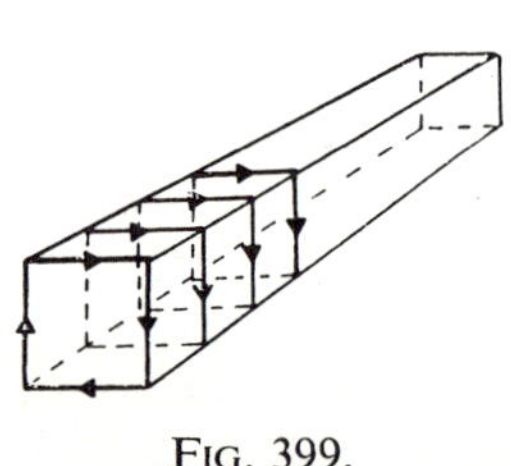

FIG. 399.

Magnetizable Materials

Only a few materials can be made into magnets. These are known as magnetic materials. Iron and steel are magnetic (i.e. magnetizable), pure nickel and cobalt to a lesser extent. Certain alloys of aluminium, nickel and cobalt can be made into very strong magnets. In the unmagnetized state we suppose that the planes of the molecular currents are pointing in all directions

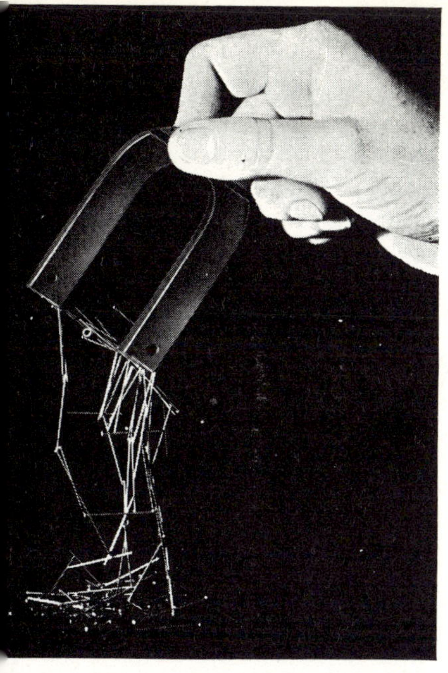

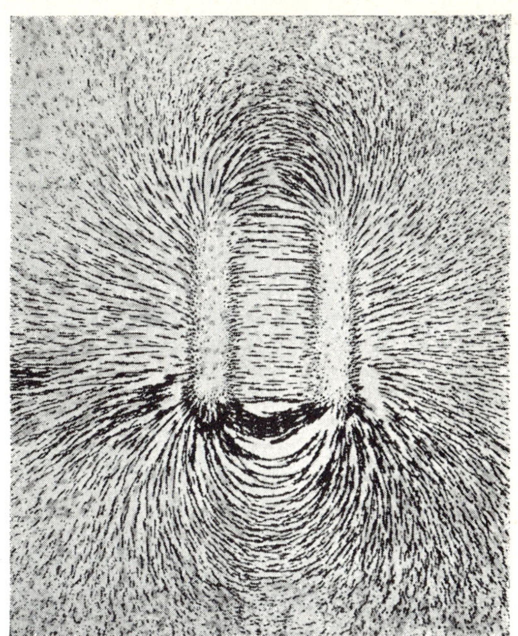

FIG. 400. *For practical purposes, magnets are often bent into the shape of a horse-shoe or a U-shape (left) so that the two poles are opposite each other: the steel pins in contact with the magnet have become magnets themselves, attracting other pins. Lines of force for this magnet are shown by the iron filings method (right); the poles of the magnet point towards the bottom of the photograph.*

and hence do not produce any external magnetic field. One way in which this could occur is shown in Fig. 401. The process of magnetization thus simply consists in "lining up" the molecular magnets. One obvious way of doing this is to place the material to be magnetized inside a solenoid round which a current flows: the molecular magnets will then experience forces to make them set with their planes perpendicular to the axis of the solenoid, and if the solenoid's field is strong enough this will occur. This is the most common way of making magnets. In the case of steel and some alloys, when the specimen is removed from the solenoid, it remains magnetized, but if iron is the substance used the molecular magnets rapidly get out of alignment and it becomes unmagnetized.

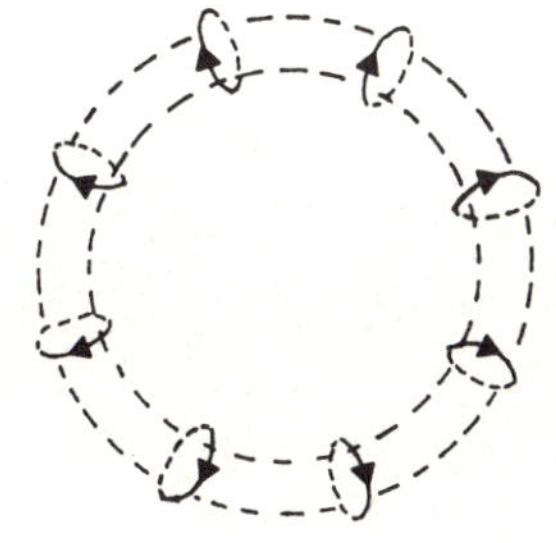

FIG. 401.

With two magnets it is easy to verify that the two north poles repel each other, that the two south poles repel each other, and that the north and south poles attract each other. The north poles can be identified by floating the magnet on a cork in a bowl of water. The pole which points north is the north pole. It is also easily verified that either pole will attract unmagnetized magnetic material such as iron or steel.

A simple experiment shows how these effects can be explained in terms of electric current circuits. A light circular coil of thin insulated wire *A* is suspended by the leads through which a current enters and leaves. Another

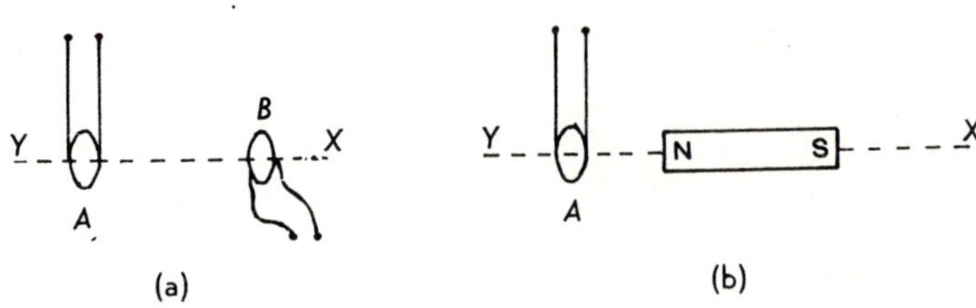

FIG. 402. *Investigating the forces exerted by electric currents and magnetic poles.*

similar coil *B*, also with a current flowing in it, is brought towards *A* (Fig. 402*a*). If the currents in *A* and *B* are in the same direction, i.e. both clockwise looking along *XY*, attraction takes place and *A* moves towards *B*. If the currents are in opposite directions, repulsion occurs and *A* swings away from *B*. The same effects occur with solenoids instead of circular coils. If the north pole of a magnet is used instead of coil *B* (Fig. 402*b*), *A* is repelled when the current in it flows anticlockwise, as seen along *XY*, and attracted when either the current *or* the magnet is reversed. When both are reversed, repulsion occurs.

Thus the forces exerted by magnetic poles on each other can be regarded as due to the forces that electric currents exert on each other. Lines of force for a magnet and its equivalent solenoid are shown as dotted lines in Fig. 403.

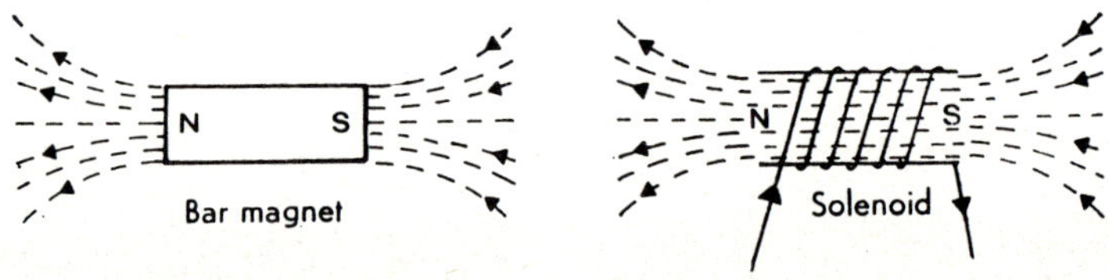

FIG. 403. *Lines of force for a bar magnet and its equivalent solenoid.*

The polarity of a solenoid can be remembered from a simple rule: *when looking along the axis of a solenoid, if the current at the near end flows clockwise, then that end is a south pole and the lines of magnetic force enter the solenoid at that end.* A diagramatic representation of this law (Fig. 404) may help you to remember it.

A body may be charged either positively or negatively, but north and south poles can never be isolated. If a magnet is cut, opposite poles are formed on either side of the gap and two complete magnets are obtained (Fig. 405). This is readily understandable if we regard magnetism in solids as a consequence of circulating charges.

FIG. 404.

FIG. 405.

Magnetization by Stroking

A rod of magnetic material can be magnetized by drawing one pole of a magnet from one end to the other several times, always using the same pole and moving it the same way along the rod (Fig. 406). An opposite pole is formed on the rod where the magnet pole leaves it. The lines of force of the stroking pole tend to align the molecular magnets as the pole moves along the bar. Only weak magnets can be made in this way.

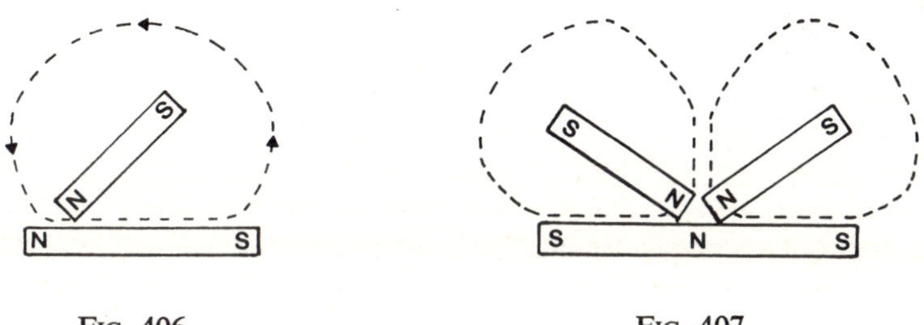

FIG. 406.

FIG. 407.

By using two similar stroking poles and beginning in the middle of the rod, the rod can be magnetized with a north pole in the middle and a south pole at each end (Fig. 407). The middle pole is known as a *consequent pole*. The same result can be achieved by winding two solenoids one on each half of the bar, and sending currents through the solenoids in appropriate directions (Fig. 408).

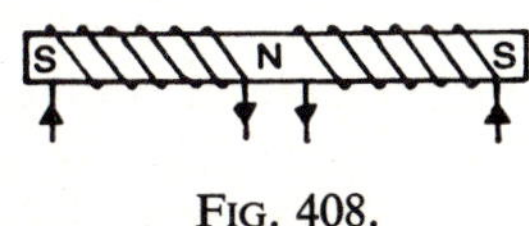

FIG. 408.

Magnetic Induction

When a magnet is brought near a piece of soft iron, the iron becomes temporarily magnetized. This can be demonstrated by bringing a strong magnet near to one end of a short piece of iron. If a paper with a heap of iron filings on it is brought up to the opposite end of the piece of iron and taken away again many filings adhere to the iron, showing that it has been magnetized. If the magnet is removed, most of these filings drop off.

If a rod of iron *A* (Fig. 409) is held with one end near the north pole of a

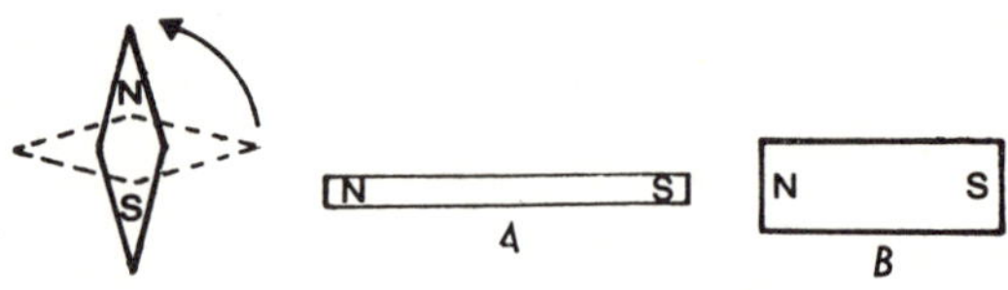

FIG. 409. *Apparatus to demonstrate magnetic induction in an iron bar.*

compass needle, there is a slight force of attraction because any pole of a magnet attracts unmagnetized iron. If the north pole of a magnet *B* is brought near the opposite end of the rod, the north pole of the compass is repelled showing that there is a north pole on that end of *A* nearest the compass pole. It might be thought that this repulsion is directly due to the north pole of the magnet, and in part it is. If the experiment is repeated with the magnet the same distance from the compass but with *A* absent, the repulsion is far less.

This phenomenon is called *magnetic induction*, and the poles in the iron are called *induced poles*. When a pole of a magnet is brought near a piece of unmagnetized iron, it induces an opposite pole on the near part of the iron and a similar pole on the more remote parts. The magnet pole will attract the nearer opposite pole more than it repels the more distant similar pole, and hence the iron is attracted. This explains why a magnet will attract unmagnetized iron. From what has already been said, the reason for the phenomenon of induction is easily given.

The lines of force of the magnet tend to align the molecular currents in the iron, thus magnetizing it.

A long rod of iron may be magnetized by induction in the earth's magnetic field. The rod is first shown to be unmagnetized by bringing each end in turn to each pole of a compass and showing that attraction occurs in each case. The rod is then held approximately north and south, preferably with the rod sloping at about 70° to the horizontal with the north end downwards. In this position the rod is approximately parallel to the earth's magnetic field in this country. The rod is then tapped with a hammer. On testing again it will be found that the end which pointed downwards repels the north pole of a compass, showing that the rod has become a weak magnet. The earth's lines of force have to some extent aligned the molecular magnets, assisted by tapping the iron. If the bar is held east and west and tapped, it loses its magnetism.

Demagnetizing a Magnet

If a magnet is knocked or banged its magnetism is reduced, probably because the molecular magnets are jolted out of alignment. Again, if the magnet is heated until it becomes red hot its magnetism is completely

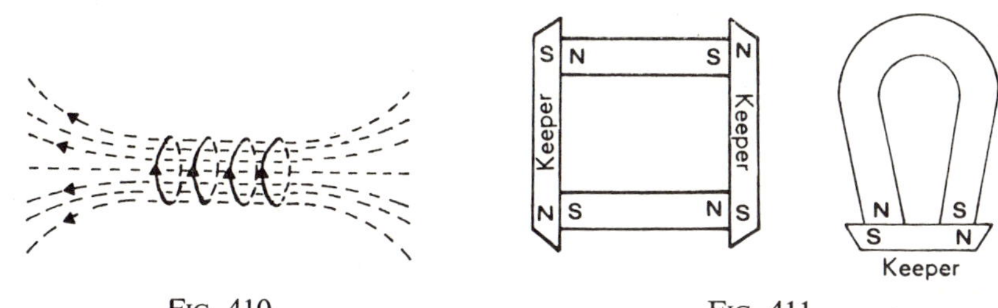

FIG. 410.

FIG. 411.

destroyed. In fact, iron above a certain temperature ceases to be a magnetic material and cannot be magnetized. This is presumably because the more energetic vibration of the atoms of the iron at these high temperatures knocks the molecular magnets out of alignment.

Even when a magnet is not ill-treated in either of these ways its magnetism gradually becomes weaker. This is more noticeable in short magnets whose poles are close together. This effect, known as self-demagnetization, can be explained as follows. In the middle of a magnet the molecular currents tend to be held in place by the field due to the molecular currents on either side (Fig. 410). At the poles these currents are exposed on one side and tend to get out of line. This effect can spread slowly towards the centre.

Self-demagnetization can be slowed down by storing bar magnets in pairs pointing opposite ways, with pieces of iron called "*keepers*" in contact with their poles (Fig. 411). Poles are induced on the "keepers" as shown, producing the effect of a ring magnet in which there are no free poles.

THE EARTH'S MAGNETIC FIELD

The fact that a horizontally suspended compass needle points roughly in the direction of geographical north has been, and still is, of great use in navigation.

In the description of the earth's field certain terms need definition. The *geographic meridian* at a point is a vertical plane passing through the point and the geographic north and south poles. The *magnetic meridian* is a vertical plane in which the magnetic axis of a compass sets itself at any place. The angle between these meridians, which is the angle between the directions of *compass north* and *true north*, is called the *declination* or *variation* at any place (Fig. 412). It may be to the east or west.

To find the direction of the magnetic meridian it is not enough to find the direction in which a pointed magnet sets, for the magnetic axis of the magnet may not lie along the line joining the points. After recording the direction of the line between the points, the magnet must be turned over and the new direction of this line recorded. The bisector of the angle between these lines is in the direction of the magnetic meridian. The finding of the direction of the geographical meridian accurately has to be done by astronomical observations. It can be found approximately by noting the direction of the shadow cast by a vertical stick at midday, which points due north from the base of the stick.

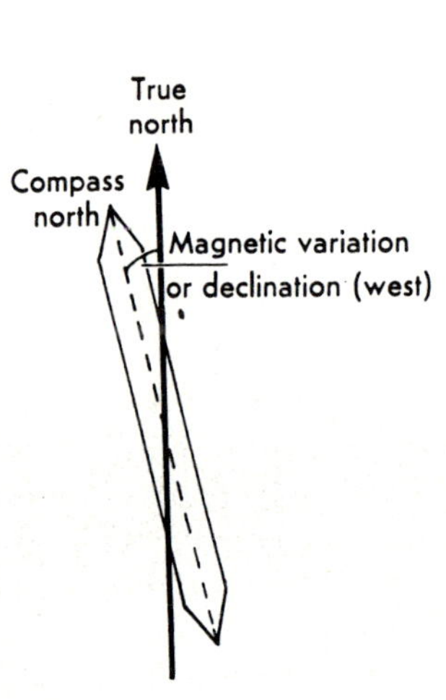

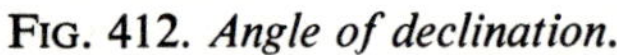
FIG. 412. *Angle of declination.*

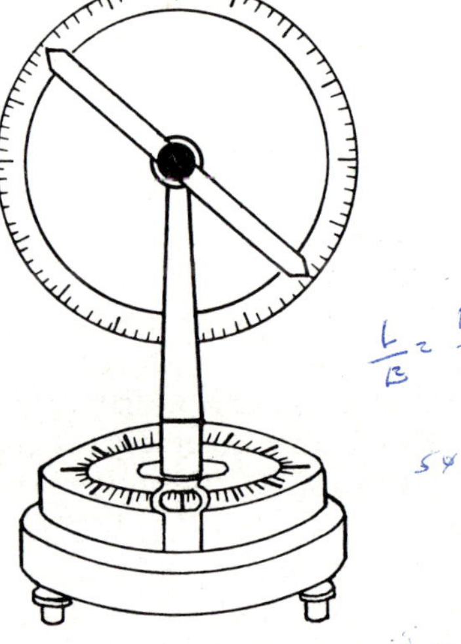

FIG. 413. *Dip circle.*

A compass needle, freely suspended at its centre of gravity, does not in general set horizontally. This shows that the earth's lines of force are usually inclined to the earth's surface. The angle between the horizontal and the earth's lines of force at any place is known as the *magnetic dip* or *inclination.* The angle of dip is approximately zero at the equator and increases towards the poles. In the northern hemisphere the north pole of a magnet dips downwards, and in the southern hemisphere the south pole dips. At the *magnetic poles* of the earth the dip is 90°.

The angle of dip at any place is measured with an instrument called a dip circle (Fig. 413). This consists of a compass needle free to rotate on a horizontal axle through its centre of gravity. A circular scale of degrees enables the angle made by the needle with the horizontal to be measured. The whole circle can be rotated about a vertical axis, and the angle turned through can be measured on a horizontal scale of degrees. The instrument is first levelled by means of a spirit level. This makes the 0-0 marks on the scale horizontal. The dip circle is then rotated until the needle points vertically downwards. The plane of rotation of the needle is then at right angles to the magnetic meridian. The dip circle is next turned through exactly 90° measured on the horizontal scale, and it is then free to rotate in the magnetic meridian. The angle of dip is read from the vertical scale.

Magnetic Maps

The general shape of the earth's magnetic field can be visualized by imagining a small but powerful magnet at the earth's centre with its axis

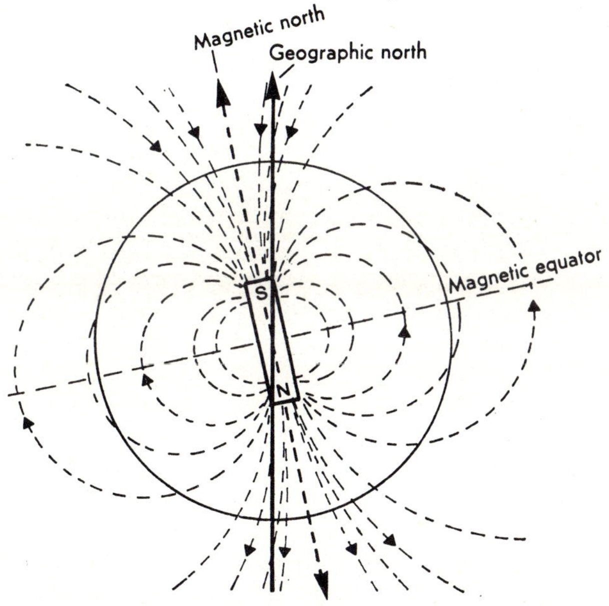

FIG. 414. *General shape of the earth's magnetic field.*

slightly inclined to the earth's axis of rotation, and its south pole pointing northwards (Fig. 414). (It must be emphasized that the existence of such a magnet is impossible.) The lines of force of such a magnet are shown dotted. At points on the great circle passing through both magnetic and geographical poles the magnetic variation will be zero (or 180°). On the magnetic equator,

FIG. 415. *The card of a mariner's compass is supported above its centre of gravity by pivoting it about the top of a small central dome. Supports for the inner gimbal can be seen to left and right of the compass box. One of the supports for the outer gimbal can be seen in the foreground.*

which is slightly inclined to the geographical equator, the dip is zero. On magnetic maps, lines drawn through places of equal dip (*isoclinic lines*) and lines drawn through places of equal variations (*isogonic lines*) are plotted. Isoclinic lines are very approximately lines of "magnetic latitude." There are many departures from this simple picture due to local deposits of magnetic rock. Isogonic lines are of great importance in the navigation of both ships and aircraft because the magnetic compass is still the standard instrument for direction finding.

In addition to varying with position on the earth's surface, the magnetic elements such as inclination and declination vary with time. The largest of these variations, the *secular variation*, is caused by the slow rotation of the magnetic poles round the geographical poles. At the present rate of change it appears that a complete cycle will take about 500 years. Besides these steady changes there are sudden and violent changes called *magnetic storms*. These are connected with sunspots and a vivid display of the Aurora Borealis.

Mariner's Compass

To eliminate the effect of pitching and rolling of the ship, the mariner's compass is mounted in gimbals, which provide two axes of rotation at right angles (Fig. 415). The compass box is thus held horizontally by its own weight. The compass itself consists of a set of magnetic needles fixed to a light aluminium "card" (Fig. 416) which is pivoted on a jewelled bearing. The card is marked in degrees. To overcome the tendency of the magnetic needles to dip, the card is supported above its centre of gravity so that its weight tends to keep it horizontal. It is mounted in a bowl containing liquid to damp its oscillations.

During the construction of a ship or an aircraft it lies in the earth's field,

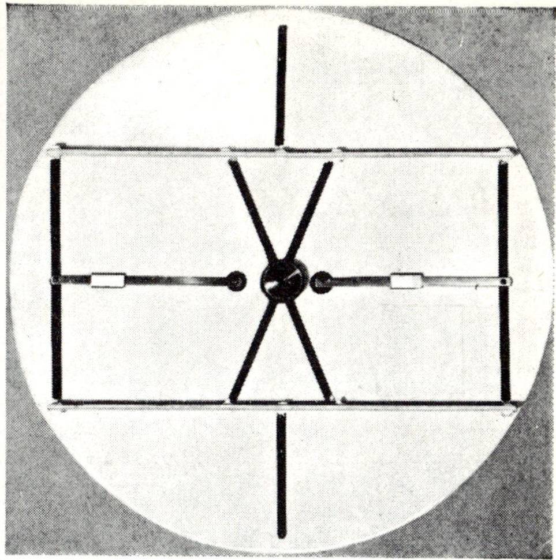

FIG. 416. *Compass card front and back (showing arrangement of magnetic needles).*

and the hammering it receives causes steel parts to develop a weak permanent magnetism. The effect of this on the compass is compensated for by suitably placed magnets in the compass mounting or *binnacle*.

QUESTIONS

1. Define the following terms: line of magnetic force; magnetic meridian; geographical meridian; magnetic variation; dip.

2. Given a wire through which an electric current flows, how would you attempt to obtain a point in space near the earth's surface where there was no magnetic field, and how would you locate this point?

3. A circular disc is magnetized parallel to a diameter. How could you determine, using the disc, (*a*) the direction of the magnetic meridian, (*b*) the direction in which the disc is magnetized?

4. A dip circle is rotated about a vertical axis at a place where the dip is 70°. Describe how the dip needle moves during one revolution.

5. A magnet, floating on a cork in a bowl of water, turns to point in the direction of the magnetic meridian but does not move either north or south. What does this show?

6. When iron filings are placed in a magnetic field they become tiny magnets by induction. How does this account for the use of filings in demonstrating the lines of force of a magnetic field?

7. When a compass is moved from the laboratory to a place outside the building, the direction in which it points often changes slightly. Why is this?

8. A current flows in a wire. How could you find its direction without breaking the wire?

9. In a compass the needle is horizontal although the field dips down into the earth. Why is this?

CHAPTER 33

ELECTRIC CURRENTS

THE various devices for the production of electricity so far discussed have mainly been of the kind which produce a static charge, and we have regarded an electric current as a flow of electrical charge through the material of a conductor. Even the most elaborate forms of this kind of device only produce minute quantities of electricity. If we require electric currents large enough to light lamps or drive machinery, batteries or dynamos must be used.

Effects of an Electric Current

If we take a torch battery and connect it to a torch bulb, the bulb lights up. This is because the wire filament inside the bulb is heated till it glows white hot. The connecting wires do not become appreciably warm, though a little heat is produced in them too. This illustrates that a fundamental property of an electric current is to *heat* a conductor through which it flows. As we shall see later, the amount of heat produced in a conductor depends on its material and on the size of the current flowing. This also illustrates the fact that an electric current can be used to transfer energy from one place to another along conducting wires. The energy liberated by the chemical changes in the battery reappears as heat and light energy in the filament of the torch bulb.

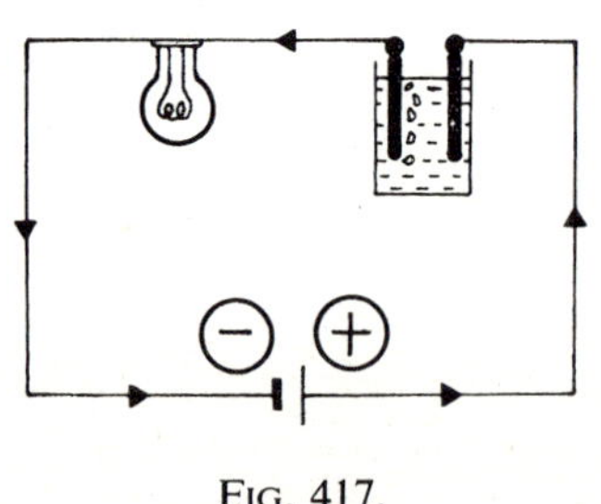

FIG. 417.

We have already seen that an electric current produces a magnetic field. This is known as its *magnetic effect*.

Electric currents also produce *chemical effects*, as is shown by the following simple experiment. If in the last experiment the circuit is broken and two copper plates dipping in a glass of water are put in between one pole of the battery and the lamp (Fig. 417), the lamp does not light, showing that little or no current can flow through the water.

If now a small handful of common salt is added to the water, as it dissolves the lamp will begin to glow showing that a current is flowing through the salt solution. After a while, bubbles of gas will appear on the plate connected

FIG. 418. *The earliest form of battery, built at the beginning of the nineteenth century, consisted of plates of zinc surrounded by plates of copper, immersed in a vessel of sulphuric acid. Positive terminal wire (right) is connected to the copper plate surrounding the first zinc plate which is connected to the second copper plate, and so on. The negative terminal is connected to the final zinc plate (left). This type of battery of simple voltaic cells suffers from two important disadvantages (polarisation and local action) which were overcome with the introduction of the Daniell cell in 1836.*

to the negative terminal of the battery and the solution becomes coloured near the plate.

It appears that when electricity flows through the salt solution, chemical decomposition occurs. Thus the chemical changes occurring in the battery to produce a current are, so to speak, producing chemical changes elsewhere in the salt solution.

Does the battery continually manufacture electric charge in some way? The answer to this question is no. Since there is no accumulation of electric charge anywhere in the circuit, as much charge flows into the battery at one pole as flows out at the other. The function of the battery is to keep the charges moving. Thus we may imagine the current in the circuit to be similar to the flow of water along a pipe. The battery would then be represented by a pump which collects the water flowing out at one end and pushes it in at the other.

SOURCES OF ELECTRICAL ENERGY

Our main source of electrical energy is the generator or dynamo. This will be dealt with in detail later (page 439). We shall now consider some simple forms of battery.

One of the earliest forms of battery consisted of plates of copper and zinc immersed in dilute sulphuric acid. The copper plate was the positive pole and the zinc negative. If both metals were pure, no action occurred until they were joined by a conducting wire. When so joined, the zinc dissolved in the acid and bubbles of hydrogen appeared on the copper plate. Such cells had a serious disadvantage in that the current they produced fell, as hydrogen bubbles developed on the copper. This phenomenon was called *polarisation.* It can be eliminated if some way can be found to eliminate the hydrogen bubbles. If, as is usually the case, the zinc used contains iron as an impurity, the zinc plate will dissolve in the acid even when there is no connection between the zinc and copper. This phenomenon, known as *local action,* is also electrical. The little pieces of iron embedded in the zinc, together with the acid, form a cell, like a simple cell, in which there is iron instead of the copper. As the iron and zinc are touching, a current flows, the zinc dissolves, and hydrogen bubbles appear on the iron. Local action can be prevented by amalgamating the surface of the zinc with mercury.

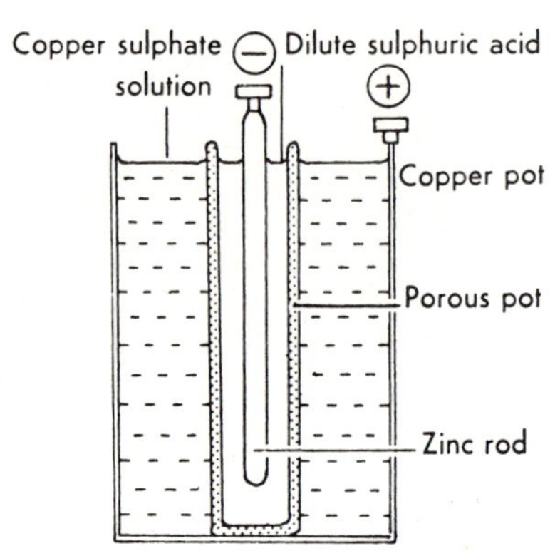

FIG. 419. *Daniell cell.*

There are two batteries in which polarisation has been eliminated in common use in elementary laboratories. These are the Daniell cell and the Leclanché cell.

Daniell Cell

A zinc rod is immersed in dilute sulphuric acid in a porous pot of earthenware. This pot is immersed in copper sulphate solution contained in a copper pot (Fig. 419). When the cell provides a current, the zinc dissolves; but copper and not hydrogen is set free at the outer copper pot so eliminating polarisation. This cell can maintain a small steady current for many hours, but cannot be left made up indefinitely as the acid and the copper sulphate gradually mix through the pores of the pot.

Leclanché Cell

A zinc rod, the negative pole, is immersed in a solution of ammonium chloride in water. The positive pole is a carbon rod which is packed round with a mixture of powdered carbon and manganese dioxide in a porous pot. This pot dips into the ammonium chloride solution (Fig. 420). When the cell is providing a current, the zinc dissolves forming zinc chloride. Ammonia and hydrogen are produced at the positive carbon pole. The ammonia escapes or dissolves in the solution which seeps through the pot and the hydrogen is oxidized to water by the manganese dioxide. If too large a current is taken continuously from this cell the manganese dioxide cannot oxidize the hydrogen fast enough and polarisation occurs. If a current is taken only occasionally and for a short while, as in a household bell circuit, the cell gives good service. It can be left for years without any attention.

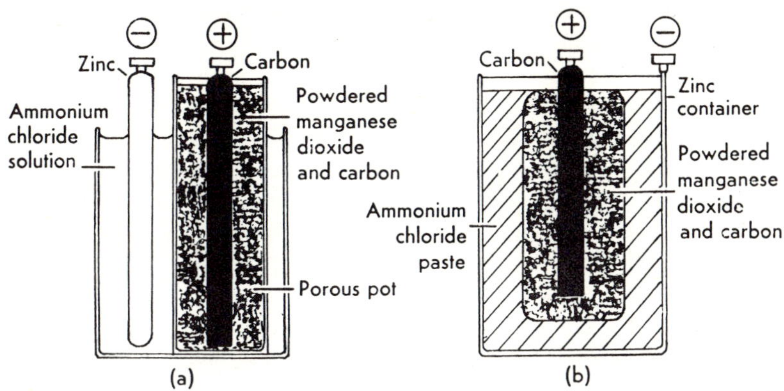

FIG. 420. *Leclanché cell.* FIG. 421. *Dry cell.*

Dry Cell

The most common form of the Leclanché cell is the "dry battery." This is used in torches and high tension batteries. This cell has a zinc container with ammonium chloride made into a paste with flour and gum inside it. The carbon rod and the powder of carbon and manganese dioxide are in a muslin or canvas bag which is inside the zinc container (Fig. 421). The cell is sealed with pitch to prevent evaporation and there is usually a small hole to permit the escape of ammonia. In an electric torch, the carbon rod, which is usually capped with brass, touches the bottom of the bulb to which one end of the filament is connected. The other end of the filament is connected to the metal case of the bulb and the circuit is completed, through the metal of the torch case to the zinc of the cell.

Lead Accumulator

The cells so far described are called *primary cells.* When the chemicals in them are exhausted, the cell cannot be recharged. Another type of cell, the lead accumulator, used in car batteries, can be recharged when it is run down by sending an electric current through it in the opposite direction to that in which it delivers current.

The lead accumulator's positive plate is a lead grid into which are compressed oxides of lead, mainly lead peroxide. The negative plate is a lead grid with spongy (i.e. porous) lead compressed into the spaces. The plates dip in dilute sulphuric acid. During discharge, lead sulphate is formed on both plates, and the density of the acid falls because it is used in the chemical reactions that go on. On recharging, the plates revert to their original state and the density of the acid rises. The state of charge or discharge of the cell can be measured by a hydrometer which measures the density of the acid, because the density rises and falls in proportion to the charge which has passed through the cell.

These cells should not be allowed to remain in a discharged state or the lead sulphate deposit becomes hard and white and the cell cannot be recharged. Owing to evaporation, the liquid level in the cell sinks. When this occurs, distilled water must be added to keep the plates covered. When the charging process is complete, bubbles of gas begin to form on the plates. The cell should not be charged beyond this point, unless the charging current is small, or the gas bubbles may loosen the deposits on the plates. If the cell is discharged too quickly, i.e. if too large a current is taken from it, heat is developed and the plates may be damaged. Similarly, the cells should not be charged with too large a current. Usually the makers specify the permissible charging and discharging currents on the label.

THE MEASUREMENT OF ELECTRICITY

The unit of electric current is called the *ampere* (symbol A) and an instrument for its measurement is called an *ammeter*. We shall see later how the unit is defined exactly (page 432) and also how an ammeter works (page 422). When a battery's poles are joined by material which is an electrical conductor, an electric current flows. This forms a circuit. The direction of the current is from the positive pole of the battery through the conducting material to the negative pole. This is the direction in which positive charges move. In many conductors, e.g. metals, it is the negative charges only which move, and these flow from the negative pole of the battery, round the circuit, to the positive pole. We must therefore distinguish between the direction of the current (the direction in which positive charges *would* flow) and the direction in which the negative charges do flow. Of course, though the current flows from positive to negative in the circuit outside the cell, it flows from negative to positive inside the cell.

Conventional Signs Used in Electrical Circuits

A single cell is represented by a long thin and a short thick line (Fig. 422*a*). The long thin line represents the positive pole, and the short thick line the negative pole.

A conducting wire that offers little or no resistance to the flow of current through it, that is one in which little or no heat is developed when a current flows and which therefore absorbs no appreciable energy, is represented by a full line. If one such wire crosses another without touching it, a little bridge can be shown in one wire (Fig. 422*b*). A more modern convention, however, is shown in Fig. 422*c* and Fig. 422*d*, for wires not connected and wires connected, respectively.

A conductor that does develop heat when a current flows is said to possess *electrical resistance.* Energy is produced as heat when a current flows in such

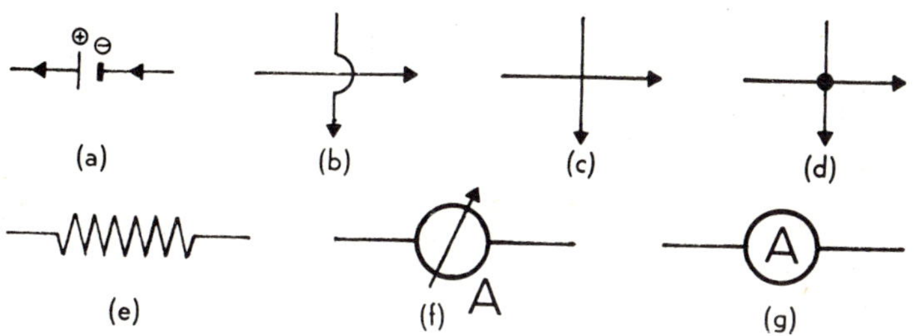

FIG. 422. *Some of the conventional signs used in electrical circuits.*

a conductor. It is represented by a zig-zag line (Fig. 422*e*). Some wires that possess this property are made from tungsten (used in electric lamp filaments), and "nichrome" (used in the heating elements of fires).

An ammeter is usually represented by a circle with an arrow through it and the letter A (Fig. 422*f*), or simply by a circle enclosing the letter A (Fig. 422*g*).

The circuit in Fig. 423 shows a cell sending a current through three resisting conductors *P*, *Q* and *R*, the connexions being made by wires of negligible resistance.

It cannot be too strongly emphasized that the current flowing through any one part of this circuit is the same as that flowing through any other. If the circuit is broken anywhere and an ammeter inserted the pointer will indicate the same current. It is clear that this must be true for when the current is flowing steadily there is no accumulation of electric charge at any part of the circuit. Hence, the charge flowing past *any* point in the circuit in a given time when a steady current flows must be the same.

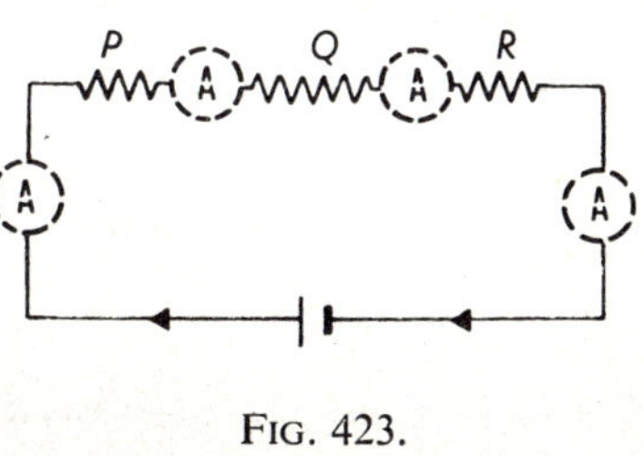

FIG. 423.

The quantity of charge passing any point in a circuit in one second when a current of one ampere is flowing has been given a special name, the *coulomb* (symbol C). If a current I is flowing in a circuit, a quantity of electricity I will flow round the circuit in one second and a quantity $I \times t$ (measured in coulombs) will flow in a time t.

Electromotive Force (e.m.f.)

If a single lead accumulator cell is connected to an appropriate torch bulb, the bulb will light fully. In this circuit practically all the energy provided by the cell is converted into heat and light in the bulb. Very little heat is developed elsewhere.

The heat energy liberated in the bulb could be measured by immersing the whole bulb in an electrically insulating liquid such as paraffin oil, and measuring the rise in temperature produced in a known time. Knowing the mass and specific heat capacity of the liquid and the heat capacity of the calorimeter containing the liquid, the rate of production of heat in joules/second, i.e. watts (symbol W), can be calculated. This is the rate at which the battery is providing energy. Suppose that this is done and that the result of the calculation is P for a single bulb attached to a single cell (Fig. 424*a*).

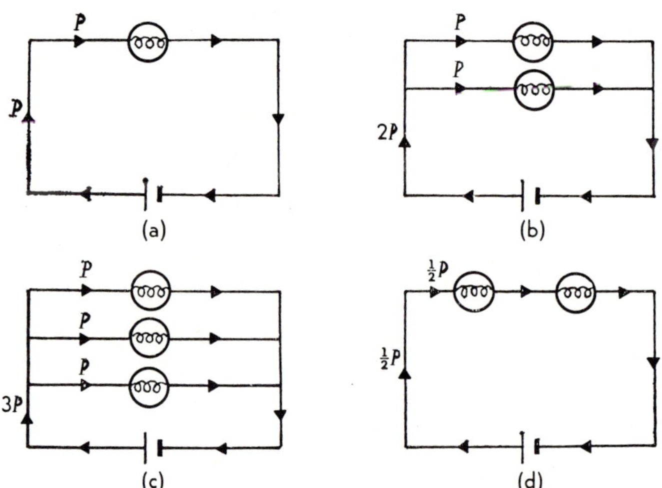

FIG. 424. *Identical light bulbs connected in series and parallel.*

Now suppose two identical bulbs are connected to the same battery as shown in Fig. 424*b*, the two bulbs are said to be connected *in parallel*. Each bulb lights up fully and now the rate of conversion of energy into heat in each bulb is P. Since each bulb uses the same current as the single bulb did, the battery is now providing twice the current that it provided when only one bulb was used. Thus the battery is providing twice the power, but since the charge it delivers every second is also doubled then the energy provided per coulomb is the same as before.

In the same way if three bulbs are connected (Fig. 424*c*) each bulb lights fully and the rate at which energy is drawn from the battery is $3P$. The current or flow of charge from the battery is now three times what it was with a single bulb, and once again the energy provided by the battery per coulomb of charge remains the same.

Two bulbs are now connected to the battery *in series* (Fig. 424*d*). Neither of the bulbs will light fully and experiment would show that the power produced in each bulb is now $P/4$. Power produced by the battery is now $P/2$, compared with P in the first experiment. By connecting an ammeter into the circuit it will be found that the current flowing is a half of that which flowed in the first experiment. Thus again the energy produced per coulomb of charge delivered is the same.

If three bulbs are now connected to the battery in series the power developed in each is $P/9$ and the power produced by the battery is $P/3$. The current flowing in the circuit is now one-third of that which flowed in the first experiment.

These somewhat idealized experiments illustrate an important characteristic of a cell. *A cell provides a fixed number of joules of energy for every coulomb of electricity it delivers to a circuit.* This characteristic is called the *electromotive force* (*e.m.f.*) of the cell. The name is unfortunate, *electromotive energy* would perhaps have been better.

This characteristic number measures the e.m.f. of the cell in *volts* (symbol V). The lead accumulator that we have been using has an e.m.f. of very nearly 2 volts. This means that for every coulomb of electricity that the cell provides, 2 joules of energy are delivered by the cell. The Daniell cell has an e.m.f. of about 1·1 volts, and the Leclanché cell an e.m.f. of nearly 1·5 volts. The e.m.f. of a cell depends on the chemicals used in construction and not on size. The advantage of having a large cell lies in the maximum current it can provide. This will be explained in more detail later (page 416).

Potential Difference

In the circuit consisting of a single cell and a lamp, heat production is almost entirely confined to the lamp filament. In the connecting wires and in the battery itself a very small amount only is developed. If we consider any two points in the circuit the energy liberated in joules when one coulomb flows between them measures what we call the *potential difference* (*p.d.*) between these points. This too is measured in volts.

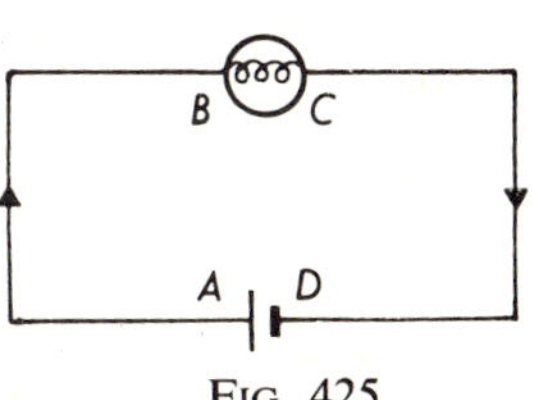

FIG. 425.

In the circuit shown in Fig. 425 the p.d. between A and B, or between C and D, is negligible. The p.d. between B and C is as near as makes no difference 2 volts, which is the e.m.f. of the cell. In Fig. 424*b* and Fig. 424*c* of the preceding section the p.d. across each lamp is 2 volts, since the same heat is developed in each lamp of the combinations as was developed in the single lamp. In Fig. 424*d*, however, the p.d. between the terminals of each lamp is 1 volt. This is because the heat developed in each lamp per coulomb of charge is only half what it was in the case of the single lamp. Similarly, in the case of three lamps in series, the p.d. is $\frac{2}{3}$ volts across each lamp.

It will be noticed that the sum of the p.d.s across all the parts into which the circuit is divided must necessarily be equal to the e.m.f. of the cell. The total energy set free in all parts of the circuit when one coulomb flows round it is the e.m.f. of the cell.

Conventionally we regard the "potential" as highest at the positive pole of the cell, and we say that the potential "falls" or "drops" as we go round the circuit to the negative pole of the cell. Suppose a 2-volt cell is connected by copper wires *AB* and *CD*, in which negligible heat is produced, to a length of uniform iron wire *BC* in which heat is produced (Fig. 426). The p.d. across *AB* and *CD* is negligible. We would express this by saying that the fall of potential or potential drop along *AB* and *CD* is negligible. The whole potential difference of 2 volts occurs between *B* and *C*, i.e. the potential falls by 2 volts from *B* to *C*. Since *BC* is a uniform wire, the same amount of heat will be liberated across each unit length and hence the fall of potential across each unit length will be the same. If *BC* is l long, the fall of potential across each unit length will be $2/l$ volts. Thus the p.d. between *B* and a point *P*, distance x from *B*, will be $2x/l$ volts.

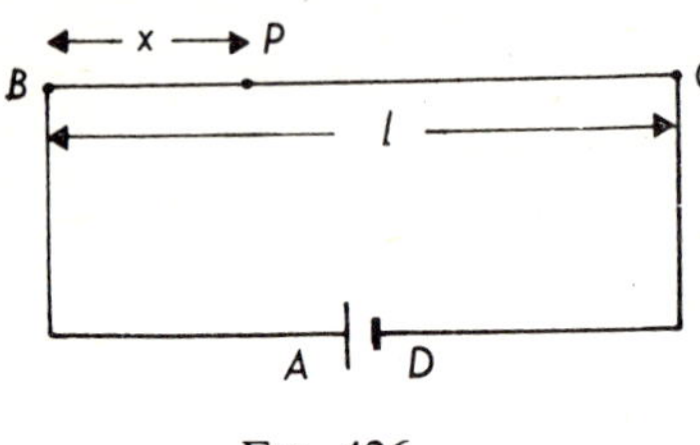

FIG. 426.

Cells in Series and Parallel

If a number of cells are connected as shown in Fig. 427*a* so that the same current flows through each cell in turn, the cells are said to be connected *in series*. When this is done, the cells act as a single cell in which the total e.m.f. is the sum of the separate e.m.f.s of the individual cells. If any cell

FIG. 427. *Three cells connected* (a) *in series and* (b) *in parallel.*

(say *B*) were to be reversed, it would act in opposition to the other two, and the resulting e.m.f. would be the sum of the e.m.f.s of *A* and *C*, minus that of *B*.

If a number of *similar* cells are connected as shown in Fig. 427*b*, with all their positive poles connected together and all their negative poles together, they are said to be connected *in parallel*. The cells then act as a single cell whose e.m.f. is the same as that of any *one* cell. The difference here is that if there are *n* such cells in parallel, each cell only contributes $1/n$ of the total current.

As we have said when dealing with the lead accumulator, there is a limit to the current which can be taken from a particular cell if damage to the cell is to be avoided. (This maximum current depends on the size of the cell.) Hence, if small currents and a large number of volts are required, cells are connected in series. If large currents are required and volts are not so important the cells are connected in parallel. If both large currents and high voltages are required several series sets can be connected in parallel (Fig. 428). If the maximum current for each cell is 5 A and if each cell has an e.m.f. of 2 V, the arrangement shown would have an e.m.f. of 6 V and could deliver 15 A.

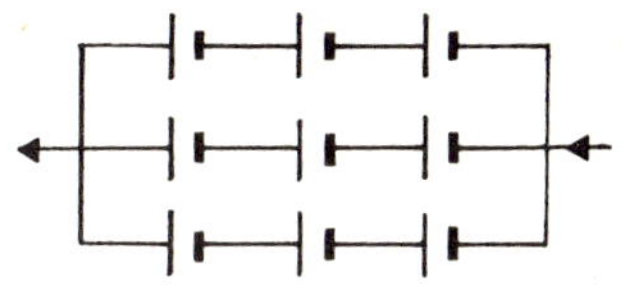

FIG. 428.

Units of Energy and Power

The unit of energy (or of work, i.e. energy transferred) is the *joule* (symbol J).

Power is the rate of supply, or transference, of energy or the energy transferred (i.e. the work done) in unit time. The rate of doing work of 1 J/s has a special name: it is called the watt (symbol W). A larger unit of power often used is the *kilowatt* (symbol kW): this is the rate of doing work of 1000 W or 1000 J/s. Another unit of power you may meet is the *horse-power*. This is equivalent to 746 W, so that a 1-kilowatt electric fire is producing heat energy at approximately 1·3 H.P.

Another unit of energy is the kilowatt-hour. This is the energy obtained from a source of power of 1 kilowatt during one hour. Since there are 3600 seconds in an hour, and a kilowatt is a rate of energy-production of 1000 joules in one second, a kilowatt-hour is a quantity of energy of 3 600 000 joules. Electric energy is marketed by this unit (which used to be called a Board of Trade Unit). Thus for the price of one unit the electricity company supplies 3 600 000 joules of energy.

Power in an Electrical Circuit

If a cell has an e.m.f. of E, then for every coulomb of charge it pushes round a circuit, an amount of energy E is provided in the circuit. If a current I is being delivered by the cell, a quantity of electricity I is flowing round the circuit every second (from the definition of a coulomb). Thus the power provided by the cell is EI. Similarly, if the p.d. between two points in a circuit is V the power given out in this part when a current I flows is VI.

Heat Produced in an Electrical Circuit

We imagine the cause of heat production when electric charges flow through a conductor to be somewhat akin to friction. When a body is pushed over a rough surface there is resistance to its motion and heat is produced. In much the same way, when electric charge flows through a conductor it experiences resistance to its motion and heat is produced. Some materials offer more resistance to the flow of charge than others; copper offers very little; thick conductors offer less resistance than thin ones of the same material as there is more "room" for the charge to flow.

Conductors in Series and Parallel

If lamps or any electrical conductors are connected end to end so that the same current flows through each in turn (Fig. 429*a*) the conductors are said to be connected *in series*. If they are connected so that the current on arriving at A splits up, part going through each conductor (Fig. 429*b*), they are said to be connected *in parallel*. This nomenclature is identical to that used in the case of cells. If the conductors are similar the current divides equally, but in general the current will not do so. When conductors are connected in parallel the p.d. is the same for each conductor. This could be verified by measuring the current in each branch with an ammeter and measuring the heat developed in each branch in a measured time. It would be found that the joules of energy obtained per coulomb of electricity was the same for all.

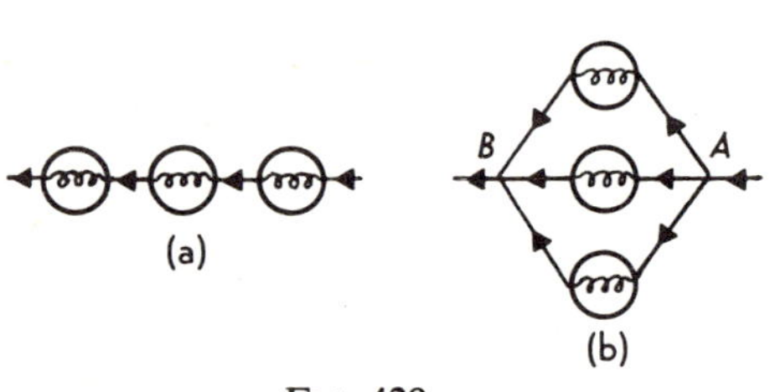

FIG. 429.

Wiring a House

In wiring a house for electricity (Fig. 430), all the electrical appliances are connected in parallel across the main supply so that each has the same p.d.,

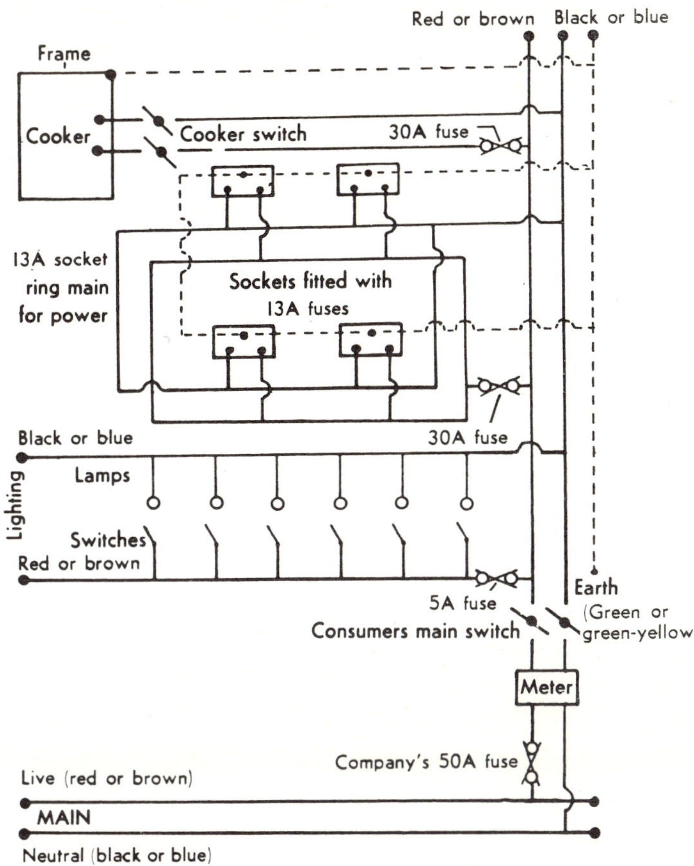

FIG. 430. *Electrical wiring diagram for a house.*

i.e. that of the main supply. Each appliance, whether it is a lamp, cooker, refrigerator, etc., is so constructed that the correct current flows when it is connected in this way. If lamps were connected in series they would not light up correctly. Each piece of electrical apparatus usually has a *fuse* attached for the sake of safety. This consists of a piece of easily-melted wire of the appropriate length and thickness, connected in series with the apparatus so that the current used by the apparatus flows through it. If through some fault the current increases beyond the safety level, that is, the point at which the heat developed in the leads becomes too large, the heat developed in the fuse wire is sufficient to melt it. This breaks the circuit and stops the current. For a lighting circuit a fuse which "blows" for a current greater than 2 A or 5 A is used. For power points a 13 A or 15 A fuse, and for heavy duty cookers a 15 A or 30 A fuse is used.

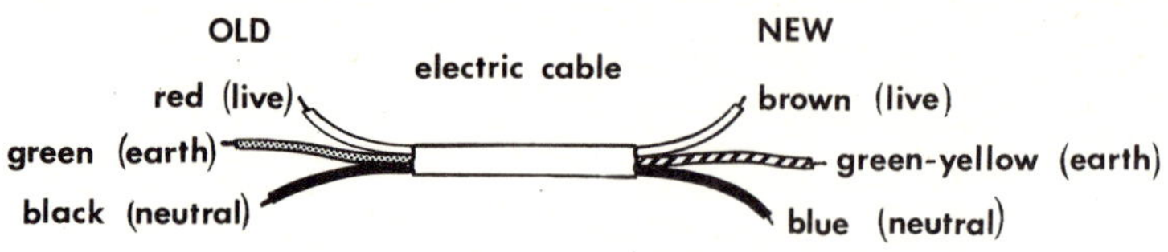

Comparison of old and new colours for electric leads.

One lead of the main supply, called the *neutral* lead, is usually "earthed". This is done by connecting it to a conductor buried in the ground. This is the blue lead of a wiring cable in the new colours. The other lead, called the *live*, is coloured brown. Since the neutral lead is earthed there is a potential difference between the live wire and the earth. Thus anyone touching the live wire experiences a current flowing through him to the earth. This experience is unpleasant and may even prove fatal if contact with earth is good. For instance, anyone in a bath makes excellent contact with earth. To minimize the risk of this, each house has an earth lead of its own. This is often obtained by connecting a wire to the metal water pipes. All metal casings of equipment are connected to this earth wire. This is the green-yellow wire in the three-wire cable used for wiring power supplies. If through some fault in insulation the metal casing becomes connected to the live wire a current flows to earth which should be big enough to blow a fuse and render the circuit dead. For lighting, only two wires are used so that light switches should not be made of metal. Nor should light switches be placed in such a position that somebody in a bath could reach them. Fuses and switches are all placed in the live wire lead or in both leads.

QUESTIONS

1. An electric toaster is labelled 400 watts, 250 volts. What current does it consume? What would be a suitable value for its fuse wire? How much would it cost to run for 20 minutes assuming that electricity were 3p per unit?

2. A dynamo supplying 10 A is driven by a 1·5 H.P. engine. Assuming perfect efficiency, calculate the e.m.f. of the dynamo. (Take 1 H.P. = 750 watts.)

3. An electric pump motor pumps 200 kg of water each minute from a well 15 m deep. If the motor is run from a 250 V supply and is perfectly efficient, calculate the current supplied to the motor. (Take $g = 10$ N/kg.)

4. A piece of apparatus requires 10 A at 60 V; a number of 2-volt cells each capable of providing 5 A is available. How many cells would be required? How would they be connected?

5. Find out from your local electrician a method of switching and wiring in order to obtain (*a*) two lamps controlled by one switch, (*b*) one lamp which can be switched on or off from two different places.

CHAPTER 34

RESISTANCE

THE circuit shown in Fig. 431 consists of a battery of lead cells in series, an ammeter "A" and a length of wire *PQ* which could conveniently be one metre of 28 S.W.G. Eureka wire. With a single cell, the current produced would be about 0·4 A but would depend somewhat on the ammeter used. Suppose it is exactly 0·4 A. It is found that if two cells are used, the current is 0·8 A. With three cells the current is 1·2 A and so on. Each cell produces a p.d. of 2 V, so that two in series produce 4 V and so on.

Ohm's Law

This experiment illustrates Ohm's Law which states that *the current through a conductor is proportional to the p.d. across the conductor provided the temperature remains constant.* In this experiment the conductor is a composite one consisting of the wire and ammeter in series. The heat produced is not enough to affect the temperature of the wires sufficiently to upset the proportionality between current and p.d. Therefore, since current and p.d. are proportional, the ratio p.d./current is constant for a particular conductor. This ratio is called the *electrical resistance* of the conductor. If the p.d. is measured in volts and the current in amperes the resistance is measured in *ohms* (symbol Ω). Thus the total resistance of the ammeter and wire in the above experiment would be $\frac{2}{0\cdot 4}$ or 5 ohms. This relationship is usually written:

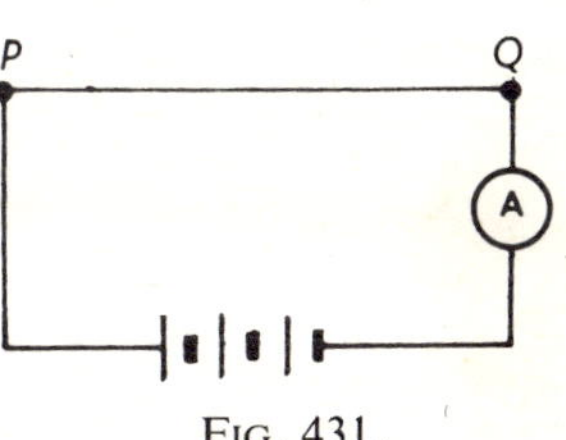

FIG. 431.

$$V = IR$$

where V is the p.d. in volts, I the current in amperes, and R the resistance in ohms of the conductor through which the current flows. It should be remembered that this relationship is *not* Ohm's Law.

Thus the p.d. across a conductor, whose resistance is known, when a known current flows through it can be calculated by multiplying current by resistance.

Resistances in Series

If in the experiment just described, another equal length of the same wire is connected in series with *PQ*, we find that the circuit resistance becomes

9·6 ohms; and if yet another similar length is added to these two the resistance becomes 14·2 ohms. Thus for each additional length of wire the resistance of the circuit increases by 4·6 ohms. This leads us to suppose that the ammeter resistance is 0·4 ohms. It also illustrates the fact that if resistors are connected in series, the total resistance is the sum of the separate resistances.

Resistors in Parallel

If two of the metre lengths of wire PQ and RS are connected in parallel (Fig. 432) the current is 0·74 A. This gives a value of $\frac{2}{0 \cdot 74}$ or 2·7 ohms for the resistance of the circuit. Allowing 0·4 ohms for the ammeter, the resistance of PQ and RS in parallel is 2·3 ohms. This is just half the resistance of a single wire. The current divides equally between the two wires, 0·37 A through each.

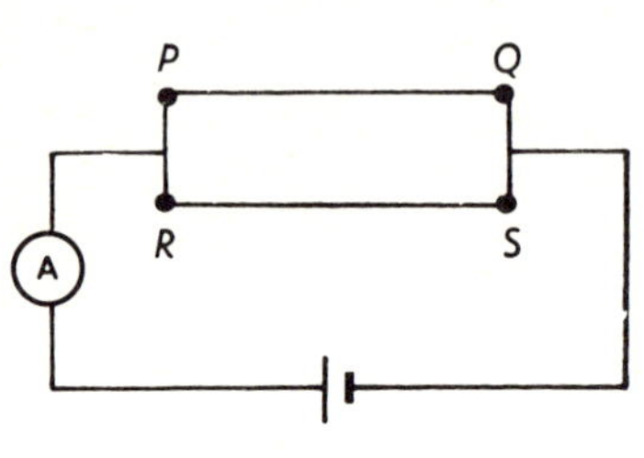

FIG. 432.

With three such wires in parallel, the effective resistance is one-third of that of a single wire and so on. In a general way it can be seen why the effective resistance should decrease as more resistors are connected in parallel, since more alternative paths are provided for the current to flow through.

If the resistors connected in parallel are not equal, the following formula can be used to calculate the effective resistance:

$$\frac{1}{R} = \frac{1}{R_1} + \frac{1}{R_2} + \frac{1}{R_3} + \text{etc.}$$

where R is the effective resistance of the resistors R_1, R_2, R_3, etc. in parallel.

Let us consider a formal proof of this equation. If I is the total current flowing into the junction X (Fig. 433a), and I_1, I_2, I_3, are the separate currents flowing in the resistances R_1, R_2, R_3 respectively, then

$$I = I_1 + I_2 + I_3$$

Let the p.d. between X and Y be V. This is the same for all three resistances. Hence:

$$\frac{V}{I_1} = R_1 \qquad \frac{V}{I_2} = R_2 \qquad \frac{V}{I_3} = R_3$$

$$\therefore I_1 = \frac{V}{R_1} \qquad I_2 = \frac{V}{R_2} \qquad I_3 = \frac{V}{R_3}$$

$$\therefore I = \frac{V}{R_1} + \frac{V}{R_2} + \frac{V}{R_3}$$

But if R is the effective resistance:

$$\frac{V}{I} = R \text{ or } I = \frac{V}{R}$$

$$\therefore \frac{V}{R} = \frac{V}{R_1} + \frac{V}{R_2} + \frac{V}{R_3}$$

$$\therefore \frac{I}{R} = \frac{I}{R_1} + \frac{I}{R_2} + \frac{I}{R_3}$$

The division of current between two resistors in parallel can easily be

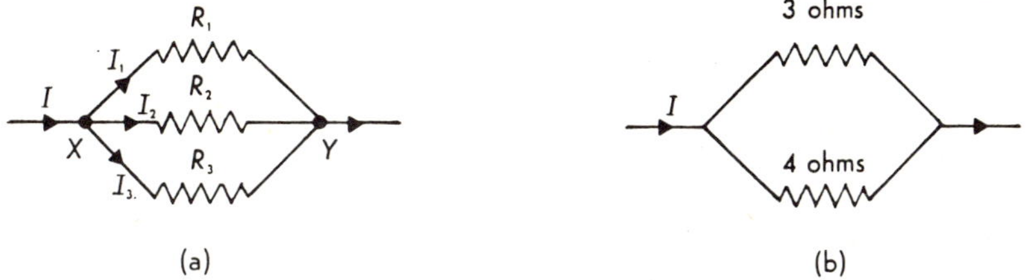

FIG. 433. *Resistances in parallel.*

calculated. As above $I^1R^1 = I^2R^2$, since the p.d. is the same for each resistor. Hence:

$$\frac{I_1}{I_2} = \frac{R_2}{R_1}$$

or the current divides in the inverse ratio of the resistances. In the example shown (Fig. 433*b*) $\frac{3}{7}$ of the total current I will flow in the 4 ohm resistor and $\frac{4}{7}$ of I in the 3 ohm resistor.

Voltmeters

Ammeters are constructed to have as low a resistance as possible so that when connected into a circuit they affect the current in the circuit as little as possible.

It is possible to construct an instrument similar to an ammeter, but with a high electrical resistance. Such instruments are used as voltmeters. Suppose that an instrument is made with a resistance of 100 ohms, and suppose it gives a full scale deflection of 10 divisions when a current of 0·1 A flows through it. To send this current through it a p.d. of $100 \times 0{\cdot}1 = 10$ volts is needed (Ohm's Law). Thus 10 volts is needed to produce a full-scale deflection of 10 divisions. If 2 volts were applied to the terminals of the instrument it would produce a deflection of 2 divisions. Thus the instrument could be used to measure p.d.s up to 10 volts.

Suppose a current is flowing through a resistor X in the part AB of a circuit (Fig. 434), and that we wish to measure the p.d. between A and B. We could of course calculate this by multiplying current by resistance if they were known. If they are not known the p.d. can be measured by connecting a voltmeter in parallel with AB. Since the p.d. across resistors in parallel is the same, the p.d. across the voltmeter will be the same as that across X, and the voltmeter will read this p.d.

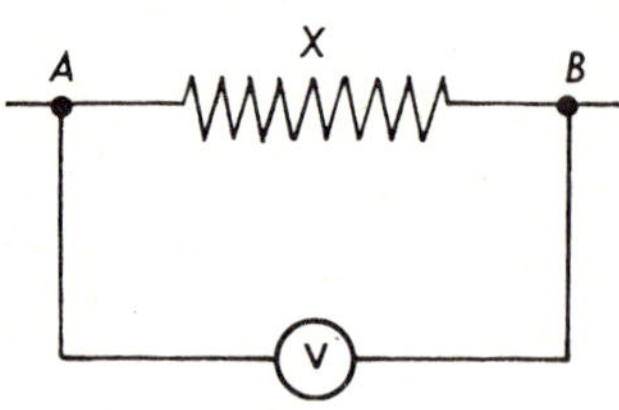

FIG. 434.

Since some current flows through it, the p.d. measured by the voltmeter will not be the same as it was across X before connecting the meter because the current through X will be changed by connecting the meter. To make this change as small as possible, the meter should have as high a resistance as possible so that very little current flows through it. This involves making a meter that will deflect for very small currents.

MEASUREMENT OF RESISTANCE

The circuit is set up as shown in Fig. 435. The resistor to be measured is R, while X is a variable resistor whose value need not be known.

The variable resistor X would normally be a *rheostat*. This consists basically of a coil of resistance wire wound on a porcelain tube as shown in the photograph (Fig. 437).

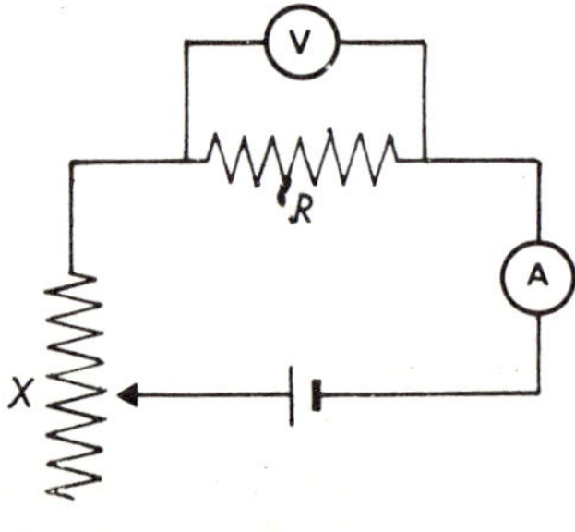

FIG. 435.

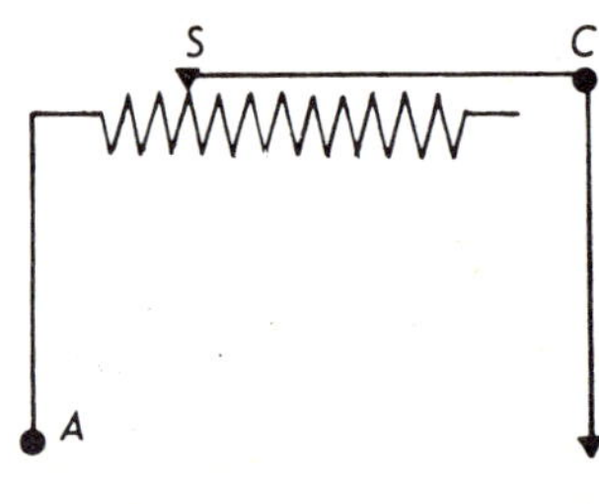

FIG. 436.

Current enters one end of the coil through a terminal A. A slider S, moving along a conducting bar, makes contact with the resistance coil. The current, which enters the coil through A, leaves through S and the terminal C of the conducting bar, so that, if S is moved to the right, the current has to flow through a greater length of wire and the resistance increases. A rheostat is

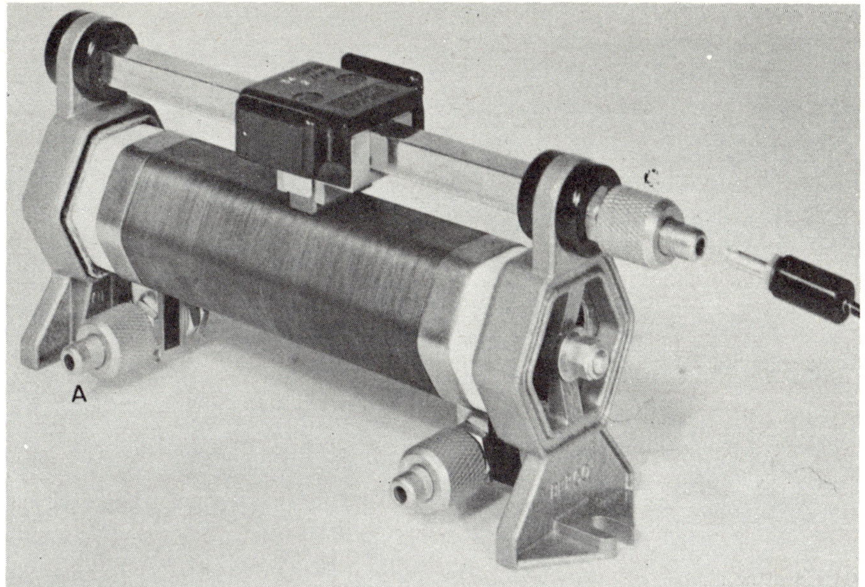

FIG. 437. *Rheostat; basically a resistance wire wound round a porcelain tube.*

diagrammatically represented in Fig. 436. Compare this with the photograph.

Returning to the circuit shown in Fig. 435, it is clear that the ammeter reading indicates the current flowing through R (if we neglect the very small current flowing through the voltmeter). The voltmeter reads the p.d. across R. The resistance of R is obtained by dividing the p.d. by the current. Several determinations can be made by varying the current by means of X, and the average found. This method is most suitable for resistors of a few ohms.

Substitution Method

This method can be used for measuring large resistances. The resistor to be measured X is connected in series with a very sensitive ammeter (galvanometer) and a cell (Fig. 438). The deflection of the galvanometer is read. X is now taken out of the circuit and replaced by a variable resistor of known value and the variable resistor is adjusted until the deflection of the galvanometer is the same as it was before. The variable resistor, which has a known value, is then the same as X.

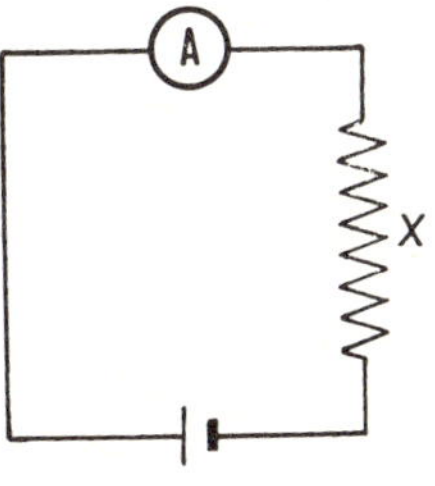

FIG. 438.

The variable resistor of known value as used in this experiment is known as a *resistance box*. Coils of wire of known resistance contained in the box have their ends connected to brass blocks on top of the box. The brass blocks can be joined by brass plugs which fit between them (Fig. 439). Current flowing in at A would have to pass through every coil before leaving at E. Hence the box would have a resistance of 10 ohms. If a plug were used to connect D with E, the current would not go through the coil of 5 ohms, it

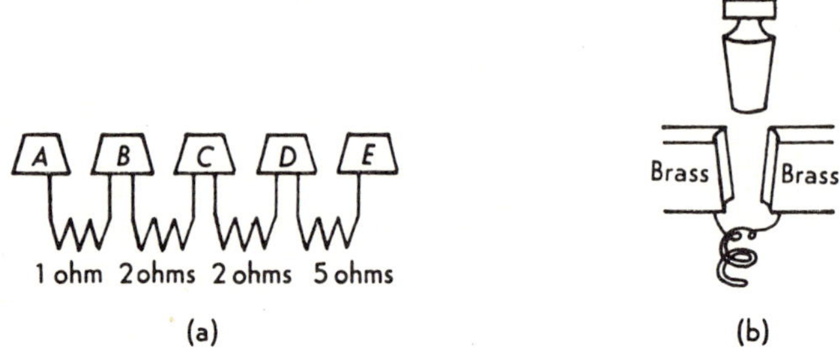

FIG. 439. *A resistance box consists of standard resistors connected to brass blocks* (*a*). *The brass blocks can be connected by brass plugs* (*b*).

would go straight from D to E. Hence the box resistance would be 5 ohms. It will be seen that any resistance in steps of 1 ohm between 0 and 10 ohms is obtainable with this box. Resistance boxes usually go up to a maximum of 10 000 ohms. The following resistance coils would be required for this: 1, 2, 2, 5, 10, 20, 20, 50, 100, 200, 200, 500, 1000, 2000, 2000, 5000 ohms.

Resistivity

Wires of the same length and thickness but made of different materials are found to have different resistances. In order to construct resistances of known value it is convenient to be able to predict what resistance a piece of wire of known length and thickness will have. This we can do if we know the *resistivity* of its material. The resistivity of a material is the resistance between opposite faces of a metre cube of the material. It is obvious that so thick a conductor will have a very low resistance and the resistivity will be a very small number.

Suppose the resistivity of a material is ρ. Since the resistance of a conductor of uniform cross section is proportional to its length, if we place l such cubes end to end (Fig. 440*a*), the resistance between end faces will be $l\,\rho$. If we now place two such sets of cubes side by side (Fig. 440*b*), since they are in parallel, the total resistance will be $l\,\rho/2$ and if A sets are in parallel the total resistance is $l\,\rho/A$. Thus the resistance R of a conductor of length l and cross sectional area A will be given by

$$R = \frac{l\,\rho}{A}$$

This enables us to calculate what the resistance of the conductor will be. In the case of wires, A will only be a small fraction of a metre^2, but the relationship still holds.

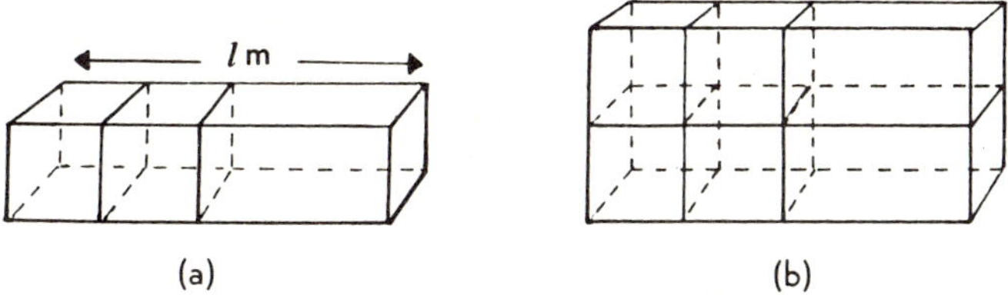

FIG. 440. *Theory of resistivity.*

The resistivity of a material is measured in the first place by rearranging the above relationship to:

$$\rho = \frac{RA}{l}$$

The resistance R in ohms of a measured length l in metres is then found. The area of cross-section A in metre^2 is calculated from the diameter of the wire and hence ρ is found. The unit of ρ is the ohm metre (Ω m).

Change of Resistance With Temperature

The resistance of most metal wires gets greater as they get hotter. This can be illustrated by the following example. The element of a 1-kilowatt electric fire consumes 1000 watts when connected to a 240-volt mains supply. Since watts are calculated by multiplying p.d. by current, the current I flowing through the element is given by:

$$240\,I = 1000$$

$$I = \frac{1000}{240}$$

$$= 4{\cdot}166 \text{ A}$$

$$\text{But } \frac{V}{I} = R$$

$$\therefore \text{ Resistance of fire} = \frac{240}{4{\cdot}166} = 57{\cdot}6\ \Omega$$

If the resistance of the fire is measured by the voltmeter and ammeter method, using a single 2-volt cell to provide the current, only a small current will flow and the element of the fire will not become hot. The resistance measured in this way will be found to be about 45 Ω. Most standard resistors are made from alloys such as "Eureka" or "Manganin". The resistance of a wire made from such alloys changes very little with its temperature.

Extension of the Range of Ammeters and Voltmeters

This is most easily understood by considering an actual example. Suppose that we have an ammeter whose scale has 15 divisions, each division corresponding to $\frac{1}{1000}$ A or 1 milliampere (symbol mA), and that the resistance of the ammeter is 5 ohms. The maximum current it can measure is 15 mA or 0·015 A. Suppose we wish to use it to measure currents up to 1·5 A, that is, we wish to use it to measure a current a hundred times as big as that which produces a full deflection.

This measurement can be made by connecting a resistor with a low resistance in parallel with the ammeter of such a value that $\frac{99}{100}$ of the current to be measured flows through the resistor and only $\frac{1}{100}$ through the

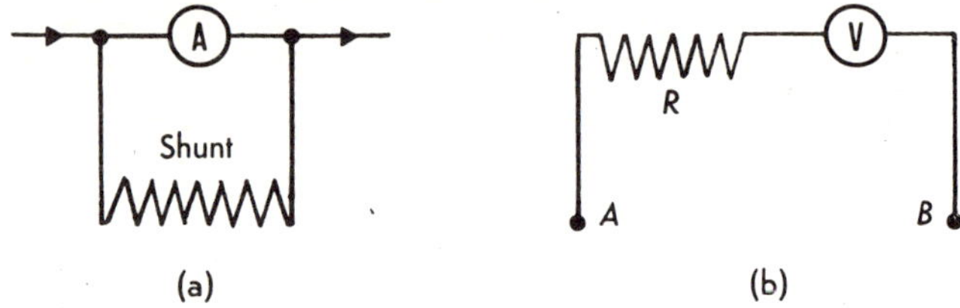

FIG. 441. *Use of resistors to extend the range of ammeters and voltmeters.*

ammeter (Fig. 441*a*). This parallel resistor is called a *shunt*. Its resistance must be therefore $\frac{1}{99}$ of the ammeter resistance, or $\frac{5}{99}$ ohms, since the current divides in the inverse ratio of the resistances. With the shunt in position, the scale readings will have to be multiplied by 100 to obtain the actual current. The effective resistance R of the ammeter with the shunt, can be calculated from

$$\frac{1}{R} = \frac{1}{5} + \frac{99}{5} = 20$$

$$R = 0{\cdot}05 \ \Omega$$

Thus the total resistance of the ammeter and shunt is much lower than that of the ammeter alone. This improves the instrument as a current measurer, for an ammeter should have as low a resistance as possible so that it does not appreciably reduce the current to be measured when it is inserted.

The original instrument used as a voltmeter would give a full-scale deflection when connected to a p.d. of 5 × 0·015 volts by Ohm's law, i.e. 0·075 volts or 75 millivolts (symbol mV). Suppose we measure p.d.'s up to 6 volts. Since a voltmeter should have as high a resistance as possible, a shunt is not used because this would reduce the resistance. We therefore put in series with the ammeter a resistor R of such a value that when 6 volts are connected across AB (Fig. 441*b*), the maximum deflection is produced. By Ohm's law, this resistance, together with the ammeter resistance, must pass 0·015 A.

when a supply of 6 volts is connected across the terminals *AB*.

$$6 = 0{\cdot}015\ (R + 5)$$
$$\therefore R = 395\ \Omega$$

Thus the resistance of the instrument is now 400 w. With the series resistor in place, each division on the scale will represent $\frac{6}{15}$ or 0·4 V.

Heat Produced in a Conductor When a Current Flows

We said earlier that the power expended in driving a current I through a conductor is VI if V is the p.d. across the conductor, and that this energy appears as heat in the conductor. If the conductor's resistance is R, by Ohm's law $V = IR$. Hence the rate of production of heat energy in the conductor is $I^2 R$. Thus *the heat produced in a wire is proportional to the square of the current passing through it.*

Again, the expression VI is equivalent to V^2/R, and thus *the heat produced in a wire is proportional to the square of the p.d. across it.* This means that if the current through a conductor is doubled, the heat produced per second becomes four times as great.

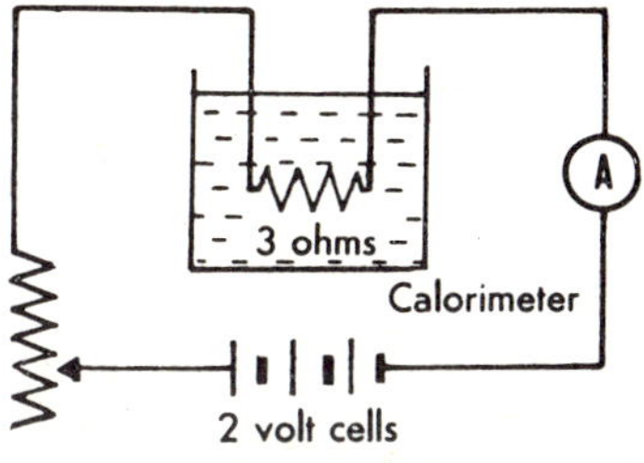

FIG. 442

This can be shown by immersing a coil of "Manganin" wire of resistance about 3 Ω, in about 50 g of water in a calorimeter (Fig. 442). The current is adjusted by the rheostat to 0·5 A and the rise in temperature in five minutes is measured by a thermometer. The water is allowed to cool, and the current increased to 1 A. Again the rise of temperature in five minutes is measured. It will be found to be four times as great as before. Similarly, if the current is raised to 1·5 A the rise of temperature in five minutes will be found to be nine times as great as in the first case. These numbers are only approximate unless the calorimeter is so well insulated that heat losses are negligible.

Internal Resistance of a Cell

The material between the poles of a cell has some resistance. This is called the *internal resistance* of the cell. Part of the e.m.f. of the cell is used up in driving the current through the cell itself. The volts which are so used (often called the "lost volts") are not available in the external circuit. Hence the p.d. of the cell is less than the e.m.f. With a lead accumulator, the internal resistance is very small, about 0·01 ohm, so the e.m.f. and p.d. are practically equal. Daniell and Leclanché cells may have internal resistances of a few ohms. The internal resistance limits the current available from the cell. If the poles are connected by a wire of negligible resistance (that is, the cell is short circuited) the only resistance in the circuit is that of the cell. The current flowing will then be the maximum possible. For example, if a Daniell cell has an e.m.f. of 1·1 volts and a resistance of 0·5 ohm, the current it produces when short circuited will be $\frac{1\cdot1}{0\cdot5} = 2\cdot2$ A. With a lead accumulator the short circuit current would be about $\frac{2}{0\cdot01}$ or 200 A. This current would be so great that the cell would suffer damage if the current flowed for more than a few seconds. Hence it is inadvisable to short circuit lead accumulators.

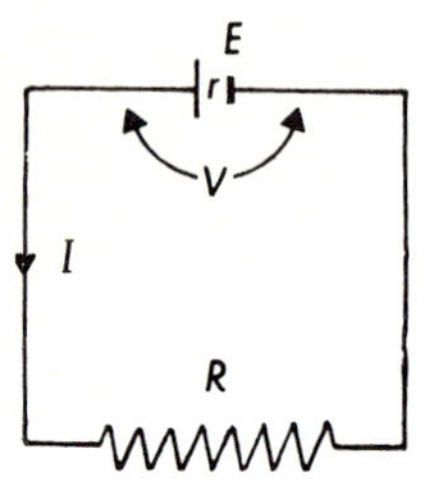

FIG. 443

In Fig. 443, a cell whose e.m.f. is E, and internal resistance r, sends a current I through an external resistance R. Since the p.d. necessary to drive a current through a resistance is obtained by multiplying current by resistance, the volts used in driving a current through the cell are Ir. Similarly if V is the p.d. of the cell, i.e. the volts used in driving a current through the external circuit $V = IR$. Hence the e.m.f. E of the cell, which is its p.d., plus the lost volts, is given by $E = IR + Ir$, or $E = I(R + r)$.

It will be noticed that the lost volts depend on the current. The smaller the current delivered by the cell, the fewer volts are lost in the battery, and the more nearly is the p.d. equal to the e.m.f. Hence the e.m.f. of a battery is often defined as the *p.d. between its terminals when no current is flowing*. If a voltmeter is connected to the terminals of the battery

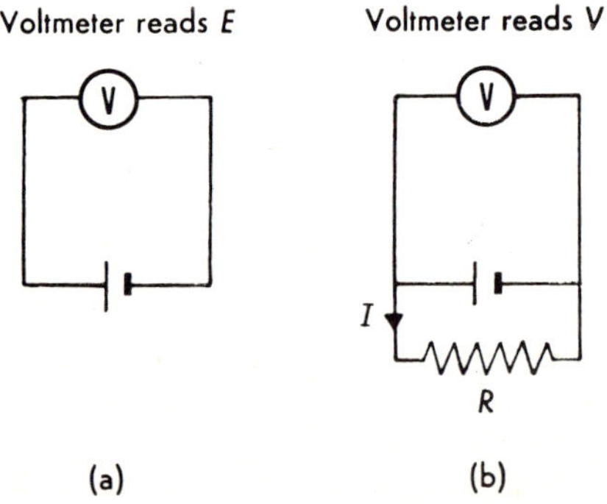

FIG. 444. *Method of measuring the internal resistance of a cell.*

it will read the p.d. If the voltmeter has a very high resistance, the current flowing will be small, and the p.d. will nearly equal the e.m.f.

It is worth mentioning that the larger the cell is in size, the smaller its internal resistance is. Thus a Daniell cell the size of a house would have an e.m.f. of 1·1 volts, but a very low internal resistance. Hence the maximum current it could produce would be much larger than the normal sized cell could produce.

The internal resistance of a cell can be measured if a high resistance voltmeter is available. This is first connected to the cell (Fig. 444*a*). Its reading gives the e.m.f. E of the cell nearly enough. A known resistor R of a few ohms resistance is now connected as an external circuit and the voltmeter again read (Fig. 444*b*). This time it records the p.d. of the cell V. The current I will be V/R. The lost volts are $E - V$, and since these are equal to Ir where r is the internal resistance, $E - V = rV/R$ from which r can be calculated.

Measurement of Resistance by the Wheatstone Bridge

The Wheatstone bridge is the most extensively used method for the accurate measurement of resistance. When a current divides between two

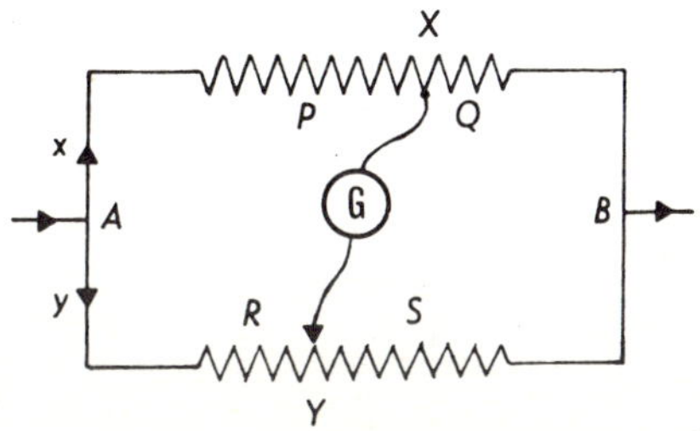

FIG. 445. *Wheatstone bridge.*

different resistors connected in parallel (Fig. 445), the p.d. is the same for each resistor. If X is a point somewhere on one of the resistors, there must be a point Y on the other resistor where the fall of potential from A to Y is equal to that between A and X. Since no p.d. exists between X and Y, a sensitive galvanometer connected between these points would show no deflection since no current flows. If x is the current flowing along AXB and y the current flowing along AYB, and if P, Q, R and S are the resistances between A and X, X and B, A and Y, and Y and B respectively, then since the p.d. across AX is the same as that across AY:

$$x \times P = y \times R$$

Again, since the p.d. across XB is the same as that across YB:

$$x \times Q = y \times S$$

Hence by division:
$$\frac{P}{Q} = \frac{R}{S}$$

OCL/PHYS—2D

This is called the principle of Wheatstone's bridge.

A simple apparatus using this principle is called the *metre bridge* (Fig. 446). AB is a uniform resistance wire 1 metre long. Either P or Q is the unknown

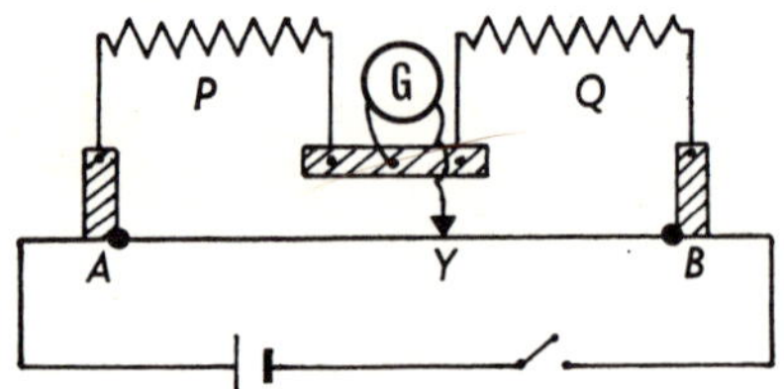

FIG. 446. *Metre bridge.*

resistance to be measured, the other is a known resistance. The contact Y is moved until the galvanometer G shows no deflection. Then:

$$\frac{P}{Q} = \frac{\text{Resistance of } AY}{\text{Resistance of } YB} = \frac{AY}{YB}$$

since the resistance of a uniform wire is proportional to its length.

For an accurate result, the known resistance should be chosen to obtain a balance point somewhere near the middle of AB.

Potentiometer

We have seen that the e.m.f. of a cell can be defined as the p.d. across its terminals when no current flows. When a voltmeter is connected to a cell, some current flows and hence the voltmeter does not measure the e.m.f.

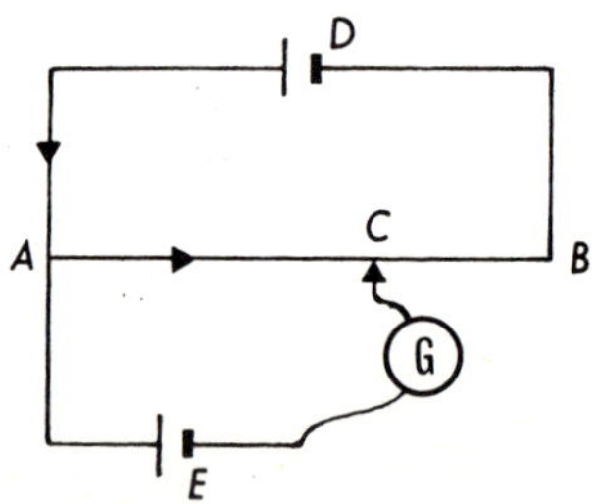

FIG. 447. *Potentiometer.*

accurately. The potentiometer principle provides a method of measuring e.m.f.s without taking any current. A lead cell D (Fig. 447) drives a steady current along the uniform resistance wire AB. If another cell E, of smaller e.m.f. than D, is connected as shown, there are two p.d.s; one due to the difference of potential between A and C trying to send a current in the direction AEC, and the other due to the cell E tending to drive a current in the opposite direction. If these are equal, the galvanometer G reads zero, and no current flows in AEC. Thus if the sliding contact C is moved so that G does not register a current, the p.d. between A and C is equal to the p.d. of the cell E. Again, since no current flows through E, the p.d. across it is equal to its e.m.f.

Since the fall of potential along a uniform wire carrying a current is proportional to the length, the e.m.f. of E is proportional to the length AC. If now the cell E is replaced by another cell S of known e.m.f. and the balance point C_1 found, the e.m.f. of S will be proportional to AC_1. Thus, provided the current through AB does not change:

$$\frac{\text{e.m.f. of } E}{\text{e.m.f. of } S} = \frac{AC}{AC_1}$$

from which the e.m.f. of E can be calculated. In practice, S is usually a standard Weston cadmium cell whose e.m.f. is 1·0183 volts at 20° C. This cell is ruined if an appreciable current is taken from it and hence a high resistance must be connected in series with it when the balance point is being found. When the balance point is known approximately, this high resistance can be removed or short circuited for the final adjustment.

QUESTIONS

1. A voltmeter of resistance 100 Ω is connected to a cell of e.m.f. 1 V and internal resistance 0·5 Ω. What will it read?

2. A battery consisting of 12 accumulators each of e.m.f. 2 V and negligible resistance is connected in series and is to be charged from 100 V main supply. If the maximum charging current permissible for the battery is 5 A, state what size of resistance must be connected in series with the battery. (The effective e.m.f. in the circuit is (100 — 24) V as the e.m.f. of the battery opposes that of the mains.)

3. An ammeter has a resistance of 1 Ω and gives its full scale deflection for a current of 2 A. State how you would adapt it to measure (*a*) currents up to 10 A, (*b*) p.d.'s up to 200 V.

4. If you wished to construct a resistor with a resistance of 5 Ω, and were provided with wire of a diameter 0·5 mm of material of resistivity 5×10^{-7} Ω m, what length of wire would be necessary?

5. A motor-car headlamp bulb is labelled 12 volts, 36 watts. What current does it take when it is switched on? What is its resistance when it is hot?

6. You have three resistors with resistances of 1, 2 and 3 Ω respectively. Show how by using some or all of these the following resistances can be obtained: 1, 2, 3, 4, 5, 6, $\frac{2}{3}$, $\frac{3}{4}$, $1\frac{1}{5}$, $3\frac{2}{3}$, $2\frac{3}{4}$, $2\frac{1}{5}$, $\frac{6}{11}$ Ω.

CHAPTER 35

APPLICATIONS OF THE ELECTROMAGNET

THE magnetic fields produced by currents have already been described, and before reading what follows it would be as well to re-read that section.

We saw that a piece of iron or steel becomes magnetized when placed inside a solenoid through which an electric current flows. If iron is used, it loses its magnetism when the current is switched off, and it is this property which leads us to use iron in the construction of electromagnets.

Since an electromagnet is used for attracting iron it is an advantage to have both poles assisting each other. Hence the magnet is made in the form of a horseshoe. Fig. 448 shows the direction of the current, the polarity of the magnet, and the induced poles on the iron armature that is being attracted.

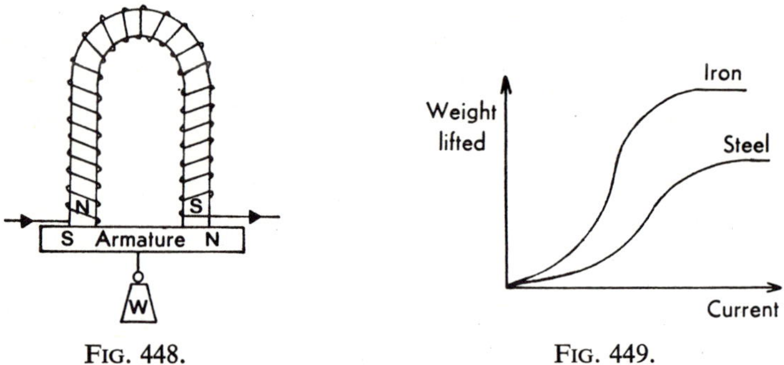

FIG. 448.

FIG. 449.

The strength of the magnet can be estimated by the total weight that can be held up by the armature. (It is necessary when making quantitative measurements to have a good fit between the poles and the iron armature, i.e. there must be no air gaps.) If gradually increasing currents are sent through the solenoid, the weight that can be held gets greater at first, but there comes a point at which further current increases produce hardly any increase in the lifting power. The magnet is then said to be saturated. Presumably the molecular magnets are then as much in line as they ever will be.

If a steel core of the same dimensions is used instead of an iron one, it does not become as strong a magnet even when saturated. Also the soft iron can be magnetized to a specified degree by a much smaller current than the steel requires. The curves (Fig. 449), a comparison of the magnetic properties of steel and iron, show the sort of behaviour that is observed.

Electric Bell

In Fig. 450, M is an electromagnet which attracts the soft iron armature A when a current flows. A is mounted on a piece of brass, fixed at P. The current flows through a fixed contact C which normally touches another contact carried by the brass strip to which A is fixed. When the current flows and M is activated, A is attracted and the contact at C broken. Thus the current cannot flow, M ceases to attract A, and the spring in the brass makes it return to its former place and so on. A clapper fixed to A strikes a gong G each time A is pulled towards the magnet. An electric buzzer works on the same principle but the attracted armature has no clapper attached and is made very light. It moves to and fro so quickly that it emits a note, in much the same way that a bee's wings emit a buzz when they vibrate quickly.

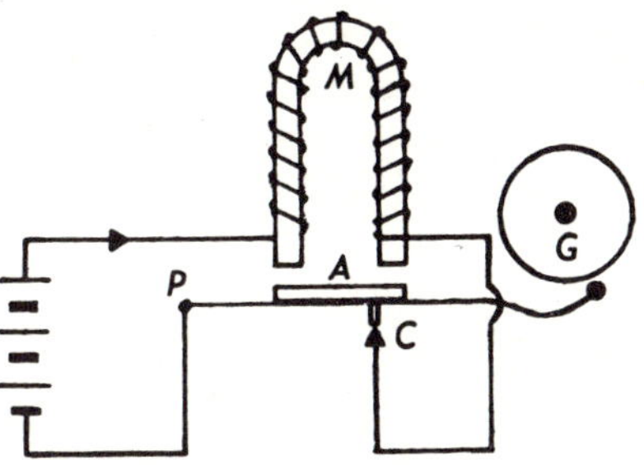

FIG. 450. *Electric bell.*

Relay

Suppose we wished to operate an electrical device requiring a considerable current by a switch placed some distance away. If the simple circuit (Fig. 451*a*)

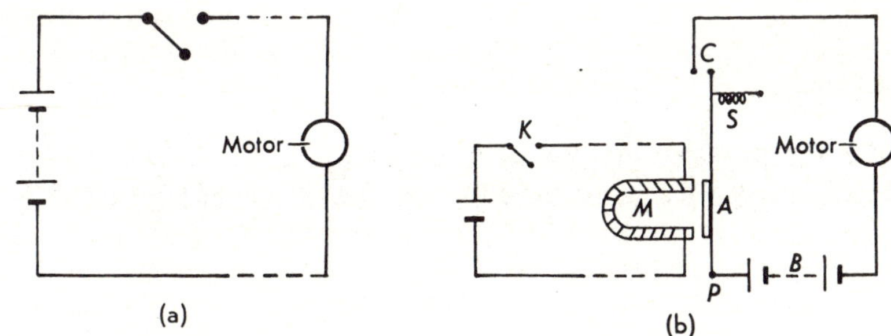

FIG. 451. *Relay circuit for switching on and off an electric motor.*

were used, the wires would have to be thick to carry the current needed, and this would be expensive. To avoid this, a *relay circuit* (Fig. 451*b*) could be used. This is a device by which a very small current is used to control a large one. The key K, when depressed, would allow a small current to flow in the coils of the magnet M. An armature A, held away from the magnet by a light spring *S* and pivoting at *P*, would be pulled to the magnet and make a contact at *C* which is normally open. This would complete the circuit of the larger battery *B* which is capable of operating the motor.

A similar device (Fig. 452) could be used to make a current cut itself off when it exceeded a certain value. The contact would normally be held closed by the spring *S*. If the current becomes great enough it pulls the armature *A* sufficiently strongly to break the contact. This device is sometimes known as a *cut-out*.

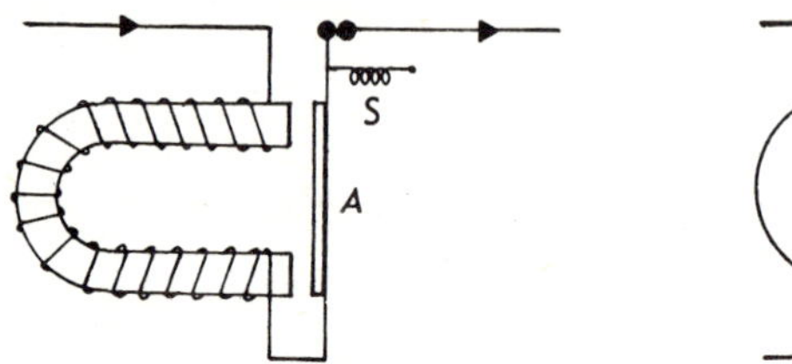

FIG. 452. *Cut-out.*

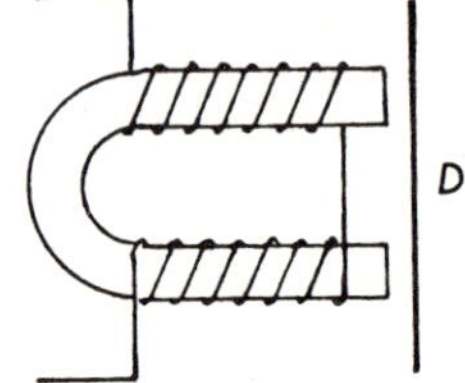

FIG. 453. *Telephone earpiece.*

Telephone Earpiece

The construction of this is best understood by taking one of a pair of wireless headphones to pieces. Coils of wire are wound on the poles of a permanent magnet (Fig. 453), and a thin iron diaphragm *D* is mounted very close to the poles. When fluctuating electric currents flow round the coils the magnet becomes stronger and weaker and its attraction for the diaphragm varies. This causes the diaphragm to move in and out, creating sound waves in the air which correspond to the variations in current. A permanent magnet is used rather than an electromagnet, but the reason for this is beyond the scope of this book.

Moving Iron Ammeters and Voltmeters

These are inexpensive instruments. They are of two distinct types. In the *attraction type*, use is made of the fact that if a piece of iron is placed with part of its length in a solenoid, when a current flows in the solenoid the iron is drawn in. If the iron is near the S-end of the solenoid (Fig. 454*a*), a north pole is induced on the end of the iron nearest the solenoid and the iron is attracted. If the current is reversed, the induced poles on the iron are reversed and the iron is still attracted. In Fig. 454*b*, the current to be measured is sent through the solenoid *S*. The iron *I* is attached to a pointer, pivoted at *P*, the other end of which moves over a scale. The moving part is made to balance about *P*, and a watch spring is used to control the movement of the pointer. The pointer comes to rest when the turning effects of the forces due to the current and the spring are balanced.

The force exerted by the solenoid on the iron depends on the strength of the current, and on the strength of the magnetization of the iron. Since the strength of magnetization of the iron depends on the strength of the current

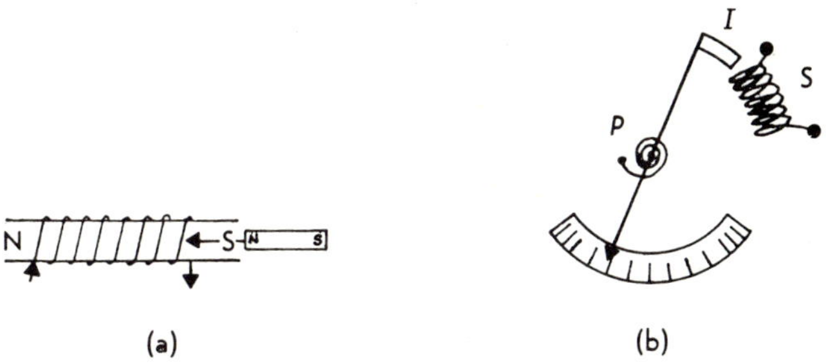

FIG. 454. *Attraction type of moving iron ammeter or voltmeter.*

also, the force of attraction depends on the square of the current. It also depends on the position of the iron in the solenoid. For this reason the scale is not uniform, i.e. equal increases in current are not represented by equal increases in deflection. The scale is cramped at each end and spread out in the middle.

In the *repulsion type* of instrument (Fig. 455), the current to be measured flows in a solenoid. A bar of iron *A* is fixed in the solenoid and another bar of iron *B*, parallel to the first, is mounted on a pivoted pointer. The pointer is counterbalanced and has a spring control as in the attraction type. When a current flows in the coil, both bars of iron become magnetized and since like poles are together, they repel each other. The force of repulsion depends on the product of the strengths of the magnets, and since the strength of each magnet depends on the current, the repulsive force again depends on the square of the current, as well as the distance apart of the bars. Hence, in this case too, the scale is not uniform.

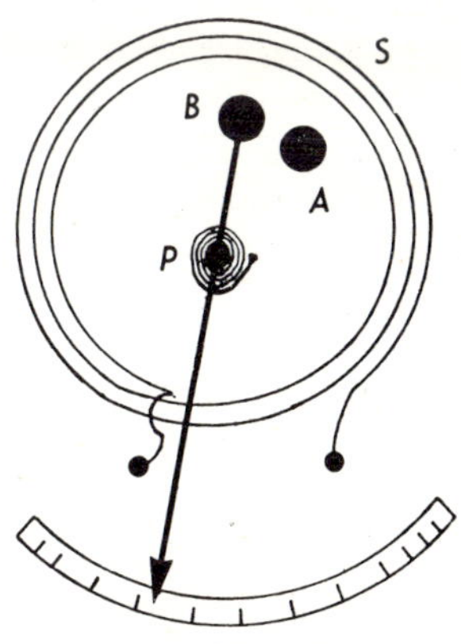

FIG. 455.

Moving iron instruments have one great advantage over the more accurate moving coil instruments to be described later; they deflect the same way for either direction of the current. This enables them to be used to measure alternating currents.

FORCES ON ELECTRIC CURRENTS IN MAGNETIC FIELDS

If a wire carrying an electric current is placed between opposite poles of a powerful horseshoe magnet, the wire experiences a force tending to move it at right angles to both the current and the field. In Fig. 456, with poles and currents as shown, the wire tends to move vertically upwards. This effect can readily be shown if instead of a wire, a strip of thin aluminium foil is used. If a 2-volt cell is used to produce the current, the wire will "belly" upwards or downwards depending on the direction of the current. The direction of the force can be obtained by the use of *Fleming's "Left-hand rule."*

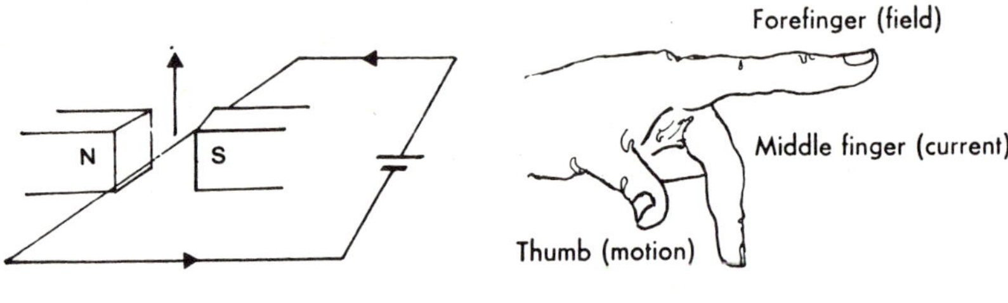

FIG. 456

FIG. 457. *Fleming's left-hand rule.*

If the thumb and first two fingers of the left hand are placed mutually at right angles (Fig. 457), and if the forefinger points in the direction of the lines of magnetic force of the magnet, and the middle finger in the direction of the current, the thumb indicates the direction of the motion or force acting on the wire. The magnitude of the force becomes greater as the size of the current is increased and as the strength of the magnet is increased. If the lines of force of the magnet are not at right angles to the current, the force decreases until, when they are parallel to the current, there is no force acting at all.

Forces on a Coil

This phenomenon is put to many practical uses. The electric motor, moving coil meters, and moving coil loudspeakers are some of them.

As many of these applications depend on the forces exerted on a coil of wire carrying a current when it is mounted in a magnetic field, we will examine this in some detail first.

Suppose that a rectangular coil *ABCD* has a current flowing in it (Fig. 458*a*) and that a magnet is placed so that there are lines of force crossing the coil in the direction shown. Application of the left hand rule shows that the side *AD* will experience a force downwards, and *BC* a force upwards. If the coil is free to turn about the axis line *EF*, it will turn until its plane is perpendicular to the field and then stop. The forces now do not tend to turn the coil, they tend to expand it (Fig. 458*b*).

It is worth noticing that the coil will come to rest when the lines of magnetic force due to the current in the coil are in the same direction as those of the magnet.

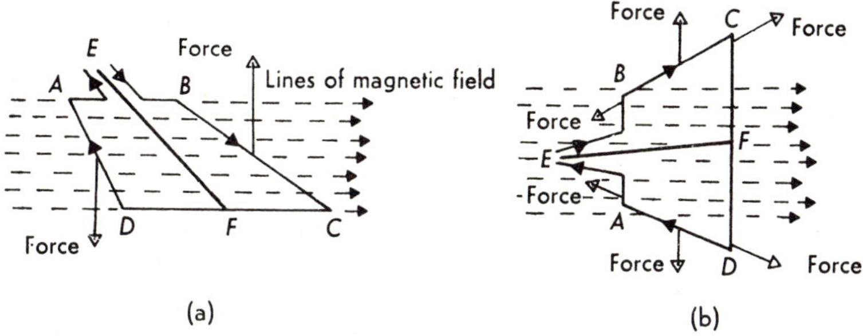

FIG. 458. *Forces on a current carrying coil in a magnetic field.*

Simple Direct Current (d.c.) Motor

In principle this consists of a coil of wire (Fig. 459) mounted on an axle (shown dotted) so that it is free to turn between the poles of a magnet. The ends of the coil are attached to the two halves, C_1 and C_2, of a split ring which are fixed to the axle and insulated from each other, and which turn

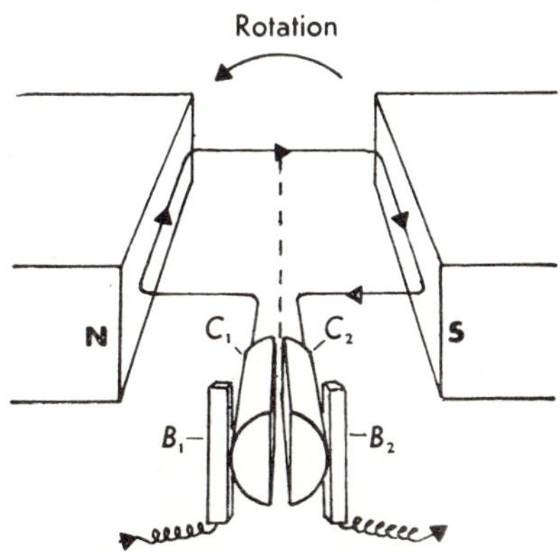

FIG. 459. *Simple direct current (d.c.) motor.*

with the coil. This is called a split ring commutator. Two fixed conductors, B_1, B_2, usually graphite, press against the ring and it is through these that the current enters and leaves the coil. These conductors are called brushes.

If the coil starts in the position shown in the diagram, it will rotate in the direction shown. When it has turned through 90°, its momentum will carry it past this position and the brushes will come into contact with the opposite halves of the ring, i.e. B_1 will make contact with C_2 and B_2 with C_1. The coil will thus go on turning and a continuous rotation is achieved. In practice, the coil which is of thick insulated wire, is wound in slots on a cylindrical iron core (Fig. 460), and the pole pieces of the magnet are hollowed out to be concentric with the core.

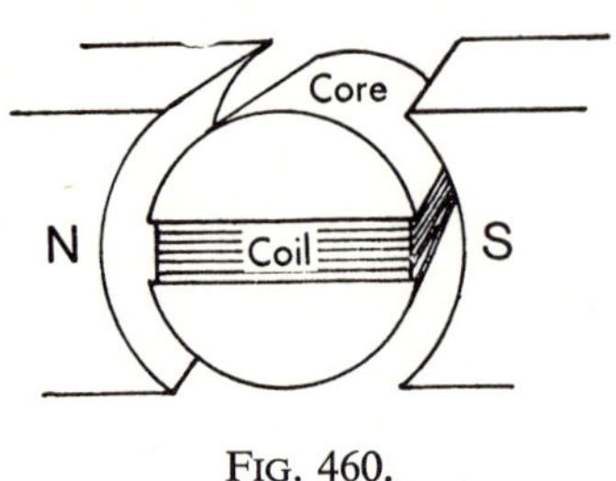

FIG. 460.

Moving Coil Ammeter and Voltmeter

These are similar to each other in construction, the only difference being that the coil of the ammeter has a low resistance and the coil of the voltmeter a high one.

The coil (Fig. 461) is fixed to a vertical spindle supported in bearings at top and bottom. The poles of the magnet are hollowed out and a soft iron core is fixed between them to produce a radial field. The production of this radial field is illustrated in Fig. 462; note that the magnetic lines of force enter the core everywhere at right angles to its surface.

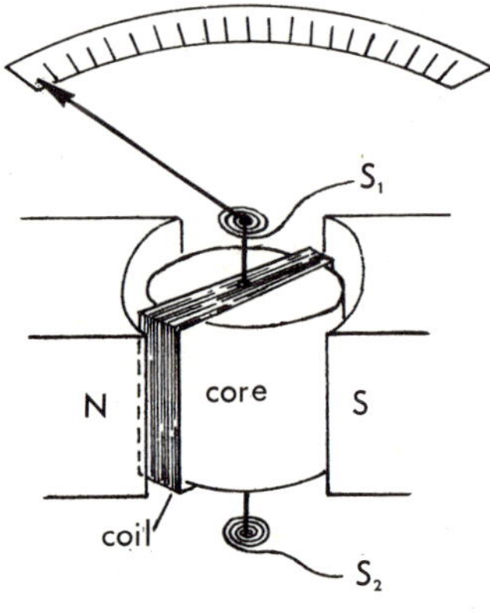

FIG. 461.

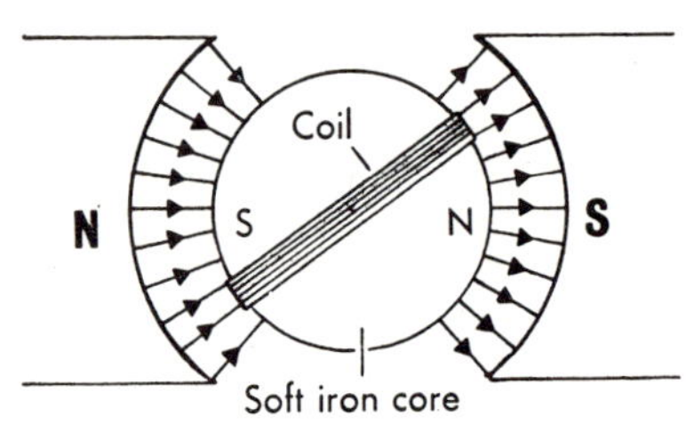

FIG. 462.

The coil, supported by its vertical spindle, is free to rotate in the gap between the poles of the magnet and the soft iron core. It does not touch either the magnet or the core. This arrangement produces a greater turning effect on the coil for two reasons: the presence of the induced poles on the

FIG. 463. *The photograph shows a moving coil ammeter. Notice the shaped poles of the magnet and the cylindrical soft iron core between them. The piece of metal screwed down across the two poles contains the top bearing for the coil. Part of the coil is visible below the left hand side of the piece of metal. The two terminals for the ammeter are on the bottom right hand side side of the photograph.*

core strengthens the magnetic field; as the coil turns it remains parallel to the field.

Springs S_1 and S_2 are fixed at top and bottom of the coil to resist its turning. One of these springs leads the current into the coil and the other leads it out. When a current flows in the coil, it turns, its turning being resisted by the springs. The angle through which it turns is proportional to the current flowing. A pointer attached to the coil moves over a scale which is uniform; equal divisions on the scale correspond to equal increases in the current.

Since the resistance of the coil is constant, the current flowing through it is proportional to the p.d. across it (Ohm's law). It follows that the deflection of the coil must also be proportional to the p.d. across it. Thus a voltmeter can be made by using a high resistance coil consisting of many turns of fine wire.

Moving Coil Galvanometer

For measuring very small currents, the coil is suspended by a phosphor-bronze or silicon-bronze wire instead of a spindle. The current enters the coil through this suspension and leaves by another fine wire at the bottom. Since the suspension is thin it exerts a very small resistance to the turning of the coil and hence very small currents will produce a deflection. The deflection is sometimes measured by a light pointer fixed to the coil, and sometimes by using a concave mirror attached to the coil. A beam of light from a lamp strikes the mirror and is reflected on to a scale. If the scale is some distance away, a very slight movement of the coil produces an observable deflection. Also the reflected beam turns through twice the angle that the mirror turns through (page 254) and this magnifies the deflection. Such instruments as these can be made to deflect for currents of less than one ten-millionth of an ampere. They are particularly useful in "null" methods (page 417) where they are used to detect the absence of a current.

Moving Coil Loudspeaker

In the output stage of a radio receiver or gramophone reproducer the sounds which are to be heard are represented by a varying electric current.

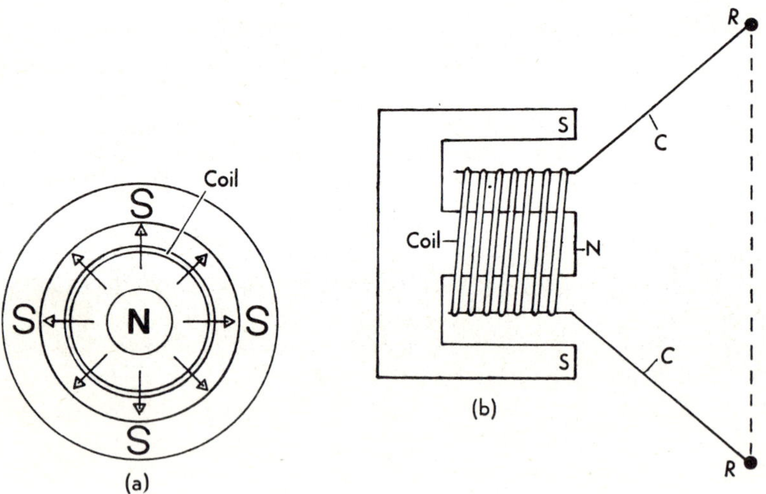

FIG. 464. *Front and side views of a moving-coil loudspeaker.*

The loudspeaker converts these currents into sound waves in the air. As we have already seen (page 422), a telephone receiver will do this. The moving coil speaker works on a different principle and is capable of producing a larger volume of sound. A cylindrical permanent magnet (Fig. 464*a*) has a pole of one kind in the centre and the opposite pole in the form of a ring round it. This produces a radial field in the gap between the poles. A coil of wire is wound on a cylindrical cardboard former and is fixed to a stiff paper cone *C* (Fig. 464*b*). The cone is fixed to a metal ring *R* which is held by stays fixed to the magnet. The coil is held in the gap between the poles and the varying currents are fed into it. If the current in the coil (Fig. 464*a*) flows clockwise, application of the left-hand rule shows that it will tend to move up out of the paper. As the current varies, the force on the coil varies and the coil moves in and out. This causes the cone to vibrate with the same frequency as the coil, and therefore at the same frequency as the current varies. The vibrating cone produces the sound.

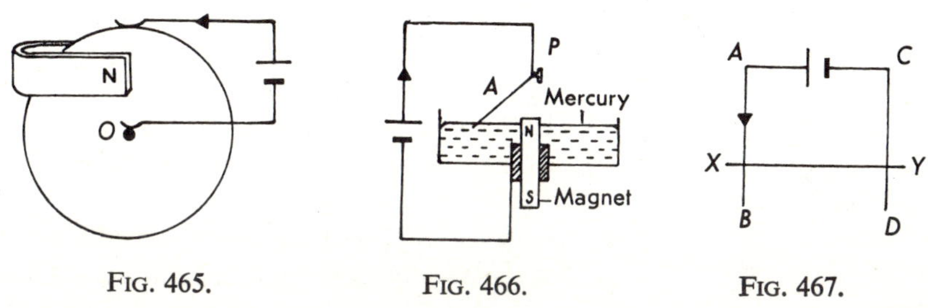

FIG. 465. FIG. 466. FIG. 467.

QUESTIONS

1. Two parallel currents flow in the same direction in wires close to each other. What forces, if any, will act on the wires? How will these forces change if (*a*) one of the currents has its direction reversed, (*b*) both of the currents have their directions reversed?

2. A solid metal disc that can rotate about an axle through its centre *O* (Fig. 465) has a current flowing along a radius through brushes that bear on the axle and the rim. The poles of a magnet are arranged as shown. Which way will the disc rotate?

3. A light wire *A* (Fig. 466), pivoted at *P*, dips into mercury, and a current is passed through it via the mercury. If a magnet is arranged in the centre of the mercury as shown, how will the light wire *A* behave?

4. A current flows vertically downwards in a wire. In what direction does the wire tend to move owing to the earth's magnetic field?

5. Two iron bars are placed side by side in a solenoid. What happens when a current flows in the solenoid?

6. In Fig. 467, *XY* is a wire sliding freely on the wires *AB*, *CD*. If a magnetic field acts vertically downwards into the paper, how will *XY* tend to move?

CHAPTER 36

CHEMICAL EFFECTS OF A CURRENT

IN THE introduction to electric currents we saw that a solution of common salt in water would conduct electricity, and that chemical action accompanied the conduction. The salt solution is an example of what is known as an *electrolyte*. An electrolyte is a liquid which will conduct electricity and which suffers chemical decomposition while doing so. Any acid, salt, or base, when dissolved in water, is an electrolyte. The metal plates by which the current enters or leaves the solution are called *electrodes*. The one at which the current enters is called the *anode*, and the other, at which the current leaves, is called the *cathode*.

Since the current passes through the salt solution there must be charged particles present in the solution to carry it. Since neither the salt nor the water was originally charged, and the solution has no electric charge as a whole, there must presumably be equal quantities of positive and negative charge in the solution. We therefore picture the presence of both positively and negatively charged particles in the solution. These particles are called *ions*.

When the electrodes are connected to a battery, the positive ions move towards the negative electrode or cathode, and the negative ions to the anode, and this is how the current flows through the solution. Before adding the salt to the water no appreciable current flowed. Hence there were very few, if any, ions present. The ions appeared when the salt dissolved. Now in all electrolytes, when a current flows, either hydrogen or a metal is set free at the cathode. Thus hydrogen and metals are associated with positive charges in solutions since they go towards the cathode. It is not so easy to generalize about what happens at the anode, but the experimental results are adequately explained by the *ionic theory*.

Crystals of sodium chloride consist of sodium (Na^+) and chloride (Cl^-) ions arranged in order, which separate when dissolved in water. Water is also slightly dissociated into hydrogen (H^+) and hydroxyl (OH^-) ions. When electrodes at different potentials are placed in the solution the sodium and hydrogen ions move to the cathode where the hydrogen ions are preferentially discharged and hydrogen gas liberated. Chloride and hydroxyl ions move to the anode; the chloride ions are preferentially discharged and the liberated chlorine saturates the surrounding liquid.

If carbon electrodes are used, the chlorine bubbles off; if metal electrodes are used, the chlorine will attack them. An excess of hydroxyl and sodium ions is formed in the original sodium chloride solution.

FARADAY'S LAWS OF ELECTROLYSIS

The *copper voltameter* consists of two copper electrodes dipping in an electrolyte of copper (II) sulphate solution. When an electric current passes through this solution nothing appears to happen, but if the electrodes are weighed before and after passing the current, the cathode is found to have gained in mass and the anode to have lost mass by the same amount. The cathode has become plated with copper on the side facing the anode, and copper has dissolved off the anode. The strength of the copper (II) sulphate solution remains unaltered.

The mass of copper deposited on the cathode depends on the strength of the current and on the time for which it is passed through the solution. If a current, flowing for a certain time, deposits a certain mass of copper, half this current flowing for twice the time will deposit the same mass of copper. Thus the mass of copper deposited depends on the total charge or quantity of electricity flowing through the solution. It does not depend on the strength of the solution.

This relation between the mass of substance liberated from solution in electrolysis and the quantity of electricity flowing through is generally true for all electrolytes. It was discovered by Faraday and the statement of it is called the *first law of electrolysis.* Thus Faraday's first law of electrolysis states that "*the mass of substance liberated in electrolysis is proportional to the quantity of electricity passed through.*"

Electrochemical Equivalent

The mass of substance liberated by 1 coulomb of electricity is called the *electrochemical equivalent*, or e.c.e., of that substance.

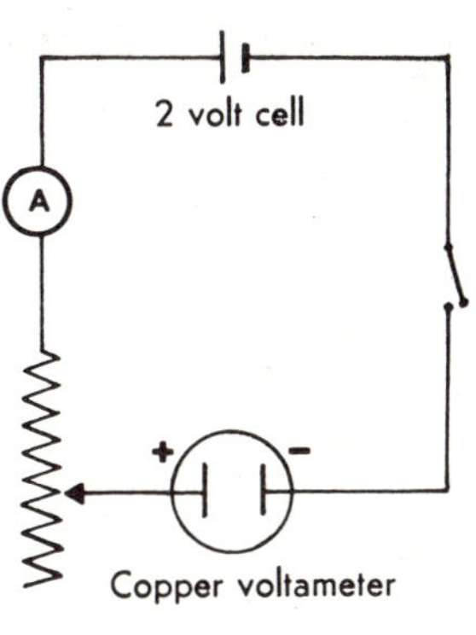

FIG. 468.

The e.c.e. of copper can be measured using a copper voltameter as follows. The circuit is as shown (Fig. 468). The current is adjusted by the rheostat to a suitable value. A suitable value of the current can be obtained by allowing 1 A for every 25 cm² of surface of the cathode to be plated. If the current is much larger than this, the copper deposit may not adhere to the cathode.

The cathode is removed from the circuit, cleaned with fine sandpaper, washed, dried in front of an electric fire, and accurately weighed. Then the cathode is replaced in the circuit and the current switched on. A quick adjustment of the current is made to its predetermined value. The current is allowed to flow for about an hour, using the rheostat to keep it constant. The time for which the current flows is noted. At the end of this time the current is switched off. The cathode is removed from the circuit, gently washed, dried, and reweighed. The drying process can be accelerated by dipping the plate in methylated spirit before drying.

From the two weighings, the mass of copper deposited can be found. By multiplying the current in amperes by the number of seconds for which it flowed, the number of coulombs of electricity flowing is calculated. Finally, by dividing the mass of copper deposited by the number of coulombs flowing the e.c.e. of copper is calculated. The answer should be 0·000 33 g/C. A more accurate value is 0·000 328 g/C. For example, if a current of 1 A is passed for 1 hour, 1·18 g of copper is deposited.

If the electrochemical equivalents of different substances are measured, it is found that they are proportional to the chemical equivalents of those substances. That is, the ratio of the electrochemical equivalents of two substances is equal to the ratio of their chemical equivalents. For example, the chemical equivalent of silver is 108 times that of hydrogen, so the electrochemical equivalent of silver is 108 times as big as that of hydrogen. This is the substance of Faraday's *second law of electrolysis*, though not the form in which he stated it. Faraday's second law is usually stated as "*the masses of different substances liberated by the same quantity of electricity are in the ratio of their chemical equivalents.*" It follows from Faraday's second law that the same quantity of electricity is required to liberate the equivalent weight in grammes of any substance. This quantity, called *a faraday*, is 96 540 coulombs. Thus 96 540 coulombs will liberate 1 g of hydrogen, 8 g of oxygen, 108 g of silver, 31·75 g of copper and so on.

The atomic weight in grammes of any element contains the same number of atoms. For example, 1 g of hydrogen and 63·5 g of copper have the same number of atoms. In electrolysis 96 540 coulombs of electricity will liberate 1 g of hydrogen, but $\frac{63 \cdot 5}{2}$ g of copper, since copper is divalent and its equivalent is half its atomic weight. This means that the copper ion must carry twice the charge that a hydrogen ion does. This illustrates how it follows from Faraday's second law that the charge on an ion is proportional to its valency. We have reason to believe that the charge on a monovalent ion is the fundamental unit of electric charge ($1 \cdot 6 \times 10^{-19}$ coulombs). No charge can be smaller than this, and every charge is a whole number multiple of this.

Silver Voltameter

The silver voltameter consists of silver electrodes immersed in silver nitrate solution. When a current flows, silver is deposited on the cathode and dissolved from the anode.

This voltameter can be used for checking the accuracy of the calibration of an ammeter. The circuit used would be similar to that used in determining the e.c.e. of copper except that the copper voltameter would be replaced by a silver one. The experiment would be carried out in a similar way, but the calculation would be different. By dividing the mass of silver deposited by the number of seconds for which the current flowed, the rate of deposition in g/s would be found. By dividing this by $\frac{108}{96\,540}$ g/C, the current in amperes could be calculated. This could be compared with the ammeter reading.

For example, if the weight of silver deposited in 1000 seconds was 2·236 g, the rate of deposition would be 2·236/1000 or 0·002 236 g/s. Hence the current would be

$$\frac{0{\cdot}002\,236}{108} \times 96\,540$$
$$= 2\text{A}$$

Water Voltameter

The water voltameter consists of platinum electrodes immersed in dilute sulphuric acid. On electrolysis, hydrogen is liberated at the cathode and oxygen at the anode.

The apparatus usually takes the form shown in Fig. 469 which enables the volumes of gas to be measured. It is found that the gases are produced at the rate of two volumes of hydrogen for one volume of oxygen. This is the proportion in which these gases combine to form water so that it is the water which suffers decomposition, the acid remaining unchanged. Hydrogen and oxygen are prepared commercially by electrolysis, but usually the electrolyte used is caustic soda solution, which produces the same two gases.

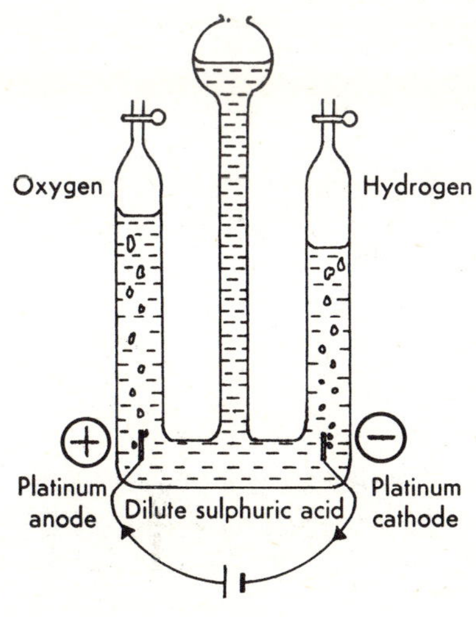

FIG. 469. *Water voltameter.*

Electroplating

This is the process of depositing a thin layer of metal by electrolysis on a baser metal.

Copper plating. Iron can be plated with copper by making the object to be plated the cathode in a cell containing copper (II) sulphate solution and a copper anode.

Silver plating. This is done by making the object the cathode in a cell with a silver anode, the electrolyte being a solution of a silver salt. Silver nitrate is not used because other silver salts are found to give a better deposit. Silver cyanide dissolved in potassium cyanide solution is used. If an attempt is made to deposit silver directly on to iron, the silver tends to flake off. Hence the iron is usually plated with copper first to which the silver deposit adheres better. In old silver electroplate, the silver deposit wears thin with constant cleaning and the copper underneath can often be seen.

Nickel plating. It is usual to plate iron first with copper as nickel does not adhere directly to iron. The electrolyte used is nickel sulphate solution and the anode is of nickel.

Chromium plating. This forms a hard and stainless coating to iron and is used for radiators, hub caps, and bumpers in cars. In making good chromium plate, the steel is first plated with copper, then nickel, and finally with chromium. The electrolyte is a strong solution of chromic acid. Chromium anodes are not used, but the strength of the solution is maintained by adding chromium trioxide. The anodes are usually lead.

Zinc plating. Iron, coated with a thin layer of zinc, resists corrosion, and is known as galvanized iron. It is often made by dipping the iron into molten zinc, but this process is being replaced by an electrolytic one in which the electrolyte is zinc sulphate solution and the anode is pure zinc.

In all electroplating, the object to be plated must be free from rust and grease. To obtain an even deposit the object is surrounded by a cylindrical anode and kept rotating. Too large a current causes a deposit which does not stick firmly to the metal underneath.

Electrolytic Metal Refining

Crude copper is refined by electrolysis. The crude copper is made the anode in a cell containing copper (II) sulphate solution. The cathode is a thin sheet of pure copper. When the current flows, pure copper deposits on the cathode and dissolves off the anode. The impurities fall to the bottom of the cell.

Aluminium and sodium are both produced electrolytically. Aluminium by the electrolysis of a melted oxide, and sodium by the electrolysis of fused caustic soda.

The e.m.f. of a Cell

In some cases it is possible to predict the e.m.f. of a cell from energy considerations. For example, when a Daniell cell is delivering a current, zinc dissolves and copper is deposited. The e.c.e. of zinc is 0·000 34 g/C, and that of copper 0·000 33 g/C. Thus, for every coulomb the cell supplies, 0·000 33 g of copper is deposited and 0·000 34 g of zinc dissolve. Measurements show that when 1 g of zinc dissolves in dilute sulphuric acid about 6800 joules are evolved, and when 1 g of copper is deposited 3700 joules are absorbed. Hence for every coulomb the cell produces, there is a net production of $(6800 \times 0{\cdot}000\,34) - (3700 \times 0{\cdot}000\,33) = 1{\cdot}1$ joules. Now when the cell is delivering a current, no heat is produced in the cell, so presumably this energy, instead of producing heat in the cell, produces energy in the circuit. The energy delivered by the cell is thus 1·1 joules/coulomb of electricity provided. Hence the e.m.f. of the cell is 1·1 volts. This calculation assumes that all the energy liberated by the chemical reactions is converted into energy in the conductors carrying the current. In the case of the Daniell cell this is very nearly true, but for other cells the agreement is not so good.

QUESTIONS

1. A steady electric current is sent through a water voltameter, a copper voltameter and a silver voltameter in series for one hour. If the mass of silver deposited is 4·025 g, calculate (*a*) the current, (*b*) the mass of copper deposited, (*c*) the mass of oxygen set free. (*d*) the mass of hydrogen set free, (*e*) the mass of water decomposed. The e.c.e. of silver can be taken as 0·001 118 g/C, and the chemical equivalents of copper, silver and oxygen as 31·5, 108 and 8 respectively.

2. Four ammeters of equal resistance are available, on each of which the scale has a mark at the 1 A calibration only. If you have a battery and a rheostat, how could you mark calibration points on the scale for 2 A, 3 A, $\frac{1}{2}$ A, $\frac{1}{3}$ A?

3. It is required to deposit a layer of copper 0·1 mm thick on 100 cm^2 of iron. The density of copper is 8·93 g/cm^3 and its e.c.e. is 0·000 33 g/C. Calculate how many coulombs will be necessary.

4. Two similar Daniell cells are used together to send a current of 1 A through a circuit for a period of one hour. What is the mass of copper deposited and the mass of zinc dissolved in each Daniell cell if the cells are (*a*) in series, (*b*) in parallel? (The e.c.e. of copper can be taken as 0·000 33 g/C, and the chemical equivalent of copper and zinc to be 31·5 and 32·5 respectively.)

CHAPTER 37

ELECTROMAGNETIC INDUCTION

WE HAVE seen that when a wire carrying a current is at right angles to a magnetic field a force acts on the wire. The direction of this force is given by the left-hand rule. For example (Fig. 470*a*), when a current flows from *A* to *B*, a force acts on the wire tending to move it upwards. Now an electric current is a flow of electric charge along a wire. Thus, although a stationary electric charge in a magnetic field experiences no force, a moving charge does experience a force.

A positive charge moving along *AB* is equivalent to a current flowing in the direction *AB*. If the positive charge moves at right angles to a magnetic field, the left-hand rule shows that there is a horizontal force acting on it (Fig. 470*b*).

Now suppose a horizontal wire *AB* is moved vertically down between the magnet poles (Fig. 470*c*). Since the wire contains electric charges and these

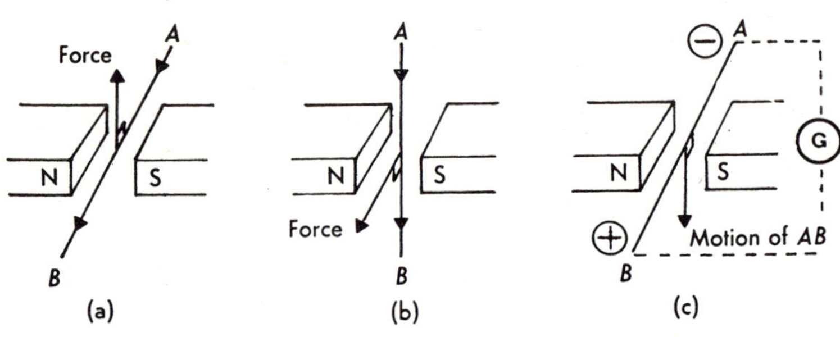

FIG. 470.

are moving *vertically* down, the positive charges will, by the left-hand rule, experience a force pushing them towards *B*. Negative charges, by a similar argument, will be pushed towards *A*. Thus *B* will become positively charged and *A* negatively charged.

Although we believe that in a metal wire only the negative charges move, the argument is unaffected. If negative charge moves from *B* to *A*, *B* is left positively charged. That is, as long as the wire is moving down through the magnetic field it will act like a battery with *B* a positive and *A* a negative pole. That this is true may be shown by connecting the ends of the wire to a galvanometer which will register a current while AB is moving across the field. This current is called an *induced current.* It is produced by the e.m.f. which is induced in the wire by its motion through the magnetic field. The e.m.f. induced is greater if the wire moves faster, and the e.m.f. can also be

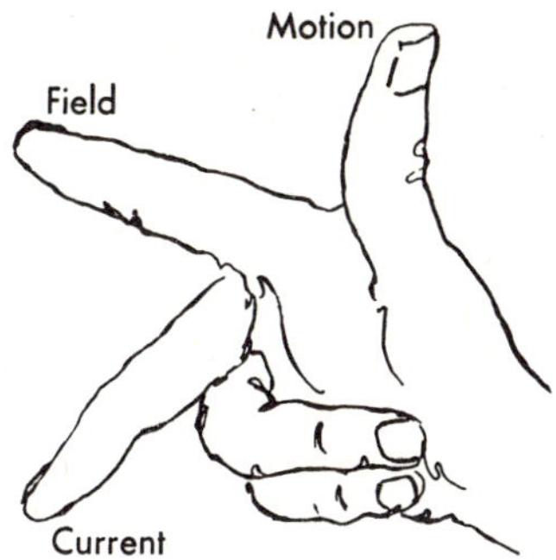

FIG. 471. *Fleming's right-hand rule.*

increased by increasing the strength of the magnet. Its direction can be obtained by the use of *Fleming's right-hand rule.* This is similar to the left-hand rule. If the forefinger points along the lines of force, and the thumb in the direction of the wire's motion (Fig. 471), the middle finger indicates the direction of the induced e.m.f.

Laws of Electromagnetic Induction

The law of electromagnetic induction states that *when a conductor moves across lines of magnetic force, an e.m.f. is induced in it which is proportional to the rate at which the conductor "cuts" the lines of force.* (We imagine that the stronger a magnetic field is, the more lines of force there are concentrated in the space in which the field acts.)

Now let us consider the same phenomenon from another angle. Suppose that we are moving a wire as in Fig. 470*c* and inducing a current in it. Reference to Fig. 470*a* will show that the wire experiences a force opposing its motion, i.e. as the wire moves down, a force acts on it trying to move it up, and this force has to be overcome in order to keep the wire moving down. Thus, work has to be done in pushing the wire down, and it is this work which is converted into electrical energy in the wire. In this way mechanical work can be directly converted into electrical energy. This illustrates another law of electromagnetic induction called *Lenz's law*, which states that *the induced e.m.f. acts in such a direction as to oppose the motion or change inducing it.*

Lenz's law shows how the principle of conservation of energy applies to the induction of currents. If the force on the wire assisted instead of opposed the motion, we should be able to obtain energy without doing work to produce it.

Another Way of Inducing Currents

Suppose a circular coil of wire with its ends connected to a galvanometer is placed between the poles of an electromagnet so that the lines of force of the magnet pass through the coil (Fig. 472). If the strength of the magnet is suddenly changed by changing the current through its magnetizing coils, the galvanometer shows a current flowing in the coil as long as the strength of the magnet is changing. When the magnet becomes stronger, the current flows one way in the coil, and when it becomes weaker, the current flows the opposite way.

In this case there is no motion to which the induced current can be attributed, and the law of induction has to be stated in a different form as follows: *When the number of lines of magnetic force through a circuit changes, there is an induced e.m.f. in the circuit.* (The number of lines of magnetic force through a circuit is often called the *magnetic flux* through the circuit.)

The two forms of the law of induction are really equivalent but it requires more advanced theory to prove this.

Lenz's law can be applied to this case even though there is no motion. When the number of lines of force through the coil increases, the current in the coil flows so as to produce lines of force in the opposite direction. When the number of lines of force through the coil decreases, the induced current flows so as to produce lines of force in the same direction. In both cases the induced current is caused by a change in the number of lines of force through the coil, and in both cases it flows in such a direction that its own lines of force *oppose the change*. Thus in Fig. 472, when the magnet's strength increases, the induced current flows in an anti-clockwise direction around the coil.

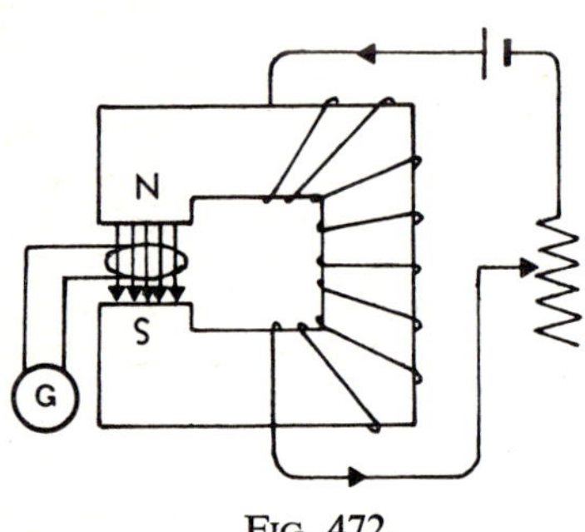

FIG. 472.

Mutual Induction between Two Coils

The principles described so far enable the results of the following experiment to be explained. A coil *B* is placed inside another coil *A* (Fig. 473). *B* is connected to a battery, rheostat, and switch. *A* is connected to a galvanometer. On switching on or increasing the current in *B*, a *momentary* current is produced in *A*, in the *opposite* direction to that in *B*. On switching off or decreasing the current in *B*, a *momentary* current is produced in *A* in the *same* direction as that in *B*. When the current in *B* is steady, there is no current in *A*. If the experiment is repeated with a rod of iron placed inside *B*, the induced currents are much greater.

If the coil *A* is replaced by another coil with more turns but of the same electrical resistance (using thicker wire), the currents in *A* are proportionally greater.

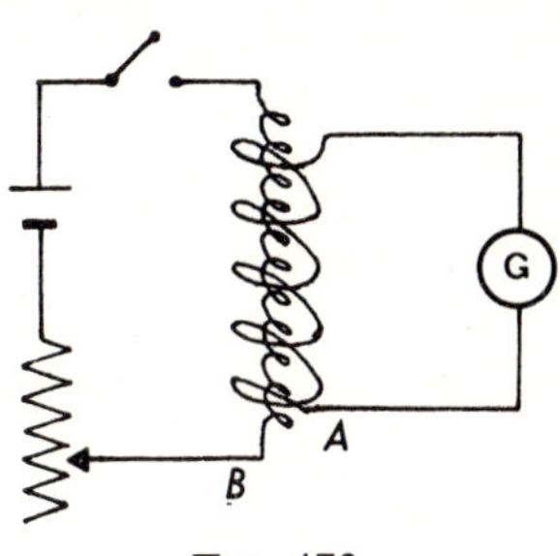

FIG. 473.

The currents in *A* are produced by changes in the magnetic flux through it due to the changing current in *B*. When a rod of iron is placed in *B*, it becomes magnetized by the current in *B* and its own lines of force are added to those of *B*. Thus the change in the number of lines of force when the current in *B* changes is much magnified. Since the change in lines of force occurs through each turn of the coil *A*, there is an e.m.f. induced in each turn.

These e.m.f.s "add up," and the total e.m.f. in *A* is proportional to the number of turns in *A*.

ELECTRIC GENERATORS

If a coil is rotated between the poles of a magnet, an e.m.f. is generated in the coil.

The direction of the e.m.f. in this case is given by applying Fleming's right-hand rule (page 437).

a.c. Dynamo

Consider the coil *ABCD* rotating between the poles of a magnet (Fig. 474). It can be seen that the e.m.f. in the right-hand conductor *CD* is in the direction *CD*, and that in the left-hand conductor *AB* the e.m.f. is in the direction *AB*. As the coil rotates, *AB* will come into the position now occupied by *CD* and the e.m.f. in it will be in the direction *BA*; similarly with *CD*. Thus the e.m.f. in the coil reverses direction when the coil is vertical, i.e. perpendicular to the lines of force. The sides of the coil *AB* and *CD* are moving across the lines of force most rapidly when the coil is in the position shown, that is, when the coil is parallel to the lines of force.

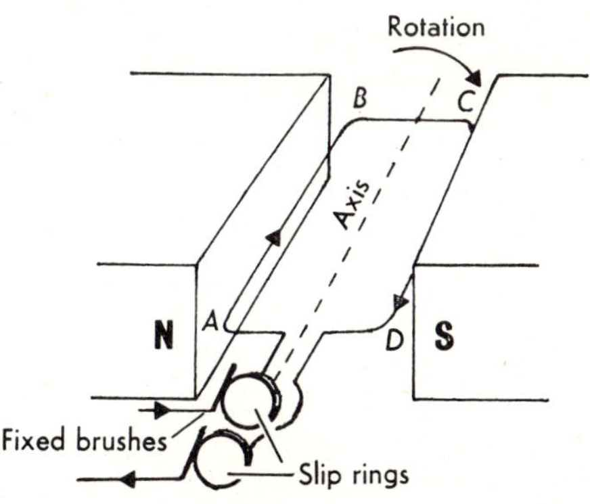

FIG. 474. *Simple a.c. dynamo.*

Hence the e.m.f. generated is a maximum when the coil is in this position. When the coil is perpendicular to the field the e.m.f. is momentarily zero. Thus a graph of the e.m.f. produced in the coil during one complete revolution would be shown as in Fig. 475. This is known as an alternating e.m.f., and Fig. 475 shows one cycle.

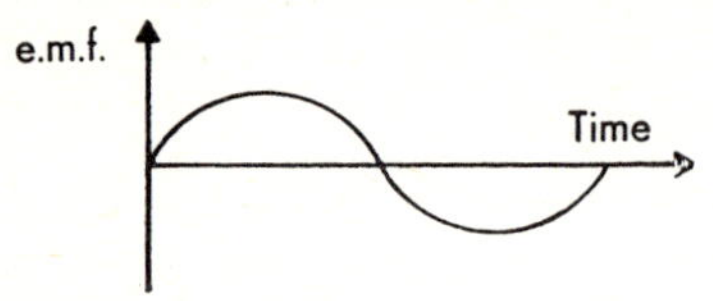

FIG. 475. *Graph for a.c. dynamo.*

In Britain the supply alternates at the rate of 50 Hz. To obtain current from the coil, its ends are each connected to a separate "slip ring", and fixed brushes bear against these rings as the coil rotates.

d.c. Dynamo

By attaching a split ring commutator to the coil instead of slip rings (Fig. 476), the e.m.f. driving the current through the *external* circuit is always in the same direction, though not steady. Its graph is shown in Fig. 477. By attaching several coils to the same shaft and using a more complicated commutator a nearly steady e.m.f. can be obtained.

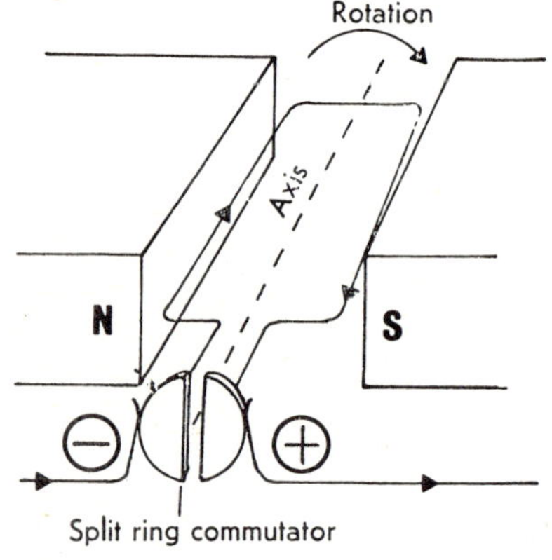

e.m.f.
Time

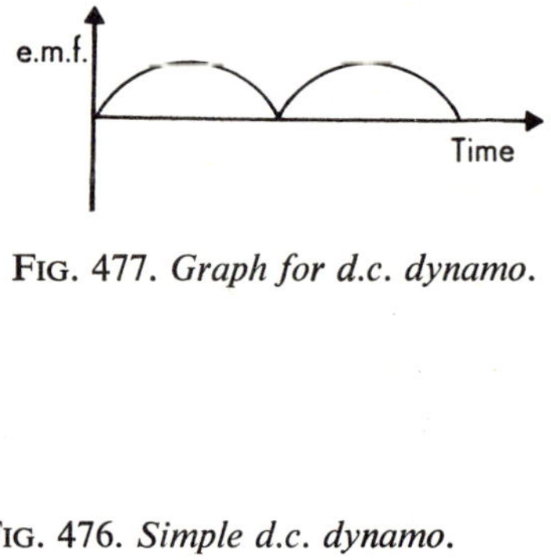

FIG. 477. *Graph for d.c. dynamo.*

FIG. 476. *Simple d.c. dynamo.*

It will be noticed that the construction of a d.c. dynamo is similar to that of a d.c. motor. The same machine can be used as either. If a current is sent through the coil, it turns and acts as a motor. If the coil is turned by an external agency it generates a current and acts as a dynamo. The two functions cannot really be separated. When the coil is being turned and generating a current, the induced current makes the coil act like a motor and tries to turn it the opposite way. The work that is done against the resisting force is converted into electrical energy. In the same way, when a d.c. motor is running, an induced e.m.f. is produced in the coils opposing the current that is being sent through it. This is called the "*back e.m.f.*" of the motor. This back e.m.f. reduces the effective voltage of the supply which is driving the current through the coils. The faster the motor turns, the greater is the back e.m.f. and hence the less current flows in the motor. When the load on the motor is increased, the coil slows down, hence the back e.m.f. drops, hence the current rises, and hence the turning force of the motor rises and the load is taken on at a reduced speed. Overloading a motor may cause the current to rise dangerously high and burn the coils with the heat produced in them.

ALTERNATING CURRENT

Electric lamps and heating appliances such as electric fires, cookers and irons will work just as well on a.c. as on d.c. because the heat produced by a current in a wire is independent of the direction of the current. The current varies all the time, and so does the rate of heat production, but it varies so rapidly that the temperature of the wire fluctuates only slightly.

Root Mean Square Values

Alternating currents and voltages are usually measured by what are called their r.m.s. values (r.m.s. stands for root mean square). The r.m.s. value of an alternating current (or voltage) is the value of the steady current (or voltage) that would produce the same average heating effect in a wire as the alternating current (or voltage) does. When the value of an alternating current (or voltage) is stated, it is the r.m.s. value that is meant. It can be shown that the r.m.s. value of an alternating current (or voltage) is the maximum or peak value divided by $\sqrt{2}$. Thus if the alternating mains voltage is quoted as 240 volts, it means that the peak value of the voltage is $240 \times \sqrt{2}$ or about 340 volts. This explains why it is more dangerous to connect oneself across an alternating supply of 240 volts than a direct supply of 240 volts. It also follows that the power consumed in watts when an alternating current flows through a wire is obtained by multiplying the r.m.s. value of the p.d. in volts, by the r.m.s. value of the current in amps. Ohm's law also holds for alternating currents, provided we are dealing only with simple resistances:

r.m.s. value of p.d. in volts = r.m.s. value in amperes × resistance in ohms

These last two statements are made with some reservations. They are true provided the frequency of alternation is not very large.

Magnetic and Chemical Effects of an Alternating Current

These are very much what we would expect from the character of alternating current. A wire carrying such a current placed near a compass needle does not cause it to deflect. This is because the direction of the magnetic field produced by the current is continually changing and the compass needle moves too slowly to be able to follow the changes. If experiments on electrolysis are attempted with alternating current, the anode and cathode are continually changing places. Thus in a copper voltameter, neither of the electrodes would gain or lose mass. In the water voltameter, a mixture of hydrogen and oxygen is obtained at each electrode. (This mixture is highly explosive.) Thus alternating current cannot be used for charging batteries.

Measuring Alternating Currents and Voltages

Moving coil meters cannot be used with an alternating supply as their deflection depends on the direction of the current. With a centre zero instrument and a very slow frequency of alternation, the pointer moves to and fro, but with a 50-cycle supply no visible motion of the pointer occurs. Instruments must be used which deflect the same way for either direction of the current. Moving iron instruments can be used. Hot wire instruments can also be used since their deflection depends on the heat produced, and this is independent of the current direction.

The principle of the hot wire ammeter is as follows. The current to be measured flows through a thin wire *AB* (Fig. 478), and heats it. This causes the wire to expand. Another thin wire attached to the middle of *AB* is attached to a silk thread which is kept under tension by the spring *S*. The silk passes round an axle *C* to which a pointer is attached. When *AB* expands, the thread takes up the slack and in doing so turns the axle and moves the pointer. The scale of this instrument, like that of a moving iron instrument, is not uniform because the heat produced depends on the square of the current. Its zero also changes with changes in room temperature. The zero can be adjusted by altering the tension in *AB*.

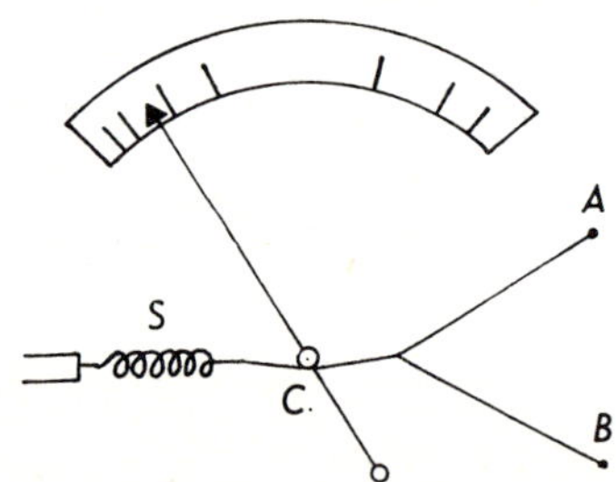

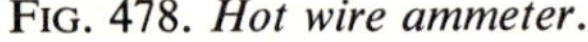

FIG. 478. *Hot wire ammeter.*

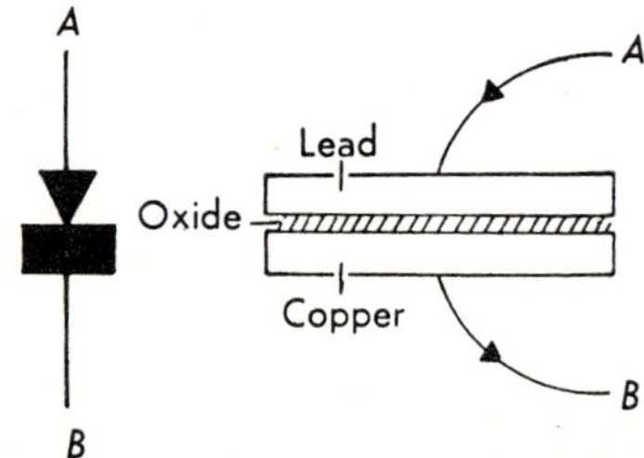

FIG. 479. *Metal rectifier.*

Rectification

A rectifier is a device which only allows current to flow one way through it, so that it can be used to convert an alternating current into direct current. Rectification is usually obtained by the use of a diode valve, but there is a type of metal rectifier which is often used. This depends on the fact that if a piece of copper is specially treated to produce a film of oxide on its surface, the resistance to a current flowing from oxide to copper is small, while the resistance to a current flowing from copper to oxide is large. A lead plate is usually put in contact with the oxide to make contact with it. Fig. 479 shows one such cell and its conventional symbol. Current flows easily from *A* to *B* but not from *B* to *A*. Heat is developed in these rectifiers when current flows, and heating impairs the efficiency of the rectification. Cooling fins are usually fitted to dissipate the heat, and if a high voltage is to be rectified a number of these elements are connected in series so that the voltage across each cell is reduced. If large currents are required a number of rectifiers are joined in parallel.

To obtain what is called "*full wave*" *rectification*, four rectifiers are connected as shown in Fig. 480*a*. It can be seen that whichever way the alternating

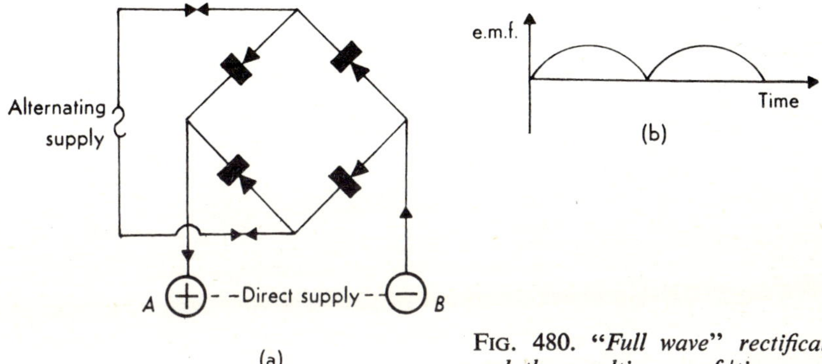

FIG. 480. "*Full wave*" *rectification and the resulting e.m.f./time graph.*

current flows, current will always flow out of *A* and into *B*. The current obtained from *AB* will not be steady, but it will always be in the same direction. The graph (Fig. 480*b*) shows how the e.m.f. across *AB* varies.

Induction Coil

This is a device which enables us to produce a high voltage from a low voltage battery. It is used in the ignition system of some cars to produce the spark which fires the mixture in the cylinders. It is also used in activating the electric fences which prevent cattle from straying.

A primary coil *B* consisting of a few layers of relatively thick wire is wound on an iron core (Fig. 481). Over this is wound the secondary coil *A* which has many thousands of turns of fine wire. When the current is switched on or off in *B*, there is a momentary e.m.f. in *A*, and the e.m.f. in *A* is large owing to the large number of turns in *A*. The switching on and off of the current in *B* is made automatic by a make-and-break similar to that in an electric bell. When current flows in *B* the iron core becomes magnetized, and attracts the iron armature. This breaks the contact *X* which is normally held closed by a steel spring and the current stops. The armature, no longer attracted by the iron core, flies back, and contact is made again and so on. In a car, this make-and-break would be operated by the engine.

Owing to factors which are too difficult to be considered here, the e.m.f. induced in *A* when the current in *B* is broken is very much larger than that when it starts. Hence it is the e.m.f. at the break which is used to produce a spark. A certain amount of sparking occurs at the contact *X*. This represents a loss of efficiency and can be reduced by connecting a capacitor *C* across the contact. The iron core is a bundle of thin rods electrically insulated from each other by a film of oxide. This is to reduce heating in the iron core which also causes a waste of electrical energy.

When an iron bar is placed inside a coil through which a varying current flows, currents are induced in the iron which heat it. These are called *eddy currents* because they eddy to and fro. Since the eddy currents flow in circles round the axis of the bar they are blocked by the insulation between the rods.

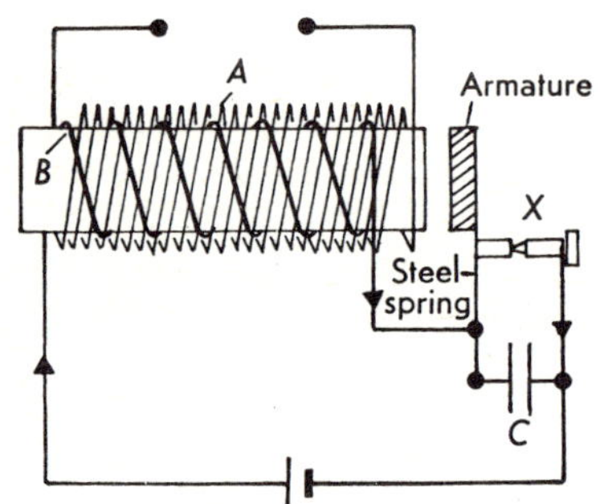

FIG. 481. *Induction coil.*

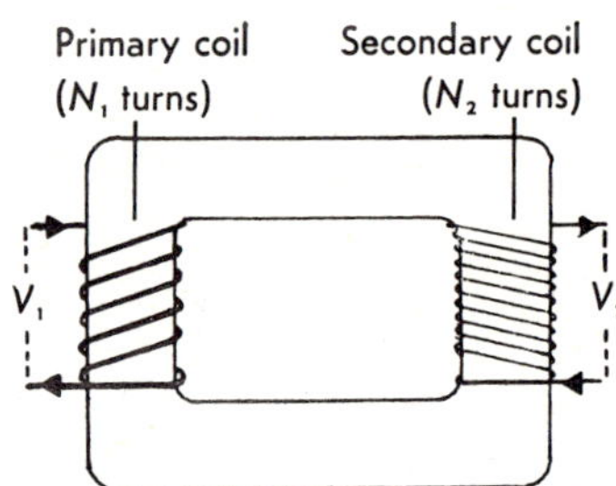

FIG. 482. *Transformer.*

Transformer

This is a device for changing the voltage of an alternating supply. In its simplest form it consists of a "closed" iron core with two coils wound on it (Fig. 482). If alternating current is sent through one coil (the primary coil) the iron core becomes magnetized first in one direction and then in the other. This produces changes in the magnetic flux through the other coil (the secondary coil) and an alternating e.m.f. is induced in it. The full theory of the transformer is complicated, but in a well-designed transformer, the ratio of the voltage developed in the secondary to that applied to the primary is

FIG. 483. *Electricity produced by generating stations at 11 000 volts is converted by transformers to 132 000 volts and fed into wires supported by pylons to be carried across the country. To tap this supply of electricity, substations are required where the voltage can be brought down to the level required by the various consumers. This is done in several stages. The substation shown transforms the 132 000 volt supply to supplies of 33 000 and 6600 volts, which are fed to other transforming and switching stations before reaching the domestic supply of 240 volts.*

the ratio of the number of turns of wire in the secondary coil to the number of turns in the primary:

$$\frac{V_1}{V_2} = \frac{N_1}{N_2}$$

Thus a transformer can be used to either "step up" or "step down" an alternating voltage. The iron core used is built up from strips or laminations insulated from each other. This is to minimize eddy current heating which would produce a loss of efficiency. The *efficiency of a transformer* is measured by the ratio of *power output/power input* or watts output from secondary coil/ watts input into primary coil.

The efficiency of a good transformer may be as great as 99 per cent. Thus if a higher voltage is obtained from a transformer, the current obtained from the secondary coil will be correspondingly less than that flowing in the primary.

Transmission of Electrical Energy

Suppose a generating station has to transmit electrical power across country to a town. The total current to be transmitted, I, which is the sum of all the currents used by all the houses, may be very large. If the transmission wires have a total resistance of R, the power wasted as heat in the transmission wires is $I^2 R$. To reduce this, R could be made very small. This would entail very thick wires which would be expensive, and also the wires would be heavy and require heavy and expensive pylons to carry them. The alternative, which is in fact adopted, is to reduce the current in the wires. The same power can be transmitted by a low current at a high voltage. The use of alternating current enables this to be done easily.

At the generating station, the voltage is raised to a high value by a transformer, power is transmitted across country at a high voltage and correspondingly low current, and transformers at the town then reduce the voltage to that required by the town. The "*grid*" system transmits power at 132 000 volts, and the voltage is reduced in several stages before it reaches the consumer.

QUESTIONS

1. A current flows in a wire. How could you determine whether it is direct or alternating?

2. A bar magnet with its north pole pointing downwards is dropped through the centre of a ring held with its plane horizontal. In what directions will current flow as the magnet passes through the ring? Where do you think the energy for the current comes from?

3. A yacht with a vertical metal mast sails due east. In which direction will an e.m.f. act in the mast? If a sensitive galvanometer is connected between top and bottom of the mast, explain why no current flows.

4. A circular metal disc is rotated about an axis through its centre and at right-angles to its plane. A magnetic field is acting at right-angles to the disc. Draw a diagram to show the direction of rotation, the direction of the magnetic field, and the direction of the induced e.m.f.

5. A metal ring is held with its plane vertical and pointing east-west and allowed to fall. Why is no current produced in the ring?

6. A pendulum made of copper is allowed to swing between the poles of a powerful magnet so that its swings are at right-angles to the lines of force: it rapidly comes to rest. Explain why this is.

7. If you were given a small bicycle dynamo, how would you determine whether it produced alternating current or direct current (*a*) by inspection, (*b*) by experiment. (You can try this out.)

8. Electric power is supplied to a factory through wires of total resistance 0·5 ohm, and the factory uses 100 kilowatts at 250 volts. Calculate (*a*) the current used, (*b*) the power lost in the transmission wires, and (*c*) the p.d. at the generating end of the wires.

CHAPTER 38

STORAGE OF ELECTRICAL ENERGY

ELECTRICAL energy can be obtained from chemical energy in cells (page 396), so that, in a sense, a cell may be said to store electrical energy. Electrical energy is also stored in charged conductors. Any charged conductor possesses electrical energy because if the charged conductor is connected to another conductor at a lower potential (e.g. the earth), a current flows and heat is produced in the connecting wire.

The electrical capacities, or *capacitances*, of conductors are compared by comparing the quantities of electricity required to raise them to the same potential. The capacitance of a conductor is *measured* by the charge required to give it unit potential. Thus the greater the charge stored when the conductor has unit potential, the greater the capacitance. So if a charge Q coulombs raises the potential of a conductor to V volts, its capacitance, C farads (symbol F), is given by:

$$C = \frac{Q}{V}$$

If a charge of one coulomb raises the potential of a conductor by one volt, the conductor is said to have a capacitance of one *farad.* But this is a very large unit of capacitance and it is customary to express capacitances in *microfarads* (symbol μF), or millionths of a farad.

The capacitance of a conductor can be increased by placing close to it a conductor at a lower potential; an earthed conductor for instance. To show this a gold leaf electroscope can be used. In most gold leaf electroscopes there is a tinfoil lining to the case which is connected to a terminal on the case. If there is a difference of potential between the leaf and this lining, the leaf deflects, and the greater this potential difference, the greater the deflection. This can be shown by connecting several H.T. batteries in series and connecting one terminal of the battery to the cap and the other to the lining. A deflection is produced, and the greater the number of batteries the greater the deflection. If the lining is connected to earth and is therefore at zero potential, the deflection of the leaves indicates the potential of the cap.

Parallel Plate Capacitor

In Fig. 484, *A* and *B* are two metal plates each fixed to an insulating stand. They are placed facing each other. *A* is connected to the cap of a gold leaf electroscope whose lining is earthed. *B* is earthed. *A* is given a charge, and the electroscope deflects. If *B* is now brought nearer to *A*, the deflection of the electroscope gets less, showing that the potential of *A* is decreased. Since its charge is not altered, its capacitance must be increased. If a sheet of insulating material such as ebonite or polythene is placed between *A* and *B* without touching either, the deflection again is decreased, showing an increase of capacitance. If *B* is moved to one side so that it is not directly facing *A*, the deflection becomes greater showing that the capacitance of *A* is decreased as the area of the plates opposite each other is reduced.

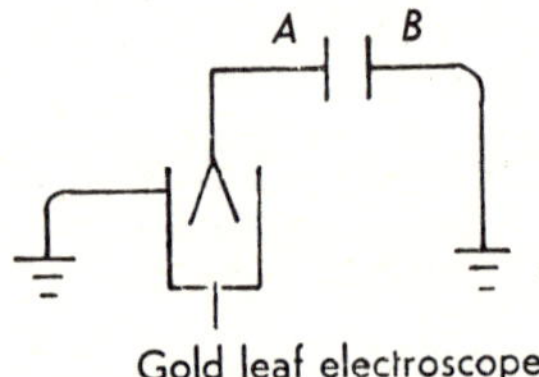

FIG. 484. *Apparatus for investigating the capacitance of a metal plate.*

Thus the capacitance of a metal plate such as *A* can be greatly increased by placing another earthed metal plate such as *B* close to it and separated from it by an insulating substance. Such an arrangement is called a parallel plate capacitor.

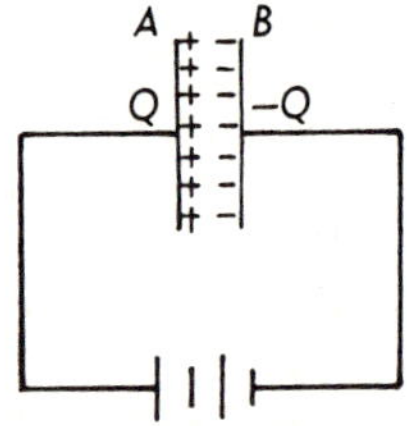

FIG. 485. *Storage of charge in a parallel plate capacitor.*

It can be shown that if such a capacitor is connected to a battery of potential difference V, the plate B being connected to the negative terminal of the battery and not being earthed (Fig. 485), the charges Q on A and B are concentrated on the faces of the plates opposite each other and are equal and opposite. The capacitance is still calculated from $C = Q/V$. The demonstration of these facts is beyond the scope of this book.

It is worth noting that when the battery is first connected to the plates, a charge flows from it on to the plates, but when the potential difference between the plates becomes equal to that of the battery, no more charge flows. If, however, the battery is replaced by an alternating voltage, the capacitor is continually charging and discharging, and thus an alternating current flows in the connecting wires. Thus an alternating current is able to "pass through" a capacitor.

Practical Capacitors

Capacitors are important in radio, television and many electrical circuits. There are four common types.

Fig. 486 shows a variable capacitor of the kind used for tuning a radio.

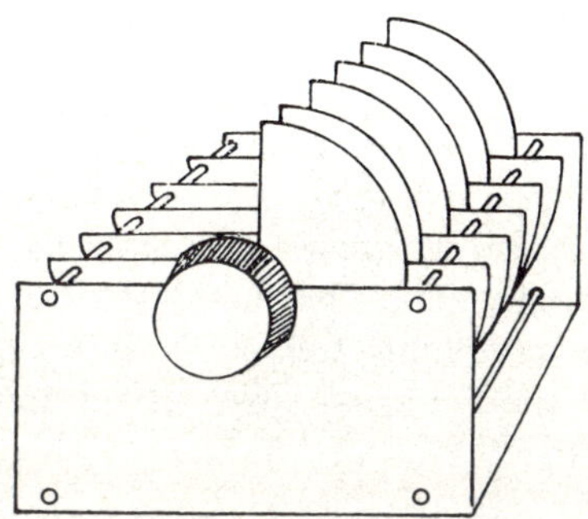

FIG. 486. *Variable capacitor of the type used in radios.*

It has two sets of semi-circular metal plates separated by air, each set consisting of several plates joined together, forming, in effect, one large plate. One set is fixed and the other rotates when the knob is turned, thus altering the

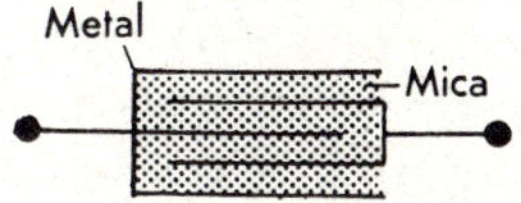

FIG. 487. *High-quality mica capacitor.*

effective area of the plates and so the capacitance. With the plates fully interleaved the maximum capacitance may be about 0·0005 μF.

In Fig. 487 a mica capacitor is illustrated, so called because mica is the insulator between the metal plates. High quality capacitors are of this type.

Fig. 488 shows a capacitor in which two tinfoil strips form the plates and waxed paper or polyester film the insulator. It is cheap and can be rolled up into a small cylinder.

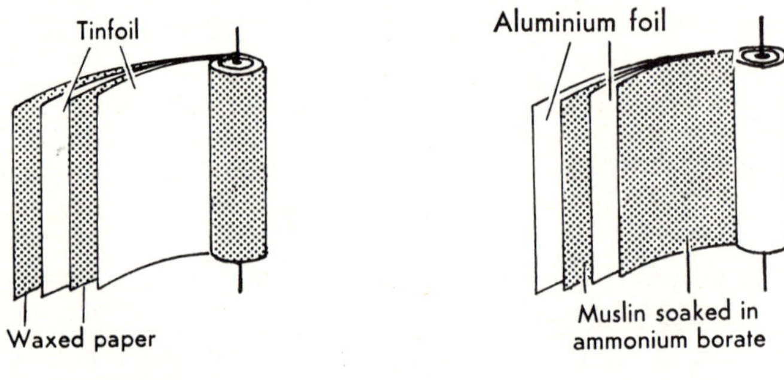

FIG. 488.

FIG. 489.

The electrolytic capacitor (Fig. 489) consists of two sheets of aluminium foil separated by muslin soaked in a solution of ammonium borate. These are rolled up, sealed in a container and the leads from the foil plates connected to a battery. Electrolysis occurs and an extremely thin film of aluminium oxide forms on the positive foil. This film is an excellent insulator and forms a capacitor with the aluminium foil. Very large capacitances are possible, 1000 μF or more, in quite a small space. It is important to ensure that in use the oxide-covered foil is always positive with respect to the other foil.

QUESTIONS

1. What is meant by the capacity of an electric conductor? How is it measured and on what factors does it depend?

2. Two similar vertical insulated plates, *A* and *B*, are placed facing each other about an inch apart. *A* is connected to the cap of a gold leaf electroscope whose lining is earthed and *A* is given a charge. State and explain what happens to the electroscope, (*a*) when *B* is earthed; (*b*) when *B*, still earthed, is brought nearer to *A*; (*c*) when *B* is moved to one side so that it is not directly facing *A*. What happens to the electroscope when a sheet of insulating material such as polythene is placed between *A* and *B*?

3. Describe how you would make a compact capacitor of large capacitance given some aluminium foil, some muslin and some solid ammonium borate. In what way does such a capacitor depend on chemical reactions between the component parts?

NUCLEAR ENERGY

CHAPTER 39

ELECTRONS

ELECTRIC sparks sometimes occur at the contacts of a switch when a circuit is broken. Each spark is a flash of light emitted by the air in the switch gap through which the current has passed. At atmospheric pressure dry air has a high resistance and normally only conducts if subjected to a large potential difference. Thus about 12 000 volts is required to produce a spark between two pointed electrodes 10 mm apart in air. A lightning flash from one charged cloud to another cloud or to the ground is a huge electric spark caused by a p.d. (potential difference) of millions of volts.

Whereas a gas at atmospheric pressure requires a high p.d. to produce a brief current flow in the form of a spark, by contrast, at low pressure a much smaller p.d. will cause a steady current. The conduction of electricity by a gas, usually termed *gas discharge*, is accompanied by some important phenomena at low pressure.

Discharge Tube Effects

The passage of electricity through air can be studied at different pressures, using a glass tube which is connected to a vacuum pump by a side tube (Fig.

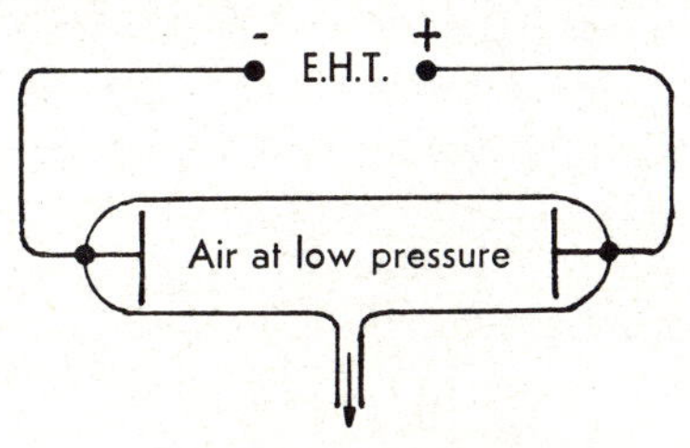

FIG. 490. *Discharge tube.*

490). A p.d. of several thousand volts d.c. is applied between two electrodes at opposite ends of the tube from the secondary of an induction coil or an extra high tension (E.H.T.) power supply. The appearance of the discharge as air is removed is shown in Fig. 491. Other gases give the same stages but different colours.

Luminous effects do not occur until the pressure has been reduced to about 20 mm of mercury when wavy violet streamers of light are seen (Fig. 491*a*).

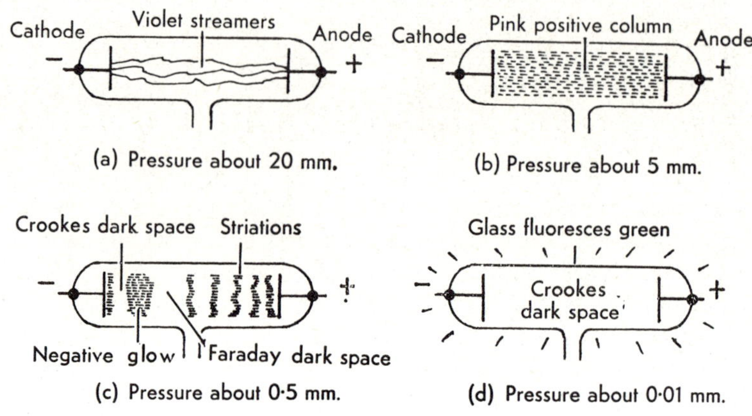

FIG. 491. *Luminous effects in discharge tube at reduced pressures.*

At a pressure of about 5 mm a pink positive column almost fills the tube (Fig. 491*b*). This is the glow used in advertising signs, its colour being dependent on the gas used: neon gives a red glow; argon gives a pale-blue glow. At a pressure of about 1 mm the bright glow in the tube begins to shrink towards the anode, when it is known as the *positive column.* At a pressure of about 0·5 mm the positive column becomes striated (Fig. 491*c*); next to it is a dark space known as the *Faraday dark space*, then a bright glow called the *negative glow*, another dark space called the *Crookes dark space*, and finally a narrow luminous glow around the cathode. As the pressure in the tube is further reduced, the positive column shrinks and the Crookes dark space expands. Eventually, at a pressure of about 0·01 mm, the Crookes dark space fills the whole tube. The gas in the tube no longer glows but the walls of the tube now fluoresce with a green light (Fig. 491*d*).

Maltese Cross Experiment

The most striking luminous effects occur in a discharge tube before the Crookes dark space fills the space between the electrodes, but it was the study of lower pressures which led to new knowledge about the atom. Although the Crookes dark space is dark, there must be something present since it makes the glass fluoresce: moreover, suitable minerals placed inside the tube also fluoresce. The tube shown in Fig. 492 has an aluminium Maltese cross as the

FIG. 492. *Maltese cross apparatus.*

anode and is evacuated to a pressure of about 0·01 mm. When the E.H.T. voltage is applied, the end of the tube fluoresces green except in the shadow of the cross. It appears that some kind of invisible radiation is being emitted by the cathode, travels in straight lines towards the anode, and causes fluorescence where it strikes the glass. The term *cathode rays* is used for this radiation.

Properties of Cathode Rays

What is the nature of cathode rays? The paddle wheel discharge tube (Fig. 493) shows that they convey energy: when the light mica vanes of the wheel are struck by cathode rays, the wheel travels along the rails. Although motion appears to be caused by the impact of the rays, it is in fact due to a secondary effect. On hitting the paddles the energy of the ray is changed into heat and raises their temperature; air molecules remaining in the tube are in turn warmed when they touch the vanes and therefore rebound with a greater

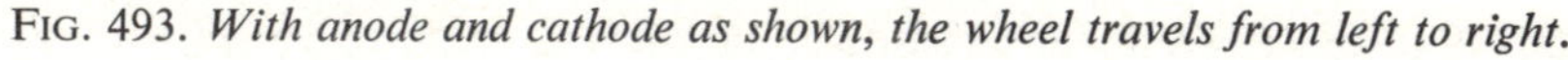
FIG. 493. *With anode and cathode as shown, the wheel travels from left to right.*

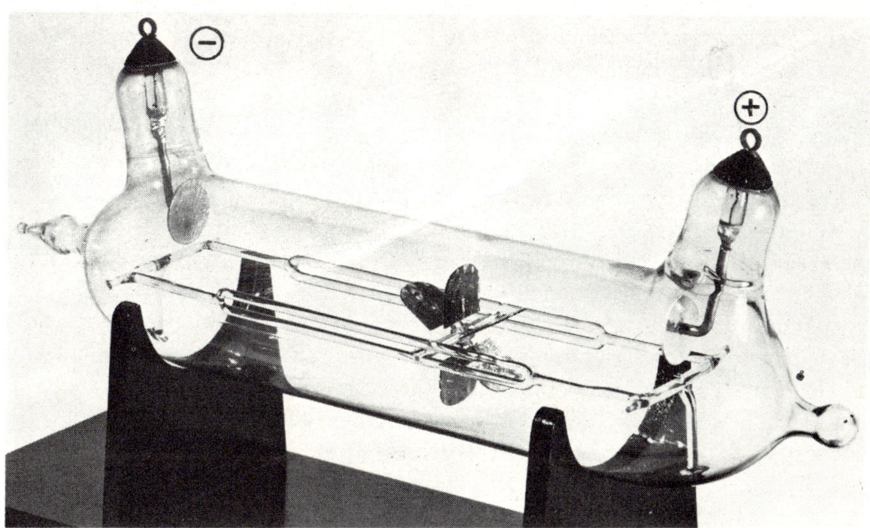

velocity. The force of reaction thus set up makes the wheel rotate. Reversing the polarity of the electrodes changes the direction of rotation of the wheel.

Energy can travel in two ways: by waves, as with light and radio signals; or by moving particles. Do cathode rays exhibit a wave-like or a particle-like behaviour? The Maltese cross experiment could be explained by either but other experimental evidence suggests that they are streams of particles of negative electricity. Thus, using a discharge tube having a screen coated with zinc sulphide which fluoresces blue and reveals the path of the cathode rays (Fig. 494), it can be seen that they are deflected by a magnetic field.

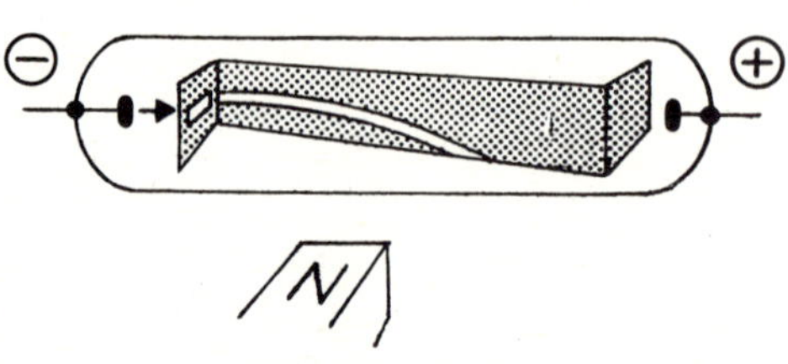

FIG. 494. *Magnetic deflection of the cathode ray.*

No one has ever succeeded in deflecting light by a magnet. If we assume that the cathode rays are negatively charged and are equivalent to a positive (conventional) current moving from anode to cathode, the direction of the deflection is that predicted by Fleming's left-hand rule (see page 424). Cathode rays may also be deflected by an electric field. If a beam of cathode rays is passed between two parallel metal plates across which there is a p.d., it is deflected towards the positive plate.

Discovery of the Electron

In 1897 Sir J. J. Thomson provided quantitative support for the belief that cathode rays are a stream of negatively charged particles, by measuring the ratio of the charge e carried by the particle to its mass m. He deflected cathode rays in a discharge tube, using electric and magnetic fields, and found that e/m always had the same value whatever the gas in the tube or the material of the electrodes. It was therefore quite reasonable to suppose that cathode ray particles were somehow or other present in all matter. Their speed in a discharge tube was about 0·1 of the speed of light.

What is the mass of a cathode ray particle? To calculate m from e/m it is necessary to know e. In 1909 Robert Millikan, an American physicist, started a series of very accurate experiments from which he found that there is a certain minimum electric charge and that a charged body either acquires this basic unit of electricity or some whole number multiple of it. A univalent ion in electrolysis also carries the basic charge and if it is assumed that a cathode ray particle does as well, then (using modern values) m is equal to $\frac{1}{1840}$ of the mass of the lightest atom, the hydrogen atom.

This is a surprising answer and showed that cathode ray particles, now called *electrons*, were very different from anything previously experienced, being much lighter than atoms. Modern work has confirmed that the electron does bear the basic unit of charge and so justifies the assumption made originally in calculating m. Sometimes e is called the fundamental unit of electricity, or the *electronic charge*; its value is $1{\cdot}602 \times 10^{-19}$ coulombs.

Thomson's Determination of e/m

To measure e/m for an electron, Sir J. J. Thomson used a gas discharge tube at cathode ray pressure (Fig. 495). A narrow electron beam is produced by a cathode C, a perforated anode A and a slit S. The beam makes a bright

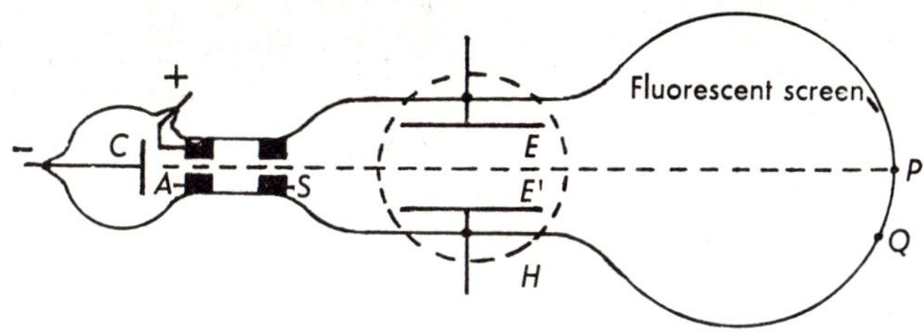

FIG. 495. *Thomson's apparatus for the determination of* e/m.

spot P on the fluorescent screen at the end of the tube. E and E' are metal plates sealed into the tube and between them an electric field can be applied from a battery. The dotted circle H represents two connected coils, one on each side of the tube, which give a magnetic field when a current is passed through them. This field is at right-angles to the plane containing the electron beam and the electric field.

First, only the magnetic field is applied, causing the beam to be deflected to Q and the distance PQ which the spot moves up or down is noted. (Reversing the direction of the magnetic field reverses the direction of the deflection.) The electric field is then applied and the p.d. adjusted so as to cancel the magnetic deflection and return the spot to its original undeflected position at P.

Knowing the magnetic and electric field strengths, the value of e/m and the velocity of the electrons can be calculated.

Millikan's Determination of e

Millikan measured the fundamental unit of electric charge e by charging very small oil drops and observing their motion under gravity and then under the action of an electric field. The principle of his apparatus is illustrated in Fig. 496. When a spray of fine oil drops is directed towards the small hole in the upper plate *A* of the cell, a few find their way through. In the space between *A* and *B* the drops are illuminated and observed as bright specks of light in a microscope. They appear to drift upwards but are actually falling slowly under gravity since the microscope inverts the image. The time taken by one drop to fall a known distance is measured by a stop watch. From this the weight of the drop can be found.

The drops usually become charged by friction in the process of spraying. Suppose a drop under observation has a negative charge, the application of a p.d. to the plates *A* and *B*, with *A* having a positive potential with respect to *B*, creates an electric field which drives the drop up to *A*. By adjusting the electric field the drop can be held stationary between the plates. In this condition the electric force on the drop equals its weight, which is already known from the first part of the experiment. The value of the charge on the drop can then be calculated.

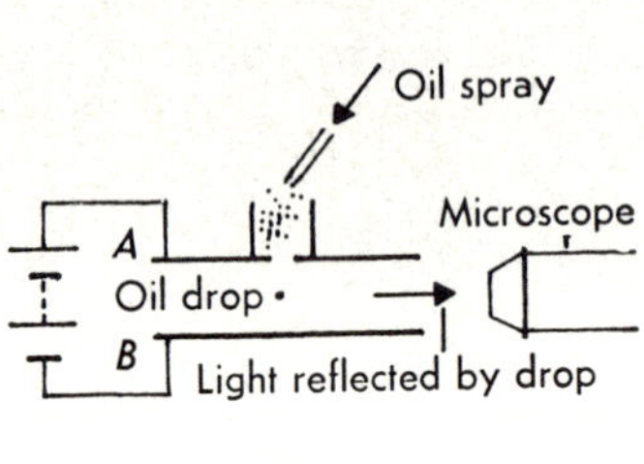

FIG. 496.

Millikan found that the size of the charge on a drop, whether positive or negative, was always the electronic charge *e* or a whole number multiple of it. Electricity, like matter, is atomic.

Electrons, Insulators and Conductors

Electrons are one of the basic constituents of all atoms and, since matter is not normally electrically charged, there must be something in the atom which carries a positive charge to balance the negative charge of the electron. We shall see later that the positive charge is associated with the anchor of the atom and cannot move as electrons can. Many of the facts of static and current electricity may now be explained in terms of the electron.

A negatively charged body is one which has more electrons than it does in the neutral state, while a positively charged body has fewer electrons than normal. Thus, when a polythene rod is rubbed electrons are transferred from cloth to polythene, making the polythene negative and leaving the cloth deficient of electrons and so positively charged. Why the electrons should pass from polythene to cloth and not in the opposite direction is not understood.

The difference between an insulator and a conductor is that in an insulator the electrons are firmly bound to their atoms whereas in a conductor they are loosely held and can travel freely from atom to atom. When an insulator is charged by rubbing, the movement of electrons is confined to the surface and no free electrons can pass through it.

Before the electron was discovered the convention had been adopted of regarding electric current as a flow of positive charges from the positive to the negative terminal of a battery. We now believe that current is a flow of negative electrons in the opposite direction, but so far as outside effects are concerned the two are equivalent. For this reason the old convention is retained and the direction of positive current considered.

Ions

In metals, electrons are entirely responsible for electrical conduction; in electrolytes and gases, the current-carriers are positive and negative ions travelling in opposite directions. A negative ion is an atom or group of atoms which has one or more electrons in excess of that required to neutralize its positive charge, while a positive ion is deficient of one or more electrons. Ions are already present in electrolytes (solutions of acids, bases and salts in water) and simply drift towards the electrodes when a p.d. is applied. On the other hand, in a gas there are normally very few ions but their number, and so the conductivity of the gas, can be increased by various ionizing agents. Three of these are:

1. *A flame*. In a flame, gas molecules are moving about rapidly and frequently colliding with one another. As a result some lose electrons and others gain them. An effective way of discharging a rod is to hold it briefly in the air above a burner; if it is negatively charged, positive ions are attracted to it and neutralize its charge.
2. *High energy radiation* such as ultra-violet radiation and X-rays.
3. *A radioactive substance*—more will be said about this later (Chapter 41).

The few ions always present in the atmosphere are caused by the so-called cosmic rays from outer space and radioactive minerals in the earth. They are responsible for a charged body, such as an electroscope, gradually losing its charge even when insulated.

In a discharge tube at cathode ray pressure, positive ions of the residual gas are the agents which, by bombardment, cause emission of electrons from the atoms of the metal cathode. Under a high p.d. and at low pressure the positive ions are accelerated towards the cathode. The ejected high-speed electrons, in their journey to the anode, are able to produce further positive ions in collisions with the neutral gas molecules they encounter. In this way the discharge is maintained.

Thermionic Emission

The operation of several electronic devices depends on the production of a stream of electrons from the atoms in which they normally reside. The method used in the gas discharge tube is neither efficient nor common today. A simpler way is to heat a metal or a metal coated with certain oxides. A pure metal like tungsten only emits a plentiful supply of electrons if it is white-hot; the oxides on the other hand release them at a very much lower temperature. This process, whereby electrons are emitted by a heated metal or oxide-coated metal, is known as *thermionic emission.* We can consider that the electrons "evaporate" from the heated surface in much the same way as vapour molecules leave a hot liquid. In general the rate of electron emission increases with temperature.

Diode Valves

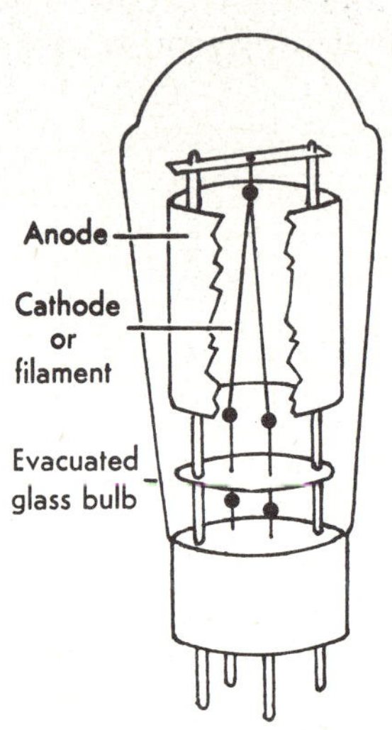

FIG. 497. *Diode valve.*

Thermionic emission is used in the diode valve (Fig. 497). It consists of a cathode or filament (usually coated with a mixture of barium and strontium oxides) heated directly by a current. Indirect heating is also used. The cathode is surrounded by a nickel anode and the whole assembly enclosed in an evacuated glass bulb having pins at the base for connecting the electrodes to an external circuit. When the anode has a positive potential with respect to the cathode, the electron stream emitted from the heated cathode is attracted to the anode and forms an electric current through the valve. If the anode is negative with respect to the cathode, electrons are repelled by the anode and therefore there is no current through the valve. Current can therefore only flow in one direction (hence the name valve), and this property enables the diode to perform certain essential operations in radio and television such as rectification (i.e. the converting of a.c. to d.c.).

Cathode Ray Tubes

A development from J. J. Thomson's e/m apparatus is the cathode ray tube now so common in television and radar sets. In it, electrons are obtained by thermionic emission from an electrode assembly similar to that in the diode and called the electron gun (Fig. 498). A narrow electron beam strikes a fluorescent screen inside the wide end of the tube and produces a bright spot of light. This beam can be deflected up and down by a p.d. applied to the *Y*-plates. If the p.d. is an alternating one, even of millions of hertz, the beam responds because the mass of the electrons is so small. In this case a vertical line of light would be seen on the screen. A p.d. on the *X*-plates gives a horizontal deflection. The brightness of the spot can be varied by controlling the number of electrons reaching the screen.

It should be noted that in cold-cathode gas discharge tubes, the emission of electrons from the cathode depends on the existence of a small amount of gas remaining in the tube and the use of high p.d's. A hot-cathode thermionic tube needs no residual gas and works no matter how highly evacuated it may be, the electrons being drawn across the vacuum by a comparatively small p.d. This smaller operating p.d. also makes hot-cathode tubes safer to work with because, as we shall see in the next chapter, there is then little danger of penetrating X-rays being produced.

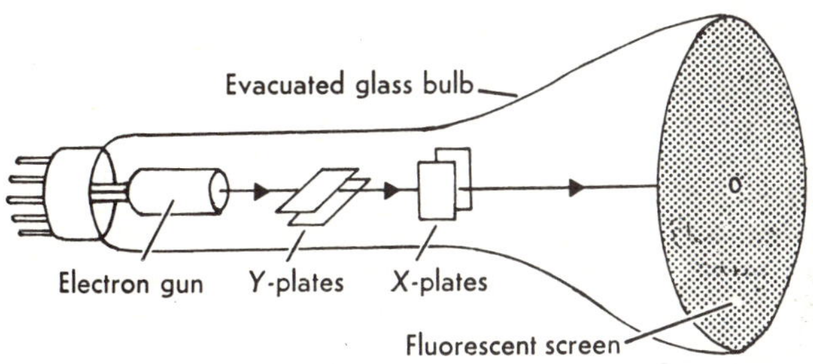

FIG. 498. *Cathode ray tube.*

QUESTIONS

1. What are cathode rays? Name three experiments which helped to establish their nature.

2. A stream of electrons travels in the direction of the arrow (Fig. 499). How will it be deflected relative to the paper if: (*a*) *P* is a N-pole and *Q* a S-pole; (*b*) *P* is joined to the positive and *Q* to the negative of a battery?

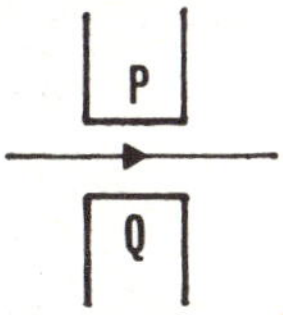

FIG. 499.

3. Describe how J. J. Thomson measured e/m for the electron. What assumption did he have to make before he was able to calculate the mass of the electron?

4. What important conclusion can be drawn from Millikan's oil drop experiment?

5. Explain in terms of electrons: (*a*) what happens when Perspex is charged by rubbing; (*b*) the difference between a conductor and an insulator; (*c*) how ions are produced in a flame.

6. What is meant by thermionic emission? Why can a diode be used as a valve? Draw a labelled diagram of a cathode ray tube and state two of its uses.

CHAPTER 40

X-RAYS

IN 1895 the German physicist Wilhelm Röntgen was experimenting in a darkened laboratory with a gas discharge tube (at cathode ray pressure) enclosed in black cardboard, when he noticed the fluorescence of a nearby screen painted with barium platinocyanide. The fluorescence only occurred when the tube was working and persisted even if the screen was two metres away from the tube. He realized that some kind of radiation was being emitted and traced its origin to that part of the glass wall of the discharge tube being struck by cathode rays (i.e. high-speed electrons). As well as visible fluorescent light the cathode rays were also causing the glass to emit an invisible radiation which was sufficiently penetrating to pass through the cardboard round the tube. In view of their unknown, mysterious nature he called them X-rays.

Gas-filled Cold-cathode X-ray Tubes

X-rays are produced whenever high-speed electrons are suddenly stopped. The earliest X-ray tubes were modified gas discharge tubes containing air at

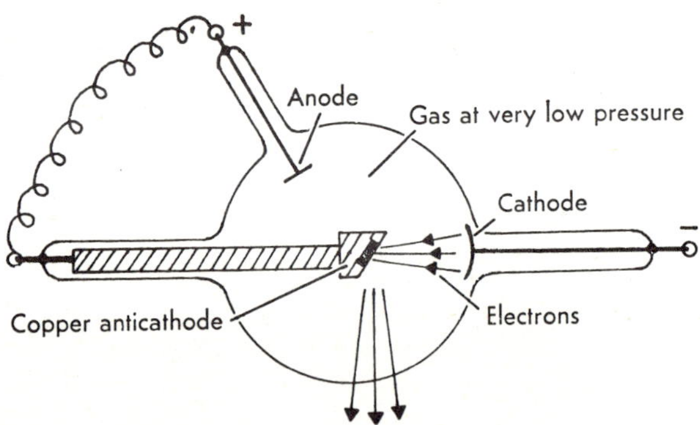

FIG. 500. *Gas-filled cold-cathode X-ray tube.*

cathode ray pressure in which a beam of electrons was focused from a concave cathode on to an anticathode (Fig. 500). The anticathode was connected to the anode and simply acted as a target to stop the electrons. A great deal of heat was generated, only a small percentage of the kinetic energy of the electrons being changed into X-rays, and so the anticathode was made from a high-melting-point metal such as tungsten embedded in a copper block. By inclining the surface of the target, the X-rays were emitted sideways from the tube. A high d.c. voltage was needed, up to 100 000 volts, and this was taken from the secondary of a large induction coil.

Evacuated Hot-cathode X-ray Tubes

Modern X-ray tubes are highly evacuated so that gas discharge does not occur, the electrons being produced by thermionic emission (Fig. 501).

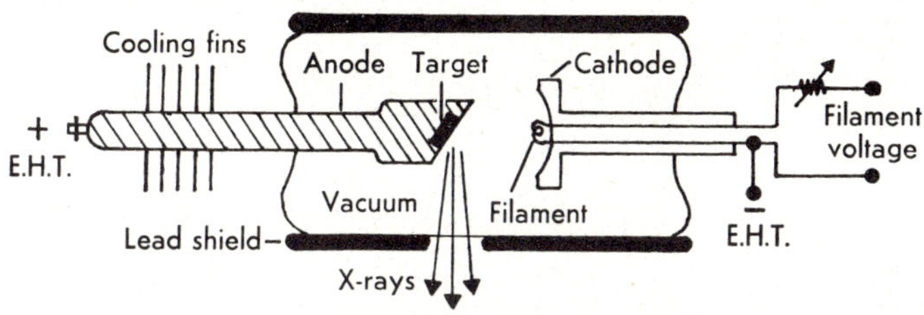

FIG. 501. *Evacuated hot-cathode X-ray tube.*

A stream of electrons is emitted by a tungsten wire filament heated electrically. The electrons are focused on to the target by the shape of the cathode and the arrangement of the filament in it.

The electrons are accelerated as in the gas-filled tube by a high p.d. between cathode and anode. This p.d. is usually obtained from a step-up mains transformer and a diode valve rectifier.

The *penetrating power of the X-rays* produced depends on the speed of the electrons, which, in turn, increases with the voltage across the tube. The terms *soft* and *hard X-rays* are used. Soft X-rays only penetrate soft objects like flesh; hard X-rays can penetrate thick, solid objects. A dentist might use about 15 000 volts to X-ray teeth, while an engineer looking for flaws in metals would require a much higher voltage, even a million volts. The *intensity of the X-rays*, that is, the quantity emitted per second, increases with the filament current; this controls the filament temperature and so the rate of emission of electrons. Control of both penetrating power and intensity was not possible in the gas-filled tube.

Properties of X-rays

The true nature of X-rays was a puzzle for many years after their discovery. Their known properties are:

1. They travel in straight lines, forming sharp shadows.

FIG. 502. *A jet engine of the type used in intercontinental airliners is here shown together with an X-ray photograph of it. The photograph shows the internal parts and their relative positions which cannot otherwise be seen except by taking the engine apart. The photograph was taken with a two-million volt X-ray camera using an exposure time of one hour.*

2. They are not deflected by magnetic or electric fields, which suggests they cannot be electrically charged particles.
3. They affect photographic plates and films in much the same way as light, making X-ray photography possible.
4. They cause substances such as barium platinocyanide and rock salt to fluoresce.
5. They penetrate more or less all matter. In general, the thicker and denser the material the less is the penetration. Thus, whilst sheets of cardboard, wood, and some metals fail to stop them, all but the most penetrating are absorbed by a 1 mm thick sheet of lead. Glass containing lead is a much better absorber than ordinary soda glass.
6. They ionize a gas, increasing its conductivity.

We now believe that X-rays are waves similar in nature to light but with a very much shorter wavelength and it was this short wavelength which made their identification as waves difficult. The longer waves are less penetrating, while the shorter ones are more penetrating.

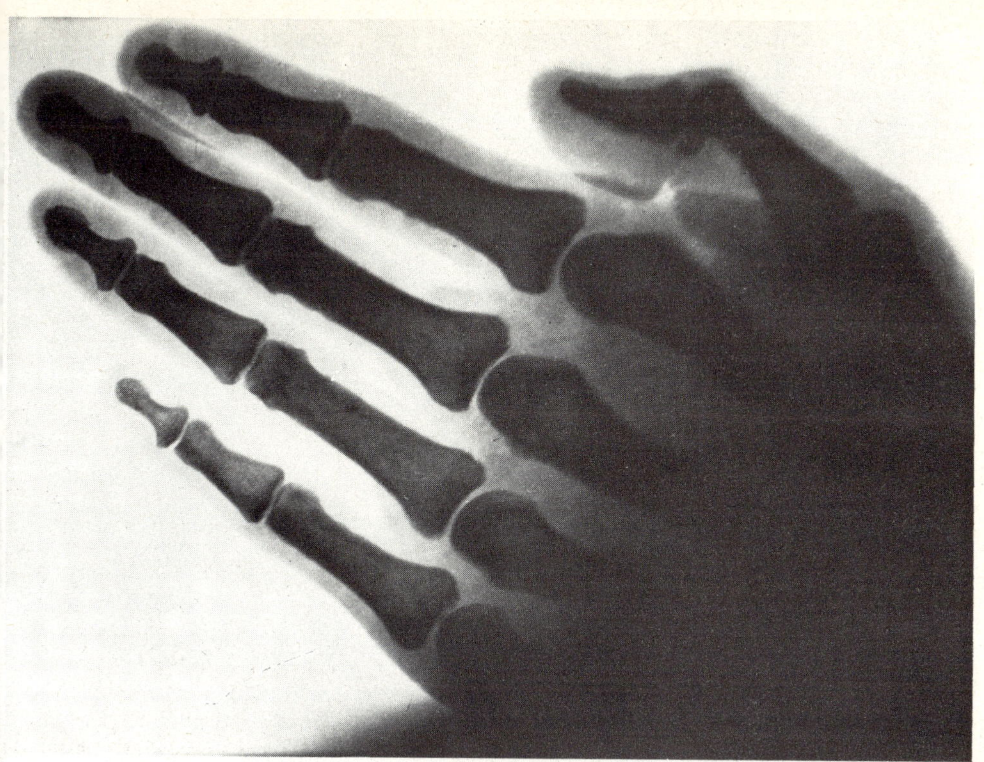

FIG. 503. *X-rays, discovered by Röntgen in 1895, rapidly found practical applications, first in medicine and later in industry. The X-ray photograph of a hand shown here was taken in 1896.*

Uses of X-rays

X-rays are useful mainly on account of their penetrating power. Radiographs or X-ray photographs have many medical applications. X-rays of a certain hardness can penetrate flesh but not bones, and cast an accurate shadow of the bones on a photographic plate or fluorescent screen. Suspected fractures can thus be investigated. In industry X-rays are used to detect imperfections in steel rails, girders for bridges, castings and welded joints.

In the detection of lung tuberculosis by mass radiography the patient stands so that X-rays cast a shadow of his chest on a fluorescent screen; this shadow is photographed by a camera. The roll of film used is easily and quickly processed and allows a large number of people to be examined in a short time. Any abnormality in a photograph may result in the patient having a full-size picture taken for further investigation.

In hospitals, very hard X-rays are used to destroy cancer cells, which cause malignant growths. Great care, however, is always taken to avoid unnecessary doses of X-rays as they have a harmful effect on normal cells which may not be apparent for several years. Today all X-ray apparatus is carefully sur-

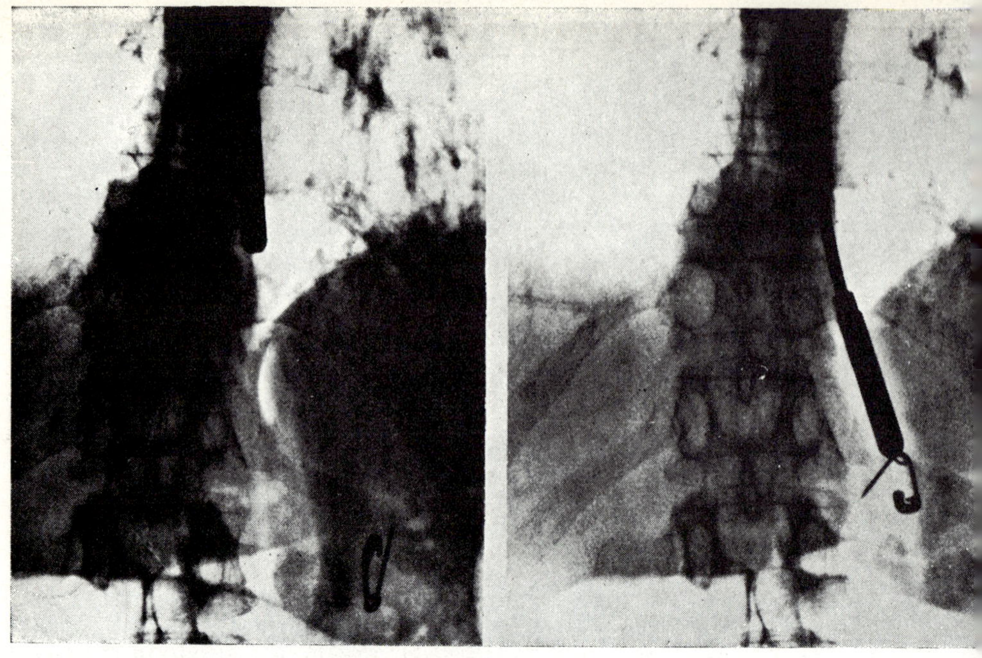

FIG. 504. *X-rays projected on to a fluorescent screen are here being used to guide a delicate operation. An open safety pin has been swallowed by a patient. A small electromagnet is inserted down the patient's throat (left). When it is seen to be in the correct position for grasping the blunt end of the pin, the electromagnet is switched on and the pin withdrawn safely (right).*

rounded by lead shields to confine the radiation to the required direction and protect the operators. Prolonged exposure to X-rays is dangerous and can cause death.

As a scientific probe X-rays have, like light, proved very helpful. The structure of crystals and the molecules of many compounds have been determined by their aid. Every element has its own characteristic X-ray spectrum which reveals much about its atomic structure.

QUESTIONS

1. Under what circumstances are X-rays produced? Draw a modern X-ray tube labelling the important parts. How do the X-rays change if the p.d. across the tube is increased? How can the intensity of the X-rays be increased?

2. State four properties of X-rays.

3. Give three uses for X-rays in medicine.

CHAPTER 41

RADIOACTIVITY

IN 1896, Henri Becquerel, Professor of Physics at the University of Paris, discovered that uranium compounds emitted a penetrating radiation which affected a photographic plate. On investigating his discovery further, Becquerel found that the rays from uranium and its compounds were similar to X-rays: not only did they affect a photographic plate, but they penetrated materials opaque to light, and ionized gases. This last fact he observed by bringing a uranium compound near a charged gold-leaf electroscope; the collapse of the leaf indicated the production of ions.

The Curies and Radium

Becquerel's discovery aroused great interest and a search for other radioactive substances was made. Foremost among the investigators were Marie Curie and her husband Pierre, who was one of Becquerel's colleagues. First, thorium compounds were found to be radioactive, but the Curies' most important discovery concerned the great activity of pitchblende, an ore containing uranium oxide. They suspected the presence of an element more radioactive than uranium itself. A preliminary analysis in 1898 led to the isolation of two new elements, which they named polonium and radium.

To confirm their claim, the Curies felt they had to work with much larger quantities in order to obtain products of greater purity. Through the co-operation of the Austrian Government they obtained from mines in Bohemia 1000 kg of pitchblende residues from which much of the uranium had been extracted (for colouring glass). After working for four years in a leaky, abandoned shed near the University of Paris they isolated 0·1 g of radium chloride in 1902, a truly great achievement for which they were later awarded a Nobel prize.

Over forty radioactive elements are known to occur naturally. They all have high atomic weights. Many artificial radioactive substances, called radioisotopes, have now been made and are used in medicine and industry.

Magnetic Deflection Experiments

The radiations from radioactive substances have many properties similar to those of X-rays but they appear to be emitted in very different circumstances. Their exact nature was only established by the work of many scientists including the Curies and Lord Rutherford. The radiations are of three types called *alpha* (α), *beta* (β) and *gamma* (γ) *rays* and some of the chief evidence for this conclusion came from studying their behaviour in magnetic and electric fields.

Alpha rays are deflected by a strong magnetic field, a fact that suggests they are charged particles with a comparatively large mass. The direction of the deflection indicates (using Fleming's left-hand rule) that they carry a positive charge. A more appropriate term for them is *alpha particles*. Their charge to mass ratio can be determined as for the electron, using both electric and magnetic fields. The value obtained, when taken in conjunction with other information, enables us to conclude that alpha particles are helium atoms which have lost two electrons and so have a positive charge twice the size of the electronic charge. They are doubly charged helium ions with a mass nearly four times that of a hydrogen atom.

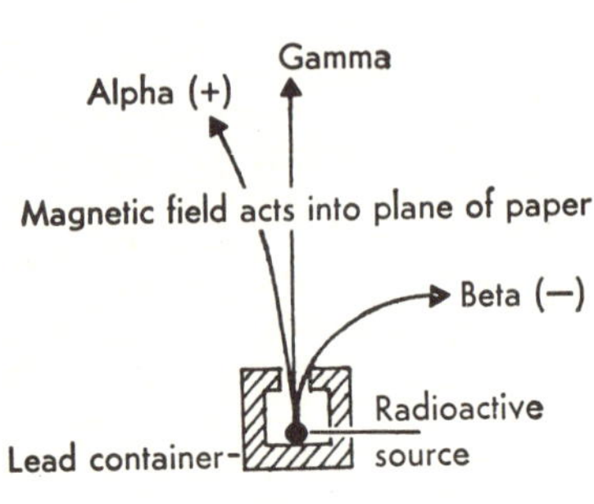

FIG. 505.

Beta rays are easily deflected by a magnetic field but in the opposite direction to alpha particles. Measurements of the ratio of charge to mass reveal that they are streams of high speed electrons similar to cathode rays. They are better called *beta particles.*

Gamma rays are unaffected by electric and magnetic fields and are in fact very short wavelength (hard) X-rays whose wavelengths can be measured.

The behaviour of the three kinds of radiation in a magnetic field is summarized in Fig. 505.

A radioactive source is shown at the bottom of a lead container with a small hole at the top to allow a narrow beam of rays to emerge. The deflections shown are produced by a magnetic field, applied at right-angles to the diagram, acting into the paper.

Naturally occurring radioactive elements emit either alpha or beta particles but rarely both. In some cases gamma rays accompany the alpha or beta particles.

Absorption of Alpha, Beta and Gamma Radiation

Alpha particles are the least penetrating, radiation being stopped by 100 mm of air, 0·1 mm of aluminium, or a thick sheet of paper. They are able to produce marked ionization in a gas, partly because of their relatively high mass and low velocity. All the alpha particles from a given radioactive element are emitted with the same velocity.

FIG. 506. *Marie and Pierre Curie in their laboratory.*

Beta particles are more penetrating than alpha particles. They produce considerably fewer ions per cm of their path length in, say, air, because of their greater speed and smaller charge. The faster they move, the less time do they spend near air molecules and so the less chance do they have of knocking an electron out of the molecule. They are emitted with various velocities approaching the velocity of light, the most energetic can penetrate several metres of air and 1 mm of aluminium.

FIG. 507.

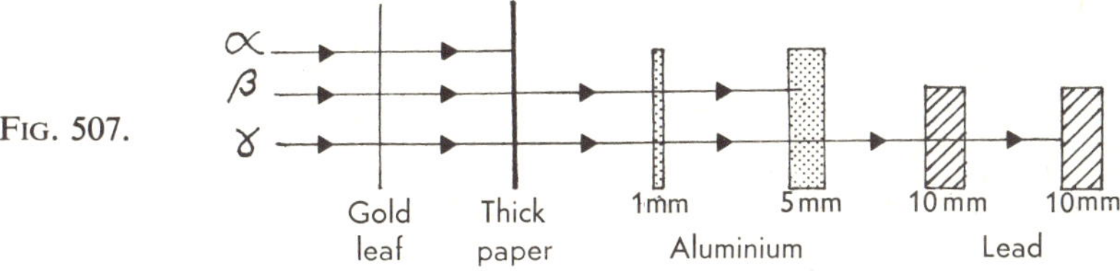

Gamma rays are highly penetrating, several centimetres of lead sometimes failing to absorb them, although they produce comparatively little ionization on passing through a gas. They travel with the speed of light. The penetrating powers of these rays and particles are summarized in Fig. 507.

Radioactive Decay and Half-life

We now know that the atoms of a radioactive element are breaking up spontaneously, emitting radiation, and changing into the atoms of another element which is itself usually radioactive. This in turn disintegrates to become something else, and so on. For example, when an atom of radium disintegrates it emits an alpha particle to become an atom of the heavy gas called radon. Radon also gives off an alpha particle forming radium *A* and eventually, after the formation of various elements by the emission of alpha and beta particles, the end-product is non-radioactive lead. A radium source is really a mixture of several elements which are themselves continually being formed and breaking up.

All radioactive elements decay at definite but different rates. Thus, measurements show that if we had one gramme of radium, half of it would have changed into radon after 1620 years. In a further 1620 years half of this would have gone, leaving a quarter of a gramme of radium. Every 1620 years the amount of radium decreases by one-half and a convenient way of expressing this fact is to say that the *half-life* of radium is 1620 years. Half-lives of different radioactive elements vary greatly; for radon it is 3·8 days and those for other members of the radium series are given in Fig. 508 which shows

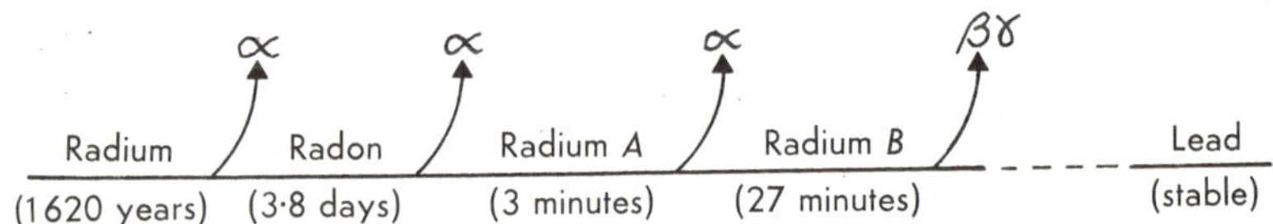

FIG. 508. *Part of the radium decay chain, showing half-lives.*

part of the radium decay chain. The rate of decay is unaffected by chemical combination, heat, or the application of pressure.

Although we can calculate what fraction of the atoms present at the start of a given interval will disintegrate during that interval, we cannot predict when a particular atom will disintegrate. Why some disintegrate rather than others we do not know. Radioactive decay is a statistical process and is rather like tossing a large number of coins. We can be sure that about half will turn up heads but we cannot be certain which they will be.

Radiation Detectors

The radiation from radioactive substances was discovered originally by its effect on a photographic plate and although this method is still common with the use of special emulsions, other devices have been invented.

The *Geiger tube* is widely used to detect ionizing radiation, especially beta

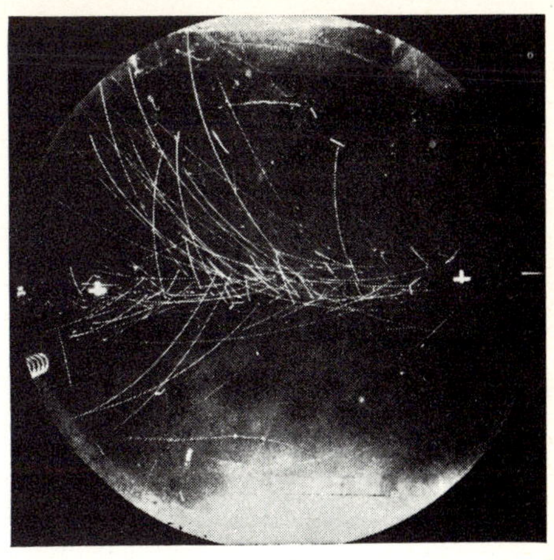

FIG. 509. *Tracks formed by atomic particles in an expansion cloud chamber. The different types of particles and rays produce distinctive tracks. By analysing and measuring such photographs as the one shown, much is learnt about the forces exerted by these particles and thus about atomic structure.*

particles and gamma rays. A common arrangement (Fig. 510) consists of a glass tube containing an easily ionized gas such as neon, at a pressure of about 10 cm Hg, along with a trace of bromine. The cathode is in the form of a metal cylinder surrounding a thin wire anode. The electrodes are connected to a suitable H.T. supply, frequently 450 volts. If beta particles of low penetrating power are to be detected, there is a very thin end-window, made of glass or mica, which allows their entry. When ionizing radiation enters the tube, gas ions are produced and are attracted to the electrodes. A very small current flows and, after amplification, is fed to a loudspeaker, which gives a click. The greater the intensity of the radiation the faster do the clicks occur. In another arrangement the tube is connected to a counting-rate meter (i.e. a *ratemeter*) which measures the rate at which radiation is being received. Alternatively, individual counts can be recorded automatically if the tube feeds an electronic device called a *scaler*.

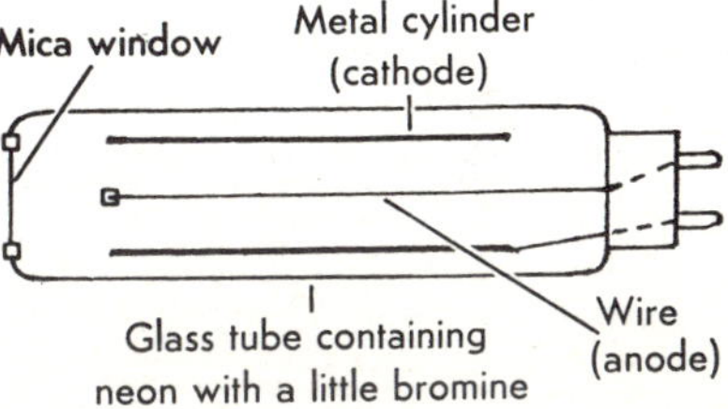

FIG. 510. *Geiger tube.*

The *expansion cloud chamber* was invented by C. T. R. Wilson in 1911 and enables the tracks of invisible particles and rays to be seen. Normally if air saturated with water vapour is cooled by a sudden expansion, drops of water condense on dust particles to form a cloud. Wilson found that gas ions can also act as condensation nuclei. If then an alpha particle, say, passes through a dust-free cloud chamber just before an expansion occurs, drops of water will condense on the positive and negative air ions it produces. With suitable illumination, a white trail marking the track of the alpha particle can be seen and photographed. The tracks are rather like the "vapour" trails of ice crystals formed by high-flying aircraft which are invisible from the ground.

Fig. 511 shows the principle of the cloud chamber, the sudden expansion being caused by downward movement of the piston. After each expansion it is desirable to remove the ions by applying a p.d. of two or three hundred

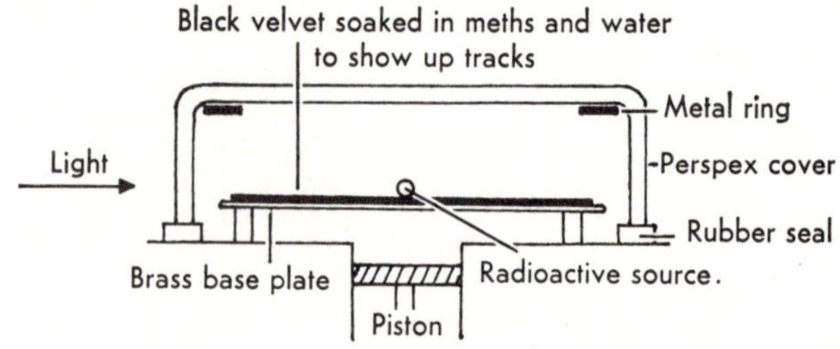

FIG. 511. *Wilson's expansion cloud chamber.*

volts between a metal ring at the top of the chamber and its base. The resulting electric field clears the ions from the chamber. Instead of using just water to saturate the air, a mixture of alcohol and water is common, since it reduces the expansion necessary for drops to form on the ions. Each kind of particle or ray can be recognized from its track. Typical tracks for alpha, beta and

FIG. 512. *Typical tracks formed in an expansion cloud chamber.*

gamma rays are shown in Fig. 512. By measuring path-lengths, the speed and kinetic energy of ionizing particles can be estimated. Work with cloud chambers has led to the discovery of new atomic particles.

Safety Precautions

Radioactive substances have to be treated with respect. Becquerel carried a few milligrammes of a radium salt in a glass tube in his pocket and after a few days it produced a sore on his body. Even small doses of radiation received over a long period have a cumulative effect and may cause leukemia (cancer of the blood) in later life. Madame Curie died because of this. The danger is all the greater because it is unseen, and one of which we may not be aware.

Alpha particles present little hazard so long as they do not get into the body, being completely stopped by the outer layers of skin. Protection from

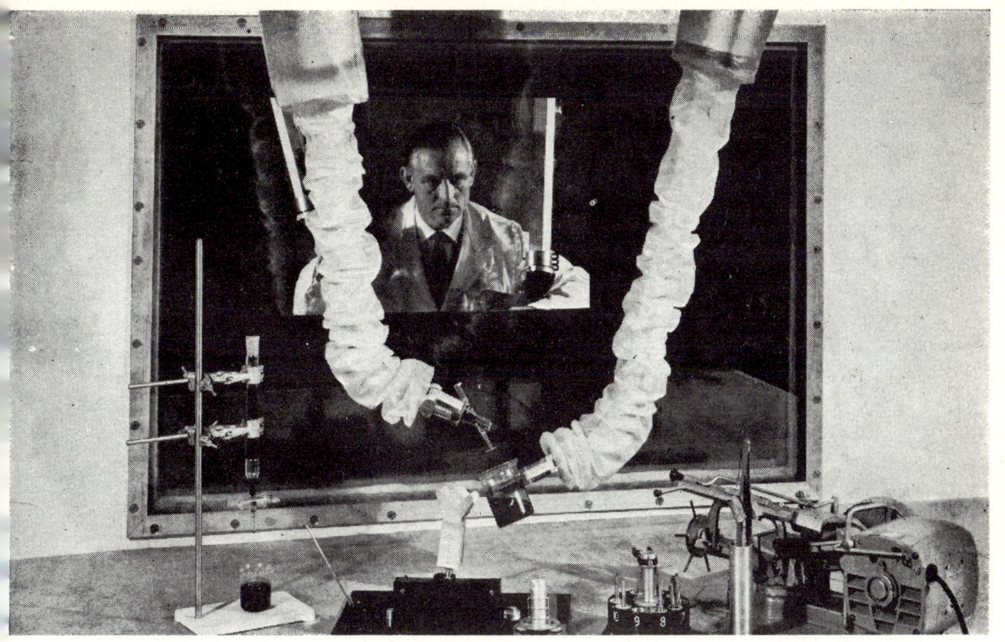

FIG. 513. *A strong radioactive source is here being handled by means of a "master-slave" manipulator. The jaws that handle the radioactive material are mechanically controlled from outside the cell by the operator. The cell has concrete walls over 1 m thick. The operator looks into the cell through a tank filled with zinc bromide which gives as much protection as the concrete.*

beta particles can be obtained from a sheet of aluminium or Perspex a few millimetres thick. Gamma rays, like X-rays, require lead shielding.

In laboratories radioactive sources are kept in thick-walled lead containers called lead castles. Eating, drinking and smoking are forbidden. Hands are washed before and after experimental work. Laboratory coats are worn and checked for radiation.

Handling precautions taken depend to some extent on the strength of the sources used. Weak sources can be lifted with forceps, but even they must not be held near the eyes or other susceptible organs. Strong sources require much more elaborate safety measures and are manipulated by mechanical tongs using remote control from behind windows of lead glass or walls of lead bricks or concrete.

Some artificial radioisotopes have half-lives of only a few days or weeks and soon become harmless after they have fulfilled their intended medical or industrial purpose. On the other hand, radium, with its long half-life of 1620 years, suffers negligible decay and is always a potential danger. The same hazard arises with radiostrontium from nuclear bombs (sometimes called atomic bombs), which can eventually get into food.

We are all subject to a certain amount of "background" radiation from

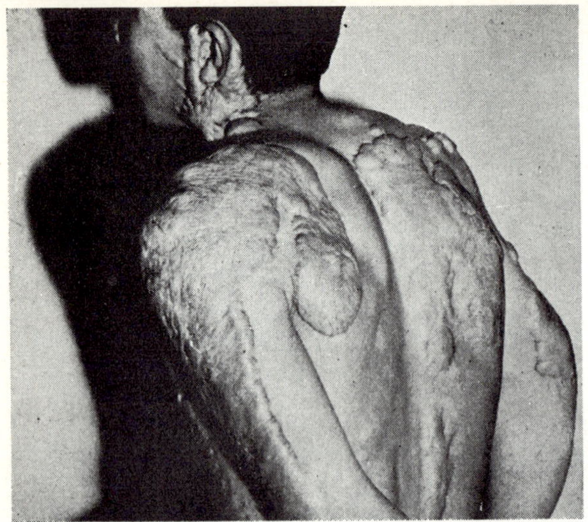

FIG. 514. *These terrible radiation burns were received as a result of the nuclear bomb dropped on Hiroshima, Japan, in 1945. They were caused by prolonged contact of radioactive fall-out from the bomb with the skin of the victim.*

radioactive minerals and from the cosmic rays entering the atmosphere from outer space. In laboratories using radioactive substances and in atomic power stations strict precautions are taken to ensure that workers do not receive more than what is considered to be the safe maximum dose.

QUESTIONS

1. Describe the nature of the three types of radiation from radioactive substances.

2. Fig. 515 shows a radioactive source inside a lead cylinder which has a small opening at the top. When a magnetic field is applied into, and perpendicular to, the paper, a photographic plate placed to the left of the opening is found to be affected. What conclusion can you draw about the radiation?

3. The half-life of cobalt 60 (a radioisotope used instead of X-rays in the treatment of cancer) is 5·3 years. Explain this statement. If a certain radioisotope has a half-life of two days, how long will it take for 1 g of it to decay to 0·25 g?

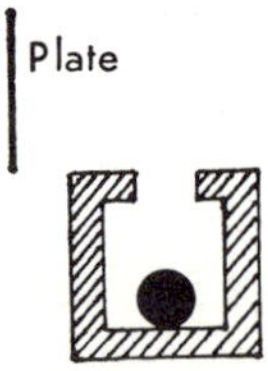

4. What is a Geiger tube used for and how does it work?

5. Explain the action of an expansion cloud chamber. What information can be deduced from the tracks?

6. A Geiger tube and scaler gives the background radiation as 1 count per second. When a watch with a luminous dial is brought close to the tube with its glass front facing the tube, 60 counts per second are recorded; when the watch is brought close to the tube with its stainless steel back nearest the tube, 2 counts per second are recorded. What conclusions can you draw regarding the kind of radiation reaching the Geiger tube from the watch in each case? Give reasons for your answers.

Plate

FIG. 515.

CHAPTER 42

THE ATOM

THERE are two possible theories regarding the structure of matter: either it is continuous, like a jelly filling a mould, and has no structure; or it consists of separate particles (i.e. atoms) with intervening spaces, rather like apples in a barrel. It is the latter view which has proved most fruitful in interpreting the observed facts, and one of the features of twentieth century science has been the growth of the atomic theory of matter.

Chemistry and the Atomic Theory

The founder of the modern atomic theory was John Dalton, an English schoolmaster who proposed it in 1803. He thought of atoms as tiny, indivisible balls. All the atoms of a given element were, he suggested, exactly alike, while the atoms of each element differed from those of other elements in both mass and behaviour: there were, therefore, as many kinds of atoms as there were elements.

According to Dalton, chemical reaction does not alter atoms but merely results in their rearrangement. Furthermore, combination to form "compound-atoms," as they were called, takes place between simple whole number ratios of atoms. Thus if elements *A* and *B* combine, they do so in the ratio of one atom of *A* to one atom of *B* or one atom of *A* to two atoms of *B* and so on. With these assumptions Dalton was able to explain the laws of conservation of mass and constant composition—the known laws of chemistry—and to predict the law of multiple proportions (page 39).

FIG. 516.

Dalton's compound-atoms are now called *molecules*, a molecule being a divisible particle containing several atoms. To enable the atomic theory to explain the chemical reactions of gases, the molecule idea has had to be extended to include some elements. Thus, the atoms of oxygen and many other gases appear to exist in pairs, forming diatomic molecules. The molecule can be thought of as the family group and the atoms as members of the family. In some families all members are alike, as in the molecule of an element, but in others the members are different as in the molecule of a compound. The water molecule contains two atoms of hydrogen and one of oxygen (Fig. 516).

Kinetic Theory

Dalton's atomic theory gives a simple explanation of the chemical behaviour of matter, while the kinetic theory deals with its physical properties. Being an extension of the atomic theory, it treats matter as molecules (containing one or more atoms) and attributes to them two further properties. First, they are in a state of continuous rapid motion and the higher the temperature the faster the motion. Second, they exert a force of attraction on one another which decreases as the distance between the molecules increases.

The theory explains in general terms many of the properties of matter and has been particularly successful with gases. By applying to gas molecules the laws of mechanics which hold for ordinary objects, the laws of Boyle and Charles can be predicted. This provides strong, though indirect, proof of the existence of molecules. One of the most direct pieces of evidence comes from Brownian motion which was discussed earlier in this book (page 51).

The kinetic theory also enables us to derive mathematical relations between some of the measurable properties of a gas and the characteristics of the individual molecules. Thus, the sizes and weights of atoms and molecules have been estimated and found to agree with the values obtained by other methods. For example, the hydrogen atom has a diameter of about 10^{-10} m.

Thomson's "Plum Pudding" Model of the Atom

By the start of the twentieth century the idea that atoms were indivisible was no longer accepted. The discovery of the electron and radioactivity had made it evident that atoms contained particles carrying positive and negative electrical charges. The question was how these sub-atomic particles were assembled inside the atom and how many were present in the atoms of a given element.

An arrangement favoured by Sir J. J. Thomson was known as the "plum pudding" model (Fig. 517). He considered that the atom was a positively charged sphere in which the negative electrons were distributed in sufficient numbers to make the atom neutral. By considering the electric forces acting on an electron, he showed that for stability the electrons had to be arranged in concentric rings in the atom.

The Russian chemist Mendeleev discovered many years previously that if the elements are arranged in order of increasing atomic weight, elements with similar chemical properties occur at regular intervals. Sir J. J. Thomson felt that this might somehow be connected with the number and arrangement of electric charges in the atom. Although some aspects of the "plum pudding" atom have proved useful, there was little enthusiasm among scientists for it.

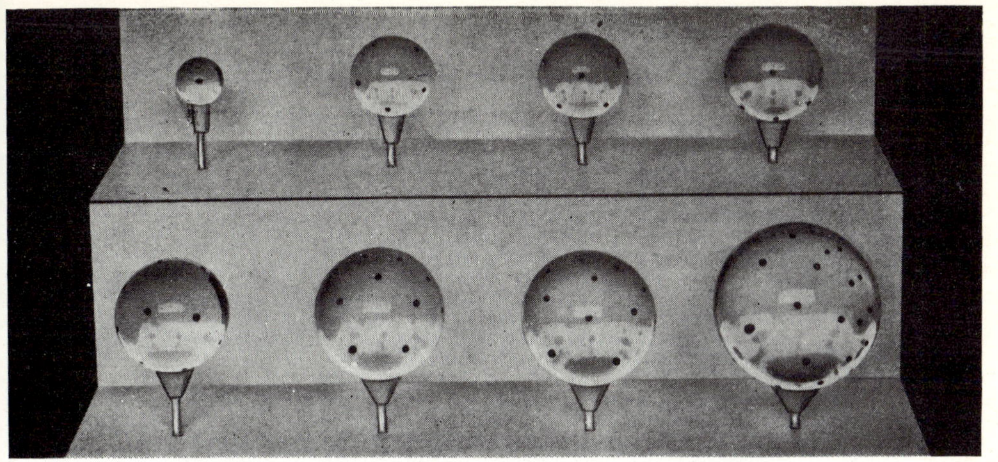

FIG. 517. *The positively-charged spheres of Thomson's "plum pudding" models of atoms are here represented by transparent spheres of perspex. The negative electrons are represented by the black spots arranged inside the perspex. In order of increasing size the models represent atoms of hydrogen (1), boron (5), carbon (6), fluorine (9), neon (10), sulphur (16), chlorine (17) and germanium (32). The numbers indicate the number of electrons present in each atom.*

Rutherford's Experiments on Alpha Particle Scattering

The next advance was made in 1911 by Lord Rutherford. Several years before, while investigating radioactivity, he had noticed that alpha particles penetrated thin foils, although some were deflected out of their straight line

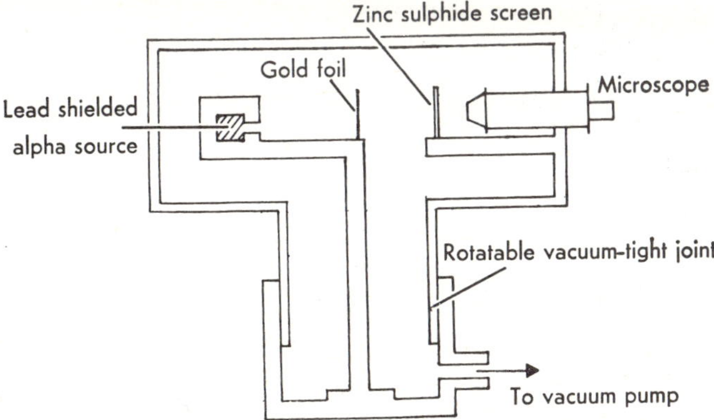

FIG. 518. *Geiger and Marsden's apparatus for investigating alpha particles.*

paths. He now set his two assistants, Geiger and Marsden, to study this further. A radioactive source fired a narrow beam of alpha particles at a very thin sheet of gold foil in an evacuated box (Fig. 518). The fate of the alpha particles was detected by a glass screen coated with zinc sulphide. The scintillations they produced were observed with a microscope which could rotate through any angle.

Geiger and Marsden spent many hours in a darkened room counting the scintillations for a wide range of angles. They found that while most of the alpha particles passed through the gold foil with little deflection, some were scattered by appreciable angles, and a few (about one in 8000) bounced back. That is, a very small percentage were reflected from the foil. This was surprising and could not be explained by the "plum pudding" model of the atom. At the time Rutherford said that it was rather like firing a shell at a piece of tissue paper and finding that it came back to hit you.

FIG. 519. *Lord Rutherford (1871-1937).*

The Nuclear Atom

To explain Geiger and Marsden's results, Rutherford proposed a nuclear model of the atom. He suggested that the positive charge, instead of being uniformly distributed throughout the atom, was concentrated in a nucleus of extremely small size compared with the whole atom. The negative electrons travelled round the nucleus rather like planets orbiting the sun.

In the scattering experiments most of the alpha particles were thus invading the empty space between the nucleus and the outermost planetary electrons of the gold atoms. In fact, they were passing *through* the atoms. However, when a positively charged alpha particle came near a positively charged gold nucleus, the two like charges repelled so violently that the alpha particle "rebounded" without necessarily striking the nucleus (Fig. 520).

Rutherford estimated that the diameter of an atom was about 10^{-10} m and that of a nucleus between 10^{-14} and 10^{-15} m, that is, ten thousand to one hundred thousand times smaller. The difference is such that the nucleus must feel like a very small fly in a very large hall.

By using X-rays, the number of positive charges on the nucleus of an atom can be found. This number, called the *atomic number*, must also be the number of planetary electrons surrounding the nucleus. The lightest element, hydrogen, has atomic number 1; for helium, the next lightest, it is 2; for lithium 3 and so on.

The heaviest naturally occurring element, uranium, with atomic number 92, has therefore 92 orbital electrons. However, since the mass of an electron is 1/1840 of the mass of a hydrogen atom, it is evident that even in the atom with the greatest number of electrons the electrons make a negligible contribution to its mass.

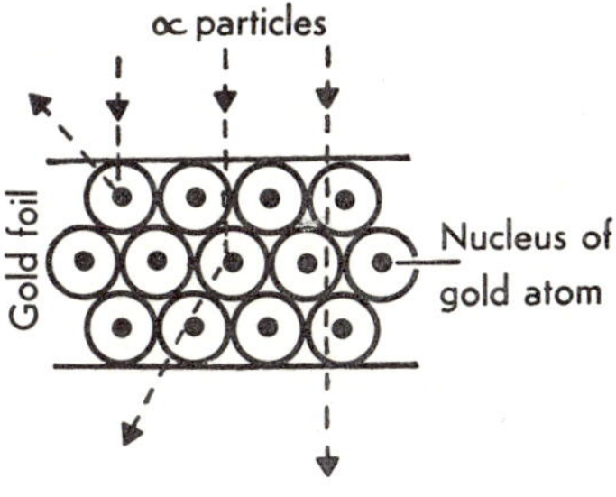

FIG. 520.

Therefore, not only does all the positive charge of an atom reside in the nucleus, but also most of its mass. The nucleus is very tiny and very dense, in fact, all the atomic nuclei in your body could in theory be packed into a grain of sand which would weigh almost as much as you do. Such very high densities have been observed among the stars, for example the "White Dwarf" stars. It appears that the atoms from which these stars are formed have been broken up and packed closely together without any waste of space. As a result, although some of the "White Dwarfs" are smaller in size than the earth, they are very much more massive.

Electron Shells

If the elements are arranged, in order of increasing atomic number, in a table of horizontal rows and vertical columns, so that elements with a similar chemical nature are in the same column, they form several periods in each of which there is a gradual change of properties. This arrangement is called the *periodic table* and the first part of it is shown in Fig. 521. It shows that the *chemical properties of the elements are periodic functions of their atomic numbers*. Thus, the third, èleventh and nineteenth elements are all alkali metals—lithium, sodium and potassium respectively.

This periodic behaviour can be explained if we assume that the orbital electrons are arranged round the nucleus in groups, called *shells*. Depending on the atomic number of an element, it can have up to seven distinct shells of different radii, denoted by the letters *K*, *L*, *M*, *N*, *O*, *P* and *Q*. Each shell has a limited capacity, thus the innermost (*K*) shell cannot have more than two electrons, the second is complete with eight electrons, the third can take eighteen and so on. The chemical properties are considered to be determined by the number of electrons in the outermost shell so that lithium, sodium and potassium are similar because they all have one outer electron (Fig. 522). Other chemical similarities can be explained in the same way.

One word of warning is necessary. Although the idea of electrons moving in certain orbits is easy to visualize and is useful for some purposes, modern atomic theory no longer considers that it is possible to pin-point electrons in this way. The picture of an atom is now one of a tiny but very dense positive nucleus surrounded by a hazy cloud of electrons.

1 Hydrogen							2 Helium
3 Lithium	4 Beryllium	5 Boron	6 Carbon	7 Nitrogen	8 Oxygen	9 Fluorine	10 Neon
11 Sodium	12 Magnesium	13 Aluminium	14 Silicon	15 Phosphorus	16 Sulphur	17 Chlorine	18 Argon
19 Potassium							

FIG. 521. *First nineteen elements of the Periodic Table.*

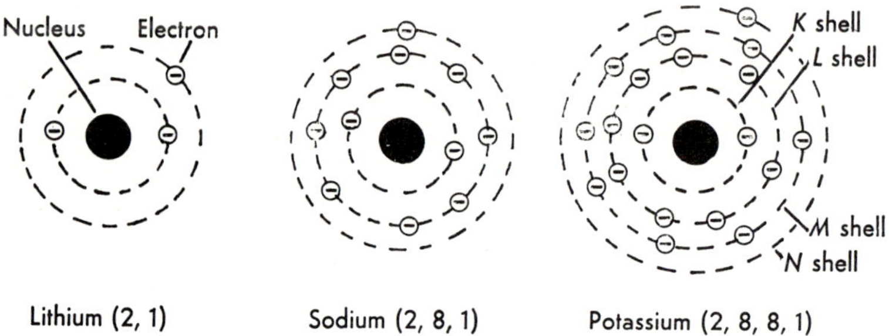

FIG. 522. *Arrangement of electrons around the nucleus in orbits, or shells.*

Structure of the Nucleus

The problem of the structure of the atom falls into two parts: first, the arrangement of the planetary electrons in the comparatively large space available to them—this we have considered; second, the structure of the small, massive, central nucleus, which forms the anchor of the atom.

The nucleus of the hydrogen atom is called a *proton*: since it carries a charge equal in size but opposite in sign to that on the electron, it is taken as the unit of positive charge. The atomic weights of many elements are almost whole numbers and therefore it was assumed that the nuclei of other atoms contained closely packed protons. Lord Rutherford verified this in 1919 when he knocked protons out of the nuclei of several elements, a point which will be discussed in the next chapter.

The proton contains practically all the mass of the hydrogen atom and so helium, with atomic weight 4, should have 4 protons in its nucleus and 4 planetary electrons. Its atomic number, however, is 2, therefore it only has 2 protons and 2 electrons. This discrepancy was only resolved in 1932 when

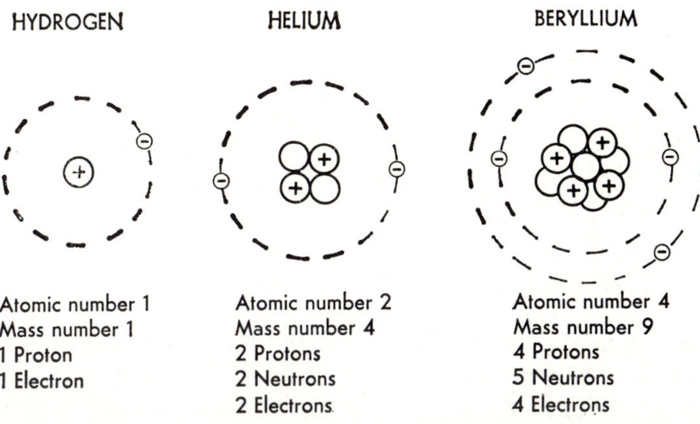

FIG. 523. *Atomic number and mass number differences explained by neutrons.*

Sir James Chadwick discovered the *neutron*, an uncharged particle with almost the same mass as the proton. It was then clear that the helium nucleus contained 2 protons and 2 neutrons, thus giving atomic weight 4 but atomic number 2. An alpha particle is a helium nucleus. Beryllium with atomic weight 9 and atomic number 4 has 4 protons and 5 neutrons in the nucleus and 4 planetary electrons (Fig. 523). Uranium has an atomic weight of 238 and an atomic number of 92 giving 92 planetary electrons, 92 protons and 146 (238 − 92) neutrons.

Protons and neutrons are called *nucleons* and the number of nucleons in the nucleus is called the *mass number* of the atom.

Isotopes

There are a few notable exceptions to the general rule that atomic weights are almost whole numbers. For example, chlorine has atomic weight 35·5. It cannot have half a proton or half a neutron; how then does this value arise?

If two elements have the same number of protons, the same number of electrons, in the same arrangement, but different numbers of neutrons, they would have the same chemical properties, the same atomic number and occupy the same position in the periodic table. But their atomic weights would be different. Such elements are called *isotopes.*

In the case of chlorine one isotope has 17 protons and 18 neutrons, giving atomic number 17 and atomic weight 35; another has 17 protons and 20 neutrons giving the same atomic number but atomic weight 37. Ordinary chlorine is a mixture of the two isotopes in the ratio of 3 to 1 so that the average atomic weight is 35·5. Chemically the two isotopes are indistinguishable, but physically one is slightly denser than the other. Using chemical symbols, they are written $^{37}_{17}Cl$ and $^{35}_{17}Cl$ where the superscript gives the mass number and the subscript the atomic number.

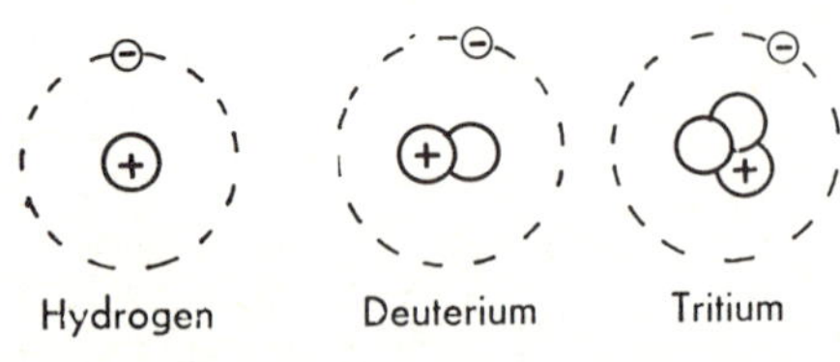

FIG. 524. *Isotopes of hydrogen.*

Most elements are mixtures of isotopes. Hydrogen has three isotopes. As well as ordinary hydrogen there is *deuterium* (heavy hydrogen) with one neutron in its nucleus, and *tritium* which has two. All three hydrogen isotopes have one proton and one planetary electron (Fig. 524). Water made from deuterium is more dense than ordinary water and is called *heavy water.*

Positive Rays

If there are holes in the cathode of a gas discharge tube at cathode ray pressure, luminous rays are seen to pass through them. Something is travelling along the tube from anode to cathode in the opposite direction to the cathode rays (Fig. 525). The rays can, like cathode rays, be deflected by electric and

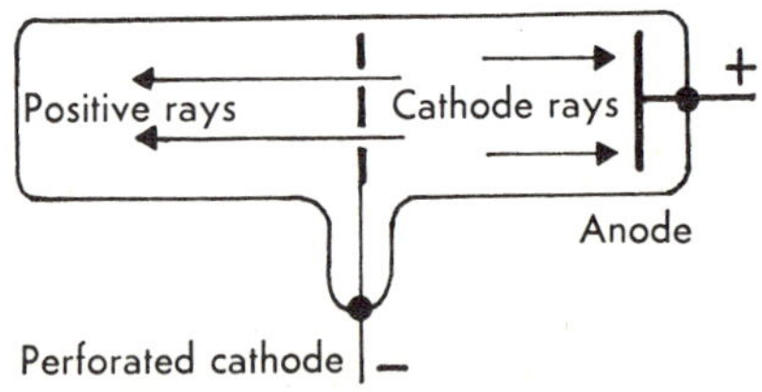

FIG. 525. *Generation of positive rays in a gas discharge tube.*

magnetic fields but in a direction which shows they are positively charged particles. Measurements indicate that the masses of the particles depend on the gas in the tube for they are always nearly equal to the masses of the gas atoms present. *Positive rays,* as they are called, are gas atoms which have lost electrons, that is, positive gas ions.

In 1913 Sir J. J. Thomson found that with neon gas in the discharge tube two kinds of neon ions, with atomic weights 20 and 22, were present. This was one of the earliest pieces of direct evidence for the existence of isotopes. It was concluded that ordinary neon with atomic weight 20·2 was a mixture of the two isotopes in the proportion which gives this value.

Effect of Alpha and Beta Decay on the Nucleus

In radioactive decay new elements are produced by the emission of alpha and beta particles from the nucleus.

An alpha particle consists of 2 protons and 2 neutrons. Hence, when a radioactive atom emits an alpha particle, its mass number decreases by 4 and since it has lost two positive charges from the nucleus, its atomic number decreases by 2. It becomes an element two places back in the periodic table. For example, when radium of mass number 226 and atomic number 88 emits an alpha particle, it decays to radon of mass number 222 and atomic number 86. The change may be written:

$$^{226}_{88}\text{Ra} \rightarrow \alpha + ^{222}_{86}\text{Rn}$$

In beta decay a neutron changes to a proton and an electron. The proton remains in the nucleus and the electron is emitted as a beta particle. The nucleus now has one more positive charge, its atomic number increases by 1 but its mass number remains the same. It becomes an element one place farther on in the periodic table. This type of decay occurs when a radioactive isotope of carbon, known as carbon 14, changes to nitrogen:

$$^{14}_{6}\text{C} \rightarrow \beta + ^{14}_{7}\text{N}$$

QUESTIONS

1. Outline briefly the evidence for our belief in the existence of atoms and molecules.

2. Describe Geiger and Marsden's experiment on the scattering of alpha particles by gold foil. What conclusion did Rutherford draw from this experiment?

3. Explain the terms atomic number and mass number. Aluminium has atomic number 13 and mass number 27, how many (*a*) protons, (*b*) neutrons, (*c*) electrons, does it have in its atom?

4. What are isotopes? How do the atoms of isotopes of a given element differ in structure? Illustrate your answer by reference to the isotopes of hydrogen.

5. Explain the meaning of the symbols $^{235}_{92}\text{U}$ and $^{238}_{92}\text{U}$.

6. What are positive rays? How did they lead to the discovery of isotopes? How do they differ from cathode rays?

CHAPTER 43

NUCLEAR ENERGY

THE first clue that the atom might be a vast store-house of energy was given by radioactive substances. One gramme of radium generates 570 joules of heat per hour and can continue to do so for a very long time. What is the source of an atom's energy?

Einstein's Equation

In 1905, while developing his theory of relativity, Einstein made the startling suggestion that matter and energy are equivalent and showed that the connection between the two was given by the equation:

$$E = mc^2$$

where, in appropriate units, E is the energy produced when a mass m disappears, c being the velocity of light. A small mass conversion would release a tremendous amount of energy, since c is so large.

The truth of Einstein's statement is no longer in doubt and all our calculations whether for atomic power stations or nuclear bombs are based on his equation. The problem is to improve on nature's performance in radioactive atoms and speed up the conversion of matter into energy by splitting nuclei.

Splitting the Atom

After the scattering experiments with metal foils had successfully established the nuclear atom, Rutherford and his team investigated the bombardment of various gases by alpha particles. In 1919 Rutherford himself studied the effect, using the apparatus shown in Fig. 526. It consisted of a brass cylinder with inlet and outlet tubes for introducing the gas, and an adjustable rod to support an alpha particle source. One end of the cylinder was closed by thin silver foil capable of stopping most of the alpha particles, but any that did penetrate fell on a zinc sulphide screen and their scintillations were observed through a microscope.

A surprising result was obtained when using nitrogen, the number of scintillations increasing markedly. More penetrating particles must have been produced to cause this effect, and by studying their deflection in a magnetic field Rutherford found they were protons (hydrogen nuclei). He assumed that they had been knocked out of nitrogen nuclei by the alpha particles. What of the disrupted nitrogen nuclei? It was not until 1925 that definite evidence of their fate emerged. In that year Blackett took photographs of the tracks of alpha particles passing through nitrogen in a cloud chamber. These made it clear that, in a disintegration, an alpha particle is absorbed by a nitrogen nucleus which, after emitting a high speed proton, becomes an isotope of

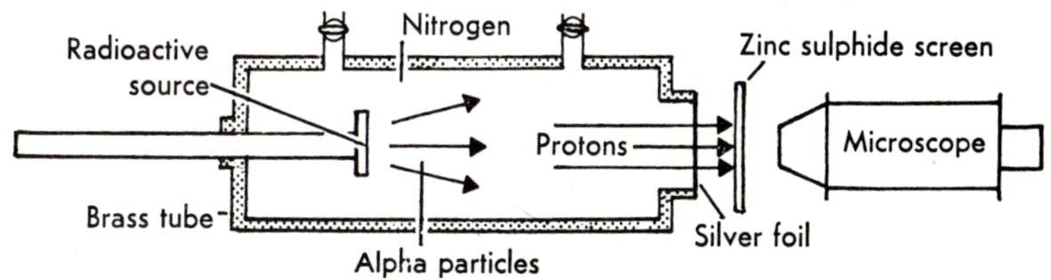

FIG. 526. *Rutherford's apparatus for splitting an atomic nucleus.*

oxygen with atomic number 8 and mass number 17. In Fig. 527, the thin track going up to the right is that of a proton and the short thick track that of a recoiling oxygen nucleus. The equation for the reaction may be written:

$$^{14}_{7}N + ^{4}_{2}He \rightarrow ^{1}_{1}H + ^{17}_{8}O$$

Nitrogen nucleus	Helium nucleus (alpha particle)	Hydrogen nucleus (proton)	Oxygen isotope nucleus

On each side of the equation the mass numbers total 18 and the atomic numbers 9.

Rutherford's suspicions were confirmed and there was now no doubt that the atomic nucleus had been split. This was a momentous discovery and realized the alchemists' ambition of turning one element into another. Although the transformation was only nitrogen to oxygen and not lead into gold, this work heralded the research which led to nuclear energy.

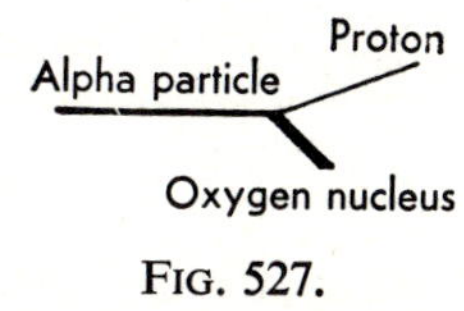

FIG. 527.

Protons were obtained from about twelve other light elements as in the nitrogen experiment. The kinetic energy of the protons was estimated from their range and in some cases was greater than that of the bombarding alpha particles, showing that additional energy had been released in the disintegration.

Cockcroft and Walton's Experiment

Alpha particles can split only the nuclei of light elements; the large positive charge on the nucleus of a heavy element exerts a powerful repulsive force and simply scatters them. By 1930 the need for a more penetrating atom-smashing missile was apparent; the proton, with only half the charge of an alpha particle, seemed promising.

Cockcroft and Walton, encouraged by Lord Rutherford, obtained protons from a hydrogen discharge tube and accelerated them to a high speed in a strong electric field. The fast protons so obtained were used to bombard a lithium target (Fig. 528). Evidence from the observation of scintillations and cloud chamber tracks revealed that occasionally a lithium nucleus absorbed a proton and then split into two alpha particles. The reaction may be represented by the equation:

$$^{7}_{3}\text{Li} + ^{1}_{1}\text{H} \rightarrow ^{4}_{2}\text{He} + ^{4}_{2}\text{He}$$

Lithium nucleus   Proton   Alpha particles

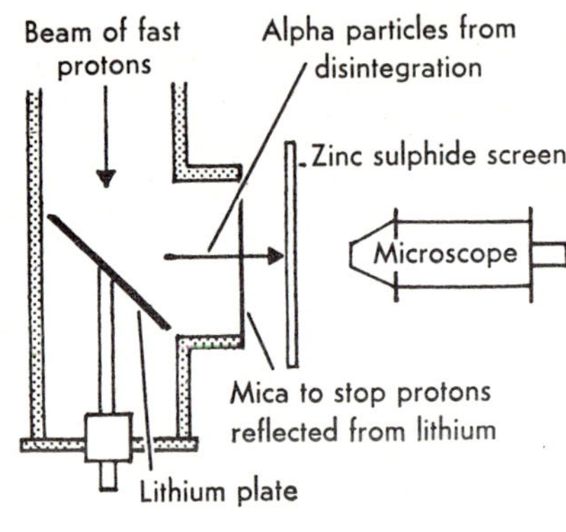

FIG. 528.

This was not only the first nuclear disintegration achieved by artificially accelerated particles but it also enabled Einstein's equation to be verified. The mass of two alpha particles is slightly less than the sum of the masses of a lithium nucleus and a proton, and according to Einstein this mass loss should appear as energy of the alpha particle fragments in a disintegration. The total kinetic energy of the two alpha particles was obtained from the distance they travelled after emission, and found to have the value predicted by $E = mc^2$.

Matter had successfully been converted to energy but the transformation was uneconomic and on a very limited scale.

Uranium Fission

When Chadwick in 1932 discovered the neutron by bombarding beryllium with alpha particles he provided physicists with an important new atomic missile.

Although the positive charge on a proton enables it to be accelerated to a high speed by an electric field, it also makes its approach to the nucleus more difficult. The neutron, being uncharged, is not subject to a repulsive force and can enter the nucleus of the heaviest atoms. Between 1934 and 1938 Enrico Fermi, an Italian physicist, split the atoms of heavy elements by neutron bombardment. But the energy supplied for the millions of shots required to produce even one disintegration far exceeded the energy released.

The outlook for nuclear energy was not bright; even Lord Rutherford had considered it an unlikely economic proposition.

In 1939 the situation changed dramatically. Two German scientists, Hahn and Strassmann, made the vital discovery that the nucleus of uranium 235 splits under neutron bombardment into two more or less equal parts,

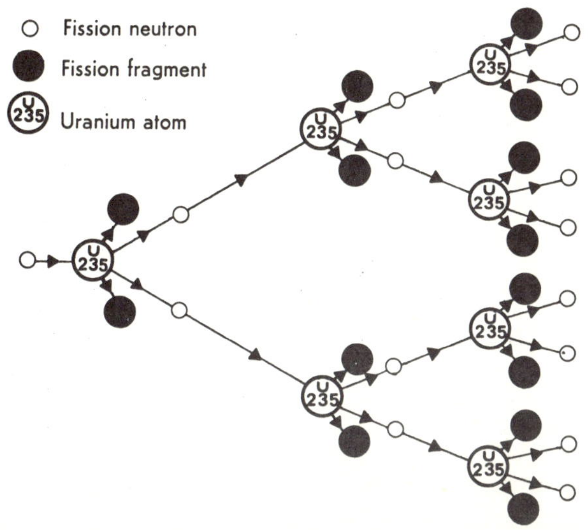

FIG. 529. *Representation of a chain reaction in uranium 235.*

frequently barium and krypton, which fly apart at great speed. This *fission* of uranium was rather different from any previous atom-splitting process in two respects. First, not only was the nucleus more deeply divided and bigger splinters obtained, but the combined mass of the disintegration products was appreciably less than that of the original uranium atom and neutron; in view of Einstein's equation, the importance of this for energy production was soon realized. Second, every uranium 235 atom split emitted one, two or sometimes three neutrons along with the two large fission fragments. The vital question now was whether the *fission neutrons* could split other atoms and set up a *chain reaction* throughout the whole mass of uranium, causing the conversion of matter into energy on a large scale.

A chain reaction does occur in uranium 235 (Fig. 529), so long as it is above a certain *critical mass.*

Fission neutrons moving at high speed through the great empty spaces in atoms are more likely to have nuclear collisions in a large mass than a small one, where escape from the surface is easier.

Nuclear Bombs

A huge release of energy can be produced by bringing together two or more pieces of uranium 235, each smaller than the critical mass. If, when combined, they exceed the critical mass, a very fast chain reaction spreads with explosive rapidity. The reaction is started in the super-critical mass of uranium, either by having a small source of neutrons in a bomb or by relying on stray neutrons which are always present in the atmosphere. There is a limit to the power of fission bombs, owing to the difficulty of getting all the sub-critical pieces together at the same instant.

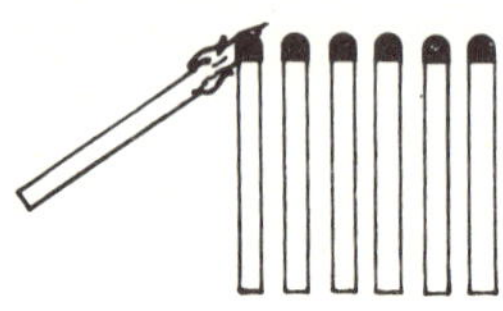

(a) A steady chain reaction

(b) An increasing chain reaction

FIG. 530.

The temperature in a nuclear bomb when it explodes is thought to be about 10 million degrees Celsius, similar to that inside the sun. As a result, the fission fragments and any unused uranium are changed into gases at very high pressure. The sudden expansion of these causes much of the blast damage accompanying the explosion.

The first nuclear bomb was exploded in the New Mexican desert on 16 July, 1945, and created such intense heat that the sand melted to form glass. One month later, the two bombs dropped on Hiroshima and Nagasaki in Japan, brought incredible devastation to life and property, spreading highly radioactive fission products over a wide area as "fall-out."

Chain Reactions

In a nuclear bomb, energy is released in a very short explosive burst by an increasing *uncontrolled* chain reaction. In a nuclear reactor, such as is used in a power station, heat has to be generated continuously by a steady, *controlled* chain reaction in which the fission neutrons are not allowed to run amok. Fig. 530*a* shows how a steady chain reaction can be demonstrated, using a row of matches; each match lights one other and the lighting rate remains constant. In Fig. 530*b* an increasing chain reaction is illustrated, each match sets fire to several others and the lighting rate increases rapidly. Control of the fission rate in a nuclear reactor is achieved by introducing an appropriate number of rods of neutron-absorbing material such as boron.

Nuclear Reactors

In a nuclear power station the heat required to produce steam is obtained from a nuclear reactor instead of a coal furnace. At present, most reactors use natural uranium; this is a mixture of two isotopes, uranium 235 and uranium 238, one in every 140 atoms being uranium 235. Uranium 238 absorbs neutrons without undergoing fission and would suffocate a uranium 235 chain reaction in natural uranium. However, uranium 238 prefers fast to slow neutrons, whereas uranium 235 prefers slow ones.

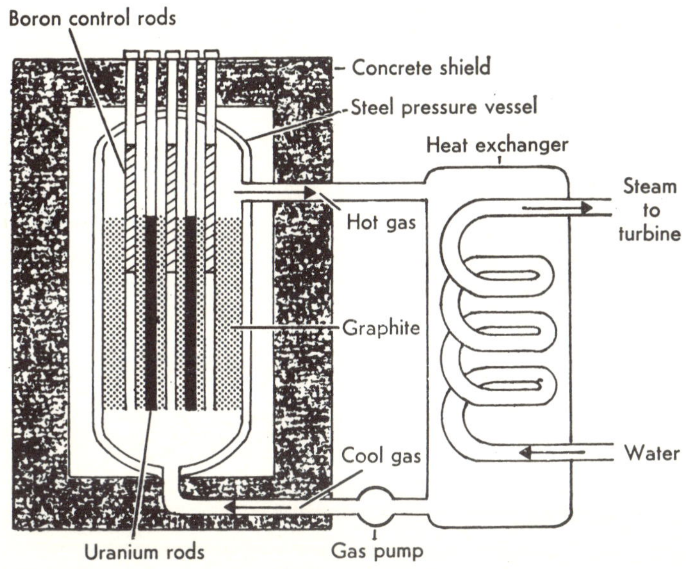

FIG. 531. *Simplified diagram of a nuclear reactor using natural uranium.*

Fission-released neutrons are fast and so are liable to capture by non-fissile uranium 238; fortunately they can be slowed down by collision with atoms of about their own mass. Such speed-reducing substances are known as *moderators*, and graphite is commonly used. Thus, with a moderator natural uranium gives a chain reaction. The alternative is to undertake the costly process of separating uranium 235 from uranium 238.

A simplified diagram of a nuclear reactor using natural uranium is shown in Fig. 531. It consists of a strong steel pressure vessel containing graphite blocks. These have a series of vertical channels, into which sufficient uranium fuel rods are inserted to make the reactor critical. Other channels contain boron steel rods, which can be raised or lowered as required. When all the control rods are in the reactor, too many neutrons are absorbed for the chain reaction to occur. Removing some rods enables the chain reaction to proceed at a steady rate.

A current of high-pressure carbon dioxide is pumped through the pressure vessel and carries away the heat produced in the reactor. This hot gas converts water to steam in a special boiler known as a heat-exchanger, and the steam is used as in a conventional power station to drive an electric turbo-generator. Thick concrete walls enclose the reactor and provide protection from neutrons and gamma rays.

Thermonuclear Fusion

Fission is not the only type of nuclear reaction in which a loss of mass yields a large amount of energy. In *fusion*, the nuclei of light elements combine to form a heavier one; thus hydrogen or deuterium (heavy hydrogen) can join to give helium. It is believed that the sun's heat is generated by this process.

For two hydrogen nuclei to combine, they must overcome the mutual repulsion which exists between them because of their positive charges. This may happen if they collide at very high speed; one way of attaining such a speed is to raise them to a temperature of several million degrees Celsius; hence the term *thermonuclear fusion.* The formation of helium by fusion is accompanied by a mass loss and a consequent release of heat energy. If this heat can maintain the high temperature necessary for further nuclei to fuse, the reaction becomes self-sustaining. Uncontrolled thermonuclear fusion has been achieved in the "hydrogen" bomb, the high temperatures being obtained by using a nuclear bomb as a detonator.

Scientists are trying to find ways of slowing down and controlling thermonuclear fusion for peaceful purposes. The task is immense, the prize enormous, and the challenge is one of the greatest man has undertaken.

QUESTIONS

1. How was the splitting of the atom first achieved? Do you think "splitting the atom" is an accurate description of what happens? If you do not, suggest a better expression. Using symbols, write the equation for the reaction.

2. State Einstein's equation; explain how Cockcroft and Walton's experiment on the bombardment of lithium by protons proved it.

3. What is meant by uranium fission? Why does it yield energy? What form does the energy take? Why is there always a chance that the reaction will sustain itself? What other condition is necessary for it to do so?

4. Why is it necessary in a nuclear reactor using natural uranium fuel to have (*a*) control rods, (*b*) a moderator? If uranium 235 only is used as fuel how could the reactor be simplified? Trace the energy changes which occur in a nuclear power station.

THIRD

REVISION

SUMMARY

THIRD
REVISION SUMMARY

ELECTRICAL ENERGY

ELECTRICITY AT REST *Page*

ELECTRIFICATION can be produced by **friction.** There are two 367
kinds of electrical charge called **positive** and **negative** respectively.

Like charges repel, unlike charges attract. 368

The force between electric charges varies inversely as the square of their separation.

The ***gold leaf electroscope*** can be used to detect electric charges 368
and to identify the sign of the charge.

All substances contain both kinds of charge. ***Conductors*** allow 369
free movement of electric charge through them, ***insulators*** resist the movement of charge.

Separation of the different kinds of charge on a conductor can be 372
produced by the presence of a charged body. This is called ***electrostatic induction.***

The ***electrophorus*** is a device for producing electric charges by 373
induction. A charged body placed inside a hollow conductor induces 375
an equal and opposite charge to its own on the inside of the conductor. If a charged conductor touches the inside of a hollow conductor, **all** its charge passes to the hollow conductor.

When a conductor is charged, **all** the charge collects on its outer 375
surface.

Two conductors are at different ***potentials*** if, on touching them 376
together, electricity flows from one to the other. *Negative charge flows from the conductor at the lower potential to that at the higher potential.*

The electrical ***capacities*** of conductors are compared by the charges 376
needed to raise them to the same potential.

The earth is taken as the practical ***zero of potential.*** 376

Electric charge collects on the **more sharply curved** parts of a con- 377
ductor and especially on points. **Lightning conductors** make use of this effect.

Page

MAGNETISM

Batteries and ***dynamos*** are devices for maintaining a continuous 379
flow of charge. This flow is called an ***electric current.***

Electric currents produce magnetic fields in their neighbourhood. This 379
can be shown by experiment.

A ***magnetic field*** is a region of space in which a compass needle 380
sets in a particular direction. The direction of a magnetic field is from the south to the north end of a compass needle when it has come to rest in the field. Magnetic fields are mapped by ***lines of force.*** These are lines drawn so that a compass needle placed anywhere on the line sets in the direction of the line.

A straight current produces lines of force in circles round the current. 381
The directions of the current and the field can be remembered by the ***"corkscrew" rule.***

A current flowing in a cylindrical coil or solenoid produces lines of 382
force inside the coil which are parallel to the axis of the coil.

Bar magnets and **current-carrying solenoids** produce similar mag- 383
netic fields.

The **magnetism of a bar magnet** can be regarded as due to rotating 384
electric charges in the atomic structure of the metal.

Magnetic materials can be magnetized by placing them inside a 385
current-carrying solenoid or by stroking.

Like poles of magnets repel each other, unlike poles attract. These 385
forces can be regarded as being due to the forces exerted by electric currents on each other.

Magnetic materials become temporarily magnetized when a 388
magnet's pole is brought near to them. This is called ***magnetic induction.***

Magnets can be **demagnetized** by heating or by rough treatment. 389

Keepers are used to minimise self-demagnetization. 389

The earth has a magnetic field. This causes a compass to set in a 389
north-south direction.

The ***geographical meridian*** is a vertical plane passing through 389
the geographical poles. The ***magnetic meridian*** is a vertical plane
in the direction of the earth's magnetic field. The angle between the 390
magnetic and geographical meridians is called the ***declination*** or
variation.

The earth's lines of force are, at most places, inclined to the earth's 390
surface. The angle between the horizontal and the earth's field at a place is called the ***dip*** or ***inclination.*** The dip can be measured with a ***dip circle.*** The strength of the earth's horizontal field, the variation,

Page

and the dip, are called the ***magnetic elements*** at a place. ***Magnetic maps*** are maps showing the *variation of the magnetic elements at different places on the earth's surface.* At the magnetic poles the dip is 90°. *The magnetic elements vary with time.* The magnetic poles are slowly rotating round the geographical poles. 391 392

Magnetic compasses are used in the navigation of ships and aircraft. 392

ELECTRIC CURRENTS

Electric currents produce **heat** in the wires along which they travel. When flowing through certain liquids they produce **chemical decomposition.** Electric currents can be produced by chemical action in cells or batteries. Examples of these are the ***simple cell,*** the ***Daniell cell,*** and the ***Leclanché cell.*** 394 395 396

In the simple cell ***polarization*** occurs due to the hydrogen bubbles formed on the copper plate. 396

The solubility of the zinc plate of a simple cell in the acid, even when no current is taken from the cell, is due to impurities in the zinc and is called ***local action.*** 396

Polarization is prevented in the Daniell cell by having copper sulphate solution in contact with the copper electrode. 396

In the Leclanché cell **polarization is minimized** by the use of manganese dioxide surrounding the positive pole. 396

Primary cells can only be recharged by the use of fresh chemicals. 397

The ***lead accumulator*** can be recharged by passing a current through it in the **opposite** direction to that in which it delivers current. 397

The density of the acid in a lead accumulator falls during discharge and rises during charge. 398

Electric currents are measured in ***amperes.*** 398

Heat is developed in a conductor through which a current flows in proportion to its electrical resistance. 399

Quantity of electricity is measured in ***coulombs.*** *A coulomb is the quantity of electricity flowing along a wire when one ampere flows for one second.* 399

A battery provides a fixed quantity of energy for each coulomb it delivers. 401

The number of **joules** of energy provided with each coulomb is called the ***electromotive force*** or ***e.m.f.*** of the battery, measured in **volts.** 401

The ***potential difference*** or ***p.d.*** between any two points in a circuit, also measured in **volts,** is *the number of joules of work done when one coulomb of charge flows between the points.* 401

Page

When cells are connected in *series*, the total e.m.f. is the sum of their 402
separate e.m.f.'s.

When similar cells are connected in *parallel*, they act as a cell 403
whose e.m.f. is the same as that of one cell, but each cell supplies
less current than if only one cell is used.

Electrical power is measured in **joules per second** or **watts.** A 403
kilowatt is a power of 1000 watts.

The ***kilowatt-hour*** measures energy and is the unit in which 403
electrical energy is marketed commercially. It is the energy obtained
from one kilowatt during one hour.

The **power** *in watts consumed in any part of an electrical circuit is* 404
obtained by multiplying the p.d. in volts by the current in amperes.

watts = volts × amperes

The rate of heat production in calories per second in a circuit can be 404
calculated by dividing the power in watts by 4·2.

Conductors are said to be connected in *series* if the same current 404
flows through each in turn.

When conductors are connected in *parallel*, the p.d. across each 404
is the same.

In a house wiring circuit all the apparatus is connected in *parallel* 405
with the main supply.

Fuses and switches are placed in the live lead of the main supply. 405

For a lighting circuit a fuse which "blows" for a current greater than 406
2 A or 5 A is used. For power points a fuse for a current greater than
13 A or 15 A is used while for heavy duty cookers a 15 A or 30 A
fuse is required.

RESISTANCE

***Ohm's law*:** *the current through a conductor is proportional to the* 407
p.d. across it, provided the temperature remains constant.

The electrical ***resistance*** of a conductor is measured by the ratio 407
p.d./current for that conductor. Resistance is measured in ***ohms.***

volts = amperes × ohms

When **resistors** are connected **in series,** *the total resistance is the* 408
sum of the separate resistances.

$$R = R_1 + R_2 + R_3$$

When **resistors** are connected **in parallel,** the effective resistance 408

Page

may be calculated by making use of the following formula: 408

$$\frac{1}{R} = \frac{1}{R_1} + \frac{1}{R_2} + \frac{1}{R_3} \ldots .$$

where R is the effective resistance **and R_1, R_2, $R_3 \ldots .$ the separate resistances.**

When a current divides through two resistors in **parallel,** the currents are in the **inverse** ratio of the resistances. 408

Ammeters **have a low resistance and are connected in series in the circuit so that the current to be measured flows through them.** 409

Voltmeters **have a high resistance and are connected in parallel with that part of the circuit across which the p.d. is to be measured.** 409

Resistances can be measured by:

Voltmeter-ammeter method. 410

Substitution method. 411

Wheatstone bridge method. 417

The ***resistivity*** of a material is measured by *the resistance between opposite faces of a metre cube of the substance*. The unit of resistivity is the ohm metre (Ω m). 412

A wire of length l and cross sectional area A made of material of resistivity ρ, has a resistance R where 412

$$R = \frac{\rho l}{A}$$

The electrical resistance of most metals **increases** as their temperature **rises.** 413

The range of currents measurable by an ammeter can be extended by connecting a suitable ***shunt*** in **parallel** with the ammeter. 414

The power in watts when a p.d. of V drives a current I through a resistance R can be calculated from 415

$$\text{Power} = VI$$

$$\text{Power} = I^2 R$$

$$\text{Power} = \frac{V^2}{R}$$

Cells offer some resistance to the flow of electricity through them. This is called the *internal resistance* of the cell. 415

The internal resistance of a cell sets a limit to the current obtainable from a cell. 416

Part of the e.m.f. of a cell is used in driving the current through the 415 cell itself. The volts so used are often called the ***"lost" volts.*** 415

If a cell of e.m.f. E sends a current I through an external resistance R: 416

$$E = I(R + r)$$

where **r** is the **internal** resistance of the cell. The p.d. V available at the terminals of the cell can be calculated from:

$$E - V = I \times r$$
$$\text{or } V = I \times R$$

The e.m.f. of a cell can be defined as *the p.d. across its terminals when no current flows.* 416

The internal resistance of a cell can be measured by the use of a high resistance voltmeter and a resistor of known resistance. 417

The ***potentiometer*** can be used to **compare** the e.m.f.s of two cells accurately. 418

APPLICATIONS OF THE ELECTROMAGNET

An ***electromagnet*** can be made by winding a ***solenoid*** on a rod of soft iron. 420

As the magnetizing current is increased, the strength of the magnet increases up to a point when no further increase of current makes it stronger. The iron is then said to be ***saturated.*** 420

Electromagnets are used in:

Electric bell. 421
Relay. 421
Telephone earpiece. 422

Moving iron ammeters and ***voltmeters*** depend for their action on the *magnetization of iron when a current flows in a coil surrounding it.* Moving iron instruments do not have a **uniform** scale, but can be used for **alternating currents.** 422 423

A wire carrying a current, when placed at **right angles** to a magnetic field, experiences a force **perpendicular** to both field and current. 424

The **direction** of the force can be obtained by the use of ***Fleming's "left-hand rule":*** 424

A current-carrying coil, when placed in a magnetic field, tends to turn until the lines of force due to the current in the coil are in line with those of the magnetic field.

This principle is made use of in the ***electric motor,*** in ***moving coil ammeters*** and ***voltmeters.*** 424

Page

CHEMICAL EFFECTS OF A CURRENT

Some liquids will conduct electricity and are chemically decomposed by the current. These liquids are called ***electrolytes*** and the process is called **electrolysis.** 430

The current is carried through the liquid by charged **ions.** These ions are **charged atoms or groups of atoms.** Hydrogen and metals form **positively** charged ions. 430

Faraday's laws of electrolysis:

1. *The mass of substance liberated in electrolysis is proportional to the quantity of electricity passed through.* 431
2. *The masses of different substances liberated by the same quantity of electricity are in the ratio of their chemical equivalents.* 432

The mass of substance liberated by **one coulomb** is called its ***electrochemical equivalent***, or ***e.c.e.*** 431

The ***e.c.e.*** of copper can be measured using a ***copper voltameter.*** 431

Faraday's laws lead us to suppose that the charge on an ion is proportional to its ***valency,*** and that all monovalent ions have the same charge. 432

When acidified water is electrolysed, hydrogen and oxygen are obtained. The apparatus often used for this is called a ***water voltameter.*** 433

Base metals can be plated with other metals by electrolysis. 434

In general, **the object to be plated is made the cathode** in an electrolytic cell, the electrolyte being the solution in water of a salt of the plating metal. Copper, silver, nickel, chromium, and zinc are all used to plate iron.

Electrolysis is used industrially in chemical manufacture. 434

The e.m.f. of some cells can be calculated from energy considerations. 435

ELECTROMAGNETIC INDUCTION

When a conductor moves across lines of magnetic force, an e.m.f. is induced in it whose magnitude is proportional to the rate at which the conductor moves across the lines of force. 437

The direction of the induced e.m.f. is obtained by ***Fleming's "right-hand rule".*** 437

Lenz's law: *the induced e.m.f. acts in such a direction as to oppose the motion or change inducing it.* 437

An alternative form of the law of induction: *when the magnetic flux through a circuit changes, an e.m.f. is induced in the circuit.* 438

If two coils are placed so that when a current flows in one of them its lines of force pass through the other, then a changing current in one coil induces an e.m.f. in the other. This is known as ***mutual induction.*** 438

The **dynamo** consists of a coil of wire rotating in a magnetic field. The method of collection of the current from the rotating coil differs in the **a.c. dynamo** and the **d.c. dynamo.** 439 440

Alternating currents and **voltages** can be measured by **moving iron instruments** and by **hot wire instruments.** 442

Alternating currents can be ***rectified*** by valves or metal rectifiers. 442

An ***induction coil*** can be used to produce high voltages from a low voltage battery. It is used in the ignition system of some motor-cars. 443

A ***transformer*** can be used to change the voltage of an alternating supply. It consists essentially of two coils wound on a laminated iron core. 444

In a transformer, the ratio of the voltages in the two coils is the same as the ratio of the turns in the coils. 445

When electricity has to be transmitted over large distances, it is more economical to transmit a **small** current at a **high** voltage. 446

STORAGE OF ELECTRICAL ENERGY

Capacitances of conductors are **compared** by comparing the quantities of electricity required to raise them to the same potential. The ***capacitance*** of a conductor is **measured** by the charge required to give it unit potential. If a charge Q coulombs raises the potential of a conductor to V volts its capacitance, C farads, is given by: 447

$$C = \frac{Q}{V}$$

The capacitance of a conductor can be increased by placing close to it a conductor at a lower potential. In the case of the parallel plate capacitor the capacitance of a metal plate A is greatly increased by placing another metal plate B, which is earthed, close to A and separated from it by an insulating substance. 447 448

NUCLEAR ENERGY

ELECTRONS

Gases conduct electricity more readily at **low pressure** and give various luminous effects. 451

Page

Cathode rays are produced in a gas discharge tube when the pressure is about 0·01 mm mercury. They are streams of **negatively charged particles** called **electrons** which can be deflected by electric and magnetic fields. 453

Electrons are constituents of all atoms, they have a mass of $\frac{1}{1840}$ of that of the hydrogen atom and carry the **fundamental unit of electricity.** 454 455

Before the discovery of the electron, electric current was regarded as a flow of **positive charges** from the positive to the negative terminal of a battery. It is now believed that, in metals, current is a flow of **negative electrons** in the opposite direction. 457

In solid **conductors** electrons are **loosely** held and can travel freely from atom to atom; in an **insulator** they are **firmly** bound to their atoms. 456

In liquids and gases the current carriers are ***ions.*** An ***ion*** is an atom or group of atoms which has lost or gained an electron. 457

In **electrolytes** (solutions of acids, bases, and salts in water) ions are already present and are attracted towards the electrodes when a p.d. is applied. In a gas very few ions arc present, but their number and the conductivity of the gas can be increased by **ionizing agents** such as high energy radiation or flames. 457

Thermionic emission: when certain metals, or oxide coated metals, are heated they emit **electrons.** Two important thermionic devices are the ***diode valve*** and the ***cathode ray tube.*** 457 458

X-RAYS

X-rays are radiation of the same nature as light with wavelengths about one thousand times smaller. They are produced when fast-moving electrons strike matter, particularly heavy metals. 462 460

Properties of X-rays:

1. They travel in straight lines. 461
2. They are **not** deflected by magnetic or electric fields. 462
3. They affect photographic film. 462
4. They cause some substances to fluoresce. 462
5. They more or less penetrate all matter. 462
6. They ionize gases. 462

The earliest X-ray tubes were modified versions of the gas discharge tube containing air at cathode ray pressure (0·01 mm). 460

Modern X-ray tubes are highly evacuated and obtain electrons by ***thermionic emission*** *The intensity of the X-ray increases with the filament current and their penetrating power with the p.d. across the tube.* 461

Page

The penetrating ability of X-rays is widely used in medicine and industry. 463

RADIOACTIVITY

Radioactive substances emit ***alpha particles, beta particles,*** and ***gamma rays.*** 465

Alpha particles are positively charged helium atoms, ***beta particles*** are electrons, and ***gamma rays*** are very short-wavelength X-rays. 466

Alpha particles are the **least penetrating** but have the **greatest ionizing power;** ***gamma rays*** have **considerable penetrating power** but cause **very little ionization** on passing through a gas. 467

The emission of an alpha or beta particle changes one radioactive element into another. 468

The ***half-life*** of a radioactive element is *the time required for half the atoms in any given amount of the element to disintegrate.* 468

Geiger tubes and ***cloud chambers*** detect particles of radiation by their ionizing effect. 469

Each kind of particle or ray produces a distinctive track in the **cloud chamber.** By studying and measuring these tracks, the speed and kinetic energy of ionizing particles can be estimated. By these means new atomic particles have been discovered. 470

About 40 natural radioactive elements exist but many hundreds of radioisotopes can be made artificially. 465

Rigid safety precautions must be taken when handling radioactive 470
substances since prolonged exposure to their radiations can cause 471
serious bodily damage which sometimes proves fatal.

THE ATOM

Atoms are much too small ever to be seen in the ordinary way but 473
Brownian motion provides some evidence for their existence. Much 474
of physics and chemistry can only be explained in terms of atoms and molecules.

The ***atom*** consists of a small, **positively** charged ***nucleus*** which accounts for most of the mass of the atom, surrounded by a cloud of orbital **electrons** sufficient in number to make the atom **electrically neutral.** Most of the atom is space. 476

The ***nucleus*** contains ***protons*** and ***neutrons. Protons*** are 478
positively charged, ***neutrons*** are **uncharged,** both have about the
same mass. The **number of protons** equals the ***atomic number*** and 479
the **number of protons plus neutrons** gives the ***mass number.***

Page

The electrons are arranged around the nucleus in ***shells*** in such a way as to account for the ***periodic variation*** of the chemical properties of the elements. 477

Isotopes of an element are atoms which have the **same number of protons** but **different numbers of neutrons.** They have the same atomic number, identical chemical properties but different mass numbers. 479

Positive rays are gas atoms which have lost electrons; that is, they are **positively charged gas ions.** The investigation into such rays in a discharge tube containing neon led to the first direct evidence for the existence of **isotopes.** 480

In radioactive decay new elements are produced by the emission of **alpha** and **beta particles** from the **nucleus.** 481

NUCLEAR ENERGY

Matter and energy are equivalent and the relation between them is given by Einstein's equation: 482

$$E = mc^2$$

Rutherford first split the **atom** in 1919 by bombarding nitrogen with **alpha particles.** Protons and an isotope of oxygen were formed. 483

Cockcroft and Walton split the **nucleus** of lithium in 1932 using artificially **accelerated protons.** Their results verified Einstein's equation. 484

In ***nuclear fission*** a **neutron** penetrates into the **nucleus** of a **uranium 235** atom, splitting it into two more or less equal parts (often barium and krypton), and producing a few more neutrons. A mass loss occurs in the reaction and is released as energy, ***nuclear energy,*** which is mostly in the form of kinetic energy (heat) of the large fission fragments. 485

A ***chain reaction*** can occur if the **mass** of uranium exceeds a **critical value.** In an ***atomic bomb*** the chain reaction increases and is uncontrolled; in a ***nuclear reactor*** the chain reaction proceeds at a steady rate. 486

Thermonuclear fusion results in the nuclei of light elements joining at a very high temperature to form the nucleus of a heavier element. The mass loss occurring appears as energy. 488

Uncontrolled nuclear fusion has been achieved in the **"hydrogen" bomb** by using a nuclear bomb as a detonator. Scientists are at present trying to find ways of slowing down and controlling thermonuclear fusion for peaceful purposes. 488

THE EXAMINATION

You may find the following suggestions useful when you come to take your examination.

Always work out the number of minutes you can take over each question, leaving time to revise your script. In the paper that follows you could allow 20 minutes over each question, 5 minutes to read through the question paper and decide which you are going to answer, and 15 minutes at the end to read through what you have written.

Make quite certain that you obey the instructions at the top of the paper (the *rubric*). You must, for instance in the paper given, answer one question from each section and not more than five altogether.

You must answer the question that is set. If the question is "State briefly . . ." then you must be brief; if it is "What experiment could you perform . . . ?" be certain that it is such an experiment and do not give an account of some historic experiment.

Illustrate your answers with diagrams, even of the simplest type. These need only be freehand, but useful, clear, freehand diagrams need practice so you must practise them well in advance.

It need hardly be said that you must be neat and that you must write legibly. But this does not mean that your writing must be large or that you spread your work over many pages.

In calculations you must put in sufficient explanation to show how you arrive at your results. A correct answer without explanation will be marked down. All answers must include units when appropriate, and they must be quoted only to a sensible number of significant figures.

SPECIMEN EXAMINATION PAPER

Attempt five questions, at least one from each section. Time allowed: 2 hours.

SECTION A

1. (*a*) State Archimedes' principle and deduce it from the laws of hydrostatic pressure.

   (*b*) A weighted wooden rod of uniform cross-sectional area floats upright in water with 80 mm of its length above the level of the water. When it floats in a liquid of density 800 kg/m^3, 60 mm of its length is exposed. If the mass of the rod is 64 g, what is its cross-sectional area?

2. (*a*) Distinguish between mass and weight. A body falls freely under gravity. How long does it take to fall 40 m? If the body (mass 60 kg) is attached to a parachute that exerts a retarding force of 240 N on it and it is again released in a vertical fall, how long will it now take to fall 40 m? (Acceleration due to gravity = 10 m/s^2; force of gravity = 10 N/kg.)

(*b*) What is the law of the conservation of energy? A mass is placed on a vertical compression spring and pressed downwards. It is then released and the spring flies upwards. It falls eventually into a bed of soft sand. Describe the various changes of energy that take place.

3. (*a*) What is meant by the term *centre of gravity*? How would you find the centre of gravity of a piece of cardboard?
(*b*) Fig. 1 shows a regular circular lamina pivoted horizontally at its centre *O*.

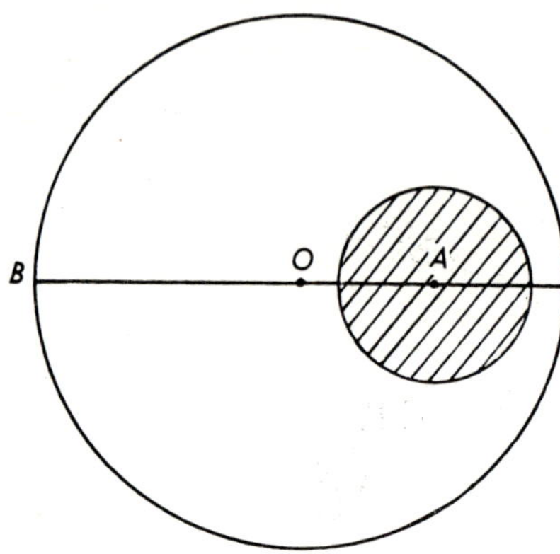

FIG. 1

A circular hole is cut in the lamina. The area of the hole is one-ninth of that of the complete lamina, and its centre *A* is 100 mm from *O*. If the mass of the complete lamina is 1 kg and its radius is 200 mm, what vertical force must act at *B* to keep *OB* horizontal? (Force of gravity = 10 N/kg.)

SECTION B

4. (*a*) Describe how you would find the specific latent heat of fusion of ice, mentioning particularly the precautions you would take to get a result as accurate as possible.
(*b*) A Thermos flask contains water at 0° C and its total heat capacity is 42 000 J/K. The water is heated electrically by a small coil in the water. How much water boils away if 50 000 J are supplied? (Specific heat capacity of water = 42 000 J/kg K; specific latent heat of vaporization of water = 2 270 000 J/kg.)

5. (*a*) How can a pure spectrum from white light be thrown on to a screen?
(*b*) Why is it that when blue paint is mixed with yellow paint the result is usually green?
(*c*) If you were given two pieces of yellow glass and told that one transmitted pure yellow light and the other a mixture of red and green how would you distinguish between them, given a third piece of glass that transmitted pure green light only?

6. (*a*) Describe how you would find the focal length of a converging lens.
(*b*) A point source of light is placed 0·8 m from a white screen. A converging

lens is placed between the two, 0·6 m from the screen, and a patch of light the same size as the lens appears on the screen surrounded by a dark area. Explain this. Leaving the lens and screen in position, how far must the source be moved to get a clear image of it on the screen?

SECTION C

7. (*a*) Describe how you would verify the statement that 96 500 coulombs will liberate 0·031 75 kg of copper when a solution of a copper salt is electrolysed.
(*b*) A battery (e.m.f. = 6 V, internal resistance = 1·5 Ω) is connected to a copper voltameter of resistance 2·5 Ω. How much copper would be deposited in 6 hours?

8. (*a*) Draw a clear diagram of an electroscope.
(*b*) Given a positive charge, how could you charge the electroscope negatively?
(*c*) How could you show that when ebonite is rubbed with fur equal amounts of positive and negative charges result?

9. (*a*) Describe the method of comparing the e.m.f.s of two cells using a potentiometer. Why is this method preferable to using a voltmeter.
(*b*) What would be the effect on your readings and result if (i) you used a potentiometer wire of double the length, and (ii) you placed a resistance in series with the first of the two cells?

10. (*a*) What are the laws of electromagnetic induction? Describe experiments that illustrate them.
(*b*) A flat coil of wire rotates at a constant rate about an axis perpendicular to a uniform magnetic field. Draw a graph of the induced e.m.f. against time, showing clearly how the graph is related to the position of the coil in the magnetic field.

ANSWERS TO TEXT QUESTIONS

CHAPTER 5

1. Cohesion and adhesion (page 53): for examples of adhesion see "Capillarity."
2. Surface tension phenomenon (page 54): the fabric of the umbrella is treated with a water-repellent substance so that the cohesion between the water molecules is greater than the adhesion between the water and the fabric. As a result the water does not penetrate the air spaces between the individual fibres.
3. Surface tension phenomenon: cohesion between the water molecules is greater than the adhesion between the water and the waxed surface of the wires. As a result the water does not penetrate the spaces between the wires and the gauze rests on the surface of the water.
4. The fabric of the tent creates the same surface tension phenomenon as the fabric of the umbrella (Question 2). If the fabric is touched, however, the surface of the film of water covering the tent may be broken and a leak would then develop.
5. "Capillarity" (pages 53-54).
6. "Diffusion of Gases" (page 58, Fig. 44).

CHAPTER 6

1. Scalars: mass, speed, work, energy, power. Vectors: displacement, force, momentum, acceleration.
2. Only (*c*) is correct.
3. Yes: a body always tends to fall towards the centre of the earth (page 79-80).
4. No: motion of the aircraft depends on action and reaction between the blades of the propeller and the air (Newton's laws of motion, page 65) and there is no air in outer space. See also "Aeroplane Changing Direction" (page 83).
5. The hammock requires the stronger ropes: for the swing see "Addition of Forces" (page 67); for the hammock see "Parallelogram of Forces" (page 69).
6. (*a*) Increases. (*b*) Decreases. (*c*) No change. Answers derived from: Newton's laws of motion (page 65); "Addition of Forces" (page 67); a spring balance measures weight and not mass.
7. Horizontal component = 260 N. Vertical component = 150 N. The horizontal component is effective.
8. Average velocity = 12 m/s. Distance travelled = 72 m. Acceleration = 4 m/s^2. Distance travelled in last second = 22 m
9. The sculler rows at 51·3° to the bank and takes 9·6 min to cross.
10. Velocity after $\frac{1}{4}$ s = 2·5 m/s; $\frac{1}{2}$ s = 5·0 m/s; 1 s = 10·0 m/s; 2 s = 20·0 m/s Average velocity during first $\frac{1}{4}$ s = 1·25 m/s; $\frac{1}{2}$ s = 2·5 m/s; 1 s = 5·0 m/s; 2 s = 10·0 m/s. Distance body falls in $\frac{1}{4}$ s = 0·31 m; $\frac{1}{2}$ s = 1·25 m; 1 s= 5·0 m; 2 s = 20·0 m.

CHAPTER 7

1. Torricelli's vacuum tube (page 98). Atmospheric pressure balances a column of mercury 760 mm high or a column of water 10 m high; it is therefore more practical to use mercury in a barometer. Another advantage of mercury is that it does not evaporate readily at room temperature so that the space above the mercury contains only minute traces of mercury vapour. Water would eva-

porate far more readily and the presence of this vapour above the column of mercury would lower the barometric reading.

2. (*a*) No effect. (*b*) Pressure = Depth × Density (page 90). (*c*) No effect.
3. Since hydrogen is less dense than air the balloon will rise. As it rises, the surrounding pressure of the atmosphere decreases (page 100) and the volume of the hydrogen increases (Boyle's law, page 102), filling the balloon and eventually bursting it. It will be shown later ("Floating Bodies," page 114) that if the balloon is made from a strong rigid material, it will stop rising when the weight of the balloon + weight of hydrogen = weight of rarefied air displaced by balloon. In this case, it will cease to rise when the same condition is true, provided the balloon does not burst.
4. The hole in the can must be large enough to allow liquid to flow out of the can and air to flow in to replace this liquid. If the air cannot enter the can, the liquid will be attempting to flow out against atmospheric pressure, creating a reduced pressure in the can (Torricellian vacuum, page 98). This it cannot do, and the liquid will remain in the can. If, however, a second hole is made in the can, in such a position that air will enter freely, then the liquid will flow from the first hole.
5. 103 000 N/m^2 above atmospheric pressure. Pressure is the same in vertical direction.
6. (*a*) 12 000 000 N. (*b*) 40 000 N/m^2 above atmospheric pressure.
7. 103 000 N/m^2. 8. 14 400 cm^3. 9. Another 400 N must be added. Atmospheric pressure could be added to give total pressure.

CHAPTER 8

1. Laws of friction (page 108). 2. Rolling friction (page 109).
3. A force a little greater than 8·48 N.
4. The beads of polystyrene replace the sliding friction between the block of wood and the surface of the table by rolling friction between the beads and the block, and the beads and the table: the block will slide more easily over the table.
5. Explain your observations with reference to "Motion in a Circle" (page 81). The coefficient of friction between the coin and the turntable will be greater than that between the coin and the shiny paper.
6. Coefficient of sliding friction = 0·2. A force of 40 N called the normal reaction (page 107).

CHAPTER 9

1. The heavier sphere will fall with the greater acceleration. Consider the resultant force P on a sphere falling through water (weight — upthrust) and substitute in the equation $P = ma$ (page 65). You will find that the greater the mass m is, the greater will be the acceleration a. (The drag due to viscosity will initially be the same for both spheres.)
2. Relative density (page 113) is lowered, and may be measured with a hydrometer (page 115). (R.D. milk = 1·03 approx.)
3. Archimedes' principle (page 112).
4. The salt increases the relative density of the water: apply Archimedes' principle and the law of floating bodies (page 114). The average density of the egg is greater than that of water but less than that of the final salt solution.
5. (*a*) R.D. metal = 2·5 (*b*) R.D. petrol = 0·75.
6. The balloon will exert a lifting force of 11×10^3 N. 7. Bernoulli's principle (page 117).

CHAPTER 10

1. So far in this study of physics two main topics have been considered: firstly, movement and changes of movement, which led on to such ideas as those of force and inertia; secondly, the idea that all matter is made of very small particles and that these particles are in constant motion (kinetic theory). These topics led on to the idea of energy (principle of the conservation of energy, page 129) which is of such importance that physics is often described as the study of energy in its various forms.
2. Both engines work at the same power.
3. No machine can produce more power than is supplied to it (principle of the conservation of energy, page 129).
4. Time required = 150 s or $2\frac{1}{2}$ min.
5. Ratio of useful P.E. to total P.E. = 5/17.
6. Average force = 55 N.
7. Velocity of rifle = 1·2 m/s. 8. Force exerted on rocket = 3600 N, i.e. 3·6 kN, in opposite direction to that in which gas is ejected.

CHAPTER 11

1. 20 000 J.
2. (*a*) 400 kJ. (*b*) 400 W.
3. Moment of a force (page 140). Weight of rule = 84 g. The 24 g mass must be hung on the 0 cm mark of the rule.
4. Consider the compression and extension of the spring attached to the base of the pogo stick and the changes in K.E. and P.E. of the child and the pogo stick in free movement: energy changes in an oscillating spring (page 136).
5. Efficiency of a machine (page 146). Base the experiment on the general procedure for finding the efficiency of a pulley system (pages 152-53); a method for finding the velocity ratio of a bicycle has already been discussed (pages 148-49).
6. An effort of 100 N is required. 7. Efficiency of machine = 35 per cent.

CHAPTER 12

1. Final temperature of water and can = 57·5° C.
2. Heat required = 537 kJ.
3. To find the specific heat capacity of a liquid (page 159).
4. Specific calorific value (page 160). To find the specific calorific value of candle wax (page 161).

CHAPTER 13

1. Relative merits of mercury and alcohol (page 164). It may be deduced that water, with its very narrow and unpractical temperature range for the liquid state, is unsuitable as a thermometric liquid.
2. Six's maximum and minimum thermometer (page 165).
3. Definition (page 166) and method of measuring (page 167) the linear expansivity.
4. The gaps would just close at a temperature of 42° C.
5. Water at 4° C. has a greater density than at any other temperature and sinking to the bottom of the pond, stays there, even when a thick ice layer has formed on the surface (pages 170-71).

6. Statement of Charles' law (page 173); method of verifying it (pages 172-73).
7. Volume of air bubble at surface = 1·29 cm^3.
8. Mass of hydrogen collected = 0·0036 g or 0·000 003 6 kg (3·6 mg)

CHAPTER 14

1. Latent heat (page 178). Heating water to boiling (pages 177-78).
2. Effects of pressure on melting and boiling points and experiments to illustrate them (pages 179-81). 3. Effect of salt on icy roads (page 179).
4. To find the specific latent heat of fusion of ice (page 182).
5. 100·4 kJ are required.
6. (*a*) 420 s. (*b*) 141 s. 7. Cooling of naphthalene (page 179).

CHAPTER 15

1. Specific heat capacity of a metal (pages 188-89).
2. The difference between the potential energy of the water at the top and bottom of the waterfall has all changed into heat energy and this is all used in raising the temperature of the water (page 187). Height of waterfall = 200 metres.
3. Specific heat capacity of a metal, using electrical energy as heat (page 190).
4. The principle of converting heat energy into mechanical energy (page 191).
5. Four-stroke petrol engine (page 193).
6. Compression-type refrigerator (page 194). The absorption-type refrigerator may also be driven by electricity.
7. Working power of engine = 0·7 watts. Overall efficiency of engine = 0·5%.

CHAPTER 16

1. Surface tension (page 198).
2. Vapour pressure and an experiment to investigate its variation with temperature (page 199).
3. A concept of boiling point in terms of kinetic theory (page 200).
4. Pressure depends on number and speed of gas molecules present (page 197).

CHAPTER 17

1. The jar loses heat by conduction, convection, and radiation: the Thermos flask (page 209) attempts to eliminate all three methods of heat transfer.
2. Comparison of the three methods of heat transfer (page 210). (*a*) Radiation. (*b*) Conduction. (*c*) Convection.
3. Thermal conductivity k (page 204). Rate of heat loss through the glass = 5 kW.
4. Land and sea breezes (page 207).

CHAPTER 18

1. Newcomen's atmospheric steam engine (page 214).
2. The radiant heat energy given out by the fire comes from electrical energy. The electrical energy comes from rotating generators (kinetic energy), which are driven either by the potential energy of water stored behind a dam, or by steam raised by burning coal or oil (a chemical-to-heat energy conversion). The potential energy in the water and the chemical energy in the coal or oil can be traced back to radiant energy from the sun (page 216). Nuclear fusion is believed to be the source of the sun's energy (page 219).

3. Chemical energy in coal (page 215).
4. Mechanical (animal, water and wind power)—thermal (Newcomen's first practical steam engine)—electrical (from chemical and mechanical sources)—nuclear.

CHAPTER 19

1. Maximum allowable error in measuring the time = 0·0002 s.
2. Average flow of water = 0·000 024 7 m^3/s (or $2{\cdot}47 \times 10^{-5}$ m^3/s).
3. Vernier scale (page 236). The vernier will measure to the nearest mm (0·1 cm). Check vernier for zero error (page 237).
4. The arrangement will work (page 236). However, the calibrations of the vernier scale must be numbered from 10 to 0 instead of 0 to 10.
5. 1·609 kilometres = 1 mile.
6. Pitch of a thread (page 238). Gap of the clamp alters by 25 mm.

CHAPTER 20

1. The sun is an extended source: consider the formation of umbra and penumbra (page 244) by rays from the sun passing the post at different heights, remembering that rays passing the top of the post must travel farther before striking the ground than those passing, say, the middle of the post.
2. Ratio of umbra/penumbra = 1/16.
3. Let length of slit = l and one side of the cardboard square = $2l$. Consider the formation of the shadow in a horizontal plane: the slit may be considered as a point source; the shadow will be an area of umbra 4 l broad. Consider the formation of the shadow in a vertical plane: the slit must now be considered as an extended source; the shadow will be an area of umbra 3 l high above and below which will be areas of penumbra, each l high. The final shadow will be an area of umbra 4 l wide and 3 l high, above and below which are bars of penumbra 4 l wide and l high.
4. Eclipses of the sun (page 245).
5. Pinhole camera (pages 245-47).
6. Sunlight filtering through trees (page 248): since the circular patches on the ground are images of the sun, they will no longer remain circular when the sun is partially eclipsed.

CHAPTER 21

1. Laws of reflection (page 252); to find the position of an image in a plane mirror (page 250); real and virtual images (page 253).
2. If the water is moving gently the lake surface acts as a distorting mirror (curved mirrors, page 257); if the water is completely ruffled, by wind or rain for example, diffuse reflection occurs (page 255). The upside down appearance of the trees is a result of lateral inversion (page 249).
3. 0·2 m of the man's legs are cut off in the reflection.
4. Images formed in a plane mirror (page 250): the object forms an image in each mirror, the image in each mirror then forms a further image in the mirror opposite, etc. In the mirror facing you, alternate images of your face and the back of your head will be formed, diminishing in size.
5. Use the piece of string to find the shortest distance from the object to your eye via the mirror; compare the path taken by the string with Fig. 246 (page 252).
6. The ray will be reflected normally; the incident ray and the reflected ray take

the same path (page 252). The angle of reflection is always equal to the angle of incidence; if the mirror is turned through an angle α the reflected ray will be turned through an angle 2α (refer to diagram of the optical lever, page 254).

CHAPTER 22

1. In drawing the rays of light travelling from the fish to the fisherman's eye, take into consideration refraction (page 262) and apparent depth (page 263).
2. Laws of refraction (page 263); refractive index (page 265). Angle of refraction $\theta_1 = 26° 13'$. $\theta_2 =$ about 24°. $\theta_3 =$ about 40° 30′. $\theta_4 = 35° 30'$.
3. Critical angle and total internal reflection (page 269). (*a*) 60° 40′. (*b*) 41° 8′. (*c*) 38° 41′.
4. The fish has a distorted view of objects outside the water due to refraction (pages 262-63). Light from the horizon falling on the surface of the water is refracted at an angle equal to the critical angle for water (page 268) and so, to the fish, the horizon appears elevated through an angle of 90°—critical angle. From the point of view of the fish, the entire field of view around the surface of the water, and all objects in it, will be compressed and distorted. The fish is also able to see parts of the pond bed twice due to internal reflection within the water (page 269). The appearance of air bubbles in water is also explicable in terms of refraction, critical angle, and total reflection from the surface of the bubble.
5. Graphical construction (page 274); worked example (pages 275-76). The screen must be placed 267 mm from the lens; the smallest screen that can be used is 33 mm high. On moving the object to its new position, the image is formed 480 mm from the lens on the object side, and it is virtual. Focal length of lens = 100 mm.
6. Application of lens formula (pages 276 and 280-81).
7. Principal axis, principal focus and focal length (page 274). (*a*) Real and inverted image, 360 mm from the lens and 40 mm high. (*b*) Virtual and erect image, 480 mm from the lens and 40 mm high.
8. Projection lantern (page 278); magnification (page 278); lens formula (page 276). The slide is 0·5 m behind the objective lens which has a focal length of 0·455 m.
9. Camera (page 279). The lens must be moved 0·0045 m (4·5 mm) away from the film.
10. Formation of image by pinhole (page 246) and by lens (pages 273-74).

CHAPTER 23

1. Structure of the eye (pages 286-87). Defects of vision (pages 290-91). (*a*) In the plane in which the curvature is greatest, the focal length is shortened. (*b*) In the plane in which the curvature is least, the focal length is increased. An eye suffering from astigmatism will see some of the spokes clearly, and the spokes at right angles to these blurred.
2. A magnifying glass (page 292) forms a virtual, erect and magnified image.
3. (*a*) The object must be 42 mm from the lens and the image will be 60 mm high. (*b*) The object must be 71 mm from the lens and the image will be 35 mm high.
4. Compound miscroscope (pages 292-93).
5. The eyepiece of the telescope must be pulled out; consider the action of an astronomical telescope (page 294, Fig. 313) and the formation of the image of a near object with a convex lens (page 275, Fig. 287). For a reflecting telescope the adjustment would be exactly the same since the action of a concave mirror is similar to that of a convex lens (pages 260-61).

CHAPTER 24

1. Formation of a pure continuous spectrum (page 297). With a green filter, the continuous spectrum disappears except for the green section, provided that the green of the filter is pure (page 301). The blue paper will appear black in all areas of the spectrum with the exception of the blue area and part of the green and violet areas (page 303).
2. The chalk acts as an indirect source of white light; consider this fact in conjunction with Fig. 320 (page 298).
3. Red flower with black leaves through a red filter; black flower with black leaves through a blue filter; red flower with green leaves through an impure yellow filter; black flower with black leaves through a pure yellow filter.
4. The red light of the strontium flame and the green light of the barium flame would mix to give a yellow appearance. Formation of a pure spectrum from such a light source would produce combined line spectra with dominant red and green lines (pages 300-301).

CHAPTER 25

1. Waves transmit energy: throwing stones at a model yacht (page 305).
2. Reflection of sound (pages 316-17). Echoes are used by ships for depth-finding. In concert halls echoes may ruin the reception of sound.
3. Measurement of the velocity of sound (page 315); the measurement will be affected by the temperature and humidity of the air, and the strength and direction of prevailing winds.
4. Depth of channel $= 160$ m.
5. The signal returns after 6×10^{-4} s.
6. Comparison of transverse and longitudinal waves (page 313). Sound-waves are longitudinal; water waves are transverse.
7. Wavelength, frequency and velocity of a wave, and their relationship (page 307). 8. Wavelength of V.H.F. station $= 3\frac{1}{3}$ m.
9. Siren (pages 308-9): adjust the speed of rotation of the siren until the frequency of the note it sounds is the same as that sounded by the tuning fork. By means of the revolution counter calculate the revolutions/s of the siren, and count the number of holes in the rotating disc. The frequency of the siren, and hence of the tuning fork, can then be calculated. Frequency of note produced $= 240$ Hz.

CHAPTER 26

1. Formation of fundamental (page 320) and of second harmonic (page 323).
2. Sonometer (page 321): (*a*) with the string under constant tension, adjust its length so that its frequency of vibration corresponds with each of the tuning forks in turn; (*b*) repeat the experiment, keeping the length of string constant, but varying the tension by adding known weights to the right-hand end of the string (Fig. 343).
3. If l is proportional to $\frac{1}{f}$, then $l \times f =$ constant in all cases. A graph of l plotted against $\frac{1}{f}$ should be a straight line through the origin.
4. If f is proportional to $\sqrt{T}$, then $\frac{f}{\sqrt{T}} =$ constant in all cases. A graph of f plotted against $\sqrt{T}$ should be a straight line through the origin.
5. Harmonics (pages 323-24) give to a note its quality or timbre (page 325).

CHAPTER 27

1. Resonance tubes (pages 329-30).
2. Closed and open tubes of the same length sounding their fundamental notes (page 328).
3. Velocity of sound in the tube = 334 m/s approx. The answer is only approximate since no allowance has been made for end correction (page 331). The second position of resonance occurs at a tube length of 0·66 m.
4. Velocity of sound in the tube (page 331) = 333 m/s. End correction = 0·01 m.
5. First position of resonance = 0·11 m. Second position of resonance = 0·36 m.
6. First position of resonance = 0·22 m. Second position of resonance = 0·47 m (end correction at both ends of the tube).

CHAPTER 28

1. Pitch, intensity and timbre (quality) of a musical note (pages 336-37).
2. Displacement-time graphs for a tuning fork and oboe (Fig. 354), and for the human voice (Fig. 355). Explanation of displacement-time graphs (pages 334-35).
3. Loudness and intensity (pages 337-38). Decibel scale of intensity levels (338).
4. Structure of the ear (pages 339-40). By opening the mouth during an explosion, an approximately equal pressure is maintained on either side of the drumskin, by means of the Eustachian tube, and the drumskin is protected from possible damage.

CHAPTER 29

1. Telephone (pages 341-42); loudspeaker (pages 342-43).
2. Microphone (page 343).
3. Chemical energy, stored in the body, is used up in moving the throat muscles; these cause a body of air to vibrate which in turn causes the diaphragm in the mouthpiece of the telephone to vibrate. The mechanical energy of the vibrating diaphragm is converted into electrical energy by electromagnetic induction, and the electric current is transmitted to the earpiece of the receiving telephone. Here, the electrical energy is once again converted into the mechanical energy of a vibrating diaphragm by means of an electromagnet; the diaphragm causes a body of air to vibrate and these vibrations reach the ear of the listener as sound waves.
4. Gramophone (pages 344-45); tape recorder (pages 345-46). Remember that the first gramophones were entirely mechanical in their reproduction of sound, so compare both the mechanical and electric gramophones.

CHAPTER 30

1. Electromagnetic spectrum (page 347).
2. The velocity of light depends on the medium in which it is travelling (page 351). It was stated earlier in the book that the refractive index of a substance = speed of light in vacuum/speed of light in that substance (page 266).
3. Light waves (page 348). The nature of light has also been discussed earlier in the book (pages 241-42). Refraction (page 351), interference and diffraction (pages 352-54) are phenomena which show light behaving like waves.
4. Interference and diffraction (page 351-54).
5. (*a*) 276 m. (*b*) 0·10 m. (*c*) 6×10^{-7} m.

CHAPTER 31

1. The charged body induces an opposite charge on the near side of the conductor and a similar charge on the far side. It attracts the opposite charge more than it repels the similar charge (page 368).
2. The strong positive charge attracts negative charge from the leaf of the electroscope to the cap until the leaf is uncharged. When brought nearer still, the strong positive charge attracts more negative charge from the leaf to the cap, leaving the leaf positively charged.
3. Charge an electroscope negatively. Bring the comb near. If the comb is negative, the deflection of the electroscope will increase. If not, charge the electroscope positively and again bring the comb near. The deflection should increase showing that the comb is positively charged.
4. The cage screens the observer from outside electrical influences (page 375). The cage is earthed as an additional safety precaution. Everything in electrical contact inside the cage is at the same potential.
5. The candle is attached to the centre of the aluminium disc to serve as an insulating handle. The gramophone record is used as the insulating plate. Charge the record by rubbing and proceed as for electrophorus (page 373).
6. (*a*) Place the can on the electroscope cap. Lower each of the charged spheres into the can in turn. The sphere that produces the greatest deflection has the greatest charge. (*b*) Lower one of the spheres into the can and note the deflection. Remove this sphere from the can, allow it to touch the other sphere and then lower it again into the can, noting the deflection. If the deflection decreases, this sphere has lost charge and, therefore, was at a higher potential.

CHAPTER 32

1. Magnetic lines of force (page 381); magnetic and geographic meridians (page 389); magnetic variation and dip (page 390).
2. If the wire is vertical and the current flows downwards, then to the east of the wire the earth's field and the current's field are in opposite directions. The current's field becomes weaker as the distance from the wire increases.
3. Consider what happens if the disc is freely suspended at its centre with its plane horizontal, and then it is turned over and again suspended in the same way.
4. During one revolution the apparent dip varies between 70° and 90°.
5. Consider what would happen if one pole was stronger than the other.
6. Magnetic induction (pages 387-88).
7. Buildings often contain iron girders, some of which may be lying approximately north-south. Magnetic induction in the earth's field (page 388).
8. Magnetic fields due to currents in a straight wire (pages 380-81).
9. The compass needle is not pivoted at its centre of gravity.

CHAPTER 33

1. The toaster consumes a current of 1·6 A, and requires a 5 A fuse wire. Cost of running toaster for 20 min $= \frac{2}{5}$p.
2. Dynamo's e.m.f. = 112·5 V.
3. Current supplied to motor = 2 A.
4. Make up two batteries of 30 cells each in series, and connect the two batteries in parallel.

CHAPTER 34

1. Reading of voltmeter $= \dfrac{100}{100{\cdot}5}$ volts.
2. A resistance of 15·2 ohms must be connected in series with the battery.
3. (*a*) Shunt of 0·25 ohms. (*b*) Series resistance of 99 ohms.
4. For a resistance of 5 ohms, approximately 1·97 m of wire would be required.
5. When switched on, the bulb takes a current of 3 A. When hot, the bulb has a resistance of 4 ohms.
6. Try various series or parallel arrangements, or combinations of series and parallel arrangements.

CHAPTER 35

1. Draw the magnetic field produced by one current (pages 380-81) and consider the effect of this field on the other current (page 424). Parallel currents in the same direction attract each other, and in the opposite direction repel each other.
2. Clockwise.
3. Viewed from above, *A* will rotate around the pole of the magnet in a clockwise direction.
4. The wire tends to move in an easterly direction.
5. When current flows in the solenoid, the two bars tend to repel each other.
5. *XY* will tend to move towards *AC*.

CHAPTER 36

1. (*a*) 1 A. (*b*) 1·174 g. (*c*) 0·298 g. (*d*) 0·037 g. (*e*) 0·336 g.
2. Connect two ammeters in parallel and connect a third in series with these two. By means of the rheostat, pass a current of 1 A through the third ammeter so that a current of $\frac{1}{2}$ A passes through each of the ammeters in parallel, allowing them to be calibrated. If the fourth ammeter is connected in parallel with the first two, then a current of $\frac{1}{3}$ A will flow through the three ammeters connected in parallel. If in the first experiment the current is adjusted so that 1 A flows through each of the two ammeters connected in parallel, then a current of 2 A will be flowing through the third ammeter and it may be calibrated. Similarly, if the current is adjusted so that a current of 1 A flows through each of the three ammeters connected in parallel, then a current of 3 A will be flowing through the remaining ammeter which is connected in series.
3. Quantity of electricity required = 27 000 coulombs approx.
4. (*a*) 1·19 g copper and 1·23 g zinc. (*b*) 0·59 g copper and 0·61 g zinc.

CHAPTER 37

1. Magnetic or electrolytic effects will determine whether the current is a.c. or d.c.
2. Viewed from above, as the north pole approaches the ring, the current is anti-clockwise. The current reverses as the magnet falls through the ring. The energy of the current comes from the kinetic energy of the moving magnet (Lenz's law, page 437).
3. The e.m.f. in the mast acts upwards. Consider what happens in the wire joining the galvanometer to the top and bottom of the mast as it moves through the earth's field.

4. Use Fleming's right-hand rule (page 437). If the disc is drawn with the magnetic field acting down into the paper, and the direction of rotation of the disc is anti-clockwise, then an e.m.f. will be induced between the edge and the centre of the disc in the direction edge→centre.
5. Laws of electromagnetic induction (page 437). Consider flux change through the ring, if any.
6. Apply Lenz's law to the pendulum (page 437).
7. (*a*) Look at the rings under the brushes. Slip rings are used for a.c. (page 439) and a split ring is used for d.c. (page 440). (*b*) Find out if the current will give a reading on a moving coil voltmeter. If it does, then the dynamo is producing d.c.
8. (*a*) Current used = 400 A. (*b*) Power lost = 80 kW. (*c*) p.d. = 450 V.

CHAPTER 38

1. Capacitance of an electric conductor (page 447).
2. (*a*) The deflection of the electroscope decreases showing that the potential of *A* is decreased and, therefore, that the capacitance of *A* has increased since its charge is unaltered. (*b*) The deflection of the electroscope decreases more; explanation as above. (*c*) The deflection of the electroscope increases showing that the capacitance of *A* has decreased. When a sheet of insulating material, such as polythene, is placed between *A* and *B*, the deflection of the electroscope decreases: explanation as in (*a*).
3. Practical capacitors (page 450, Fig. 489).

CHAPTER 39

1. Experimental evidence suggests that cathode rays are streams of high speed electrons (page 454): paddle wheel experiment (page 453); deflection in electric and magnetic fields (page 454).
2. (*a*) Out of the paper perpendicularly. (*b*) Up towards *P*.
3. Discovery of the electron and Thomson's determination of e/m (pages 454-55). Thomson assumed that the electron carried the fundamental unit of charge.
4. The conclusion that electricity is atomic can be drawn from Millikan's experiment (page 455-56).
5. Electrons, insulators, and conductors (pages 456-57). Production of ions in a flame (page 457).
6. Thermionic emission and the diode valve (pages 457-58). Cathode ray tubes (pages 458-59) are used in television and radar.

CHAPTER 40

1. X-rays are produced when high speed electrons are stopped by matter. If the p.d. across an X-ray tube (page 461) is increased, the penetrating powers of the rays are increased. The intensity of the X-rays is increased by increasing the cathode current.
2. Properties of X-rays (pages 461-62).
3. Uses of X-rays (pages 463-64).

CHAPTER 41

1. Alpha and beta particles, and gamma rays (page 466).
2. Application of Fleming's left-hand rule (page 424) shows that the radiation consists of positively charged alpha particles.

3. Radioactive half-life (page 468): after 5·3 years half the cobalt 60 originally present will have decayed. It will take 4 days for 1 g of the radioisotope to decay to 0·25 g.
4. Geiger tube (pages 468-69).
5. Expansion cloud chamber (pages 469-70).
6. When the glass of the watch is facing the tube, 59 counts per second are due to beta particles and gamma rays, since they can both penetrate glass whereas alpha particles cannot. When the stainless steel back of the watch faces the tube, 1 count per second is due to gamma radiation, since alpha and beta particles are stopped by the steel.

CHAPTER 42

1. Growth of the atomic theory of matter (pages 473-74).
2. Geiger and Marsden's experiment (pages 475-76). Rutherford's explanation of the results (pages 476-77).
3. Atomic number = number of protons. Mass number = number of protons + neutrons. (*a*) 13 protons. (*b*) 14 neutrons. (*c*) 13 electrons.
4. Isotopes (pages 479-80).
5. $^{235}_{92}U$ stands for the uranium isotope with mass number 235 and atomic number 92.
6. Positive rays (pages 480-81).

CHAPTER 43

1. Splitting of the atom or, more accurately, splitting of the nucleus, was first achieved by bombarding nitrogen with alpha particles (pages 482-83).
2. $E = mc^2$. Cockcroft and Walton's experiment (pages 483-84): E was calculated from the distances travelled by the alpha particles produced; mc^2 was calculated from the mass loss that occurs and found to equal E.
3. Uranium fission (page 485). The splitting of a uranium nucleus yields energy with an equivalent loss in mass. The energy is in the form of radiation, and the kinetic energy of fission products which is changed into heat by collision with other atoms. Neutrons are also produced which can sustain the reaction, provided the amount of uranium present exceeds the critical size.
4. Nuclear reactors (pages 487-88). Control rods are required to give a controlled reaction while a moderator is required to slow down fast fission neutrons and prevent their capture by uranium 238. If uranium 235 only is used as a fuel, no moderator is required. Energy changes in a nuclear power station may be summarised as follows: uranium—kinetic energy of fission products—heat, first in carbon dioxide and then in steam—kinetic energy of turbine—electrical.

ANSWERS TO QUESTIONS IN THE SPECIMEN EXAMINATION PAPER

(*Page numbers in brackets refer to the text.*)

SECTION A

1. (*a*) Archimedes' principle (page 112). There are several methods of approaching the deduction of Archimedes' principle from the laws of hydrostatic pressure.

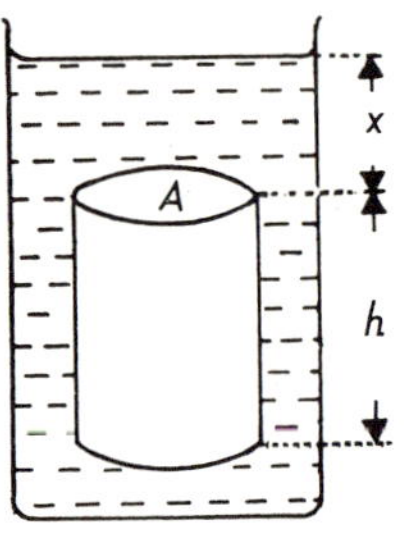

FIG. 1

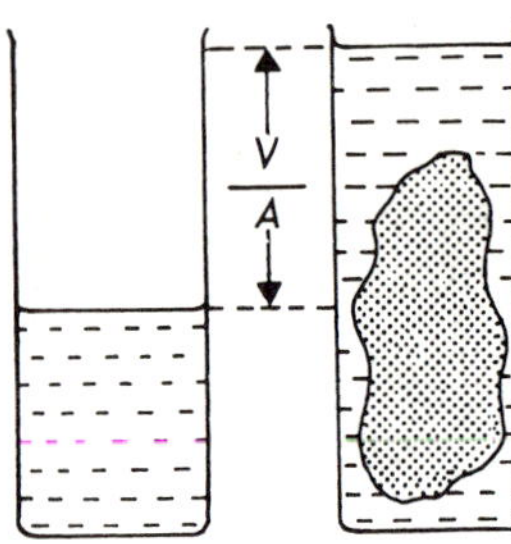

FIG. 2

First method. Imagine a cylinder (height h and cross-sectional area A) immersed in a liquid so that the top surface is a distance x below the surface of the liquid (Fig. 1). If d is the density of the liquid:

Pressure at top surface of cylinder $= x\,d\,g$

Force downwards on top of cylinder $= A\,x\,d\,g$

Similarly, force upwards on bottom of cylinder $= A(h + x)d\,g$

$\therefore$ Resultant force upwards on cylinder $= A(h + x)d\,g - A\,x\,d\,g$
$= A\,h\,d\,g$

Volume of cylinder $= A\,h$

But volume of cylinder = Volume of liquid "displaced"

$\therefore$ Weight of liquid "displaced" $= A\,h\,d\,g$

$\therefore$ Weight of liquid "displaced" = Resultant force upwards on cylinder

Although true for a cylinder this proof is not a general one.

Second (*general*) *method.* A body of volume V is lowered into a vessel of cross-sectional area A containing liquid (Fig. 2). The level of the liquid in the vessel will rise by a height V/A. The pressure on the bottom of the vessel will also increase by an amount Vdg/A.

$$\text{Thrust on bottom surface of vessel (downwards)} = \frac{VdgA}{A} = Vdg$$

But thrust downwards on bottom of vessel = Thrust upwards on body

$$\therefore \text{ Upthrust on body} = Vdg$$
$$\text{Weight of liquid of volume } V = Vdg$$
$$\therefore \text{ Upthrust on body} = \text{Weight of liquid displaced by body}$$

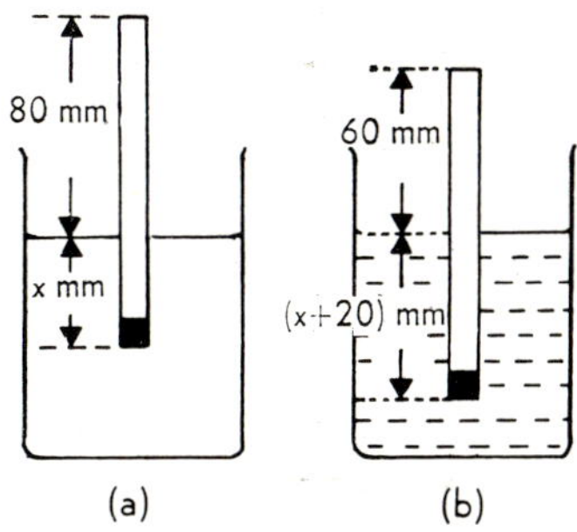

FIG. 3

(*b*) *Calculation:*

(You must remember that though you can measure quantities in convenient units, you must use kilogrammes, metres, seconds in calculations.)

$$\text{Let area of cross-section} = A$$
$$\text{Let length of rod immersed in water} = \frac{x}{1000}$$
$$\text{Volume of water displaced by rod} = \frac{Ax}{1000}$$
$$\text{Density of water} = 1000 \text{ kg/m}^3$$
$$\text{Mass of water displaced by rod} = 1000 \times \frac{Ax}{1000}$$
$$\text{Weight of water displaced by rod} = Axg$$

where g is the force of gravity on 1 kg

$$\text{Mass of rod} = 0{\cdot}064 \text{ kg}$$
$$\text{Weight of rod} = 0{\cdot}064\, g$$

By Archimedes' principle these two weights are equal.

$$Axg = 0{\cdot}064g$$
$$Ax = 0{\cdot}064 \text{ m}^3$$

Similarly in the second case:

$$\text{Volume of liquid displaced} = \frac{A\,(x+20)}{1000}$$
$$\text{Weight of liquid displaced} = 800 \times \frac{A\,(x+20)}{1000}g$$

This equals weight of rod.

$$\frac{800A\,(x+20)}{1000}g = 0{\cdot}064\, g$$
$$A\,(x+20) = 0{\cdot}080$$
$$Ax + 20A = 0{\cdot}080$$
$$\text{But } Ax = 0{\cdot}064$$
$$\therefore 20A = 0{\cdot}016$$
$$A = 0{\cdot}0008$$
$$\therefore \text{ cross-sectional area} = 0{\cdot}0008 \text{ m}^2$$

2. (*a*) The terms *mass* and *weight* are often used as if they were the same but in physics *mass* is a measure of the quantity of matter in a body and is constant for that body; *weight* is a force and is the force that the earth exerts on the body and depends on the distance of the body from the centre of the earth.

Calculation for a body falling freely under gravity:

Acceleration $= 10$ m/s²
∴ Velocity after time $t = 10 \times t$ m/s
Velocity at start of fall $= 0$ m/s
∴ Average velocity $= 5 \times t$ m/s
Distance travelled in time $t = 5 \times t^2$ m
But we know that in time t the body falls 40 m
$\therefore 5\,t^2 = 40$
$t^2 = 8$
∴ Time taken to fall 40 m $= 2\sqrt{2}$ s $= 2{\cdot}8$ s

Calculation for a body falling attached to a parachute:

Mass of body $= 60$ kg
Force of earth's attraction on body $= 60 \times 10$ N
Opposing force of parachute on body $= 240$ N
Resultant force on body $= 360$ N

$$\text{But acceleration} = \frac{\text{Force}}{\text{Mass}} = 6 \text{ m/s}^2$$

Average velocity in time $t = 3\,t$ m/s
∴ Distance travelled in time $t = 3\,t^2$
$3\,t^2 = 40$
$t^2 = 13{\cdot}3$
$t = 3{\cdot}6$
∴ Time taken to fall 40 m $= 3{\cdot}6$ s

2. (*b*) Law of conservation of energy (page 129). When the spring is compressed (Fig. 4*a*) it has a great deal of potential energy. The mass has a small amount

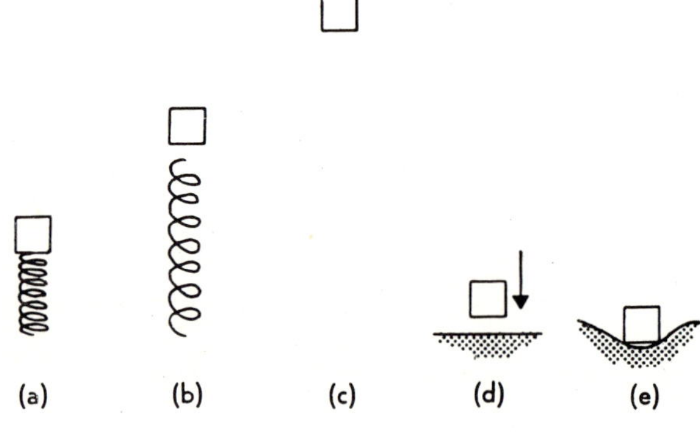

FIG. 4

of gravitational energy if the zero of this energy is taken to be when the mass is on the ground. When the spring is released its potential energy is transformed mainly into kinetic energy of the mass and slightly into the increase of the potential energy of the mass (the mass is now farther from the ground) (Fig. 4*b*). The mass flies upwards and momentarily comes to rest at its highest point (Fig. 4*c*). All its kinetic energy has now been transformed into gravitational potential energy. As it falls, (Fig. 4*d*) this energy is reconverted into kinetic energy and just before it hits the sand it will have no potential energy and maximum kinetic energy. Immediately after impact all this energy is transformed into heat, and the mass and the sand will be at a slightly higher temperature. The sand will not gain any potential energy since it does not compress (Fig. 4*e*).

3. (*a*) Definition of centre of gravity (page 142). Method for finding the centre of gravity of a card (page 142).

(*b*) The mass of the circle cut from the lamina is $\frac{1}{9}$ kg. If this were placed back in the hole, the whole lamina would balance about O. It follows that the moment about O of the force F required at B must be the same as the moment about O of the weight of a mass of $\frac{1}{9}$ kg placed at A. The moment of the weight at A would act in a clockwise direction so that the force required at B must also act in a clockwise direction. Taking moments about O:

$$F \times 200 = \tfrac{1}{9} \times 10 \times 100$$
$$F = \tfrac{5}{9}\text{ N}$$

Force required at B to keep OB horizontal is $\frac{5}{9}$ N in an upward direction.

SECTION B

4. (*a*) Method to find the specific latent heat of fusion of ice (page 182).

(*b*) *Calculation:*

$$\text{Heat supplied to flask and water} = 5 \times 10^5\text{ J}$$
$$\text{Heat required to raise flask and water to } 100^\circ\text{ C} = 4{\cdot}2 \times 10^3 \times 10^2\text{ J}$$
$$\text{Heat available to boil water away} = 0{\cdot}8 \times 10^5\text{ J}$$
$$\text{Mass of water boiled away} = \frac{0{\cdot}8 \times 10^5}{2{\cdot}27 \times 10^6} = 0{\cdot}035\text{ kg}$$

5. (*a*) The formation of a pure spectrum from a source of white light using a convex lens and a triangular glass prism (page 297).

(*b*) Blue pigments reflect some green and violet light as well as blue, while yellow pigments reflect red and green light as well as yellow. Therefore, when yellow and blue pigments are mixed the only colour not absorbed by one or other of the pigments is green (page 303).

(*c*) When one of the yellow filters Y and the green filter G are placed in front of a source of white light (Fig. 5*a*) no light comes through. *Y* must therefore be the filter which transmits only pure yellow light. All other light having been subtracted, the pure yellow light is then absorbed by the green

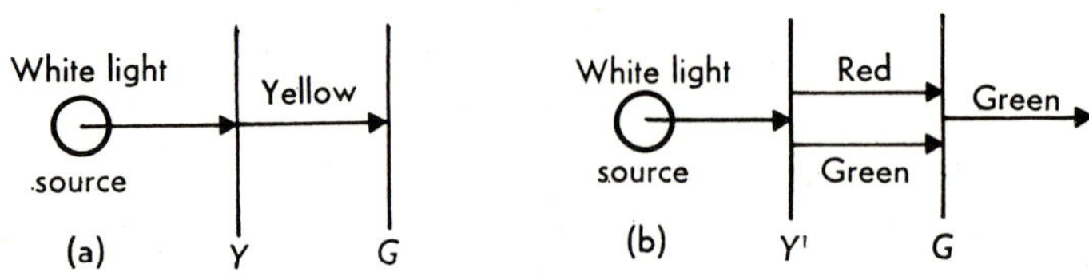

FIG. 5

filter. If the experiment is repeated with the other yellow filter Y^1, both red and green light will be transmitted to the green filter which will then subtract the red light and transmit the green (Fig. 5*b*). The source would still be visible through the two filters.

6. (*a*) An approximate value of the focal length is found by focusing an image of a distant window on to a white screen. Using a half-metre rule, the distance between the lens and the screen is measured when the image is as clear as possible. Suppose this distance is *d*. The converging lens is then placed on a horizontal plane mirror and an optical pin is mounted above it so that its point is vertically over the centre of the lens (Fig. 6*a*). The pin is held in such a

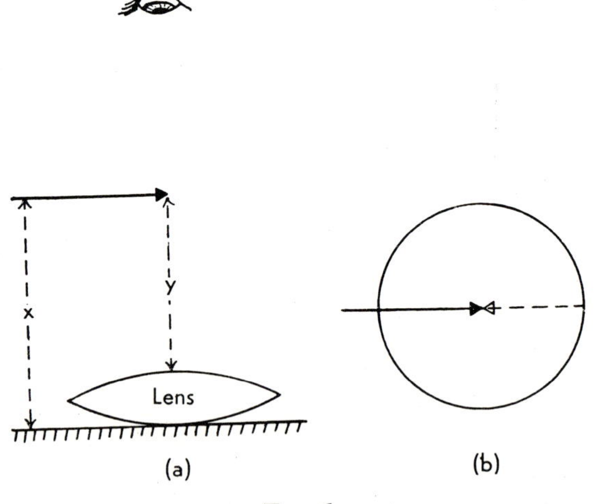

FIG. 6

way that it can be moved in a vertical line. It is adjusted to be at a distance of about *d* from the lens.

On looking down on the pin (the eye must be at a distance greater than the least distance of distinct vision above the pin) the pin and its image can be seen (Fig. 6*b*). The pin is moved up and down until there is no parallax between it and its image. That is, when you move your head from side to side there is no relative motion between the two. The distance y from the top of the lens to the point of the pin and the distance x from the pin to the plane mirror are measured and the average D_1 taken. The pin is then put out of adjustment and the process is repeated several times to get values D_2, D_3, etc. The focal length of the lens is the average of these values.

(*b*) The diverging beam of light from the source O falls partly on the lens (Fig. 7). Since the light falling on the lens forms a patch of light the same size

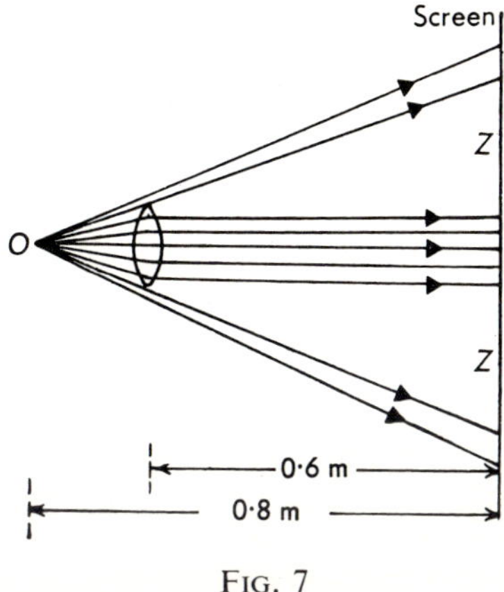

FIG. 7

as the lens on the screen, it follows that the lens refracts the light into a parallel beam. The part of the screen Z receives no light from the source and therefore appears dark.

Calculation: from the definition of the principal focus we know that a ray of light passing through the focus emerges from the lens parallel to the principal axis (page 274). Since the beam of light ,emerging from the lens in Fig. 7 is parallel it follows that O is at the principal focus of the lens hence the focal length of the lens is 0·2 m.

$$f = +0{\cdot}2\text{ m},\ v = +0{\cdot}8\text{ m},\ u \text{ is unknown.}$$

$$\frac{1}{u}+\frac{1}{v}=\frac{1}{f}$$

$$\frac{1}{u}+\frac{1}{0{\cdot}8}=\frac{1}{0{\cdot}2}$$

$$\frac{1}{u}=\frac{4}{0{\cdot}8}-\frac{1}{0{\cdot}8}=\frac{3}{0{\cdot}8}$$

$$u = +0{\cdot}27\text{ m}$$

The source must be moved 0·07 m further away from its original position in order to obtain a clear image on the screen (Fig. 8).

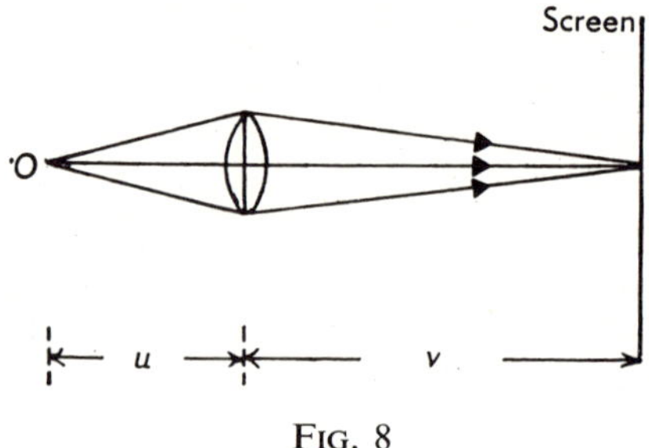

FIG. 8

Alternative (graphical) method: at the right hand side of the page draw a vertical line XY to represent the screen and 150 mm away from it a parallel line PQ to represent the lens (Fig. 9). Draw a horizontal line cutting these two at L and I (the principal axis). Make a point A on XY some distance below I.

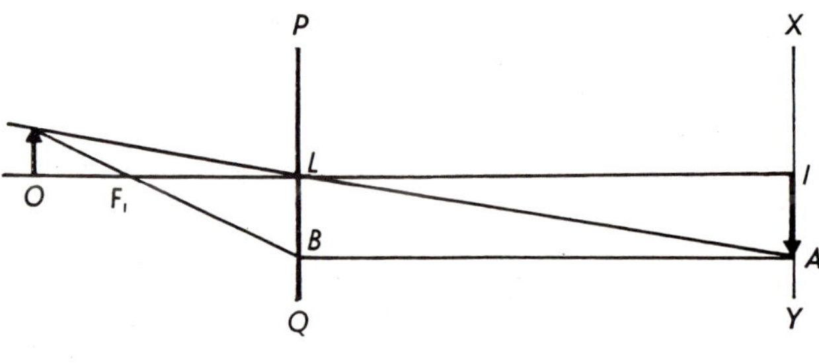

FIG. 9

Join AL and produce it on the far side of the lens. Draw AB parallel to LI. Mark F_1 50 mm (i.e. 200 mm on full scale) from L. Join BF_1 and produce backwards until it meets AL produced. This gives the position of the source O. Measure LO. This distance multiplied by 4 gives the distance the source must be from the lens.

SECTION C

7. (*a*) Determination of the electrochemical equivalent of copper (page 431).
(*b*) *Calculation:*
Total resistance of battery and voltameter $= 1{\cdot}5 + 2{\cdot}5 = 4\,\Omega$

$$\therefore \text{ Current flowing in circuit } = \frac{6}{4} = 1{\cdot}5 \text{ A}$$

Quantity of electricity flowing through circuit in 6 h = Current (A) × Time (s)
$= 1{\cdot}5 \times 6 \times 60 \times 60$ coulombs

But 96 500 coulombs will deposit 0·031 75 kg copper (page 432).

$$\therefore\ 1{\cdot}5 \times 6 \times 3600 \text{ coulombs will deposit } \frac{31{\cdot}75 \times 1{\cdot}5 \times 6 \times 3600}{1000 \times 96\,500}$$

$$= 0{\cdot}0106 \text{ kg copper}$$

8. (*a*) The electroscope (page 368).

(*b*) To charge an electroscope negatively using a positively charged object (pages 372-73).

(*c*) To show that when ebonite is rubbed with fur equal amounts of positive and negative charge result (page 375).

9. (*a*) Method of comparing the e.m.f.s of two cells using a potentiometer (page 418).

(*b*) The result of comparing e.m.f.s by means of a potentiometer depends on only two things: the current in the wire remains constant while the measurements are made; the wire is of uniform resistance per unit length. (i) Doubling the length of the potentiometer wire affects neither of these conditions so that the ratio of E_1/E_2 remains the same. It is clear therefore that if the length of the wire were doubled then the length of wire for balancing each cell would also be doubled. (ii) No current flows through a cell while its e.m.f. is being measured by the potentiometer method, therefore neither the resistance of the cell nor any resistance placed in series with the cell has any effect on the result.

10. (*a*) Laws of electromagnetic induction and experiments to illustrate them (pages 437-38).

(*b*) The points 1, 2, 3, 4, on the e.m.f./time graph (Fig. 10) are directly related

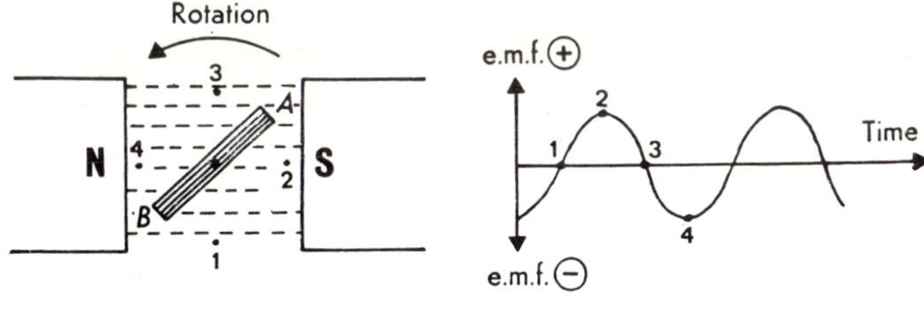

FIG. 10

to the positions 1, 2, 3, 4, of point *A* on the coil rotating in a magnetic field. A negative e.m.f. would produce a current in the coil in the direction out of the paper at *B* into the paper at *A*.

ABBREVIATIONS USED IN THE TEXT

A	ampere
a.c.	alternating current
°C	degree Celsius (Centigrade)
cm	centimetre
d.c.	direct current
e.c.e.	electrochemical equivalent
(E.)H.T.	(extra) high tension
e.m.f.	electromotive force
g	gramme
H.P.	horse power
h	hour
Hz	hertz
J	joule
K	kelvin
K.E.	kinetic energy
kg	kilogramme
kJ	kilojoule
km	kilometre
km/h	kilometre per hour
kN	1000 newton.
kW	kilowatt
m	metre
M.A.	mechanical advantage
min	minute
mm	millimetre
N	newton
p.d.	potential difference
P.E.	potential energy
R.D.	relative density
rev	revolution
r.m.s.	root mean square
s	second
S.H.C.	specific heat capacity
s.t.p.	standard temperature and pressure
V.H.F.	very high frequency
V.R.	velocity ratio
V	volt
W	watt
Ω	ohm

INDEX

ACKNOWLEDGEMENTS

The publishers wish to thank the following for their permission to use the photographs on the pages indicated: Aerofilms and Aeropictorial Ltd., 180. British Electric Resistance Co. Ltd., 411. Central Office of Information (Crown Copyright), 305. Central Press Photos Ltd., 47 (bottom). Fox Photos Ltd., 48, 56, 66, 82, 96, 110, 126, 132, 135, 144, 149, 152, 187, 196, 217, 241, 255, 260, 285, 318, 322, 333. Geo. Salter and Co. Ltd., 25. Greater London Council, Architects Department, 49. Henry Hughes and Son Ltd. (held by the Science Museum), 392. Keystone Press Agency Ltd., 47 (top), 472. Mansell Collection, 33 (bottom). Middle East Archive, 9. Ministry of Technology (Hydraulics Research Station), 349, 351, 353. Natural Rubber Producers Research Association, 89. Paul Popper Ltd., 33 (top), 45, 78, 211, 261, 269. Pictorial Press Ltd., 14. *Radio Times* Hulton Picture Library, 10, 11 (left), 13, 467, 476. Ronan Picture Library, 247, 253, 256, 264, 268, 270, 283, 300, 316, 336, 337, 341, 344, 346, 369, 385, 445. Science Museum, 8 (top), 11 (right), 12, 17, 22, 50, 120, 121, 135, 191, 213, 291, 378; (Crown Copyright), 16, 24, 38, 271, 311, 393 (right), 395, 453, 475. S. Smith and Sons (England) Ltd. (Kelvin Hughes Division), 393 (left). Sport and General Press Agency Ltd., 46. United Kingdom Atomic Energy Authority, 21, 41, 220, 469, 471. United Press International, 76. United States Information Service, 8 (bottom), 19, 83, 115, 208, 218, 272, 462, 464. Wellcome Trustees, 463. Weston Electrical Installation Co. (held by the Science Museum), 427.